ORDER IN CHAOS

Order in Chaos

Proceedings of the International Conference on
Order in Chaos held at the Center for Nonlinear Studies
Los Alamos, New Mexico 87545, USA
24-28 May 1982

Editors:

David Campbell
Harvey Rose

*Center for Nonlinear Studies
and
Theoretical Division
Los Alamos National Laboratory
Los Alamos, New Mexico 87545, USA*

1983

NORTH-HOLLAND AMSTERDAM

© North-Holland Publishing Company, 1983

All rights reserved. No part of this publication may be reproduced, stored in a retrieval system, or transmitted, in any form or by any means, electronic, mechanical, photocopying, recording or otherwise, without the prior permission of the copyright owner.

PUBLISHERS:

NORTH-HOLLAND PUBLISHING COMPANY
AMSTERDAM · NEW YORK · OXFORD

SOLE DISTRIBUTORS FOR THE U.S.A. AND CANADA:

ELSEVIER SCIENCE PUBLISHING COMPANY, INC.
52 VANDERBILT AVENUE
NEW YORK, N.Y. 10017

REPRINTED FROM PHYSICA D VOLUME 7D (1983)

Library of Congress Cataloging in Publication Data

```
International Conference on Order in Chaos (2nd :
   1982 : Center for Nonlinear Studies)
   Order in chaos.

   Sponsored by Applied Mathematical Sciences Program,
United States Dept. of Energy, and Center for Non-
linear Studies, Los Alamos National Laboratory.
   "Reprinted from Physica, D volume 7D (1983)"--T.p.
verso.
   1. Chaotic behavior in systems--Congresses.
2. Dynamics--Congresses.  3. Nonlinear theories--
Congresses.  I. Campbell, David, 1944 July 23-
II. Rose, Harvey, 1947-     .  III. United States.
Dept. of Energy.  Applied Mathematical Sciences

Program.  IV. Center for Nonlinear Studies (Los Alamos
National Laboratory)  V. Title.
QA843.I57  1982      003        83-11659
ISBN 0-444-86727-9  (Elsevier Science)
```

Printed in The Netherlands

PREFACE

To the uninitiated, the phrase 'Order in Chaos' may seem self-contradictory: chaos is, after all, conventionally viewed as the complete absence of order. Yet precisely this title attracted two hundred and ten scientists from fourteen countries to Los Alamos from May 24–28, 1982, to attend the second annual international conference of the Center for Nonlinear Studies. The purposes of the conference were to survey the rapid recent developments and to anticipate the trends for future research in the area 'chaos in deterministic systems'. The breadth of scientific interest in this topic was reflected in the variety of subjects discussed at the meeting. Presentations ranged from abstract mathematics through numerical simulations to experimental studies of fluid mechanics, chemistry, and biology. Even weather prediction made an appearance.

To appreciate the appeal of the conference title – and the importance of the field of research it describes – requires a closer look at the apparently contradictory components. The concepts of 'order' and 'determinism' in the natural sciences recall the predictability of the motion of simple physical systems obeying Newton's laws: the rigid plane pendulum, a block sliding down an inclined plane, or motion in the field of a central force are all examples familiar from elementary physics. In contrast, the concept of 'chaos' recalls the erratic, unpredictable behavior of elements of a turbulent fluid or the 'randomness' of Brownian motion as observed through a microscope. For such chaotic motions, knowing the state of the system at a given time does not permit one to predict it for all later times. In place of the determinism of the orderly systems, one has only probabilistic estimates and statistical averages.

Thus, in some sense, the possibility that chaos exists in deterministic systems runs directly counter to normal intuition. To understand that this possibility is nonetheless real, we can refer to the deeper insight of Henri Poincaré, one of the founders of modern dynamical systems theory. Writing in the pre-quantum era of pure Newtonian determinism, Poincaré noted [as translated in *Science and Method*].

> *A very small cause which escapes our notice determines a considerable effect that we cannot fail to see, and then we say that the effect is due to chance. If we knew exactly the laws of nature and the situation of the universe at the initial moment, we could predict exactly the situation of that same universe at a succeeding moment. But even if it were the case that the natural laws had no longer any secret for us, we could still only know the initial situation* approximately. *If that enabled us to predict the succeeding situation* with the same approximation, *that is all we require, and we should say that the phenomenon had been predicted, that it is governed by laws. But it is not always so; it may happen that small differences in the initial conditions produce very great ones in the final phenomena. A small error in the former will produce an enormous error in the latter. Prediction becomes impossible, and we have the fortuitous phenomenon.*

Hence, the crucial ingredient in deterministic chaos is a very sensitive dependence on initial conditions. Motions that start close to each other develop in time in dramatically different ways, and uncertainties in the initial values develop rapidly – exponentially, in fact – in time. Although the motion from instant to instant can be predicted, over macroscopic times it becomes no more predictable than a random sequence.

At first it might appear that the distinction between orderly and chaotic motions is merely one of the complexity of the system involved. In the parlance of dynamical systems theory, the orderly motions described above involve just one 'degree of freedom', whereas the chaotic fluid involves many – in conventional hydrodynamics, infinitely many – degrees of freedom. It is thus tempting to associate simple systems with order and complicated ones with chaos.

In fact, this naïve association is wrong for several fundamental reasons, some obvious and some subtle. First, everyday experience tells us that complicated systems with many degrees of freedom can undergo very orderly motion. For example, a fluid in smooth (laminar) flow moves in a regular, totally predictable manner.

Second, it is less familiar but nonetheless true that very simple physical systems can exhibit chaotic behavior, with all the associated randomness and unpredictability. Numerical experiments show that the motion of a rigid plane pendulum, if damped and driven, becomes truly chaotic. This result illustrates strikingly that a completely deterministic system can produce chaos without the addition of any external random noise. In other words, one does not have to put randomness in to get it out. The existence of deterministic motions that produce chaos is a clear example of order in chaos.

Third, it is now well established that, at least in some cases, the chaos observed in very complicated systems can be understood quantitatively in terms of simple models that involve very few degrees of freedom. This profound result is of great potential significance for understanding chaotic behavior in the physical world.

These three, and many other, important aspects of 'Order in Chaos' were discussed at the conference and are reflected in the ensuing Proceedings. At a rough level, the articles in this collection divide into two major areas. First, there are attempts to identify the essential qualitative and quantitative features of deterministic chaos, with the aim of describing and modelling it more accurately. Second, there are discussions of the transition from regular motion to chaos in both experimental systems and theoretical models. But as perhaps befits the subject of chaos, the contributions do not fall neatly into more detailed separate categories; rather, they overlap in context and technique in a variety of interesting, nonlinear ways. To provide some order in this chaos, we have chosen to collect the articles into the seven loose-knit groupings shown in the Table of Contents. We urge the dedicated reader to seek other conceptual links among the articles.

In a very real sense, this conference represented the 'end of the beginning' of the field of deterministic chaos. Many of the fundamentals of low dimensional chaos are now both theoretically modeled and experimentally verified, and a variety of intriguing questions seem ripe for answering. We hope that this volume conveys some of the excitement, vitality, and anticipation that those of us who attended the meeting felt.

Finally, it is indeed with pleasure that we express on behalf of the organizing committee our gratitude to those who worked so hard to insure the success of the conference; the pedagogical efforts of the speakers, the active involvement of the participants, the generous financial support of the Applied Mathematical Sciences Program of the U.S. Department of Energy, and the tremendous administrative support provided by the staff of Los Alamos National Laboratory and the Center for Nonlinear Studies are all gratefully acknowledged. We are particularly indebted to Mary Frances Gomez and Marian Martinez for their unflagging attention to detail and for their efficient resolution of the inevitable crises. Lastly, as editors, we wish to express our sincere thanks to our colleague Doyne Farmer for his contributions both to our understanding of the subject and to this introduction and to the authors and the editorial staffs of Physica D and North-Holland Publishing Company for enabling us to assemble this permanent tribute to 'Order in Chaos'.

David CAMPBELL
Harvey ROSE

Editors

ORGANIZING COMMITTEE

David K. CAMPBELL
Mitchell J. FEIGENBAUM
Harvey A. ROSE
Alwyn C. SCOTT

SPONSORS

Applied Mathematical Sciences Program
   United States Department of Energy
Center for Nonlinear Studies
   Los Alamos National Laboratory

CONTENTS

Preface   vii
Committees – Sponsors   ix
Contents   x

CHAPTER 1: OVERVIEWS OF ORDER IN CHAOS

SWINNEY, H.L., Observations of order and chaos in nonlinear systems   3
FEIGENBAUM, M.J., Universal behavior in nonlinear systems   16
RUELLE, D., Five turbulent problems   40

CHAPTER 2: EXPERIMENTAL OBSERVATIONS OF CHAOTIC SYSTEMS

EPSTEIN, I.R., Oscillations and chaos in chemical systems   47
ROUX, J.-C., Experimental studies of bifurcations leading to chaos in the Belousof–Zhabotinsky reaction   57
HAUCKE, H. and MAENO, Y., Phase space analysis of convection in a ^{3}He-superfluid ^{4}He solution   69
LIBCHABER, A., FAUVE, S. and LAROCHE, C., Two-parameter study of the routes to chaos   73
SMITH, C.W. and TEJWANI, M.J., Bifurcation and the universal sequence for first sound subharmonic generation in superfluid helium-4   85
GLASS, L., GUEVARA, M.R. and SHRIER, A., Bifurcation and chaos in a periodically stimulated cardiac oscillator   89

CHAPTER 3: MATHEMATICAL PROPERTIES AND MODEL SYSTEMS

GUCKENHEIMER, J., Persistent properties of bifurcations   105
HOLMES, P. and WHITLEY, D., On the attracting set for Duffing's equation, II: A geometrical model for moderate force and damping   111
LANFORD, O.E., III, Period doubling in one and several dimensions   124
FOWLER, A.C., GIBBON, J.D. and McGUINESS, M.J., The real and complex Lorenz equations and their relevance to physical systems   126
MOON, H.T., HUERRE, P. and REDEKOPP, L.G., Transitions to chaos in the Ginzburg–Landau equation   135

CHAPTER 4: DIMENSION, FRACTAL STRUCTURES, AND COHERENCE VERSUS CHAOS

FARMER, J.D., OTT, E. and YORKE, J.A., The dimension of chaotic attractors   153
GREBOGI, C., OTT, E. and YORKE, J., Crises, sudden changes in chaotic attractors, and transient chaos   181
CRUTCHFIELD, J.P. and PACKARD, N.H., Symbolic dynamics of noisy chaos   201
MANDELBROT, B.B., On the quadratic mapping $z \to z^2 - \mu$ for complex μ and z: The fractal structure of its $\mathfrak{M}$ set, and scaling   224
AUBRY, S., The twist map, the extended Frenkel–Kontorova model and the devil's staircase   240

BISHOP, A.R., FESSER, K., LOMDAHL, P.S. and TRULLINGER, S.E., The infuence of solitons in the initial state on chaos in the driven damped sine-Gordon chain 259

CHAPTER 5: TRANSITION TO CHAOS IN MAPS OF THE CIRCLE

MACKAY, R.S., A renormalization approach to invariant circles in area preserving maps 283
SHENKER, S.J., Quasiperiodicity in dissipative systems: A renormalization group analysis *(Abstract)* 301
SIGGIA, E.D., A universal transition from quasi-periodicity to chaos *(Abstract)* 302

CHAPTER 6: FLUIDS AND VORTICES

MARSDEN, J. and WEINSTEIN, A., Coadjoint orbits, vortices, and Clebsch variables for incompressible fluids 305
PALMORE, J.I., Bifurcation of stationary vortex configurations, II: Topology and integrability 324
HOLM, D.D. and KUPERSHMIDT, B.A., Noncanonical formulation of ideal magnetohydrodynamics 330
KUPERSHMIDT, B.A., On dual spaces of differential Lie algebras 334

CHAPTER 7: QUANTUM CHAOS

GUTZWILLER, M.C., Stochastic behavior in quantum scattering 341
HELLER, E.J., Quantum-classical correspondence for the Fourier spectrum of a trajectory 356

LIST OF CONTRIBUTORS 362

CHAPTER 1

OVERVIEWS OF ORDER IN CHAOS

OBSERVATIONS OF ORDER AND CHAOS IN NONLINEAR SYSTEMS

Harry L. SWINNEY

Department of Physics, University of Texas, Austin, Texas 78712, USA

Experiments on nonlinear electrical oscillators, the Belousov–Zhabotinskii reaction, Rayleigh–Bénard convection, and Couette–Taylor flow have revealed several common routes to chaos that have also been found in numerical studies of models with a few degrees of freedom. Experimental results are presented illustrating the following transition sequences: period doubling and the U-sequence, intermittency, the periodic–quasiperiodic–chaotic sequence, frequency locking, and an alternating periodic–chaotic sequence.

1. Introduction

We will describe some recent experimental studies of order and chaos in nonlinear systems. Although no attempt at completeness will be made, we will mention most of the transition sequences that have been found to be common to diverse systems.

Noisy (nonperiodic) behavior arising from stochastic driving forces such as thermal fluctuations and fluctuations in a system's environment has long been studied in laboratory experiments, but the experiments to be discussed here concern the nonperiodic (chaotic) behavior that arises primarily from the nonlinear nature of the systems rather than stochastic driving forces. The distinction between stochastic and deterministic noise in experiments is difficult, but the papers of Guckenheimer [55], and Farmer, Ott, and Yorke [53] in this volume suggest that the distinction can be made in systems with a few active degrees of freedom. We will show that the Poincaré sections and maps obtained in experiments on some rather complex nonlinear systems indicate that these systems (for some control parameter ranges) exhibit a dynamical behavior that can be described accurately by deterministic models with a few degrees of freedom.

Four well-studied nonlinear systems are described in section 2, and methods used to characterize their dynamical behavior are outlined in section 3. Some transition sequences that have been observed for a number of different systems are described in section 4. Section 5 is a discussion.

2. Four nonlinear systems

Nonlinear electrical circuits. The characteristic frequencies of electrical circuits can easily be made about 10^7 times higher than the typical oscillation frequencies of the chemical and hydrodynamic systems described in the following paragraphs. Such high information production rates make nonlinear electrical circuits (analog computers) ideal for examining different types of dynamical behavior, developing methods of data analysis, and studying the dependence of behavior on several control parameters [1–4, 51, 57, 63, 64, 68, 70]. An example of a simple nonlinear circuit is shown in fig. 1a; this series circuit has three degrees of freedom–q (the charge across the varactor), $\dot{q}$, and the angle θ, where the driving voltage is $V(t) = V_0 \sin\theta$ with $\theta = \omega t$. The behavior of this circuit is usually studied as a function of V_0, but it can also be studied as a function of other control parameters–ω, R, L, C_0, and β, where the nonlinear capacitance under reverse voltage is given by $C \approx C_0/[1 + \beta V_c]^{1/2}$.

The Belousov–Zhabotinskii reaction [5–15]. This reaction, the most thoroughly studied oscillating

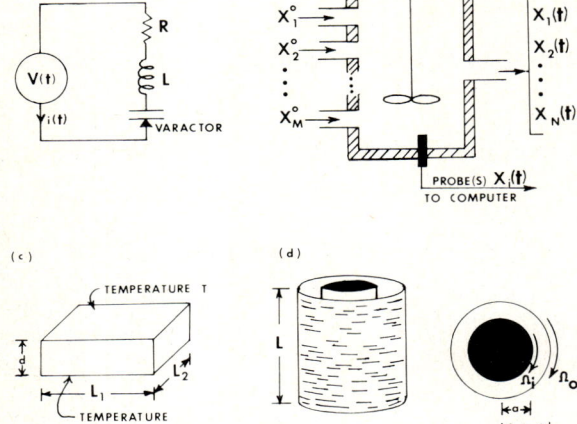

Fig. 1. Four nonlinear systems. (a) A series circuit with a varactor diode which conducts for a forward voltage and has a nonlinear capacitance for a reverse voltage [3, 4]. (b) A stirred flow chemical reactor. In the Belousov–Zhabotinskii reaction $M = 4$ (malonic acid, potassium bromate, cerium sulfate, and sulfuric acid) and $N > 30$. (c) Rayleigh–Bénard convection in a finite box. (d) Couette–Taylor system.

chemical system, involves the cerium-catalyzed bromination and oxidation of malonic acid by a sulfuric acid solution of bromate. (See the papers of Roux [9] and Epstein [5] in this volume.) The reaction can be maintained in a steady state away from equilibrium by continuously pumping the chemicals into a stirred flow reactor, as shown in fig. 1b. In a vigorously stirred reactor the system is essentially homogeneous so the reaction can be modeled by a set of coupled nonlinear ordinary differential equations. For example, the reaction

$$A + B \underset{k_b}{\overset{k_f}{\rightleftharpoons}} C$$

is described by the equations

$$\frac{dA}{dt} = -k_f AB + k_b C - r(A - A^0),$$

$$\frac{dB}{dt} = -k_f AB + k_b C - r(B - B^0),$$

$$\frac{dC}{dt} = k_f AB - k_b C - rC,$$

where A^0, B^0, and C^0 (with $C^0 = 0$) are the concentrations of the chemicals in the input to the reactor and r is the flow rate. Generalizing, the reactions among N chemical species of concentration $X_i(t)$ are described by

$$\frac{dX_i}{dt} = g_i(X_j) - r(X_i - X_i^0) \quad [i, j = 1, \ldots, N],$$

where the functions $g_i(X_j)$ involve nonlinear terms of the form X_i^2 and $X_i X_j$ (i.e., three-body interactions can be neglected). Transitions in the dynamical behavior are studied as a function of the flow rate: as $r \to 0$, the system approaches thermodynamic equilibrium, while for large r the chemicals have no time to react as they pass through the reactor; the interesting dynamics occurs for r between these extremes. The behavior can also be studied as a function of other control parameters—the reactor temperature and the input concentrations X_i^0.

Rayleigh–Bénard convection [16–30]. In contrast to the nonlinear oscillator and stirred flow reactor, which presumably have a well-defined finite number of degrees of freedom, the next two examples, the Rayleigh–Bénard and Couette–Taylor systems, are continuum hydrodynamic systems which can in principle have an infinite number of degrees of freedom (although just beyond the onset of chaos there are presumably only a few degrees of freedom that are excited).

In a Rayleigh–Bénard system a fluid is contained between parallel plates heated from below, as shown in fig. 1c. (Also see the papers of Libchaber [26] and Maeno and Haucke [28] in this volume.) The behavior is usually studied as function of the (dimensionless) Rayleigh number $R_a = (g\alpha d^3/\kappa\nu)\Delta T$, where g is the gravitational acceleration, α the thermal expansion coefficient, d the separation between the plates, κ the thermal diffusivity, and ν the kinematic viscosity. Other control variables are the Prandtl number, $P = \nu/\kappa$, the aspect ratios, $\Gamma_1 = L_1/d$ and $\Gamma_2 = L_2/d$, and the boundary conditions at the side walls.

Couette–Taylor system [31–41]. In this system a fluid is contained between concentric cylinders that rotate independently with angular velocities Ω_i (inner) and Ω_o (outer); see fig. 1d. The (dimensionless) Reynolds numbers are then $R_i = (b - a)a\Omega_i/\nu$ and $R_o = (b - a)b\Omega_o/\nu$, where a and b are the radii of the inner and outer cylinders, respectively. Most experiments including those to be described here have been conducted with $R_o = 0$. The behavior is quite different and much richer when both cylinders are rotated (Andereck, Liu, and Swinney [31]), because the instabilities do not depend simply on the differential rotation rate of the cylinders, but on a subtle interplay between the radial pressure gradient and the centrifugal force $r(\Omega_{\text{fluid}})^2$. (There is no equivalence principle for rotating reference frames!) Other control parameters for this system are the radius ratio a/b, the aspect ratio $\Gamma = L/(b - a)$, and the boundary conditions at the ends.

Other systems. Some results from experiments on a few other systems [42–50] will be mentioned in section 4.

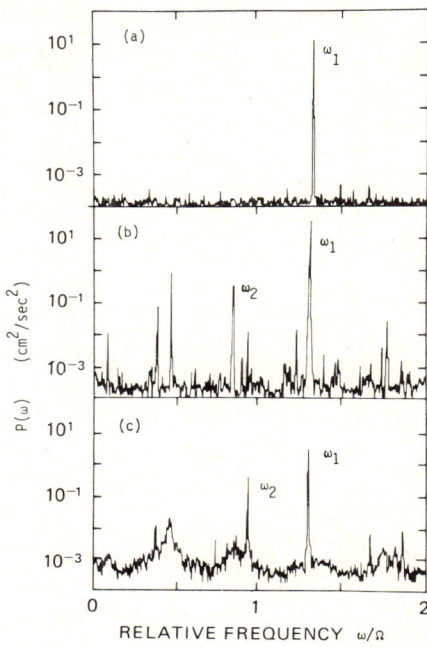

Fig. 2. Power spectra for different dynamical regimes in the Couette–Taylor system (from [35, 41]). (a) $R/R_c = 9.6$, periodic; the spectrum consists of a single fundamental frequency, ω_1. (b) $R/R_c = 11.0$, quasiperiodic; the spectrum consists of two fundamental frequencies, ω_1 and ω_2, and integer combinations. (c) $R/R_c = 18.9$, chaotic; the spectrum contains broadband noise in addition to the sharp components ω_1 and ω_2. The noise in (a) and (b) is instrumental, while in (c) the fluid noise is well above the instrumental noise level. These spectra illustrate the periodic–quasiperiodic–chaotic transition sequence discussed in section 4.3.

3. Analysis of dynamical behavior [51–72]

In experiments the time dependence of a dynamical variable $V(t)$ is determined in sequential time intervals $t_k = k(\Delta t)$, where $k = 1, \ldots, n$ (typically $n = 8192$). The time series $V(t_k)$ is recorded in a computer and its power spectral density $P(\omega)$ (the modulus squared of the Fourier transform) is calculated using the Cooley–Tukey fast Fourier transform algorithm.

Power spectra make it possible to distinguish between periodic, quasiperiodic, and chaotic regimes, as fig. 2 illustrates. However, the broadband noise or broadened spectral lines that indicate nonperiodic (chaotic) behavior could arise from stochastic as well as deterministic processes. A better method of analysis is needed to determine if the nonperiodic behavior is characteristic of a deterministic nonlinear system.

Before the turn of the century Poincaré showed that much can be learned about dynamical behavior from an analysis of trajectories in a multidimensional phase space in which a single point characterizes the entire system at an instant of time. The experimenters' dilemma has been that for a system with N degrees of freedom it seemed that it would be necessary to measure N independent variables, an almost impossible chore for complex systems.

A much simpler alternative was suggested several years ago by Ruelle [65] and Packard et al. [63]. Their idea, which is justified by embedding theorems [69, 71], was that a multi-dimensional phase portrait can be constructed from measurements of a *single* variable, as follows: For almost every observable $V(t)$ and time delay T the m-dimensional portrait constructed from the vectors $\{V(t_k), V(t_k+T),\ldots,V(t_k+(m-1)T)\}$, $k=1,\ldots,\infty$, will have many of the same properties (strictly speaking, will give an embedding of the original manifold) as one constructed from measurements of the N independent variables, if $m \geq 2N+1$. In practice m is increased by one at a time until additional structure fails to appear in the phase portrait when an extra dimension is added. Phase portraits constructed for a periodic state and a chaotic state in the Belousov–Zhabotinskii reaction are shown in fig. 3 [11, 14].

Rather than analyze the phase portrait directly it is easier to analyze Poincaré sections and maps. A Poincaré section is formed by the intersection of "positively" directed orbits with a $(m-1)$-dimensional hypersurface. For example, fig. 4(a) shows a Poincaré section constructed for the 3-dimensional phase portrait in fig. 3b. The orbits for this chaotic attractor clearly lie essentially along a sheet. Thus intersections of this sheet-like attractor with a plane lie to a good approximation along a parameterizable curve, not on a higher dimensional set. (Actually, the Poincaré section must have a dimension at least slightly greater than unity because of the fractal nature of the attractor [52, 53, 61]). The parameter values at successive intersections provide a sequence $\{X_n\}$ which defines a one-dimensional map, $X_{n+1}=f(X_n)$, as shown in fig. 4b. The data appear to fall on a single-valued curved. This indicates that the system is deterministic: for any X_n, the map *determines* X_{n+1}.

The power spectrum for the data in fig. 3b contains broadband noise [11, 14], indicating that the state is nonperiodic, but the phase portrait must be analyzed to determine if the system is really characterized by a *strange attractor*. To

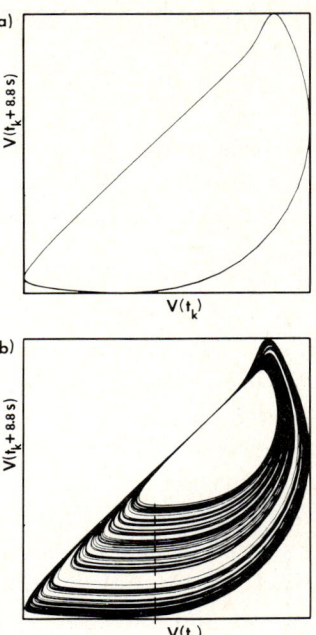

Fig. 3. (a) A two-dimensional phase portrait for a periodic state observed in experiments on the Belousov–Zhabotinskii reaction; the corresponding power spectrum has a single sharp fundamental component and its harmonics. (b) A two-dimensional projection of a three-dimensional phase portrait [with the third axis, $V(t_k+17.6\,\text{s})$, normal to the page] for a chaotic state observed in the Belousov–Zhabotinskii reaction; the corresponding power spectrum contains broadband noise. The attractor in (a) is a limit cycle and in (b) a strange attractor. (From [11, 14].)

demonstrate that the phase space trajectories define a strange attractor it must be shown that the post-transient subset described by the trajectories is:

(1) An *attractor*–orbits rapidly return to this subset after finite perturbations. However, perturbations too large could send the orbit out of the basin of attraction for the attractor; see [59, 66, 67].)

(2) *Strange*–nearby orbits diverge exponentially on the average ("sensitive dependence on initial conditions" [66]) [59, 62, 67, 68].

Studies of the effect of perturbations on the state characterized by the phase portrait in fig. 3b show that the trajectories lie on an *attractor* [11, 12]). In addition, an analysis of the corresponding map,

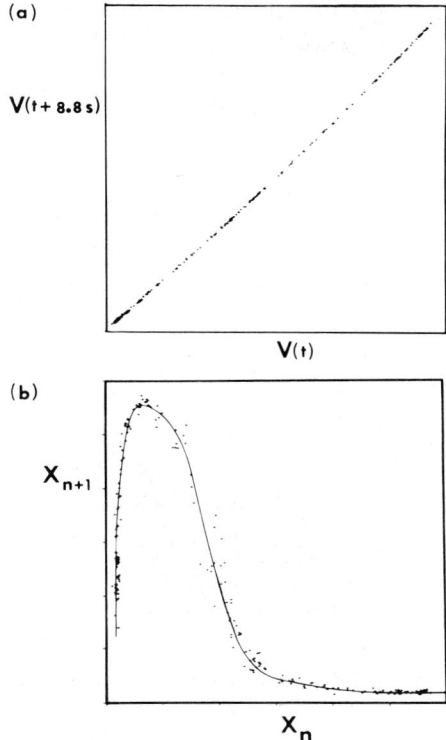

Fig. 4. (a) A Poincaré section formed by the intersection of trajectories in a three-dimensional phase space with the plane (normal to the page) passing through the dashed line in fig. 3b. (b) A one-dimensional map constructed from the data in fig. 4a. (From [11, 14].)

and folding has been directly observed by analyzing Poincaré sections through the different parts of the attractor [12]; in fig. 3b the folding occurs in the part of the attractor where the orbits appear (at the resolution of this figure) to narrow down to a line.

Other methods of analysis of phase portraits include the determination of the following properties (see the papers in this volume by Farmer, Ott and Yorke [53], Guckenheimer [55], Mandelbrot [61], Packard [63], and Shaw [68]); (1) attractor dimension (the terms capacity, Hausdorff, fractal, information or Renyi, and Lyapunov dimension correspond to different definitions of dimension); (2) entropy (topological, metric or Kolmogorov–Sinai); (3) the spectrum of Lyapunov exponents; and (4) probability distribution functions for the Poincaré sections and maps.

4. Transition sequences

4.1. Intermittency

Some systems exhibit a transition from periodic behavior (for $R < R_T$) to a chaotic behavior (for $R > R_T$) characterized by occasional bursts of noise [73, 74]. For R only slightly greater than R_T there are long intervals of periodic behavior between the short bursts, but with increasing R the intervals between the bursts decrease; it becomes more and more difficult and finally impossible to recognize the regular oscillations of the periodic state. Examples of intermittency transitions are shown in fig. 5.

Pomeau and Manneville [74] have shown that intermittency appears at a tangent bifurcation where a stable fixed point of a map disappears. Direct evidence of the tangent bifurcation has been observed in the experiments by Pomeau et al. [8] on the Belousov–Zhabotinskii reaction; by Jeffries and Perez [2] on a nonlinear oscillator; and by Bergé et al. [17] on convection. Also, the predicted behavior of the mean time T between bursts, $T \propto (R - R_T)^{-1/2}$, has been observed in the experiments of Jeffries et al.

fig. 4b, shows that the attractor is *strange*–the largest Lyapunov exponent, given by

$$\lambda = \int_0^1 P(X) \left[\ln \left| \frac{df}{dX} \right| \right] dX,$$

where $P(X)\,dX$ is the probability of finding an iterate of the map in the interval $(X, X + dX)$, is positive [7, 11, 12].

The one-dimensional map [fig. 4b] indicates that the attracting sheet seen in cross-section in fig. 4a must exhibit the stretching and folding that is characteristic of strange attractors. This stretching

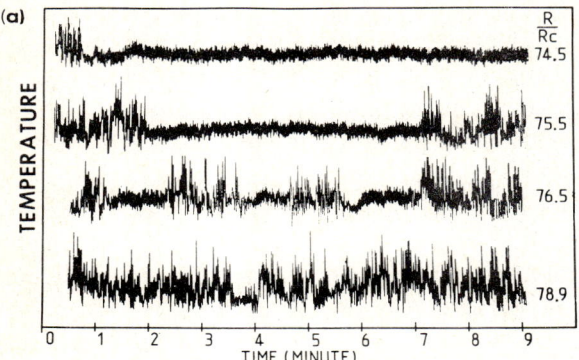

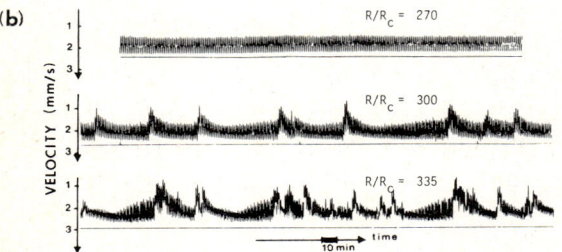

Fig. 5. Intermittency in convection. The turbulent bursts occur with increasing frequency with increasing Rayleigh number. (a) Temperature measurements of Maurer and Libchaber [30]; Prandtl number, 0.62, and aspect ratios $\Gamma_1 = 2.4$ and $\Gamma_2 = 2.0$. (b) Velocity measurements of Bergé et al. [17]; Prandtl number, 130, and aspect ratios $\Gamma_1 = 2.0$ and $\Gamma_2 = 1.2$.

4.2. Frequency locking

In some experiments a transition from a quasiperiodic state to a frequency-locked (periodic) state has been observed with increasing control parameter. The periodic state persists for some range in control parameter and then, in some cases, there is a well-defined transition to a chaotic state, as illustrated by data from a Rayleigh–Bénard experiment shown in fig. 6. This quasiperiodic→locked→chaotic sequence has been discussed theoretically [75–78].

4.3. Periodic–quasiperiodic–chaotic sequence

This transition sequence, first suggested by Ruelle and Takens [82] more than a decade ago, has been observed in many experiments since it was first observed by Gollub and Swinney in 1975 [36]. The sequence is illustrated by data for the Couette–Taylor system in fig. 2.

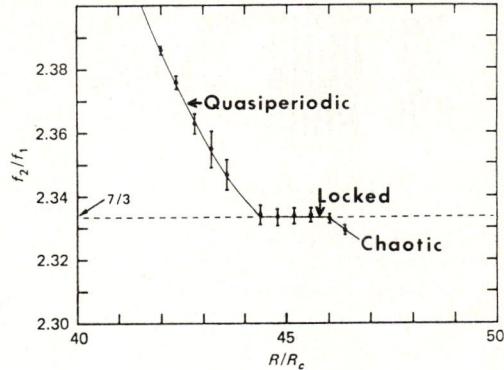

Fig. 6. Frequency locking in the convection experiments of Gollub and Benson [23]. The Prandtl number was 2.5 and the aspect ratios were $\Gamma_1 = 3.5$ and $\Gamma_2 = 2.0$. The curve through the data is drawn to guide the eye.

In the Ruelle–Takens picture [80, 82], when a system makes a transition from a quasiperiodic state with two incommensurate frequencies (a flow on a 2-torus) to quasiperiodic state with three incommensurate frequencies (a flow on a 3-torus), there is in every suitably differentiable neighborhood of the vector field on the 3-torus a vector field which has a strange attractor. Since chaos could thus arise from infinitesimal perturbations of a 3-frequency state, states with three independent frequencies would usually not be observed. (Three independent frequencies have, in fact, been seen in only a few experiments; see [20] and [24].)

An alternative theoretical picture of the quasiperiodic–chaotic transition has recently been developed by Rand et al. [81], Shenker [84], and Feigenbaum et al. [79], and is described in the papers of Shenker [83] and Siggia [85] in this volume. In this theory the 2-torus develops wrinkles as the onset of chaos is approached, and the corresponding power spectrum has a self-similar structure, at least for a system with the ratio of frequencies near the Golden Mean, $(5^{1/2} - 1)/2$. Although the detailed predictions have been developed for frequencies in the ratio of the Golden Mean, the breakdown of the torus is predicted to occur for other irrational frequency ratios. Thus far there have been no experimental observations of the predicted self-similar spectrum near the

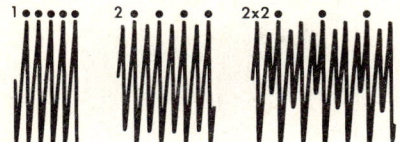

Fig. 7. Period doubling sequence time series with periods $\tau(115\,\text{s})$, 2τ, and $2^2\tau$, obtained in experiments on the Belousov–Zhabotinskii reaction; period $2^3\tau$ was also observed. The quantity measured was the bromide ion potential. The dots above the time series are separated by one period. (From [13].)

onset of chaos, but experiments are underway in several laboratories on periodically driven oscillating systems where the frequency ratio can be adjusted to the Golden Mean.

4.4 Period doubling

The period doubling route to chaos [87–90] has been observed in experiments on Rayleigh–Bénard convection [21, 24, 26, 29], nonlinear electrical oscillators [3, 4], acoustics [48–50], shallow water waves [46–47], a hybrid optical system [45], and the Belousov–Zhabotinskii reaction [13]. At least two or three period doublings were observed in each of these experiments; for example, see fig. 7. The measured values of Feigenbaum's universal number δ [87–90] (which describes asymptotically the ratio of successive intervals in the bifurcation parameter between period doubling transitions) and the scaling parameter α are consistent with the theory for one-dimensional maps with a single extremum. However, the experimental values of δ and α are accurate to only about 5% at best [4] because the rapid convergence rate of the doubling sequence makes it very difficult to observe many doublings.

In systems with many active degrees of freedom, departures from the period doubling sequence are observed. In Rayleigh–Bénard convection Arneodo et al. [86] have shown that this departure can be understood in terms of a two-dimensional Hénon-like map.

4.5. The U-sequence

Universality in the period doubling sequence for one-dimensional maps is now well known. Perhaps less well known is the U (universal)-sequence that occurs beyond the accumulation point (2^∞-cycle) of the 2^n-sequence. Metropolis, Stein, and Stein [93] found, several years before the universal scaling properties of one-dimensional quadratic maps were discovered by Feigenbaum, that one-dimensional maps with a single extremum (not necessarily quadratic) exhibit universal dynamics as a function of the bifurcation parameter. Beyond the period doubling sequence, which is an infinite sequence of doublings of a periodic state with one oscillation per period, periodic states with K oscillations per period appear for all natural numbers K, and each of these "K-cycles" undergoes its own infinite period doubling sequence, $2^n K$ [87, 92, 93]. Fig. 8a shows examples of a fundamental 5-cycle, 6-cycle, and 3-cycle (and the first doubling of the 3-cycle) observed in the Belousov–Zhabotinskii reaction.

The order in which the periodic states appear as a function of bifurcation parameter and the iteration patterns of the corresponding maps are all deduced in the theory using only the single-extremum property of the one-dimensional map [92]. Table I shows all U-sequence states with $K \leqslant 6$. The full U-sequence consists of the (infinitely long) extension of table I to include all the periodic states allowed by the theory. The larger the fundamental

Table I
The U-sequence states with periods up to 6 (in order of occurrence as a function of bifurcation parameter) [93]

Period	Map iteration pattern
1	0
2	0–1
2 × 2	2–0–3–1
6	2–0–4–3–5–1
5	2–0–3–4–1
3	2–0–1
2 × 3	2–5–3–0–4–1
5	2–3–0–4–1
6	2–3–0–4–5–1
4	2–3–0–1
6	2–3–4–0–5–1
5	2–3–4–0–1
6	2–3–4–5–0–1

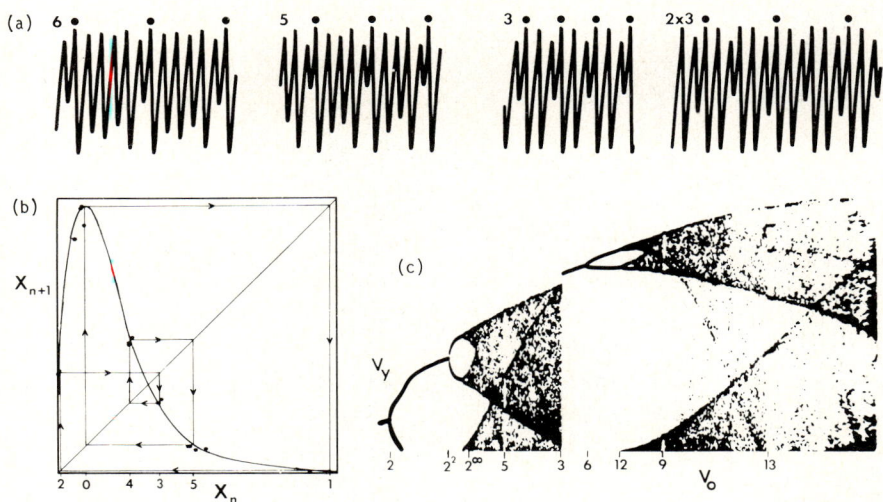

Fig. 8. Observations of the U-sequence. (a) Time series with periods 6τ (where $\tau = 115$ s), 5τ, 3τ, and $2 \times 3\tau$, observed in the Belousov–Zhabotinskii reaction [13]. The dots above the time series are separated by one period. (b) The one-dimensional map constructed from data for the state with period 6τ shown in (a); the iteration pattern is 2–0–4–3–5–1, as predicted (see table I) [13]. (c) A bifurcation diagram obtained in experiments of Testa et al. [4] on a nonlinear electrical oscillator. The vertical axis is the voltage across a varactor [see fig. 1a] and the horizontal axis is the control parameter, the amplitude V_o of the driving voltage. The onsets of some of the U-sequence states are indicated at the bottom of the diagram.

period K, the larger the number of allowed states; there are three distinct allowed 5-cycles, four distinct 6-cycles (see table I), and 27 distinct 9-cycles.

Table I also shows the predicted map iteration patterns—the order of visitation of points on the X-axis—for periodic states with $K \leq 6$. Each iteration pattern occurs only once, and for a given value of the bifurcation parameter not more than one periodic state is stable. An experimentally determined map illustrating the iteration pattern for a period-six state is shown in fig. 8b.

Beyond the $2^\infty K$-cycle of each period doubling sequence there is a chaotic reverse bifurcation sequence, as discussed by Lorenz [60]; although the chaotic states do not exist for intervals in bifurcation parameter, the set of bifurcation parameter values for which the behavior is chaotic has positive measure [56]. Both chaotic and periodic states can be seen in the bifurcation diagram obtained for a nonlinear oscillator shown in fig. 8c.

Many states of the U-sequence have been observed in experiments on nonlinear electrical oscillators [4], the Belousov–Zhabotinskii reaction [13], and Rayleigh–Bénard convection in a magnetic field [26]. The observed iteration patterns and ordering of the states are in accord with the theory for one-dimensional maps.

4.6. Alternating periodic–chaotic sequences

Fig. 9a shows an alternating periodic–chaotic transition sequence observed in an experiment on the Belousov–Zhabotinskii reaction [11, 14]; time series for the first three periodic states (P_1^0, P_1^1, and P_1^2) are shown in figs. 9b–d, respectively, and the time series for the third chaotic state ($C_1^{2,3}$), which occurs between P_1^2 and P_1^3, is shown in fig. 9e. [Notation: P = periodic, C = chaotic. The subscript (superscript) is the number of large (small) amplitude oscillations per period; see fig. 9.] Alternating periodic–chaotic transition sequences similar to that in fig. 9 have been observed in other experiments on the Belousov–Zhabotinskii reaction [6, 7, 10, 15] (for rather different control parameters) and in an experiment on a driven Josephson junction [42]. In addition, alternating

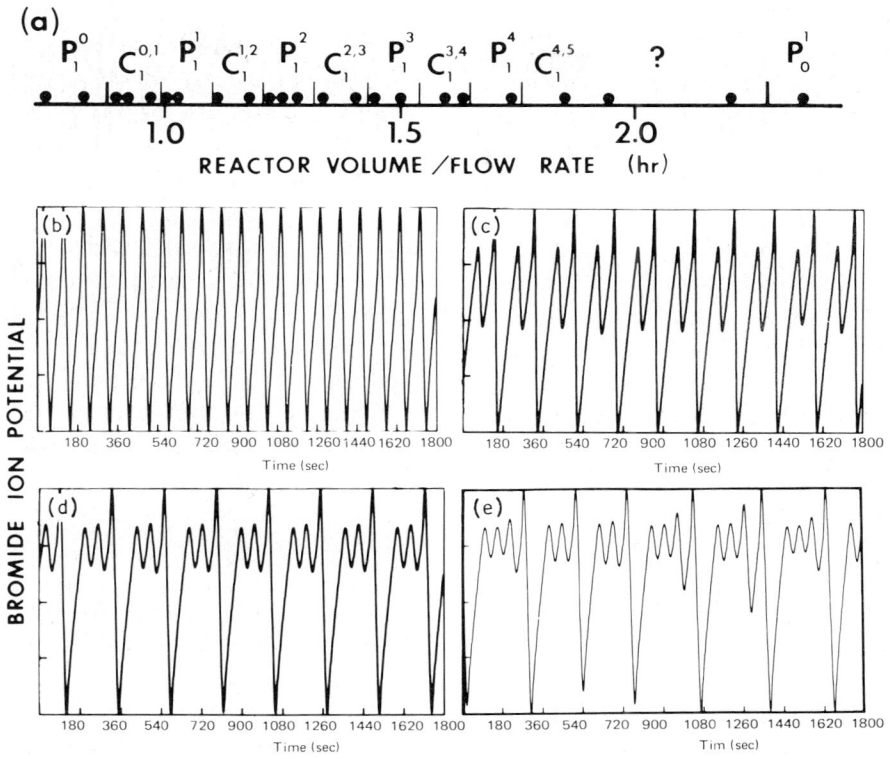

Fig. 9. (a) An alternating periodic-chaotic sequence observed in the Belousov–Zhabotinskii reaction. (b) A time series for the first periodic state, P_1^0. (c) P_1^1. (d) P_1^2. (e) A time series for the third chaotic state, $C_1^{2,3}$, where the number of small amplitude oscillations following each large amplitude oscillation is either two or three but is unpredictable. (From [11, 14].)

periodic–chaotic sequences have been found in studies of models of the Belousov–Zhabotinskii reaction [14, 97, 98] and the Josephson junction [94] (or, equivalently, the forced pendulum [95]) and in a symbolic dynamics analysis of a driven van der Pol oscillator [96]. A one-dimensional map that has an alternating periodic–chaotic sequence is described in the paper by Roux [9] in this volume.

Perhaps these alternating periodic–chaotic sequences have different mathematical descriptions; nevertheless, several common features can be noted: (1) The sequences are finite, not infinite; successive states exist for comparable ranges in control parameter. (2) Successive periodic states are simply related; for example, the states P_1^0, P_1^1, and P_1^2 in figs. 9b–d have (in each period) one large amplitude oscillation and, respectively 0, 1, and 2 small amplitude oscillations. (3) The chaotic states are mixtures of nearby periodic states; for example, $C_1^{n,n+1}$ is a nonperiodic mixture of states P_1^n and P_1^{n+1} (and perhaps occasional cycles of P_1^{n-1} and P_1^{n+2}), as fig. 9e illustrates. (4) The route by which a periodic state becomes chaotic has not been established in most cases, but presumably the transition occurs through period doubling or intermittency (see [9]). For the data in fig. 9 the P_1^0 to $C_1^{0,1}$ transition occurs through period doubling, as fig. 7 illustrates. (5) Each "chaotic" regime can contain many subintervals that are periodic. For example, the U-sequence states shown in fig. 8 occur within the $C_1^{0,1}$ regime.

4.7. Soft mode instability

Langford et al. [101] predicted that an instability associated with a low frequency mode could result from the nonlinear competition between a symmetry-breaking linear instability and oscillatory instability (see also [99, 100]); just above the instability the soft-mode frequency would increase linearly as a function of the bifurcation parameter. Such an instability has been observed in the convection experiments of Libchaber [26] described in this volume; he found that the system gradually became chaotic with increasing Rayleigh number beyond the instability.

5. Discussion

In section 4 we have considered some common features of transitions observed in experiments on diverse systems. While it is natural to focus on common features, it should be emphasized that the range of dynamical behavior that has been observed is quite large. We consider now some observations not mentioned in section 4.

Aspect ratio dependence. In experiments on Rayleigh–Bénard convection Ahlers and co-workers [16] found that as the aspect ratio was increased, nonperiodic behavior occurred at lower and lower Rayleigh numbers. A similar dependence on aspect ratio was subsequently observed in other hydrodynamic experiments [27, 32, 34, 37]. The number of accessible modes and the equilibration time both increase rapidly with increasing aspect ratio; therefore, a large aspect ratio system is especially susceptible to small external perturbations and apparently never settles down into an ideal ordered state. Thus the nonperiodic behavior observed in large aspect ratio systems at small Rayleigh or Reynolds numbers may not correspond to deterministic chaos.

Nonuniqueness. It is widely recognized that nonlinear systems can have two or more stable states at a given set of values of the control parameters, but the extreme degree of nonuniqueness in real systems is not often appreciated. For example, experiments in our laboratory indicate that a Couette–Taylor system with $R_i = 10R_c$, $R_o = 0$, $a/b = 0.88$ and $\Gamma = 30$ has more than 100 different stable states, some periodic, some quasiperiodic, and some chaotic! Each of these states corresponds to a phase space attractor which has its own basin of attraction (set of initial conditions for which the system will asymptotically approach that attractor). There is no systematic way to determine if all basins of attraction have been discovered, even at particular values of the control parameters. In fact, two independent investigators working on the same kind of system at the same control parameters could observe quite different phenomena because of different Reynolds number histories.

Multiple control parameters. Transition sequences are usually investigated as a function of a single control parameter (e.g., voltage, flow rate, Rayleigh number, or Reynolds number; see section 1). New kinds of bifurcations are possible when the dependence on two control parameters is considered [56]. The bifurcations that can occur with more than two control parameters have not been classified, but the experiments of Andereck et al. [31] and King and Swinney [38] on the Couette–Taylor system as a function of R_i, R_o, Γ, and a/b reveal an incredible richness in dynamical behavior.

Summary. In view of the great variety of behavior observed in experiments on nonlinear systems, it would be premature at this time to make sweeping generalizations about routes to chaos. Nevertheless, it is encouraging that a small number of common transition scenarios, as described in section 4, are beginning to emerge from theory and experiment.

Acknowledgements

I am happy to acknowledge that this research was conducted in collaboration with the University of Texas nonlinear dynamics group; the individual collaborations are cited in the text. This research is supported by National Science Foundation Grants MEA79-09585 and CHE79-23627 and by The Robert A. Welch Foundation Grant F-805.

References (Alphabetical under each subheading)

Experiments

Nonlinear oscillators (Analog computer studies of nonlinear oscillations are also described in [51, 57, 63, 64, 70])

[1] J.P. Gollub, E.J. Romer and J.E. Socolar, "Trajectory divergence for coupled relaxation oscillators: measurements and models", J. Stat. Phys. 23 (1980) 321.

[2] C. Jeffries and J. Perez, "Observation of a Pomeau-Manneville intermittent route to chaos in a nonlinear oscillator", Phys. Rev. A26 (1982) 2117.

[3] P.S. Linsay, "Period doubling and chaotic behavior in a driven anharmonic oscillator," Phys. Rev. Lett. 47 (1981) 1349.

[4] J.S. Testa, J. Pérez and C. Jeffries, "Evidence for universal chaotic behavior of a driven nonlinear oscillator", Phys. Rev. Lett. 48 (1982) 714; "Evidence for bifurcation and universal chaotic behavior in nonlinear semiconducting devices", LBL Report 13719, Dec 1981.

Belousov–Zhabotinskii reaction

[5] I.R. Epstein, "Oscillations and chaos in chemical systems", Physica 7D (1983) 47.

[6] J.L. Hudson, M. Hart and D. Marinko, "An experimental study of multiple peak periodic and nonperiodic oscillations in the Belousov–Zhabotinskii reaction", J. Chem. Phys. 71 (1979) 1601.

[7] J.L. Hudson and J.C. Mankin, "Chaos in the Belousov–Zhabotinskii reaction", J. Chem. Phys. 74 (1981) 6171.

[8] Y. Pomeau, J.C. Roux, A. Rossi, S. Bachelart and C. Vidal, "Intermittent behavior in the Belousov–Zhabotinsky reaction", J. Phys. Lett. 42 (1981) L271.

[9] J.C. Roux, "Experimental studies of bifurcations leading to chaos in the Belousov–Zhabotinsky reaction", Physica 7D (1983) 57.

[10] J.C. Roux, A. Rossi, S. Bachelart and C. Vidal, "Experimental observations of complex dynamical behavior during a chemical reaction", Physica 2D (1981) 395.

[11] J.C. Roux, J.S. Turner, W.D. McCormick and H.L. Swinney, "Experimental observations of complex dynamics in a chemical reaction", in Nonlinear Problems: Present and Future, A.R. Bishop, D.K. Campbell and B. Nicolaenko, eds (North-Holland, Amsterdam, 1982), p. 409; J.C. Roux and H.L. Swinney, "Topology of chaos in a chemical reaction", in Nonlinear Phenomena in Chemical Dynamics, C. Vidal and A. Pacault, eds (Springer, Berlin, 1981), p. 81.

[12] J.C. Roux, R.H. Simoyi and H.L. Swinney, "Observation of a strange attractor", Physica 8D (1983) 257.

[13] R.H. Simoyi, A. Wolf and H.L. Swinney, "One-dimensional dynamics in a multi-component chemical reaction", Phys. Rev. Lett. 49 (1982) 245.

[14] J.S. Turner, J.C. Roux, W.D. McCormick and H.L. Swinney, "Alternating periodic and chaotic regimes in a chemical reaction–experiment and theory", Phys. Lett. 85A (1981) 9.

[15] C. Vidal, J.C. Roux, S. Bachelart, "Experimental study of the transition to turbulence in the Belousov–Zhabotinskii reaction", N.Y. Acad. Sci. 357 (1980) 377.

Rayleigh–Bénard convection

[16] G. Ahlers and R. Behringer, "Evolution of turbulence from the Rayleigh–Bénard instability", Phys. Rev. Lett. 40 (1978) 712; G. Ahlers and R.W. Walden, "Turbulence near onset of convection", Phys. Rev. Lett. 44 (1980) 445.

[17] P. Bergé, M. Dubois, P. Manneville and Y. Pomeau, "Intermittency in Rayleigh–Bénard convection", J. Phys. Lett. 41 (1980) L341.

[18] F.H. Busse, "Transition to turbulence in Rayleigh–Bénard convection", in Hydrodynamic Instabilities and the Transition to Turbulence, H.L. Swinney and J.P. Gollub, eds (Springer, Berlin, 1981), p. 97.

[19] M. Dubois and P. Bergé, "Experimental evidence for the oscillators in a convective biperiodic regime", Phys. Lett. 76A (1980) 53.

[20] S. Fauve and A. Libchaber, "Rayleigh–Bénard experiment in a low Prandtl number fluid, mercury", in Synergetics Conference, H. Haken, ed. (Springer, Berlin, 1981).

[21] M. Giglio, S. Musazzi and U. Perini, "Transition to chaotic behavior via a reproducible sequence of period doubling bifurcations", Phys. Rev. Lett. 47 (1981) 243.

[22] J.P. Gollub, "What causes noise in a convecting fluid", Physica 118A (1983) 28.

[23] J.P. Gollub and S.V. Benson, "Phase locking in the oscillations leading to turbulence", in Pattern Formation and Pattern Recognition, H. Haken, ed (Springer, Berlin, 1979).

[24] J.P. Gollub and S.V. Benson, "Many routes to turbulent convection", J. Fluid Mech. 100 (1980) 449.

[25] J.P. Gollub, A.R. McCarriar and J.F. Steinman, "Convective pattern evolution and secondary instabilities", J. Fluid Mech., 125 (1982) 259; J.P. Gollub and A.R. McCarriar, "Convection Patterns in Fourier Space," Phys. Rev. A 26 (1982) 3470.

[26] A. Libchaber, S. Fauve, and C. Laroche, "Two parameter study of the routes to chaos", Physica 7D (1983) 73; A. Libchaber, C. Laroche & S. Fauve, "Period doubling cascade in mercury, a quantitative measurement", J. Phys. Lett 43 (1982) L211.

[27] A. Libchaber and J. Maurer, "Local probe in a Rayleigh–Bénard experiment in liquid helium", J. Phys. Lett. (Paris) 39 (1978) L369.

[28] H. Haucke and Y. Maeno, "Phase space analysis of convection in ^{3}He-superfluid ^{4}He solution", Physica 7D (1983) 69.

[29] J. Maurer and A. Libchaber, "Rayleigh–Bénard experi-

ment in liquid helium; frequency locking and the onset of turbulence", J. Phys. Lett. (Paris) 40 (1979) L419.
[30] J. Maurer and A. Libchaber, "Effect of the Prandtl number on the onset of turbulence in liquid helium", J. Phys. Lett. 41 (1980) L515. See also [86].

Couette–Taylor system

[31] C.D. Andereck, S.S. Liu and H.L. Swinney, "Flow between independently rotating concentric cylinders", to be published.
[32] T.B. Benjamin and T. Mullin, "Anomalous Modes in the Taylor experiment", Proc. R. Soc. Lond. A377 (1981) 221.
[33] R.C. Di Prima and H.L. Swinney, "Instabilities and transition in flow between concentric rotating cylinders", in Hydrodynamic Instabilities and the Transition to Turbulence, H.L. Swinney and J.P. Gollub, eds. (Springer, Berlin, 1981), p. 139.
[34] R.J. Donnelly, K. Park, S. Shaw and R.W. Walden, "Early nonperiodic transitions in Couette flow", Phys. Rev. Lett. 44 (1980) 987.
[35] P.R. Fenstermacher, H.L. Swinney and J.P. Gollub, "Dynamical instabilities and the transition to chaotic Taylor vortex flow", J. Fluid Mech. 94 (1979) 103.
[36] J.P. Gollub and H.L. Swinney, "Onset of turbulence in a rotating fluid", Phys. Rev. Lett. 35 (1975) 927.
[37] A. Lorenzen, G. Pfister and T. Mullin, "End effects on the transition time-dependent motion in the Taylor experiment", Phys. Fluids 26 (1983) 10.
[38] G. King and H.L. Swinney, "Limits of stability and defects in wavy vortex flow" Phys. Rev. A27 (1983) 1240.
[39] V.S. L'vov and A.A. Predtechensky, "On Landau and stochastic attractor pictures in the problem of transition to turbulence", Physica 2D (1981) 38; V.S. L'vov, A.A. Predtechenskii and A.I. Chernykh, "Bifurcation and chaos in a system of Taylor vortices: a natural and numerical experiment", Soviet Physics JETP 53 (1981) 562.
[40] R. Shaw, C.D. Andereck, L.A. Reith and H.L. Swinney, "Superposition of traveling waves in the circular Couette system", Phys. Rev. Lett. 48 (1982) 1172.
[41] H.L. Swinney and J.P. Gollub, "The transition to turbulence", Physics Today 31, No. 8 (August 1978) 41.

Other systems

[42] J. Clarke and R. Koch, private communication (1982).
[43] D. Farmer, J. Hart and P. Weidman, "A phase space analysis of a baroclinic flow", Phys. Lett. 91A (1982) 22.
[44] M. Guevara, L. Glass and A. Shrier, "Phase locking, period-doubling bifurcations, and irregular dynamics in periodically stimulated cardiac cells", Science 214 (1981) 350. L. Glass, M. R. Guevara, and A. Shrier "Bifurcation and chaos in a periodically stimulated cardiac oscillator", Physica 7D (1983) 89.
[45] H.M. Gibbs, F.A. Hopf, D.L. Kaplan and R.L. Shoemaker, "Observation of chaos in optical bistability", Phys. Rev. Lett. 46 (1981) 474.
[46] J.P. Gollub and C.W. Meyer, "Symmetry-breaking instabilities on a fluid surface", Physica A, to appear.

[47] R. Keolian, L.A. Turkevich, S.J. Putterman, I. Rudnick and J. Rudnick, "Subharmonic sequences in the Faraday experiment: departures from period doubling", Phys. Rev. Lett. 47 (1981) 1133.
[48] W. Lauterborn and E. Cramer, "Subharmonic route to chaos observed in acoustics", Phys. Rev. Lett. 47 (1981) 1445.
[49] C.W. Smith and M.J. Tejwani, "Bifurcation and the universal sequence for first sound subharmonic generation in superfluid helium-4", Physica 7D (1983) 85.
[50] C.W. Smith, M.J. Tejwani and D.A. Farris, "Bifurcation universality for first-sound subharmonic generation in superfluid helium-4", Phys. Rev. Lett. 48 (1982) 492.

Theory

Phase portrait reconstruction, strange attractors, maps, dimension, etc.

[51] J. Crutchfield, D. Farmer, N. Packard, R. Shaw, G. Jones and R. Donnelly, "Power spectral analysis of a dynamical system", Phys. Lett. 76A (1980) 1.
[52] J.D. Farmer, "Dimension, fractal measures, and chaotic dynamics", in Evolution of Ordered and Chaotic Patterns in Systems Treated by the Natural Sciences and Mathematics, H. Haken, ed. (Springer, Berlin 1983).
[53] J.D. Farmer, E. Ott and J. Yorke, "The dimension of chaotic attractors", Physica 7D (1983) 153.
[54] H.S. Greenside, A. Wolf, J. Swift and T. Pignaturo, "Impracticality of a box-counting algorithm for calculating the dimensionality of strange attractors", Phys. Rev. A 25 (1982) 3453.
[55] J. Guckenheimer, "Persistent properties of bifurcations", Physica 7D (1983) 105. J. Guckenheimer, "On a codimension two bifurcation", in Lecture Notes in Mathematics 898, D.A. Rand and L.S. Young, eds. (Springer, Berlin, 1981), p. 99.
[56] B.-L. Hao and S.-Y. Zhang, "Hierarchy of chaotic bands", J. Stat. Phys. 28 (1982) 769.
[57] P. Holmes, "A nonlinear oscillator with a strange attractor", Phil. Trans. Roy. Soc. A292 (1979) 419; P. Holmes and D. Whitley, "On the attracting set for Duffing's equation II: A geometrical model for moderate force and damping", Physica 7D (1983) 111.
[58] M.V. Jakobson, "Absolutely continuous invariant measures for one-parameter families of one-dimensional maps", Commun. Math. Phys. 81 (1981) 39.
[59] O.E. Lanford, "Strange attractors and turbulence", in Hydrodynamic Instabilities and the Transition to Turbulence, H.L. Swinney and J.P. Gollub, eds. (Springer, Berlin, 1981), p. 12.
[60] E. Lorenz, Ann. N.Y. Acad. Sci. 357 (1980) 282.
[61] B. Mandelbrot, Fractuals, Form, Chance, and Dimension (Freeman, San Francisco, 1977).
[62] E. Ott, "Strange attractors and chaotic motions of dynamical systems", Rev. Mod. Phys. 53 (1981) 655.
[63] N.H. Packard, J.P. Crutchfield, J.D. Farmer and R.S. Shaw, "Geometry from a time series", Phys. Rev. Lett. 45

(1980) 712. J.P. Crutchfield and N.H. Packard, "Symbolic dynamics of noisy chaos", Physica 7D (1983) 201.
[64] O. Rössler, Z. Naturforsch 31a (1976) 1664.
[65] D. Ruelle, private communication.
[66] D. Ruelle, "Sensitive dependence on initial condition and turbulent behavior of dynamical systems", Ann. N.Y. Acad. Sci. 316 (1979) 408.
[67] D. Ruelle, "Strange attractors", The Mathematical Intelligencer 2 (1980) 126.
[68] R. Shaw, "Strange attractors, chaotic behavior, and information flow", Z. Naturforsch 36a (1981) 80.
[69] F. Takens, "Detecting strange attractors in turbulence", in Lecture Notes in Mathematics 898, D.A. Rand and L.S. Young, eds. (Springer, Berlin, 1981), p. 366.
[70] Y. Ueda, "Randomly transitional phenomena in the system governed by Duffing's equation", J. Stat. Phys. 20 (1979) 181.
[71] H. Whitney, "Differentiable manifolds", Ann. Math. 37 (1936) 645.
[72] J.A. Yorke and E.D. Yorke, "Chaotic behavior and fluid dynamics", in Hydrodynamic Instabilities and the Transition to Turbulence, H.L. Swinney and J.P. Gollub, eds. (Springer, Berlin, 1981), p. 77.

Intermittency (see experiments [2, 8, 17, 24, 30])
[73] J.E. Hirsch, B.A. Huberman and D.J. Scalapino, "Theory of intermittency", Phys. Rev. A25 (1982) 519.
[74] Y. Pomeau and P. Manneville, "Intermittent transition to turbulence in dissipative dynamical systems", Commun. Math. Phys. 74 (1980) 189.

Frequency locking–chaotic transition (see experiments [23, 24, 29])
[75] D.G. Aronson, M.A. Chory, G.R. Hill and R.P. McGehee, "Bifurcations from an invariant circle for two-parameter families of maps of the plane", Communications in Mathematical Physics 83 (1982) 303; "Resonance phenomena for two parameter families of maps of the plane", in Nonlinear Dynamics and Turbulence, G.I. Barrenblatt, G. Iooss and D.D. Joseph, eds. (Pittman, London, 1982).
[76] L. Glass and R. Perez, "Fine structure of phase locking", Phys. Rev. Lett. 48 (1982) 1772.
[77] P.H. Steen and S.H. Davis, "Quasiperiodic bifurcation in nonlinearly coupled oscillators near a point of strong resonance", SIAM J. Appl. Math., 42 (1982) 1345.
[78] E.C. Zeeman and D.A. Rand, to be published.

Quasiperiodic–chaotic transition (see experiments [24, 27, 30, 33, 35, 39])
[79] M.J. Feigenbaum, L.P. Kadanoff and S.J. Shenker, "Quasiperiodicity in dissipative systems: a renormalization group analysis", Physica 5D (1982) 370.
[80] S. Newhouse, D. Ruelle and F. Takens, "Occurrence of strange Axiom A attractors near quasiperiodic flows on $T^m, m \geq 3$", Commun. Math. Phys. 64 (1978) 35.
[81] D. Rand, S. Ostlund, J. Sethna and E. Siggia, "A universal transition from quasi-periodicity to chaos in dissipative systems", Phys. Rev. Lett., 49 (1982) 132. See also [85].
[82] D. Ruelle and F. Takens, "On the nature of turbulence", Commun. Math. Phys. 20 (1971) 167.
[83] S. Shenker, "Quasiperiodicity in dissipative systems: a renormalization group analysis", Physica 7D (1983) 301 (*Abstract*).
[84] S.J. Shenker, "Scaling behavior in a map of a circle onto itself: empirical results", Physica 5D (1982) 405.
[85] E. Siggia, "A universal transition from quasiperiodicity to chaos," Physica 7D (1983) 302 (*Abstract*).

Period doubling (see experiments [3, 4, 13, 21, 24, 26, 44–50])
[86] A. Arneodo, P. Coullet, C. Tresser, A. Libchaber, J. Maurer and D. d'Humieres, "About the observation of the uncompleted cascade in Rayleigh–Bénard experiment", Physica 6D (1983) 385. See also [28].
[87] P. Collet and J.P. Eckmann, Iterated Maps of the Interval as Dynamical Systems (Birkhaüser, Boston, 1980).
[88] O. Lanford, "Period doubling in one and several dimensions", Physica 7D (1983) 124.
[89] M.J. Feigenbaum, "Quantitative universality for a class of nonlinear transformations", J. Stat. Phys. 19 (1978) 25.
[90] M.J. Feigenbaum, "Universal behavior in nonlinear systems", Physica 7D (1983) 16.

U-sequence (see experiments [4, 13])
[92] J. Guckenheimer, "On the bifurcation of maps of the interval", Inventiones Mathematicae 39 (1977) 165.
[93] N. Metropolis, M.L. Stein and P.R. Stein, "On finite limit sets for transformations on the unit interval", J. Comb. Theory A 15 (1973) 25; see also P.J. Myrberg, Ann. Akad. Sc. Fennicae A, I, No. 336/3 (1963).

Alternating periodic–chaotic sequence (see experiments [6, 7, 9, 10, 11, 14, 15, 42])
[94] E. Ben-Jacob, I. Goldhirsch, Y. Imry, and S. Fishman, "Intermittent chaos in Josephson Junctions", Phys. Rev. Lett. 49 (1982) 1599.
[95] D. D'Humieres, M.R. Beasley, B.A. Huberman and A. Libchaber, "Chaotic states and routes to chaos in the forced pendulum", Phys. Rev. A26 (1982) 3483.
[96] M. Levi, Qualitative Analysis of the Periodically Forced Relaxation Oscillations, Memoirs Am. Math. Soc. No. 244 (1981).
[97] A.S. Pikovsky and M.I. Rabinovich, "Stochastic oscillations in dissipative systems", Physica 2D (1981) 8.
[98] K. Tomita and I. Tsuda, "Towards the interpretation of Hudson's experiment on the Belousov–Zhabotinsky reaction–chaos due to delocalization", Prog. Theor. Phys. 64 (1980) 1138.

Soft-mode instability (see experiment [26])
[99] P. Holmes, "Unfolding a degenerate nonlinear oscillator", Ann. N.Y. Acad. Sci. 357 (1980) 473.
[100] G. Iooss and W.F. Langford, "Conjectures on the routes to turbulence via bifurcations", Ann. N.Y. Acad. Sci. 357 (1980) 489.
[101] W.F. Langford, A. Arnedo, P. Coullet, C. Tresser and J. Coste, "A mechanism for a soft mode instability", Phys. Lett. 78A (1980) 11.

UNIVERSAL BEHAVIOR IN NONLINEAR SYSTEMS*

Mitchell J. FEIGENBAUM

Center for Nonlinear Studies, Los Alamos National Laboratory, Los Alamos, NM 87545, USA

A semipopular account of the universal scaling theory for the period doubling route to chaos is presented.

1. Introduction

There exist in nature processes that can be described as complex or chaotic and processes that are simple or orderly. Technology attempts to create devices of the simple variety: an idea is to be implemented, and various parts executing orderly motions are assembled. For example, cars, airplanes, radios, and clocks are all constructed from a variety of elementary parts each of which, ideally, implements one ordered aspect of the device. Technology also tries to control or minimize the impact of seemingly disordered processes, such as the complex weather patterns of the atmosphere, the myriad whorls of turmoil in a turbulent fluid, the erratic noise in an electronic signal, and other such phenomena. It is the complex that interest us here.

When a signal is noisy, its behavior from moment to moment is irregular and has no simple pattern of prediction. However, if we analyze a sufficiently long record of the signal, we may find that signal amplitudes occur within narrow ranges a definite fraction of the time. Analysis of another record of the signal may reveal the same fraction. In this case, the noise can be given a *statistical* description. This means that while it is impossible to say what amplitude will appear next in succession, it is possible to estimate the probability or likelihood that the signal will attain some specified range of values. Indeed, for the last hundred years disorderly processes have been taken to be statistical (one has given up asking for a precise causal prediction), so that the goal of a description is to determine what the probabilities are, and from this information to determine various behaviors of interest – for example, how air turbulence modifies the drag on an airplane.

We know that perfectly definite causal and *simple* rules can have statistical (or random) behaviors. Thus, modern computers possess "random number generators" that provide the statistical ingredient in a simulation of an erratic process. However, this generator does nothing more than shift the decimal point in a rational number whose repeating block is suitably long. Accordingly, it is possible to predict what the nth generator number will be. Yet, in a list of successive generated numbers there is such a seeming lack of order that all statistical tests will confer upon the numbers a pedigree of randomness. Technically, the term "pseudorandom" is used to indicate this nature. One now may ask whether the various complex processes of nature themselves might not be merely pseudorandom, with the full import of randomness, which is untestable, a historic but misleading concept. Indeed our purpose here is to explore this possibility. What will prove altogether remarkable is that some very simple schemes to produce erratic

---

* Reprinted with minor additions and with permission from *Los Alamos Science*, Vol. I, No. 1, p. 4–27 (1980).

numbers behave *identically* to some of the erratic aspects of natural phenomena. More specifically, there is now cogent evidence that the problem of how a fluid changes over from smooth to turbulent flow can be solved through its relation to the simple scheme described in this article. Other natural problems that can be treated in the same way are the behavior of a population from generation to generation and the noisiness of a large variety of mechanical, electrical, and chemical oscillators. Also, there is now evidence that various Hamiltonian systems – those subscribing to classical mechanics, such as the solar system – can come under this discipline.

The feature common to these phenomena is that, as some external parameter (temperature, for example) is varied, the behavior of the system changes from simple to erratic. More precisely, for some range of parameter values, the system exhibits an orderly *periodic* behavior; that is, the system's behavior reproduces itself every *period* of time T. Beyond this range, the behavior fails to reproduce itself after T seconds; it almost does so, but in fact it requires *two* intervals of T to repeat itself. That is, the period has *doubled* to $2T$. This new periodicity remains over some range of parameter values until another critical parameter value is reached after which the behavior *almost* reproduces itself after $2T$, but in fact, it now requires $4T$ for reproduction. This process of successive period doubling recurs continually (with the range of parameter values for which the period is $2^n T$ becoming successively smaller as n increases) until, at a certain value of the parameter, it has doubled ad infinitum, so that the behavior is no longer periodic. Period doubling is then a characteristic route for a system to follow as it changes over from simple periodic to complex aperiodic motion. All the phenomena mentioned above exhibit period doubling. In the limit of aperiodic behavior, there is a unique and hence *universal* solution common to all systems undergoing period doubling. This fact implies remarkable consequences. For a given system, if we denote by Λ_n the value of the parameter at which its period doubles for the nth time, we find that the values Λ_n converge to Λ_∞ (at which the motion is aperiodic) *geometrically* for large n. This means that

$$\Lambda_\infty - \Lambda_n \propto \delta^{-n} \tag{1}$$

for a fixed value of δ (the *rate* of onset of complex behavior) as n becomes large. Put differently, if we define

$$\delta_n \equiv \frac{\Lambda_{n+1} - \Lambda_n}{\Lambda_{n+2} - \Lambda_{n+1}}, \tag{2}$$

δ_n (quickly) approaches the constant value δ. (Typically, δ_n will agree with δ to several significant figures after just a few period doublings.) What is quite remarkable (beyond the fact that there is always a geometric convergence) is that, for all systems undergoing this period doubling, the value of δ is *predetermined* at the universal value [1, 2]

$$\delta = 4.6692016\ldots \tag{3}$$

Thus, this definite number must appear as a natural rate in oscillators, populations, fluids, and all systems exhibiting a period-doubling route to turbulence! In fact, most measurable properties of *any* such system in this aperiodic limit now can be determined, in a way that essentially bypasses the details of the equations governing each specific system because the theory of this behavior is universal over such details. That is, so long as a system possesses certain *qualitative* properties that enable it to undergo this route to complexity, its *quantitative* properties are determined. (This result is analogous to the results of the modern theory of critical phenomena, where a few qualitative properties of the system undergoing a phase transition, notably the dimensionality, determine *universal* critical exponents. Indeed at a *formal* level the two theories are identical in that they are fixed-point theories, and the number δ, for example, can be viewed as a critical exponent.) Accordingly, it is sufficient to study the simplest system exhibiting this phenomenon to comprehend the general case.

2. Functional iteration

A random number generator is an example of a simple iteration scheme that has complex behavior. Such a scheme generates the next pseudorandom number by a definite transformation upon the present pseudorandom number. In other words, a certain function is reevaluated successively to produce a sequence of such numbers. Thus, if f is the function and x_0 is a starting number (or "seed"), then $x_0, x_1, \ldots, x_n, \ldots$, where

$$x_1 = f(x_0),$$
$$x_2 = f(x_1),$$
$$\vdots$$
$$x_{n+1} = f(x_n), \qquad (4)$$
$$\vdots$$

is the sequence of generated pseudorandom numbers. That is, they are generated by *functional iteration*. The nth element in the sequence is

$$x_n = f(f(\ldots f(f(x_0))\ldots)) \equiv f^n(x_0), \qquad (5)$$

where n is the total number of applications of f. [$f^n(x)$ is not the nth power of $f(x)$; it is the nth *iterate* of f.] A property of iterates worthy of mention is

$$f^n(f^m(x)) = f^m(f^n(x)) = f^{m+n}(x), \qquad (6)$$

since each expression is simply $m+n$ applications of f. It is understood that

$$f^0(x) = x. \qquad (7)$$

It is also useful to have a symbol, $\circ$, for functional iteration (or composition), so that

$$f^n \circ f^m = f^m \circ f^n = f^{m+n}. \qquad (8)$$

Now f^n in eq. (5) is itself a definite and computable function, so that x_n as a function of x_0 is known in principle.

If the function f is *linear* as, for example,

$$f(x) = ax \qquad (9)$$

for some constant a, it is easy to see that

$$f^n(x) = a^n x, \qquad (10)$$

so that, for this f,

$$x_n = a^n x_0 \qquad (11)$$

is the solution of the *recurrence relation* defined in eq. (4),

$$x_{n+1} = ax_n. \qquad (12)$$

Should $|a| < 1$, then x_n geometrically converges to zero at the rate $1/a$. This example is special in that the linearity of f allows for the explicit computation of f^n.

We must choose a *nonlinear* f to generate a pseudorandom sequence of numbers. If we choose for our nonlinear f

$$f(x) = a - x^2, \qquad (13)$$

then it turns out that f^n is a polynomial in x of order 2^n. This polynomial rapidly becomes unmanageably large; moreover, its coefficients are polynomials in a of order up to 2^{n-1} and become equally difficult to compute. Thus even if $x_0 = 0$, x_n is a polynomial in a of order 2^{n-1}. These polynomials are nontrivial as can be surmised from the fact that for certain values of a, the sequence of numbers generated for almost all starting points in the range $(a - a^2, a)$ possess *all* the mathematical properties of a random sequence.

Put differently, applying the simplest of *nonlinear* iteration schemes to itself sufficiently many times can create vastly complex behavior. Yet, precisely because the same operation is reapplied, it is conceivable that only a select few self-consistent patterns might emerge where the consistency is determined by the key notion of iteration and *not* by the particular function performing the iterates.

3. The fixed-point behavior of functional iterations

Let us now make a direct onslaught against eq. (13) to see what it possesses. We want to know the behavior of the system after many iterations. As we already know, high iterates of f rapidly become very complicated. One way this growth can be prevented is to have the first iterate of x_0 be precisely x_0 itself. Generally, this is impossible. Rather this condition *determines* possible x_0's. Such a self-reproducing point is called a *fixed point* of f. The sequence of iterates is then $x_0, x_0, x_0, \ldots$ so that the behavior is *static*, or if viewed as periodic, it has period 1.

It is elementary to determine the fixed points of eq. (13). For future convenience we shall use a modified form of eq. (13) obtained by a translation in x and some redefinitions:

$$f(x) = 4\lambda x(1-x), \tag{15}$$

so that as λ varied, $x = 0$ is always a fixed point. Indeed, the fixed-point condition for eq. (15),

$$x^* = f(x^*) = 4\lambda x^*(1 - x^*), \tag{16}$$

gives as the two fixed points

$$x^* = 0, \quad x_0^* = 1 - 1/4\lambda. \tag{17}$$

The maximum value of $f(x)$ in eq. (15) is attained at $x = \frac{1}{2}$ and is equal to λ. Also, for $\lambda > 0$ and x in the interval $(0, 1)$, $f(x)$ is always positive. Thus, if λ is anywhere in the range $[0, 1]$, then any iterate of any x in $(0, 1)$ is also always in $(0, 1)$. Accordingly, in all that follows we shall consider only values of x and λ lying between 0 and 1. By eq. (16) for $0 \leq \lambda < \frac{1}{4}$, only $x^* = 0$ is within range, whereas for $\frac{1}{4} \leq \lambda \leq 1$, both fixed points are within the range. For example, if we set $\lambda = \frac{1}{2}$ and we start at the fixed point $x_0^* = \frac{1}{2}$ (that is, we set $x_0 = \frac{1}{2}$), then $x_1 = x_2 = \cdots = \frac{1}{2}$; similarly if $x_0 = 0$, $x_1 = x_2 = \cdots = 0$, and the problem of computing the nth iterate is obviously trivial.

What if we choose an x_0 *not* at a fixed point? The easiest way to see what happens is to perform a graphical analysis. We graph $y = f(x)$ together with $y = x$. Where the lines intersect we have $x = y = f(x)$, so that the intersections are precisely the fixed points. Now, if we choose an x_0 and plot it on the x-axis, the ordinate of $f(x)$ at x_0 is x_1. To obtain x_2, we must transfer x_1 to the x-axis before reapplying f. Reflection through the straight line $y = x$ accomplishes precisely this operation. Altogether, to iterate an initial x_0 successively,

1) move *vertically* to the graph of $f(x)$,
2) move *horizontally* to the graph of $y = x$, and
3) repeat steps 1, 2, etc.

Fig. 1 depicts this process for $\lambda = \frac{1}{2}$. The two fixed points are circled, and the first several iterates of an arbitrarily chosen point x_0 are shown. What should be obvious is that if we start from any x_0 in $(0, 1)$ ($x = 0$ and $x = 1$ excluded), upon continued iteration x_n will converge to the fixed point at $x = \frac{1}{2}$. No matter how close x_0 is to the fixed point at $x = 0$, the iterates diverge away from it. Such a fixed point is termed *unstable*. Alternatively, for almost all x_0 near enough to $x = \frac{1}{2}$ [in this case, all x_0 in $(0, 1)$], the iterates converge towards $x = \frac{1}{2}$. Such a fixed point is termed *stable* or is referred to as an *attractor* of period 1.

Now, if we don't care about the *transient* behav-

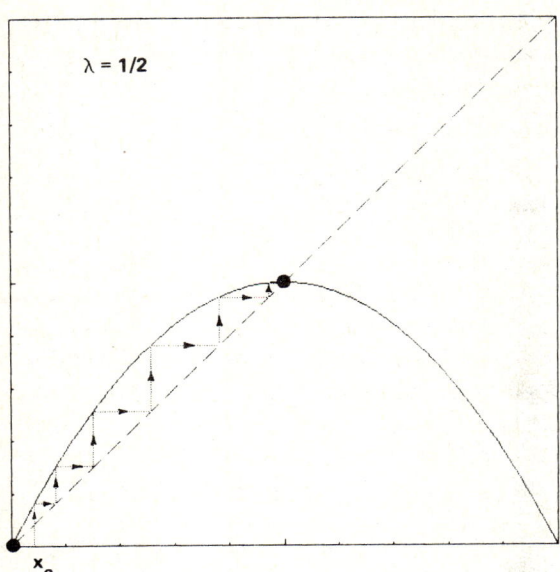

Fig. 1. Iterates of x_0 at $\lambda = 0.5$.

$$= 4\lambda(1-2x) \text{ so that}$$

$$f'(0) = 4\lambda \tag{18}$$

and

$$f'(x_0^*) = 2 - 4\lambda. \tag{19}$$

For $0 < \lambda < \frac{1}{4}$, only $x^* = 0$ is stable. At $\lambda = \frac{1}{4}$, $x_0^* = 0$ and $f'(x_0^*) = 1$. For $\frac{1}{4} < \lambda < \frac{3}{4}$, x^* is unstable and x_0^* is stable, while at $\lambda = \frac{3}{4}$, $f'(x_0^*) = -1$ and x_0^* also has become unstable. Thus, for $0 < \lambda < \frac{3}{4}$, the eventual behavior is known.

ior of the iterates of x_0, but only about some regular behavior that will emerge eventually, then knowledge of the stable fixed point at $x = \frac{1}{2}$ satisfies our concern for the *eventual* behavior of the iterates. In this restricted sense of eventual behavior, the existence of an attractor determines the solution *independently* of the initial condition x_0 provided that x_0 is within the *basin of attraction* of the attractor; that is, that it *is* attracted. The attractor satisfies eq. (16), which is explicitly independent of x_0. This condition is the basic theme of universal behavior: if an attractor exists, the eventual behavior is independent of the starting point.

What makes $x = 0$ unstable, but $x = \frac{1}{2}$ stable? The reader should be able to convince himself that $x = 0$ is unstable because *the slope of $f(x)$ at $x = 0$ is greater than* 1. Indeed, if x^* is a fixed point of f and the derivative of f at x^*, $f'(x^*)$, is smaller than 1 in absolute value, then x^* is stable. If $|f'(x^*)|$ is greater than 1, then x^* is unstable. Also, only *stable* fixed points can account for the eventual behavior of the iterates of an arbitrary point.

We now must ask, "For what values of λ are the fixed points attracting?" By eq. (15) $f'(x)$

4. Period 2 from the fixed point

What happens to the system when λ is in the range $\frac{3}{4} < \lambda < 1$, where there are no attracting fixed points? We will see that as λ increases slightly beyond $\lambda = \frac{3}{4}$, f undergoes period doubling. That is, instead of having a stable cycle of period 1 corresponding to one fixed point, the system has a stable cycle of period 2; that is, the cycle contains two points. Since these two points are fixed points of the function f^2 (f applied twice) and since stability is determined by the slope of a function at its *fixed* points, we must now focus on f^2. First, we examine a graph of f^2 at λ just below $\frac{3}{4}$. Figs. 2a and 2b show f and f^2, respectively, at $\lambda = 0.7$.

To understand fig. 2b, observe first that, since f is symmetric about its maximum at $x = \frac{1}{2}$, f^2 is also symmetric about $x = \frac{1}{2}$. Also, f^2 must have a fixed point whenever f does because the second iterate of a fixed point is still that same point. The main ingredient that determines the period-doubling behavior of f as λ increases is the relationship of the slope of f^2 to the slope of f. This relationship is a consequence of the chain rule. By definition

$$x_2 = f^2(x_0),$$

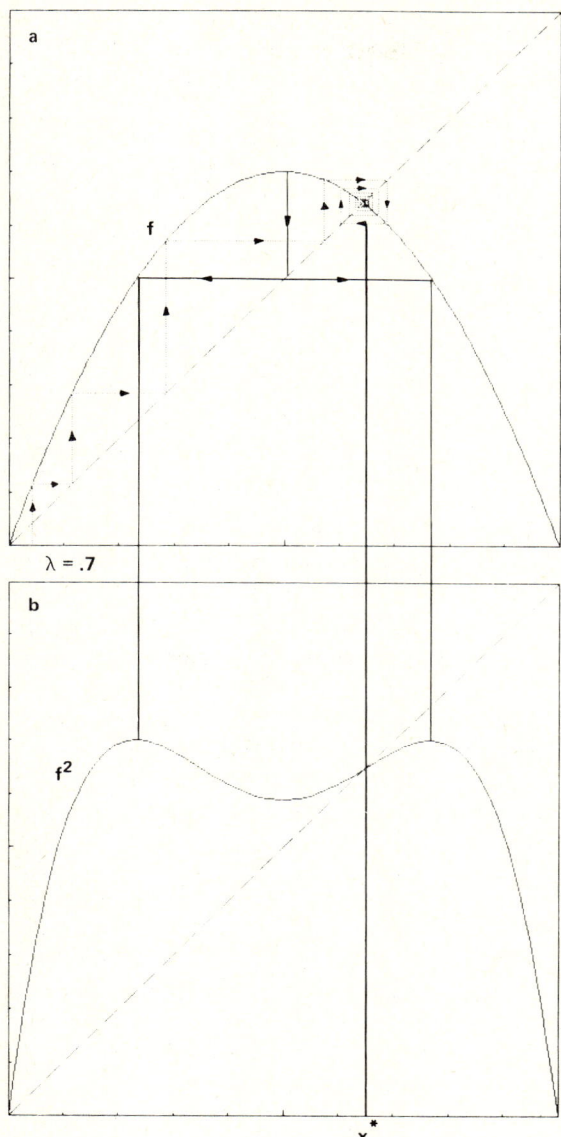

Fig. 2. $\lambda = 0.7$. x^* is the stable fixed point. The extrema of f^2 are located in (a) by constructing the inverse iterates of $x = 0.5$.

where

$$x_1 = f(x_0), \quad x_2 = f(x_1).$$

We leave it to the reader to verify by the chain rule that

$$f^{2'}(x_0) = f'(x_0)f'(x_1) \tag{20}$$

and

$$f^{n'}(x_0) = f'(x_0)f'(x_1)\ldots f'(x_{n-1}), \tag{21}$$

and elementary result that determines period doubling. If we start at a fixed point of f and apply eq. (20) to $x_0 = x^*$, so that $x_2 = x_1 = x^*$, then

$$f^{2'}(x^*) = f'(x^*)f'(x^*) = [f'(x^*)]^2. \tag{22}$$

Since at $\lambda = 0.7$, $|f'(x^*)| < 1$, it follows from eq. (22) that

$$0 < f^{2'}(x^*) < 1.$$

Also, if we start at the extremum of f, so that $x_0 = \frac{1}{2}$ and $f'(x_0) = 0$, it follows from eq. (21) that

$$f^{n'}(\tfrac{1}{2}) = 0 \tag{23}$$

for all n. In particular, f^2 is extreme (and a minimum) at $\frac{1}{2}$. Also, by eq. (20), f^2 will be extreme (and a maximum) at the x_0 that will iterate under f to $x = \frac{1}{2}$, since then $x_1 = \frac{1}{2}$ and $f'(x_1) = 0$. These points, the *inverses* of $x = \frac{1}{2}$, are found by going *vertically* down along $x = \frac{1}{2}$ to $y = x$ and then *horizontally* to $y = f(x)$. (Reverse the arrows in fig. 1, and see fig. 2a.) Since f has a maximum, there are *two* horizontal intersections and, hence, the two maxima of fig. 2b. *The ability of f to have complex behaviors is precisely the consequence of its double-valued inverse*, which is in turn a reflection of its possession of an extremum. A monotone f, one that always increases, *always* has simple behaviors, whether or not the behaviors are easy to compute. A *linear* f is always monotone. The f's we care about always fold over and so are *strongly* nonlinear. This folding nonlinearity gives rise to universality. Just as linearity in any system implies a definite method of solution, folding nonlinearity in any system also implies a definite method of solution. In fact folding nonlinearity in the aperiodic limit of period doubling in any system is solvable, and many systems, such as various coupled nonlinear differential equations, possess this nonlinearity.

To return to fig. 2b, as $\lambda \to \frac{3}{4}$ and the maximum value of f increases to $\frac{3}{4}$, $f'(x^*) \to -1$ and $f^{2\prime}(x^*) \to +1$. As λ increases beyond $\frac{3}{4}$, $|f'(x^*)| > 1$ and $f^{2\prime}(x^*) > 1$, so that f^2 must develop two new fixed points beyond those of f; that is, f^2 will cross $y = x$ at two more points. This transition is depicted in figs. 3a and 3b for f and f^2, respectively, at $\lambda = 0.75$, and similarly in fig. 4a and 4b at $\lambda = 0.785$. (Observe the exceptionally slow convergence to x^* at $\lambda = 0.75$, where iterates approach the fixed point not geometrically, but rather with deviations from x^* inversely proportional to the square root of the number of iterations.) Since x_1^* and x_2^*, the new fixed points of f^2, are *not* fixed points of f, it must be that f sends one into the other:

$$x_1^* = f(x_2^*)$$

and

$$x_2^* = f(x_1^*).$$

Such a *pair of points*, termed a *2-cycle*, is depicted by the limiting unwinding circulating square in fig. 4a. Observe in fig. 4b that the slope of f^2 is in excess of 1 at the fixed point of f and so is an unstable fixed point of f^2, while the two new fixed points have slopes smaller than 1, and so are *stable*; that is, every two iterates of f will have a point attracted toward x_1^* if it is sufficiently close to x_1^* or toward x_2^* if it is sufficiently close to x_2^*. This means that the sequence under f,

$$x_0, x_1, x_2, x_3, \ldots,$$

eventually becomes arbitrarily close to the sequence

$$x_1^*, x_2^*, x_1^*, x_2^*, \ldots,$$

so that this is a stable 2-cycle, or an *attractor of*

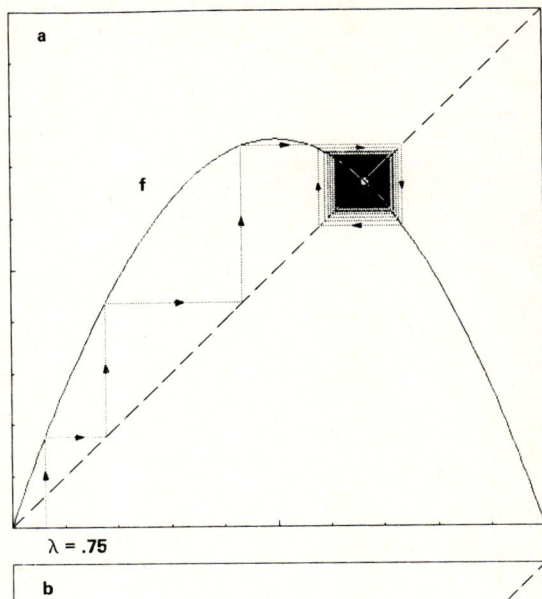

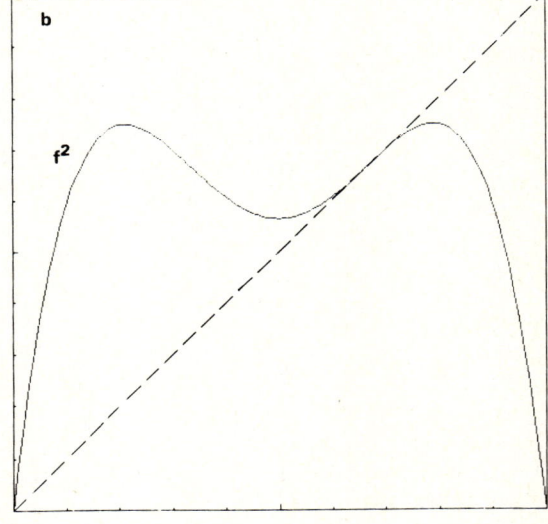

Fig. 3. $\lambda = 0.75$. (a) depicts the slow convergence to the fixed point. f^2 osculates about the fixed point.

period 2. Thus, we have observed for eq. (15) the first period doubling as the parameter λ has increased.

There is a point of paramount importance to be observed; namely, f^2 has the same slope at x_1^* and at x_2^*. This point is a direct consequence of eq. (20), since if $x_0 = x_1^*$, then $x_1 = x_2^*$, and vice versa, so that the product of the slopes is the same. More

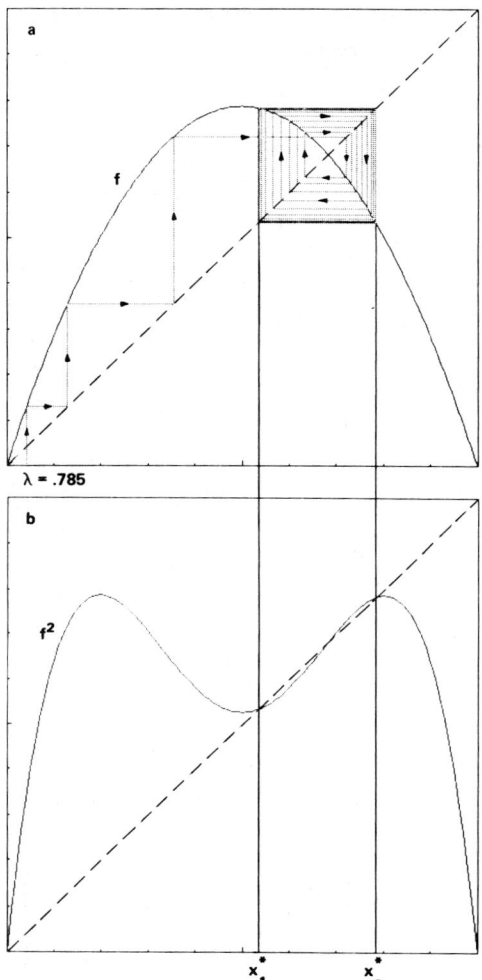

Fig. 4. $\lambda = 0.785$. (a) shows the outward spiralling to a stable 2-cycle. The elements of the 2-cycle, x_1^* and x_2^*, are located as fixed points in (b).

generally, if $x_1^*, x_2^*, \ldots, x_n^*$ is an n-cycle so that

$$x_{r+1}^* = f(x_r^*), \quad r = 1, 2, \ldots, n-1,$$

and

$$x_1^* = f(x_n^*), \tag{24}$$

then *each* is a fixed point of f^n with identical slopes:

$$x_r^* = f^n(x_r^*), \quad r = 1, 2, \ldots, n, \tag{25}$$

and

$$f^{n\prime}(x_r^*) = f'(x_1^*) \ldots f'(x_n^*). \tag{26}$$

From this observation will follow period doubling *ad infinitum*.

As λ is increased further, the minimum at $x = \tfrac{1}{2}$ will drop as the slope of f^2 through the fixed point of f increases. At some value of λ, denoted by λ_1, $x = \tfrac{1}{2}$ will become a fixed point of f^2. Simultaneously, the right-hand maximum will also become a fixed point of f^2. [By eq. (26), both elements of the 2-cycle have slope 0.] Figs. 5a and 5b depict the situation that occurs at $\lambda = \lambda_1$ [3].

5. Period doubling *ad infinitum*

We are now close to the end of this story. As we increase λ further, the minimum drops still lower, so that both x_1^* and x_2^* have negative slopes. At some parameter value, denoted by Λ_2, the slope at *both* x_1^* and x_2^* becomes equal to -1. Thus at Λ_2 the same situation has developed for f^2 as developed for f at $\Lambda_1 = \tfrac{3}{4}$. This transitional case is depicted in figs. 6a and 6b. Accordingly, just as the fixed point of f at Λ_1 issued into being a 2-cycle, so too does *each* fixed point of f^2 at Λ_2 create a 2-cycle, which in turn is a 4-cycle of f. That is, we have now encountered the second period doubling.

The manner in which we were able to follow the creation of the 2-cycle at Λ_1 was to anticipate the presence of period 2, and so to consider f^2, which would resolve the cycle into a pair of fixed points. Similarly, to resolve period 4 into fixed points we now should consider f^4. Beyond being the fourth iterate of f, eq. (8) tells us that f^4 can be computed from f^2:

$$f^4 = f^2 \circ f^2.$$

From this point, we can abandon f itself, and take

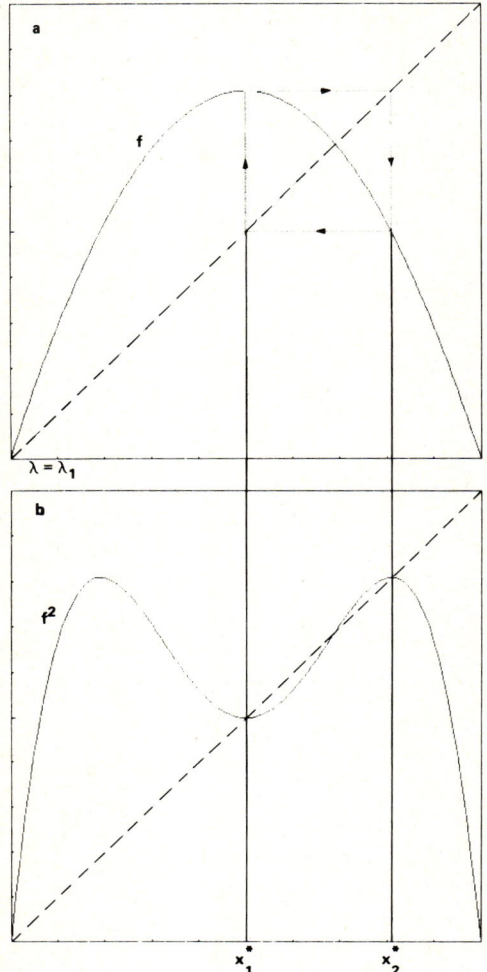

Fig. 5. $\lambda = \lambda_1$. A superstable 2-cycle. x_1^* and x_2^* are at extrema of f^2.

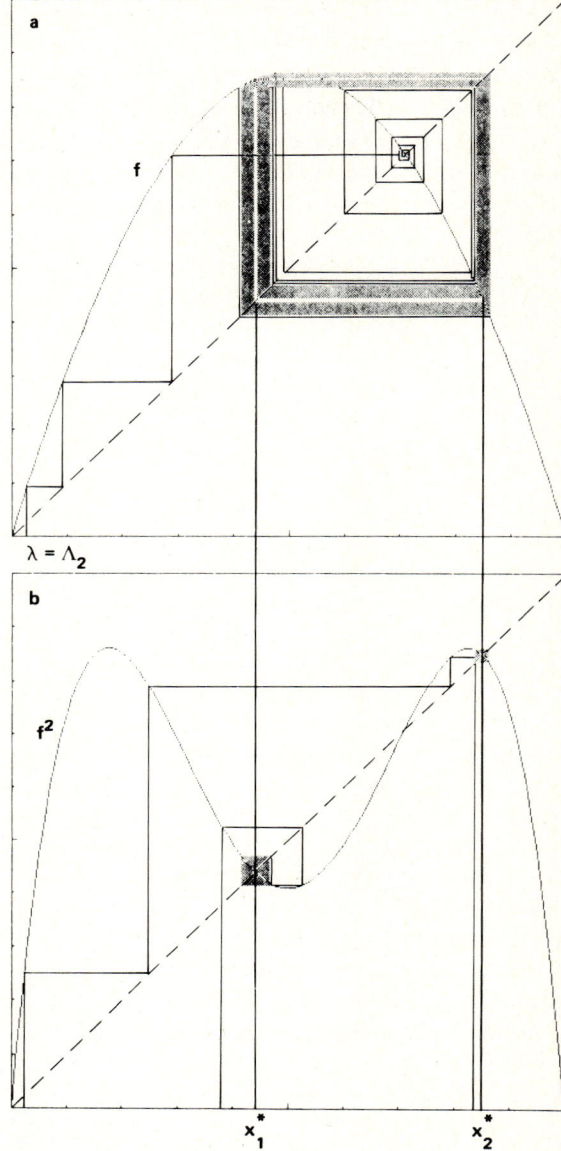

Fig. 6. $\lambda = \Lambda_2$. x_1^* and x_2^* in (b) have the same slow convergence as the fixed point in fig. 3a.

f^2 as the "fundamental" function. Then, just as f^2 was constructed by iterating f with itself we now iterate f^2 with itself. The manner in which f^2 reveals itself as being an iterate of f is the slope equality at the fixed points of f^2, which we saw imposed by the chain rule. Since the operation of the chain rule is "automatic", we actually needed to consider only the fixed point of f^2 nearest to $x = \frac{1}{2}$; the behavior of the other fixed point is slaved to it. Thus, at the level of f^4, we again need to focus on only the fixed point of f^4 nearest to $x = \frac{1}{2}$: the

other *three* fixed points are similarly slaved to it. Thus, a recursive scheme has been unearthed. We now increase λ to λ_2, so that the fixed point of f^4 nearest to $x = \frac{1}{2}$ is again at $x = \frac{1}{2}$ with slope 0. Figs. 7a and 7b depict this situation for f^2 and f^4, respectively. When λ increases further, the maximum of f^4 at $x = \frac{1}{2}$ now moves up, developing a fixed point with negative slope. Finally, at Λ_3 when the slope of this fixed point (as well as the other three) is again -1, each fixed point will split into a pair giving rise to an 8-cycle, which is now stable. Again, $f^8 = f^4 \circ f^4$, and f^4 can be viewed as fundamental. We define λ_3 so that $x = \frac{1}{2}$ again is a fixed point, this time of f^8. Then at Λ_4 the slopes are -1, and another period doubling occurs. Always,

$$f^{2^{n+1}} = f^{2^n} \circ f^{2^n}. \tag{27}$$

Provided that a constraint on the range of λ does not prevent it from decreasing the slope at the appropriate fixed point past -1, this doubling must recur ad infinitum.

Basically, the mechanism that f^{2^n} uses to period double at Λ_{n+1} is the same mechanism that $f^{2^{n+1}}$ will use to double at Λ_{n+2}. The function $f^{2^{n+1}}$ is constructed from f^{2^n} by eq. (27), and similarly $f^{2^{n+2}}$ will be constructed from $f^{2^{n+1}}$. Thus, there is a definite operation that, by acting on functions, creates functions; in particular, the operation acting on f^{2^n} at Λ_{n+1}, (or better, f^{2^n} at λ_n) will determine $f^{2^{n+1}}$ at λ_{n+1}. Also, since we need to keep track of f^{2^n} only in the interval including the fixed point of f^{2^n} closest to $x = \frac{1}{2}$ and since this interval becomes increasingly small as λ increases, the part of f that generates this region is also the restriction of f to an increasingly small interval about $x = \frac{1}{2}$. (Actually, slopes of f at points farther away also matter, but these merely set a "scale", which will be eliminated by a rescaling.) The behavior of f away from $x = \frac{1}{2}$ is immaterial to the period-doubling behavior, and in the limit of large n only the *nature of f's maximum* can matter. This means that in the infinite period-doubling limit, all functions with a quadratic extremum will have identical behavior. $[f''(\frac{1}{2}) \neq 0$ is the generic circumstance.] Therefore, the operation on functions will have a *stable fixed point* in the space of functions, which will be the common universal limit [2] of high iterates of any specific function. To determine this universal limit we must enlarge our scope vastly, so that the role of the starting point, x_0, will be played by an arbitrary *function*; the attracting fixed point will become a universal function obeying an equation implicating only itself. The role of the function in the equation $x_0 = f(x_0)$ now must be played by an *operation* that yields a new function when it is performed upon a function. In fact, the heart of

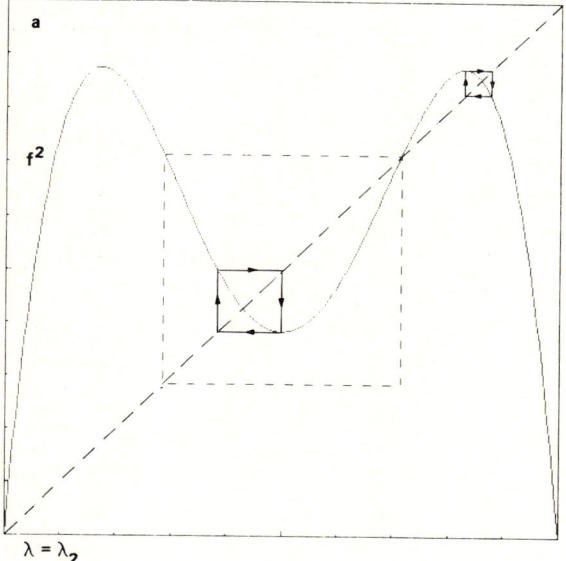

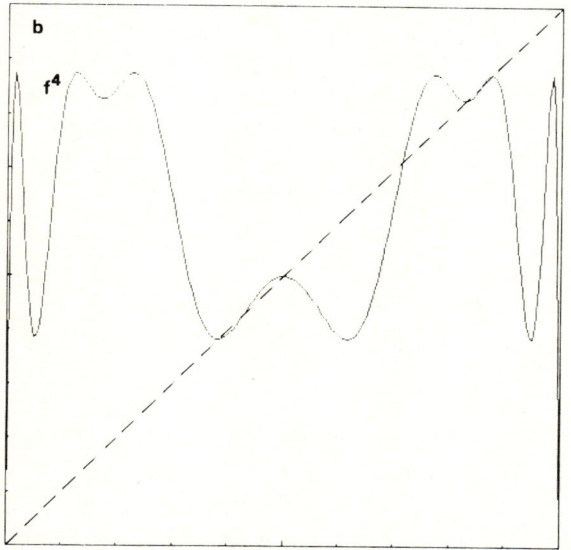

Fig. 7. $\lambda = \lambda_2$. A superstable 4-cycle. The region within the dashed square in (a) should be compared with all of fig. 5a.

this operation is the functional composition of eq. (27). If we can determine the exact operator and actually can solve *its* fixed-point problem, we shall understand why a special number, such as δ of eq. (3), has emerged independently of the specific system (the starting function) we have considered.

6. The universal limit of high iterates

In this section we sketch the solution to the fixed-point problem. In fig. 7a, a dashed square encloses the part of f^2 that we must focus on for all further period doublings. This square should be compared with the unit square that comprises all of fig. 5a. If the fig. 7a square is reflected through $x = \frac{1}{2}$, $y = \frac{1}{2}$ and then *magnified* so that the circulation squares of figs. 4a and 5a are of equal size, we will have in each square a piece of a function that has the same kind of maximum at $x = \frac{1}{2}$ and falls to zero at the right-hand lower corner of the circulation square. Just as f produced this second curve of f^2 in the square as λ increased from λ_1 to λ_2, so too will f^2 produce another curve, which will be similar to the other two when it has been magnified suitably and reflected twice. Fig. 8 shows

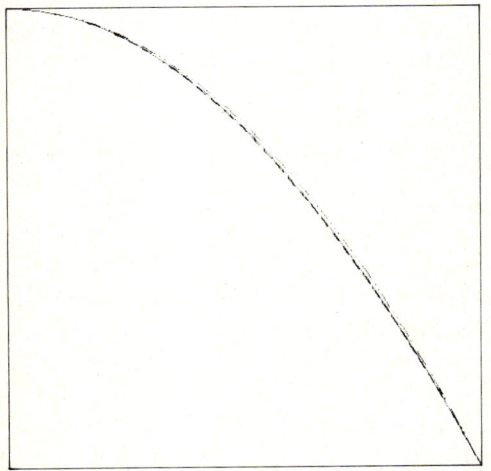

Fig. 8. The superposition of the suitably magnified dotted squares of $f^{2^{n-1}}$ at λ_n (as in figs. 5a, 7a, ...).

this superposition for the first *five* such functions; at the resolution of the figure, observe that the last three curves are coincident. Moreover, the scale reduction that f^2 will determine for f^4 is based solely on the functional composition, so that if these curves for f^{2^n}, $f^{2^{n+1}}$, *converge* (as they obviously do in fig. 8), the scale reduction from level to level will *converge to a definite constant*. But the width of each circulation square is just the distance between $x = \frac{1}{2}$ when it is a fixed point of f^{2^n} and the fixed point of f^{2^n} next nearest to $x = \frac{1}{2}$ (figs. 7a and 7b). That is, asymptotically, *the separation of adjacent elements of period-doubled attractors is reduced by a constant value from one doubling to the next*. Also from one doubling to the next, this next nearest element *alternates* from one side of $x = \frac{1}{2}$ to the other. Let d_n denote the algebraic distance from $x = \frac{1}{2}$ to the nearest element of the attractor cycle of period 2^n, in the 2^n-cycle at λ_n. A positive number α scales this distance down in the 2^{n+1}-cycle at λ_{n+1} [1, 2]

$$\frac{d_n}{d_{n+1}} \sim -\alpha. \qquad (28)$$

But since rescaling is determined only by functional composition, there is some function that composed with itself will *reproduce* itself reduced in scale by $-\alpha$. The function has a quadratic maximum at $x = \frac{1}{2}$, is symmetric about $x = \frac{1}{2}$, and can be scaled by hand to equal 1 at $x = \frac{1}{2}$. Shifting coordinates so that $x = \frac{1}{2} \to x = 0$, we have

$$-\alpha g(g(x/\alpha)) = g(x). \qquad (29)$$

Substituting $g(0) = 1$, we have

$$g(1) = -\frac{1}{\alpha}. \qquad (30)$$

Accordingly, eq. (29) is a definite equation for a function g depending on x through x^2 and having a maximum of 1 at $x = 0$. There is a unique smooth solution to eq. (29), which determines

$$\alpha = 2.502907875\ldots, \tag{31}$$

Knowing α we can predict through eq. (28) a definite scaling law binding on the iterates of any scheme possessing period doubling. The law has, indeed, been amply verified experimentally. By eq. (29), we see that the relevant operation upon functions that underlies period doubling is functional composition followed by magnification, where the magnification is determined by the fixed-point condition of eq. (29) with the function g the fixed point in this space of functions. However, eq. (29) does not describe a stable fixed point because we have not incorporated in it the parameter increase from λ_n to λ_{n+1}. Thus, g is not the limiting function of the curves in the circulation squares, although it is intimately related to that function. The full theory is described in the next section. Here we merely state that we can determine the limiting function and thereby can *determine the location of the actual elements of limiting 2^n-cycles*. We also have established that g is an unstable fixed point of functional composition, where the rate of divergence away from g is precisely δ of eq. (3) and so is computable. Accordingly, there is a full theory that determines, in a precise quantitative way, the aperiodic limit of functional iterations with an *unspecified* function f [4].

7. Some details of the full theory

Returning to eq. (28), we are in a position to describe theoretically the universal scaling of high-order cycles and the convergence to a universal limit. Since d_n is the distance between $x = \frac{1}{2}$ and the element of the 2^n cycle at λ_n nearest to $x = \frac{1}{2}$ and since this nearest element is the 2^{n-1} iterate of $x = \frac{1}{2}$ (which is true because these two points were coincident before the nth period doubling began to split them apart), we have

$$d_n = f^{2^{n-1}}(\lambda_n, \tfrac{1}{2}) - \tfrac{1}{2}. \tag{32}$$

For future work it is expedient to perform a coordinate translation that moves $x = \frac{1}{2}$ to $x = 0$. Thus, eq. (32) becomes

$$d_n = f^{2^{n-1}}(\lambda_n, 0). \tag{33}$$

Eq. (28) now determines that the rescaled distances,

$$r_n \equiv (-\alpha)^n d_{n+1}$$

will converge to a definite finite value as $n \to \infty$. That is,

$$\lim_{n \to \infty} (-\alpha)^n f^{2^n}(\lambda_{n+1}, 0) \tag{34}$$

must exist if eq. (28) holds.

However, from fig. 8 we know something stronger than eq. (34). When the nth iterated function is *magnified* by $(-\alpha)^n$, it converges to a definite function. Eq. (34) is the value of this function at $x = 0$. After the magnification, the convergent functions are given by

$$(-\alpha)^n f^{2^n}(\lambda_{n+1}, x/(-\alpha)^n).$$

Thus,

$$g_1(x) \equiv \lim_{n \to \infty} (-\alpha)^n f^{2^n}(\lambda_{n+1}, x/(-\alpha)^n) \tag{35}$$

is the limiting function inscribed in the square of fig. 8. The function $g_1(x)$ is, by the argument of the restriction of f to increasingly small intervals about its maximum, the *universal* limit of all iterates of all f's with a quadratic extremum. Indeed, it is numerically easy to ascertain that g_1 of eq. (35) is always the same function independent of the f in eq. (32).

What is this universal function good for? Fig. 5a shows a crude approximation of $g_1[n = 0$ in the limit of eq. (35)], while fig. 7a shows a better approximation ($n = 1$). In fact, the extrema of g_1 near the fixed points of g_1 support circulation squares each of which contains two points of the cycle. (The two squares shown in fig. 7a locate the four elements of the cycle.) That is, g_1 determines the location of elements of high-order 2^n-cycles near $x = 0$. Since g_1 is *universal*, we now have the amazing result that the location of the actual elements of highly doubled cycles is universal! The reader might guess this is a *very* powerful result. Fig. 9 shows g_1 out to x sufficiently large to have 8 circulation squares, and hence locates the 15 elements of a 2^n-cycle nearest to $x = 0$. Also, the universal value of the scaling parameter α, obtained numerically, is

$$\alpha = 2.502907875\ldots. \tag{36}$$

Like δ, α is a number that can be *measured* [through an experiment that observes the d_n of eq. (28)] in any phenomenon exhibiting period doubling.

If g_1 is universal, then of course its iterate g_1^2 also is universal. Fig. 7b depicts an early approximation to this iterate. In fact, let us define a new universal function g_0, obtained by scaling g_1^2:

$$g_0(x) \equiv -\alpha g_1^2(-x/\alpha). \tag{37}$$

(Because g_1 is universal and the iterates of our quadratic function are all symmetric in x, both g_1 and g_0 are symmetric functions. Accordingly, the minus sign within g_1^2 can be dropped with impunity.) From eq. (35), we now can write

$$g_0(x) = \lim_{n \to \infty} (-\alpha)^n f^{2^n}(\lambda_n, x/(-\alpha)^n). \tag{38}$$

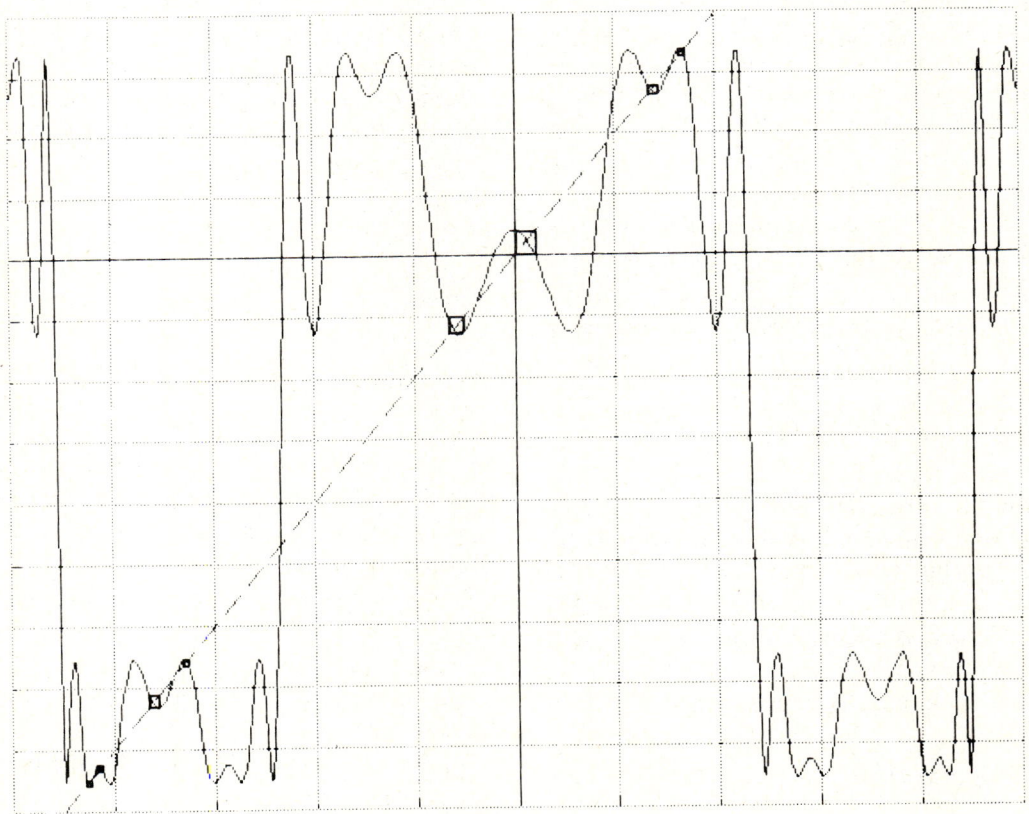

Fig. 9. The function g_1. The squares locate cycle elements.

[We introduced the scaling of eq. (37) to provide one power of α per period doubling, since each successive iterate of f^{2n} reduces the scale by α.]

In fact, we can generalize eqs. (35) and (38) to a *family* of universal functions g_r:

$$g_r(x) = \lim_{n\to\infty} (-\alpha)^n f^{2n}(\lambda_{n+1}, x/(-\alpha)^n). \tag{39}$$

To understand this, observe that g_0 locates the cycle elements as the fixed points of g_0 at extrema; g_1 locates the same elements by determining two elements per extremum. Similarly, g_r determines 2^r elements about each extremum near a fixed point of g_r. Since each f^{2n} is always magnified by $(-\alpha)^n$ for each r, the scales of all g_r are the same. Indeed, g_r for $r > 1$ looks like g_1 of fig. 9, except that each extremum is slightly higher, to accommodate a 2^r-cycle. Since each extremum must grow by convergently small amounts to accommodate higher and higher 2^r-cycles, we are led to conclude that

$$g(x) = \lim_{r\to\infty} g_r(x) \tag{40}$$

must exist. By eq. (39),

$$g(x) = \lim_{n\to\infty} (-\alpha)^n f^{2n}(\lambda_\infty, x/(-\alpha)^n). \tag{41}$$

Unlike the functions g_r, $g(x)$ is obtained as a limit of f^{2n}'s at a *fixed value* of λ. Indeed, this is the special significance of λ_∞; it is an isolated value of λ at which repeated iteration and magnification lead to a convergent function.

We now can write the equation that g satisfies. Analogously to eq. (37), it is easy to verify that all g_r are related by

$$g_{r-1}(x) = -\alpha g_r(g_r(-x/\alpha)). \tag{42}$$

By eq. (40), it follows that g satisfies

$$g(x) = -\alpha g(g(x/\alpha)). \tag{43}$$

The reader can verify that eq. (43) is invariant under a magnification of g. Thus, the theory has nothing to say about absolute scales. Accordingly, we must fix this by hand by setting

$$g(0) = 1. \tag{44}$$

Also, we must specify the nature of the maximum of g at $x = 0$ (*for* example, quadratic). Finally, since g is to be built by iterating $a - x^2$, it must be both smooth and a function of x through x^2. With these specifications, eq. (43) has a *unique* solution. By eqs. (44) and (43),

$$g(0) = 1 = -\alpha g(g(0)) = -\alpha g(1),$$

so that

$$\alpha = -1/g(1). \tag{45}$$

Accordingly, eq. (43) determines α together with g.

Let us comment on the nature of eq. (43), a so-called functional equation. Because g is smooth, if we know its value at a finite number of points, we know its value to some approximation on the interval containing these points by any sufficiently smooth interpolation. Thus, to some degree of accuracy, eq. (43) can be replaced by a finite coupled system of nonlinear equations, exactly then, eq. (43) is an infinite-dimensional, nonlinear vector equation. Accordingly, we have obtained the solution to one-dimensional period doubling through our infinite-dimensional, explicitly universal problem. Eq. (43) must be infinite-dimensional because it must keep track of the infinite number of cycle elements demanded of any attempt to solve the period-doubling problem. Rigorous mathematics for equations like eq. (43) is just beyond the boundary of present mathematical knowledge.

At this point, we must determine two items. First, where is δ? Second, how do we obtain g_1, the real function of interest for locating cycle elements? The two problems are part of one question. Eq. (42) is itself an iteration scheme. However, unlike the elements in eq. (4), the elements acted on in eq. (42) are *functions*. The analogue of the function of

f in eq. (4) is the operation in function space of functional composition followed by a magnification. If we call this operation T, and an element of the function space ψ, eq. (42) gives

$$T[\psi](x) = -\alpha \psi^2(-x/\alpha). \qquad (46)$$

In terms of T, eq. (42) now reads

$$g_{r-1} = T[g_r], \qquad (47)$$

and eq. (43) reads

$$g = T[g]. \qquad (48)$$

Thus, g is precisely the fixed point of T. Since g is the limit of the sequence g_r, we can obtain g_r for large r by linearizing T about its fixed point g. Once we have g_r in the linear regime, the exact repeated application of T by eq. (47) will provide g_1. Thus, we must investigate the stability of T at the fixed point g. However, it is obvious that T is *unstable* at g: for a large enough r, g_r is a point arbitrarily close to the fixed point g; by eq. (47), successive iterates of g_r under T move away from g. How unstable is T? Consider a one-parameter family of functions f_λ, which means a "line" in the function space. For each f, there is an isolated parameter value λ_∞, for which repeated applications of T lead to convergence towards g [eq. (41)]. Now, the function space can be "packed" with all the lines corresponding to the various f's. The set of all the points on these lines specified by the respective λ_∞'s determines a "surface" having the property that repeated applications of T to any point on it will converge to g. This is the surface of stability of T (the "stable manifold" of T through g). But through each point of this surface issues out the corresponding line, which is one-dimensional since it is parametrized by a single parameter, λ. Accordingly, T is *unstable* in only *one* direction in function space. Linearized about g, this line of instability can be written as the one-parameter family

$$f_\lambda(x) = g(x) - \lambda h(x), \qquad (49)$$

which passes through g (at $\lambda = 0$) and deviates from g along the unique direction h. But f_λ is just one of our transformations [eq. (4)]! Thus, as we vary λ, f_λ will undergo period doubling, doubling to a 2^n-cycle at Λ_n. By eq. (41), λ_∞ for the family of functions f_λ in eq. (49) is

$$\lambda_\infty = 0. \qquad (50)$$

Thus, by eq. (1)

$$\lambda_n \sim \delta^{-n}. \qquad (51)$$

Since applications of T by eq. (47) iterate in the opposite direction (diverge away from g), it now follows that the rate of instability of T along h must be precisely δ.

Accordingly, we find δ and g_1 in the following way. First, we must linearize the operation T about its fixed point g. Next, we must determine the stability directions of the linearized operator. Moreover, we expect there to be precisely one direction of instability. Indeed, it turns out that infinitesimal deformations (conjugacies), of g determine *stable* directions, while a unique unstable direction, h, emerges with a stability rate (eigenvalue) precisely the δ of eq. (3). Eq. (49) at λ_r is precisely g_r for asymptotically large r. Thus g_r is known asymptotically, so that we have entered the sequence g_r and can now, by repeated use of eq. (47), step down to g_1. All the ingredients of a full description of high-order 2^n-cycles now are at hand and evidently are universal.

Although we have said that the function g_1 universally locates cycle elements near $x = 0$, we must understand that it doesn't locate all cycle elements. This is possible because a finite distance of the scale of g_1 (for example, the location of the element nearest to $x = 0$) has been magnified by α^n for n diverging. Indeed, the distances from $x = 0$ of all elements of a 2^n-cycle, "accurately" located by g_1, are reduced by $-\alpha$ in the 2^{n+1}-cycle. However, it is obvious that some elements have no such scaling: because $f(0) = a_n$ in eq. (13), and $a_n \to a_\infty$, which is a definite nonzero number, the distance from the origin of the element of the 2^n-cycle farthest to the right certainly has not been reduced by $-\alpha$ at each period doubling. This suggests that we must measure locations of elements on the far right with respect to the farthest right point. If we do this, we can see that these distances scale by α^2, since they are the images through the quadractic maximum of f at $x = 0$ of elements close to $x = 0$ scaling with $-\alpha$. In fact, if we image g_1 through the maximum of f (through a quadratic conjugacy), then we shall indeed obtain a new universal function that locates cycle elements near the right-most element. The correct description of a highly doubled cycle now emerges as one of universal local clusters [2].

We can state the scope of universality for the location of cycle elements precisely. Since $f(\lambda_1, x)$ exactly locates the two elements of the 2^1-cycle, and since $f(\lambda_1, x)$ is an approximation to g_1 [$n = 0$ in eq. (35)], we evidently can locate both points exactly by appropriately sealing g_1. Next, near $x = 0$, $f^2(\lambda_2, x)$ is a better approximation to g_1 (suitably scaled). However, in general, the more accurately we scale g_1 to determine the smallest 2-cycle elements, the greater is the error in its determination of the right-most elements. Again, near $x = 0$, $f^4(\lambda_3, x)$ is a still better approximation to g_1. Indeed, the suitably scaled g_1 now can determine several points about $x = 0$ accurately, but determination of the right-most elements is still worse. In this fashion, it follows that g_1, suitably scaled, can determine 2^r points of the 2^n cycle near $x = 0$ for $r \ll n$. If we focus on the neighborhood of one of these 2^r points at some definite distance from $x = 0$, then by eq. (35) the larger the n, the larger the *scaled* distance of this region from $x = 0$, and so, the poorer the approximation of the location of fixed points in it by g_1. However, just as we can construct the version of g_1 that applies at the right-most cycle element, we also can construct the version of g_1 that applies at this chosen neighborhood. Accordingly, the universal description is set through an acceptable tolerance: if we "measure" f^{2^n} at some definite n, then we can use the actual location of the elements as foci for 2^n versions of g_1, each applicable at one such point.

For all further period doubling, we determine the new cycle elements through the g_1's. In summary, the *more accurately we care to know the locations* of arbitrarily high-order cycle elements, the *more parameters we must measure* (namely, the cycle elements at some chosen order of period doubling). This is the sense in which the universality theory is asymptotic. Its ability to have serious predictive power is the fortunate consequence of the high convergence rate $\delta (\approx 4.67)$. Thus, typically after the first two or three period doublings, this asymptotic theory is already accurate to within several percent. If a period-doubling system is *measured* in its 4- or 8-cycle, its behavior throughout and symmetrically beyond the period-doubling regime also is determined to within a few percent.

To make precise dynamical predictions, we do not have to construct all the local versions of g_1; all we really need to know is the local *scaling* everywhere along the attractor. The scaling is $-\alpha$ at $x = 0$ and α^2 at the right-most element. But what is it at an arbitrary point? We can determine the scaling law if we order elements not by their location on the x-axis, but rather by their order as iterates of $x = 0$. Because the time sequence in which a process evolves is precisely this ordering, the result will be of immediate and powerful predictive value. It is precisely this scaling law that allows us to compute the spectrum of the onset of turbulence in period-doubling systems [5].

What must we compute? First, just as the element in the 2^n-cycle nearest to $x = 0$ is the element halfway around the cycle from $x = 0$, the element nearest to an arbitrarily chosen element is precisely the one halfway around the cycle from it. Let us denote by $d_n(m)$ the distance between the mth cycle element (x_m) and the element nearest to it in a 2^n-cycle. [The d_n of eq. (28) is $d_n(0)$]. As just explained,

$$d_n(m) = x_m - f^{2^{n-1}}(\lambda_n, x_m). \tag{52}$$

However, x_m is the mth iterate of $x_0 = 0$. Recalling from eq. (6) that powers commute, we find

$$d_n(m) = f^m(\lambda_n, 0) - f^m(\lambda_n, f^{2^{n-1}}(\lambda_n, 0)). \tag{53}$$

Let us, for the moment, specialize to m of the form 2^{n-r}, in which case

$$d_n(2^{n-r}) = f^{2^{n-r}}(\lambda_n, 0) - f^{2^{n-r}}(\lambda_n, f^{2^{n-1}}(\lambda_n, 0))$$
$$= f^{2^{n-r}}(\lambda_{(n-r)+r}, 0)$$
$$- f^{2^{n-r}}(\lambda_{(n-r)+r}, f^{2^{n-1}}(\lambda_n, 0)). \quad (54)$$

For $r \ll n$ (which can still allow $r \gg 1$ for n large), we have, by eq. (39),

$$d_n(2^{n-r}) \approx (-\alpha)^{-(n-r)}[g_r(0) - g_r((-\alpha)^{n-r} f^{2^{n-1}}(\lambda_n, 0))]$$

or

$$d_n(2^{n-r}) \approx (-\alpha)^{-(n-r)}[g_r(0) - g_r((-\alpha)^{-r+1} g_1(0))]. \quad (55)$$

The object we want to determine is the local scaling at the mth element, that is, the ratio of nearest separations at the mth iterate of $x = 0$, at successive values of n. That is, if the scaling is called σ,

$$\sigma_n(m) \equiv \frac{d_{n+1}(m)}{d_n(m)}. \quad (56)$$

[Observe by eq. (28), the definition of α, that $\sigma_n(0) \approx (-\alpha)^{-1}$.] Specializing again to $m = 2^{n-r}$, where $r \ll n$, we have by eq. (55)

$$\sigma(2^{n-r}) \approx \frac{g_{r+1}(0) - g_{r+1}((-\alpha)^{-r} g_1(0))}{g_r(0) - g_r((-\alpha)^{-r+1} g_1(0))}. \quad (57)$$

Finally, let us rescale the axis of iterates so that all 2^{n+1} iterates are within a unit interval. Labelling this axis by t, the value of t of the mth element in a 2^n-cycle is

$$t_n(m) = m/2^n. \quad (58)$$

In particular, we have

$$t_n(2^{n-r}) = 2^{-r}. \quad (59)$$

Defining σ along the t-axis naturally as

$$\sigma(t_n(m)) \approx \sigma_n(m) \quad (\text{as } n \to \infty),$$

we have by eqs. (57) and (59),

$$\sigma(2^{-r-1}) = \frac{g_{r+1}(0) - g_{r+1}((-\alpha)^{-r} g_1(0))}{g_r(0) - g_r((-\alpha)^{-r+1} g_1(0))}. \quad (60)$$

It is not much more difficult to obtain σ for all t. This is done first for rational t by writing t in its binary expansion:

$$t_{r_1 r_2 r_3 \ldots} = 2^{-r_1} + 2^{-r_2} + \cdots.$$

In the 2^n-cycle approximation we require σ_n at the $2^{n-r_1} + 2^{n-r_2} + \cdots$ iterate of the origin. But, by eq. (8),

$$f^{2^{n-r_1} + 2^{n-r_2} + \cdots} = f^{2^{n-r_1}} \circ f^{2^{n-r_2}} \circ \cdots.$$

It follows by manipulations identical to those that led from eq. (54) to eq. (60) that σ at such values of t is obtained by replacing the individual g_r terms in eq. (60) by appropriate iterates of various g_r's.

There is one last ingredient to the computation of σ. We know that $\sigma(0) = -\alpha^{-1}$. We also know that $\sigma_n(1) \approx \alpha^{-2}$. But, by eq. (59),

$$t_n(1) = 2^{-n} \to 0.$$

Thus σ is discontinuous at $t = 0$, with $\sigma(0 - \epsilon) = -\alpha^{-1}$ and $\sigma(0 + \epsilon) = \alpha^{-2} (\epsilon \to 0^+)$. Indeed, since $x_{2^{n-r}}$ is always very close to the origin, each of these points is imaged quadratically. Thus eq. (60) actually determines $\sigma(2^{-r-1} - \epsilon)$, while $\sigma(2^{-r-1} + \epsilon)$ is obtained by replacing each numerator and denominator g_r by its square. The same replacement also is correct for each multi-g_r term that figures into σ at the binary expanded rationals [6].

Altogether, we have the following results. $\sigma(t)$ can be computed for all t, and it is *universal* since its explicit computation depends only upon the universal functions g_r. σ is *discontinuous* at all the rationals. However, it can be established that the

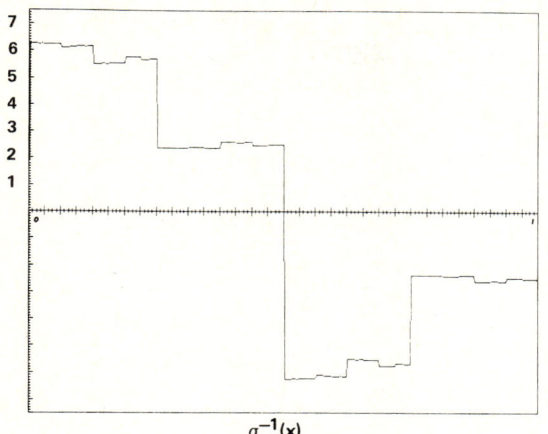

Fig. 10. The trajectory scaling function. Observe that $\sigma(x + \tfrac{1}{2}) = -\sigma(x)$.

larger the number of terms in the binary expansion of a rational t, the smaller the discontinuity of σ. Lastly, as a finite number of iterates leaves t unchanged as $n \to \infty$, σ must be *continuous* except at the rationals. Fig. 10 depicts $1/\sigma(t)$. Despite the pathological nature of σ, the reader will observe that basically it is constant half the time at α^{-1} and half the time at α^{-2} for $0 < t < \tfrac{1}{2}$. In a succeeding approximation, it can be decomposed in each half into two slightly different quarters, and so forth. [It is easy to verify from eq. (52) that σ is periodic in t of period 1, and has the symmetry

$$\sigma(t + \tfrac{1}{2}) = -\sigma(t).$$

Accordingly, we have paid attention to its first half $0 < t < \tfrac{1}{2}$.] With σ we are at last finished with one-dimensional iterates per se.

8. Universal behavior in higher dimensional systems

So far we have discussed iteration in *one* variable; eq. (15) is the prototype. Eq. (14), an example of iteration in two dimensions, has the special property of preserving areas. A generalization of eq. (14),

$$x_{n+1} = y_n - x_n^2$$

and

$$y_{n+1} = a + bx_n, \tag{61}$$

with $|b| < 1$, contracts areas. Eq. (61) is interesting because it possesses a so-called *strange attractor*. This means an attractor (as before) constructed by folding a curve repeatedly upon itself (fig. 11) with the consequent property that two initial points very near to one another are, in fact, very far from each other when the distance is measured along the folded attractor, which is the path they follow upon iteration. This means that after some iteration, they will soon be far apart in actual distance as well as when measured along the attractor. This general mechanism gives a system highly sensitive dependence upon its initial conditions and a truly statistical character: since very small differences in initial conditions are magnified quickly, unless the initial conditions are known to *infinite precision*, all known knowledge is eroded rapidly to future ignorance. Now, eq. (61) enters into the early stages of statistical behavior through period doubling. Moreover, δ of eq. (3) is *again* the rate of onset of complexity, and α of eq. (31) is again the rate at which the spacing of adjacent attractor points is vanishing. Indeed, the one-dimensional theory determines all behavior of eq. (61) in the onset regime.

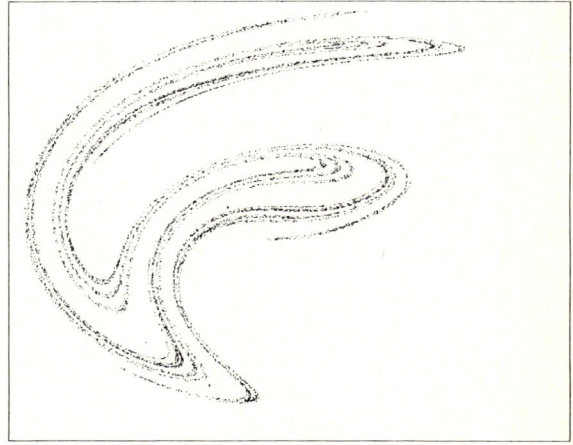

Fig. 11. The plotted points lie on the "strange attractor" of Duffing's equation.

In fact, dimensionality is irrelevant. The same theory, the same numbers, etc. also work for iterations in N dimensions, provided that the system goes through period doubling. The basic process, wherever period doubling occurs *ad infinitum*, is functional composition from one level to the next. Accordingly, a modification of eq. (29) is at the heart of the process, with composition on functions from N dimensions to N dimensions. Should the specific iteration function contract N-dimensional volumes (a dissipative process), then in general there is one direction of slowest contraction, so that after a number of iterations the process is effectively one-dimensional. Put differently, the one-dimensional solution to eq. (29) is always a solution to its N-dimensional analogue. It is the relevant fixed point of the analogue if the iteration function is contractive [7].

9. Universal behavior in differential systems

The next step of generalization is to include systems of differential equations. A prototypic equation is Duffing's oscillator, a driven damped anharmonic oscillator.

$$\ddot{x} + k\dot{x} + x^3 = b \sin 2\pi t. \tag{62}$$

The periodic drive of period 1 determines a natural time step. Fig. 12a depicts a period 1 attractor, usually referred to as a *limit cycle*. It is an attractor because, for a range of initial conditions, the solution to eq. (62) settles down to the cycle. It is period 1 because it repeats the same curve in every period of the drive. Fig. 12b and 12c depict attractors of periods 2 and 4 as the friction or damping constant k in eq. (62) is reduced systematically. The parameter values $k = \lambda_0, \lambda_1, \lambda_2, \ldots$, are the damping constants corresponding to the most stable 2^n-cycle in analogy to the λ_n of the one-dimensional functional iteration. Indeed, this oscillator's period doubles (at least numerically!) *ad infinitum*. In fact, by $k = \lambda_5$, the δ_3 of eq. (2) has converged to 4.69. Why is this? Instead of considering the entire

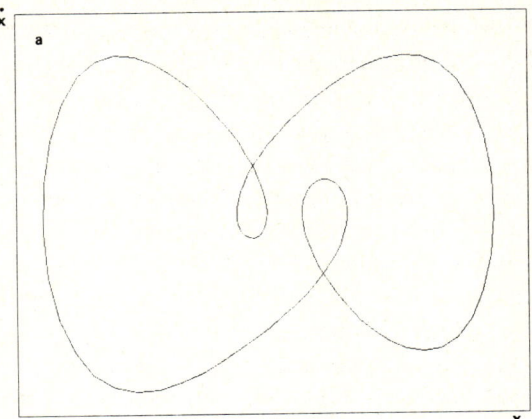

Fig. 12a. The most stable 1-cycle of Duffing's equation in phase space $(x, \dot{x})$.

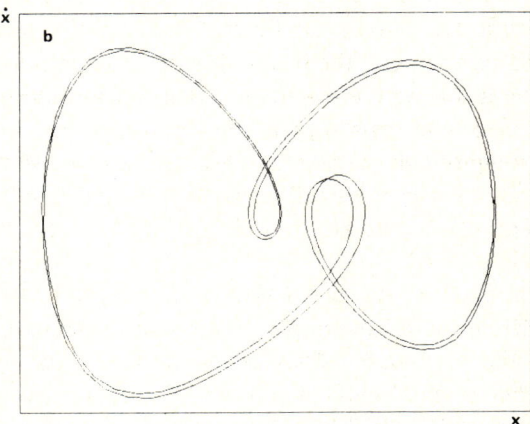

Fig. 12b. The most stable 2-cycle of Duffing's equation. Observe that it is two displaced copies of fig. 12a.

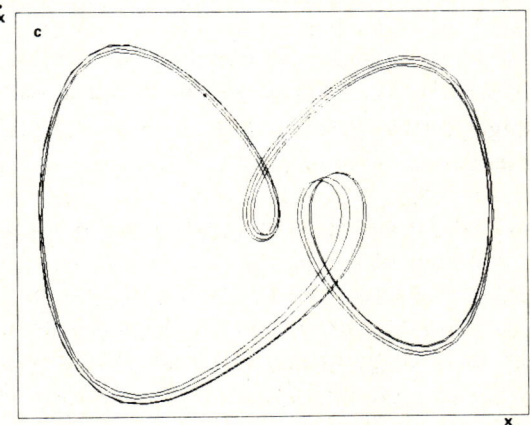

Fig. 12c. The most stable 4-cycle of Duffing's equation. Observe that the displaced copies of fig. 12b have either a broad or a narrow separation.

trajectories as shown in fig. 12, let us consider only where the trajectory point is located every 1 period of the drive. The 1-cycle then produces only one point, while the 2-cycle produces a pair of points, and so forth. This *time-one map* [if the trajectory point is $(x, \dot{x})$ now, where is it one period later?] is by virtue of the differential equation a smooth and invertible function in two dimensions. Qualitatively, it looks like the map of eq. (61). In the present state of mathematics, little can be said about the analytic behavior of time-one maps; however, since our theory is universal, it makes no difference that we don't know the explicit form. We still can determine the complete quantitative behavior of eq. (62) in the onset regime where the motion tends to aperiodicity. If we already know, by measurement, the precise form of the trajectory after a few period doublings, we can compute the form of the trajectory as the friction is reduced throughout the region of onset of complexity by carefully using the full power of the universality theory to determine the spacings of elements of a cycle.

Let us see how this works in some detail. Consider the time-one map of the Duffing's oscillator in the superstable 2^n-cycle. In particular, let us focus on an element at which the scaling function σ (fig. 10) has the value σ_0, and for which the next iterate of this element also has the scaling σ_0. (The element is not at a big discontinuity of σ.) It is then intuitive that if we had taken our time-one examination of the trajectory at values of time displaced from our first choice, we would have seen the same scaling σ_0 for this part of the trajectory. That is, the differential equations will extend the map-scaling function continuously to a function along the entire trajectory so that, if two successive time-one elements have scaling σ_0, then the entire stretch of trajectory over this unit time interval has scaling σ_0. In the last section, we were motivated to construct σ as a function of t along an interval precisely towards this end.

To implement this idea, the first step is to define the analogue of d_n. We require the spacing between the trajectory at time t and at time $T_n/2$ where the period of the system in the 2^n-cycle is

$$T_n \approx 2^n T_0. \tag{63}$$

That is, we define

$$d_n(t) \equiv x_n(t) - x_n(t + T_n/2). \tag{64}$$

(There is a d for each of the N variables for a system of N differential equations.) Since σ was defined as periodic of period 1, we now have

$$d_{n+1}(t) \approx \sigma(t/T_{n+1})d_n(t). \tag{65}$$

The content of eq. (65), based on the n-dependence arising solely through the T_n in σ, and not on the detailed form of σ, already implies a strong scaling prediction, in that the ratio

$$\frac{d_{n+1}(t)}{d_n(t)},$$

when plotted with t scaled so that $T_n = 1$, is a function *independent* of n. Thus if eq. (65) is true for *some* σ, whatever it might be, then knowing $x_n(t)$, we can compute $d_n(t)$ and from eq. (65) $d_{n+1}(t)$. As a consequence of periodicity, eq. (64) for $n \to n+1$ can be solved for $x_{n+1}(t)$ (through a Fourier transform). That is, if we have measured any chosen coordinate of the system in its 2^n-cycle, we can compute its time dependence in the 2^{n+1}-cycle. Because this procedure is recursive, we can compute the coordinate's evolution for all higher cycles through the infinite period-doubling limit. If eq. (65) is true and σ now known, then by measurement at a 2^n-cycle and at a 2^{n+1}-cycle, σ could be *constructed* from eq. (65), and hence all higher order doublings would again be determined. Accordingly, eq. (65) is a very powerful result. However, we know much more. The universality theory tells us that period doubling is universal and that there is a *unique* function σ which, indeed, we have computed in the previous section. Accordingly, by *measuring* $x(t)$ in some chosen 2^n-cycle (the higher the n, the more the number of effective parameters to be determined empirically, and the more precise are the predictions), we now can compute the entire evolution of the system on its route to turbulence.

How well does this work? The empirically determined σ [for eq. (62)] of eq. (65) is shown for $n = 3$ in fig. 13a and $n = 4$ in fig. 13b. The figures were constructed by plotting the ratios of d_{n+1} and d_n scaled respective to $T = 16$ in fig. 13a and $T = 32$ in fig. 13b. Evidently the scaling law eq. (65) is being obeyed. Moreover, on the same graph fig. 14 shows the empirical σ for $n = 4$ and the recursion

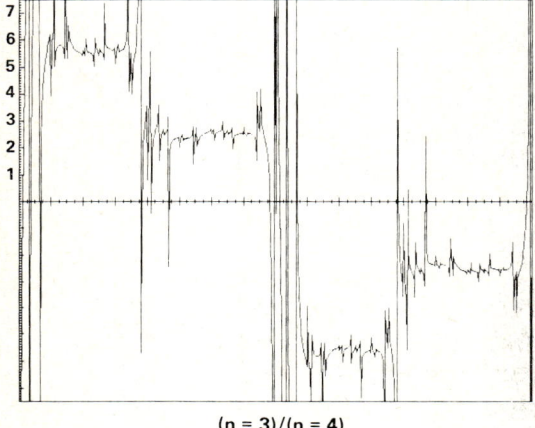

Fig. 13a. The ratio of nearest copy separations in the 8-cycle and 16-cycle for Duffing's equation.

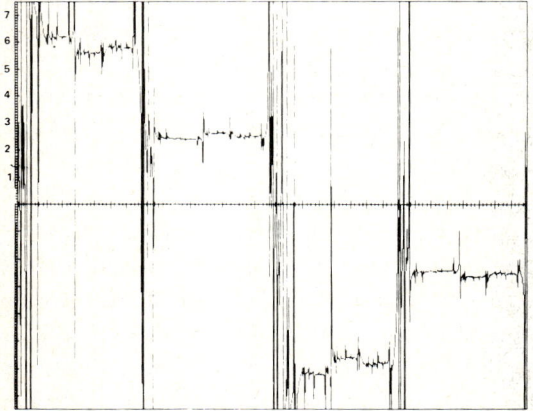

Fig. 13b. The same quantity as in fig. 13a, but for the 16-cycle and 32-cycle. Here, the time axis is twice as compressed.

theoretical σ of fig. 10. The reader should observe the detail-by-detail agreement of the two. In fact, if we use eq. (65) and the theoretical σ with $n = 2$ as empirical input, the $n = 5$ frequency spectrum agrees with the empirical $n = 5$ spectrum to within 10%. (The $n = 4$ determines $n = 5$ to within 1%.) Thus the asymptotic universality theory is correct *and* is already well obeyed, even by $n = 2$! [5].

Eqs. (64) and (65) are solved, as mentioned above, through Fourier transforming. The result is a recursive scheme that determines the Fourier coefficients of $x_{n+1}(t)$ in terms of those of $x_n(t)$ and the Fourier transform of the (known) function $\sigma(t)$. To employ the formula accurately requires knowledge of the entire spectrum of x_n (amplitude *and* phase) to determine each coefficient of x_{n+1}. However, the formula enjoys an approximate local prediction, which roughly determines the amplitude of a coefficient of x_{n+1} in terms of the amplitudes (alone) of x_n near the desired frequency of x_{n+1}.

What does the spectrum of a period-doubling system look like? Each time the period doubles, the fundamental frequency halves; period doubling in the continuum version is termed half-subharmonic bifurcation, a typical behavior of coupled nonlinear differential equations. Since the motion *almost* reproduces itself every period of the drive, the amplitude at this original frequency is high. At the first subharmonic halving, spectral components of the odd halves of the drive frequency come in. On the route to aperiodicity they saturate at a certain amplitude. Since the motion more nearly reproduces itself every two periods of drive, the next saturated subharmonics, at the odd fourths of the original frequency, are smaller still than the first ones, and so on, as each set of odd 2^nths comes into being. A crude approximate prediction of the theory is that whatever the system, the saturated amplitudes of each set of successively lower half-frequencies define a smooth interpolation located 8.2 dB *below* the smooth interpolation of the previous half-frequencies. [This is shown in fig. 15 for eq. (62).] After subharmonic bifurcations *ad infinitum*, the system is now no longer periodic; it

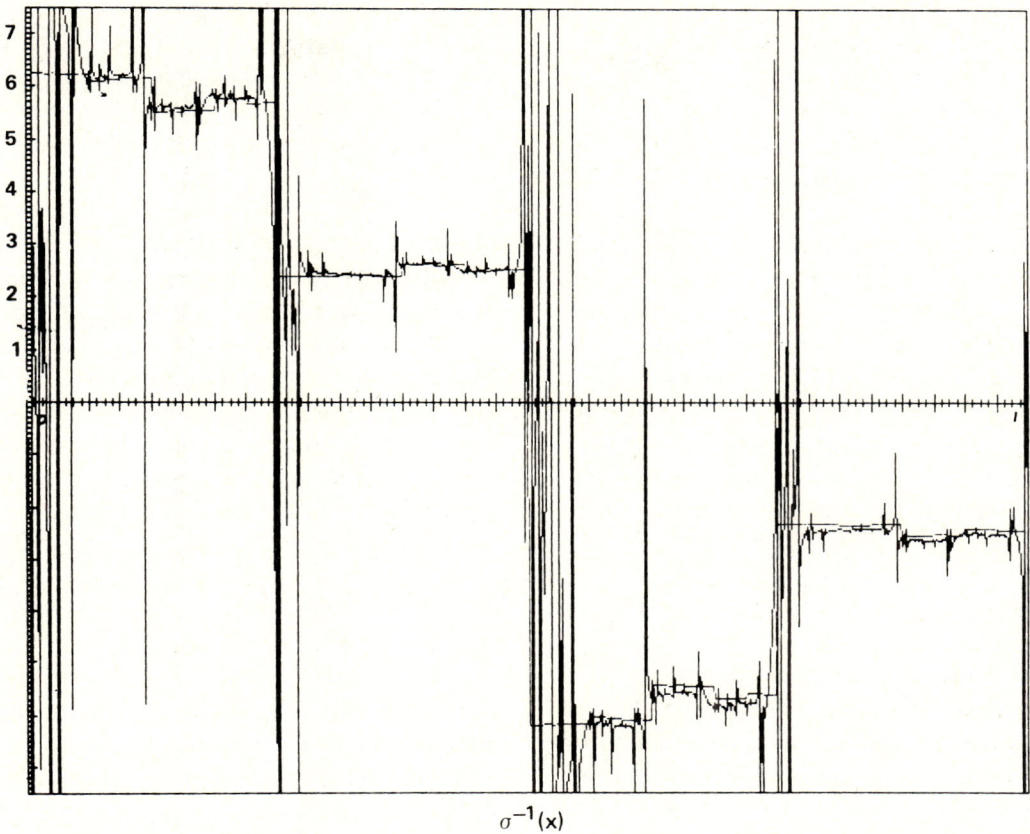

Fig. 14. Figure 13b overlayed with fig. 10 compares the universal scaling function σ with the empirically determined scaling of nearest copy separations from the 16-cycle to the 32-cycle for Duffing's equation.

has developed a continuous broad spectrum down to zero frequency with a definite internal distribution of the energy. That is, the system emerges from this process having developed the beginnings of broad-band noise of a determined nature. This process also occurs in the onset of turbulence in a fluid.

10. The onset of turbulence

The existing idea of the route to turbulence is Landau's 1941 theory. The idea is that a system becomes turbulent through a succession of instabilities, where each instability creates a new degree of freedom (through an indeterminate phase) of a time-periodic nature with the frequencies successively higher and incommensurate (*not* harmonics); because the resulting motion is the superposition of these modes, it is quasi-periodic.

In fact, it is experimentally clear that quasi-periodicity is incorrect. Rather, to produce the observed noise of rapidly decaying correlation the spectrum must become *continuous* (broad-band noise) down to zero frequency. The defect can be eliminated through the production of successive half-subharmonics, which then emerge as an allowable route to turbulence. If the general idea of a succession of instabilities is maintained, the new modes do *not* have indeterminate phases. However, only a small number of modes need be excited to produce the required spectrum. (The number of modes participating in the transition is, as of now,

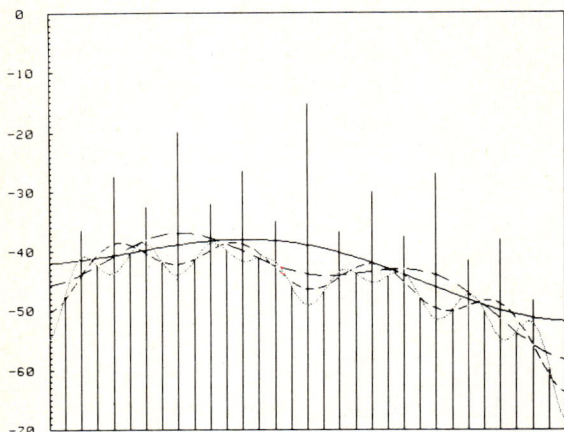

Fig. 15. The subharmonic spectrum of Duffing's equation in the 32-cycle. The dotted curve is an interpolation of the odd 32nd subharmonics. The shorter dashed curve is constructed similarly for the odd 16th subharmonics, but lowered by 8.2 dB. The longer dashed curve of the 8th subharmonics has been dropped by 16.4 dB, and the solid curve of the 4th subharmonics by 24.6 dB.

an open experimental question.) Indeed, knowledge of the phases of a small number of amplitudes at an early stage of period doubling suffices to determine the phases of the transition spectrum. What is important is that a purely causal system can and does possess essentially statistical properties. Invoking *ad hoc* statistics is unnecessary and generally incompatible with the true dynamics.

A full theoretical computation of the onset demands the calculation of successive instabilities. The method used traditionally is perturbative. We start at the static solution and add a small time-dependent piece. The fluid equations are linearized about the static solution, and the stability of the perturbation is studied. To date, only the first instability has been computed analytically. Once we know the parameter value (for example, the Rayleigh number) for the onset of this first time-varying instability, we must determine the correct form of the solution after the perturbation has grown large *beyond* the linear regime. To this solution we add a new time-dependent perturbative mode, again linearized (now about a time-varying, nonanalytically available solution) to discover the new instability. To date, the second step of the analysis has been performed only numerically. This process, in principle, can be repeated again and again until a suitably turbulent flow has been obtained. At each successive stage, the computation grows successively more intractable.

However, it is just at this point that the universality theory solves the problem; it works only after enough instabilities have entered to reach the asymptotic regime. Since just two such instabilities already serve as a good approximate starting point, we need only a few parameters for each flow to empower the theory to complete the hard part of the infinite cascade of more complex instabilities.

Why should the theory apply? The fluid equations make up a set of coupled field equations. They can be spatially Fourier-decomposed to an infinite set of coupled ordinary differential equations. Since a flow is viscous, there is some smallest spatial scale below which no significant excitation exists. Thus, the equations are effectively a finite coupled set of nonlinear differential equations. The number of equations in the set is completely irrelevant. The universality theory is generic for such a dissipative system of equations. Thus it is possible that the flow exhibits period doubling. If it does, then our theory applies. However, to prove that a given flow (or any flow) actually should exhibit doubling is well beyond present understanding. All we can do is experiment.

Fig. 16 depicts the experimentally measured spectrum of a convecting liquid helium cell at the onset of turbulence [8]. The system displays measurable period doubling through four or five levels; the spectral components at each set of odd half-subharmonics are labelled with the level. With $n = 2$ taken as asymptotic, the dotted lines show the crudest interpolations implied for the $n = 3$, $n = 4$ component. Given the small amount of *amplitude* data, the interpolations are perforce poor, while ignorance of higher odd multiples prevents construction of any significant interpolation at the right-hand side. Accordingly, to do the crudest test, the farthest right-hand amplitude was dropped, and the oscillations were smoothed

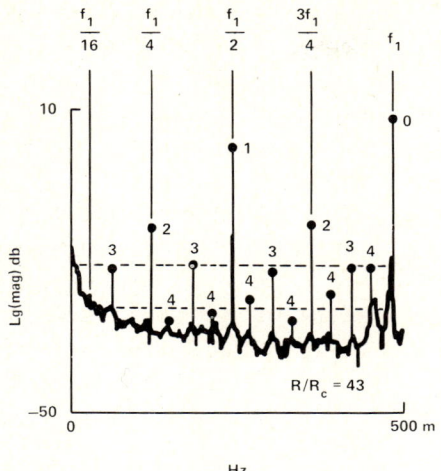

Fig. 16. The experimental spectrum (redrawn from Libchaber and Maurer) of a convecting fluid at its transition to turbulence. The dashed lines result from dropping a horizontal line down through the odd 4th subharmonics (labelled 2) by 8.2 and 16.4 dB.

away by averaging [9]. The experimental results, -8.3 dB and -8.4 dB, are in surprisingly good agreement with the theoretical 8.2!

From this good experimental agreement and the many period doublings as the clincher, we can be confident that the measured flow has made its transition according to our theory. A measurement of δ from its fundamental definition would, of course, be altogether convincing. (Experimental resolution is insufficient at present [10].) However, if we work backwards, we find that the several percent agreement in 8.2 dB is an *experimental observation* of α in the system to the same accuracy. Thus, the present method has provided a theoretical calculation of the actual dynamics in a field where such a feat has been impossible since the construction of the Navier–Stokes equations. In fact, the scaling law eq. (65) transcends these equations, and applies to the *true* equations, whatever they may be.

References

[1] M.J. Feigenbaum, "Universality in Complex Discrete Dynamical Systems", in Los Alamos Theoretical Division Annual Report, July 1975–September 1976, LA-6816-PR (1977) pp. 98–102.
[2] M.J. Feigenbaum, J. Stat. Phys. 19 (1978) 25.
[3] See R.M. May, Nature 261 (1976) 459, for an early account of their properties.
[4] M.J. Feigenbaum, J. Stat. Phys. 21 (1979) 669.
[5] M.J. Feigenbaum, Phys. Lett. A74 (1979) 375.
[6] M.J. Feigenbaum, Commun. Math. Phys. 77 (1980) 65.
[7] P. Collet and J.P. Eckmann, Iterated Maps on an Interval As Dynamical System (Birkhäuser, Boston, 1980).
[8] A. Libchaber and J. Maurer, J. Phys. Colloq. 41 (1980) pp. 3–51.
[9] For a fuller account of the power spectrum than that in refs. 5 and 6 see M.J. Feigenbaum, in Nonlinear Phenomena in Chemical Dynamics (Springer, Berlin 1981) pp. 95–102.
[10] Since the time when this article was originally written new and more accurate fluid experiments have also measured δ. See M. Giglio, S. Muzzati and U. Perini, Phys. Rev. Lett. 47 (1981) 243; and A. Libchaber, C. Laroche and S. Fauve, J. Physique Lett. 43 (1982) L211.

FIVE TURBULENT PROBLEMS

David RUELLE
Institut des Hautes Etudes Scientifiques, 35 route de Chartres, 91440 Bures-sur-Yvette, France

Five questions which involve strange attractors in one way or another are discussed. These are: the choice of probability measures to describe turbulence, the Hausdorff dimension of Julia sets for rational maps of the Riemann sphere, the existence of nonperiodic "turbulent" crystal structure, the mechanism generating intermittency in developed hydrodynamic turbulence, the distribution of characteristic exponents in a turbulent fluid.

1. Introduction

In recent years we have acquired some understanding of deterministic chaos in differentiable dynamical systems and of the onset of turbulence in dissipative physical systems. While there are certainly many details to fix, it may be that the study of the onset of turbulence will not yield real surprises anymore. Whether this is true or not, I think, that it is necessary to go back to the fundamental theoretical problems of the subject and to test our ideas in situations which may be unconventional. In what follows I shall show more precisely what I have in mind by discussing several questions in which I have been recently interested.

2. What is the measure describing turbulence?

The time average which is produced in a physical or computer experiment corresponds to a probability measure ρ invariant under time evolution (in statistical mechanics, this would be called an ensemble). For a given differentiable dynamical system, like the Hénon map, there are however many invariant probability measures, and one has to decide which one is selected in experiments. My belief is that the choice is produced by the smoothing influence of the small levels of noise present in physical experiments (roundoff errors in computer

---

* See Ruelle [17] for a discussion of this point, with reference to earlier work.
† Numerical estimates of the Hausdorff dimension of an attractor have been performed by Russel, Hanson, and Ott [22].

studies). From this one can obtain a condition on the probability measure ρ, namely

$$h(\rho) = \sum \text{positive characteristic exponents,} \quad (1)$$

where $h(\rho)$ is the entropy (Kolmogorov–Sinai invariant)*. In some cases one can show that (1) completely determines ρ. Unfortunately, it is difficult to obtain $h(\rho)$ numerically. The calculations of Shimada [23] for the Lorenz attractor, and of Curry [5] for the Hénon attractor are compatible with (1), but do not support the conjecture (1) decisively. Recently, L.-S. Young [25] has proved the following interesting identity for any invariant probability measure σ for an invertible map (diffeomorphism) in two dimensions:

$$\text{HD}(\sigma) = h(\sigma)\left(\frac{1}{\lambda_1} + \frac{1}{|\lambda_2|}\right). \quad (2)$$

In this formula, λ_1, λ_2 are the characteristic exponents associated with σ (for definiteness, let $\lambda_1 > 0 > \lambda_2$, and assume σ to be ergodic); $\text{HD}(\sigma)$ is the Hausdorff dimension of σ (see below). In the present situation, (1) reduces to $h(\sigma) = \lambda_1$, and therefore (1) is – the view of (2) – equivalent to

$$\text{HD}(\rho) = 1 + \frac{\lambda_1}{|\lambda_2|}. \quad (3)$$

This is precisely the formula conjectured by Kaplan and Yorke [10], but where HD is the Hausdorff dimension of a particular measure instead of the Hausdorff dimension of an attractor†. The quantity $\text{HD}(\sigma)$ is by definition the infimum of the Hausdorff dimension of sets Y such that

$\sigma(Y) = 1$. One can however replace $HD(\sigma)$ by more computable quantities: the condition $\lambda_1 > 0 > \lambda_2$ implies that, for σ-almost every x,

$$\lim_{r \to 0} \frac{\log \sigma B_x(r)}{\log r} = HD(\sigma),$$

where $B_x(r)$ is the ball of radius r centered at x. In particular, one can replace $HD(\sigma)$ by the Rényi dimension

$$\lim_{\epsilon \to 0} \left[\left(\log \frac{1}{\epsilon} \right)^{-1} \inf_{\text{diam} \leq \epsilon} H(\sigma) \right],$$

where H is the entropy $-\Sigma_i \sigma(A_i) \log \sigma(A_i)$ computed with respect to the partition*.

3. What is the Hausdorff dimension of Julia sets?

The Hausdorff dimension is a most natural concept for sets which are self-similar (see Mandelbrot [12]). Self-similarity is ensured by invariance under conformal transformations and occurs thus naturally in the study of polynomial (or rational) transformations of the Riemann sphere. For instance the map $z \to z^2$ has the unit circle $\{z: |z| = 1\}$ as invariant set of dimension 1. The perturbed $z \to z^2 + \lambda$ has an invariant curve near the unit circle, and one can show that this curve J has dimension > 1 for small $\lambda \neq 0$. (The curve J is the Julia set of the map $z \to z^2 + \lambda$. A review of the Julia–Fatou theory of the iteration of rational transformations is given by Brolin [3]). One can show that the Hausdorff dimension of J is a real analytic function for λ of small λ (see Ruelle [19]). In fact a non-rigourous (but presumably exact) calculation yields

$$HD(J) = 1 + \frac{|\lambda|^2}{4 \log 2} + \text{higher order}. \tag{4}$$

This agrees with numerical results of L. Garnett

*I am indebted to L.-S. Young for correspondence on this question.

(unpublished). It would be quite possible to compute more terms in (4) and compare them with numerical estimates of $HD(J)$.

Incidentally, let me mention that some polynomial maps preserve a smooth measure, and also give rise to fractal curves, which have been studied by Nauenberg (private communication).

4. Are there turbulent crystals?

The time dependence of differentiable dynamical systems presents characteristic phenomena, which are now well documented in dissipative physical systems: occurrence of steady states, periodic states, quasi-periodic states, frequency locking, and finally chaos or turbulence. The space dependence of equilibrium states (or rather *Gibbs states*) in statistical mechanics shows many of the same phenomena. In particular, modulated crystals correspond to quasi-periodic states, and the occurrence of supercells to frequency locking (see Janner and Janssen [8] for a review). It is therefore natural to ask if turbulent crystals will also occur. One-dimensional systems at zero temperature can be associated with area preserving differentiable dynamical systems (see Aubry [1], Bak [2], Janssen and Tjon [9]), but realistic systems seem much more difficult to handle. In the case of differentiable dynamical systems there is an argument to the effect that a quasi-periodic system with more than two periods can give turbulent (or chaotic, or noisy) behavior under small perturbation (see Ruelle and Takens [21], Newhouse, Ruelle and Takens [13]). It seems that this argument may also apply to a three-dimensional system of statistical mechanics at temperature $\neq 0$ (see Ruelle [20]). The result would be that trying to modulate a crystal by more than two new frequencies might lead to a turbulent crystal. This would be an equilibrium structure, not invariant under translations, but having fuzzy diffraction peaks (some of them at least). A turbulent crystal would differ from an amorphous solid (which is out of equilibrium) and a fluid (which is trans-

5. What is the mechanism of intermittency in developed turbulence?

The work *intermittency* has been used to describe rather different phenomena. For instance, the intermittency of Pomeau and Manneville [16] may occur at the onset of turbulence. We want to discuss here a different phenomenon, which consists in the concentration of dissipation and vorticity in a small part of physical space for a turbulent fluid at high Reynolds number. "Physical space" is taken to be two- or three-dimensional. Dissipation and vorticity correspond to the symmetric and antisymmetric part of the velocity gradient tensor. So, the velocity gradient is large on only a small part of physical space. This should bring some corrections to the otherwise rather successful Kolmogorov theory of turbulence, which assumes more homogeneity. Scaling ideas give a reasonable model of intermittency (see Frisch, Sulem and Nelkin [7] and reference given there to earlier work). The mechanism of intermittency is however more obscure. It is tempting to attack the problem in two-dimensions, where the concentration of vorticity in individual vortices is easily observed at the surface of a river for instance. Already Poincaré [15] was aware of the problem, and gave a discussion based on stability of the motion arguments. It is interesting that Poincaré's discussion uses the fact that the fluid is viscous. A very different idea was proposed by Onsager [14]. He considered an inviscid two-dimensional fluid where the vorticity if concentrated in a finite (but large) number of points. One can show that the motion of these vortices is Hamiltonian, and one can think of applying statistical mechanics to them. As it turns out the Hamiltonian of vortices is the same thing as the potential energy of a system of electric charges in two-dimensions. Since the Hamiltonian of vortices has no kinetic energy, Onsager argued that negative temperature states could occur. In such a state, vortices of the same sign would attract each other, leading to big blobs of positive and negative vorticity – and this would explain intermittency. Unfortunately, the modern theories on the statistical mechanics of systems of electric charges in two-dimensions show that no negative temperature states exist, and it therefore does not seem to be possible to explain intermittency by Onsager's mechanism. The origin of intermittency remains thus rather obscure. As relevant to the discussion one may mention, besides Poincaré's discussion (two-dimensional), the ideas of Frisch and Morf [6] (rather general, and independent of dimensionality) and the computer work of Chorin [4] in three dimensions. Also, since vorticity in a viscous fluid is normally created at the boundary, this geometric fact may largely be responsible for intermittency.

6. What is the spectrum of characteristic exponents in a turbulent fluid?

If several systems with independent dynamics occupy disjoint regions $\Lambda_1, \ldots, \Lambda_N$ of space, the spectrum of characteristic exponents of the joint system is simply the union of the spectra of the subsystems (each characteristic exponent being given a suitable multiplicity). If an interaction between subsystems is introduced, one may hope that this does not affect the spectrum too much and that, in the large volume limit, a number of characteristic exponents per unit volume may thus be defined. A proof of this for non-trivial dynamical systems seems beyond reach for the moment. Otherwise, very interesting questions could be asked about the limiting density of exponents. For instance, is the density finite or infinite at 0, i.e., for infinite relaxation times?

Even though one cannot prove the existence of a limiting distribution of characteristic exponents for large volumes, one can prove the existence of certain bounds which are uniform in the volume. Sinai [24] has obtained such a lower bound for the sum of positive characteristic exponents divided by the volume for a Hamiltonian system of hard spheres. For a viscous fluid in d dimensions, let ρ be a probability measure on the space of velocity fields, which is invariant under the Navier–Stokes time evolution. We make suitable smoothness assumptions and, if $d=3$, we suppose that time evolution is well defined near the support of ρ. The characteristic exponents are obtained by studying a linearization of the Navier–Stokes equation around a solution. This linearization is similar to a Schrödinger equation. Using estimates for negative eigenvalues of the Schrödinger operator due to Lieb and Thirring [11], one can prove bounds of the type

$$\sum_{k:\mu^{(k)}(\rho)\geq 0} (\mu^{(k)}(\rho))^\gamma$$

$$\leq K_{\gamma,d} \nu^{-\gamma/2 - 3d/4} \int \rho(\mathrm{d}\boldsymbol{v}) \int_\Omega \mathrm{d}x \epsilon_v(x)^{\gamma/2 + d/4}, \qquad (5)$$

where

$$\epsilon_v(x) = \frac{\nu}{2} \sum_{i,j} \left(\frac{\partial v_i}{\partial x_j} + \frac{\partial v_j}{\partial x_i}\right)^2$$

is the rate of energy dissipation per unit volume, ν is the kinematic viscosity, γ any number ≥ 1, $\mu^{(1)}(\rho) \geq \mu^{(2)}(\rho) \geq \cdots$ are the characteristic exponents associated with ρ (assumed to be ergodic for simplicity), and the $K_{\gamma,d}$ are universal constants.

We may consider a turbulent viscous fluid as an information source (random number generator). The rate of information production is the Kolmogorov–Sinai invariant (also called entropy – but this is not the thermodynamic entropy of the fluid). The Kolmogorov–Sinai invariant is bounded above by the sum of the positive characteristic exponents (see Ruelle [18]), and therefore by (5) with $\gamma = 1$. In particular, for $d = 2$ we see that *the rate of information production is bounded above by a constant multiple of the rate of energy dissipation*. For $d = 3$ one has for the Hausdorff dimension $\mathrm{HD}(A)$ of an attractor A an estimation of the form

$$\limsup_{\Omega \to \infty} \frac{\mathrm{HD}(A)}{\mathrm{vol}\,\Omega} \geq K\nu^{-9/4} \sup_{\rho:\mathrm{supp}\,\rho \subset A}$$

$$\times \left(\int \rho(\mathrm{d}\boldsymbol{v}) \frac{\mathrm{d}x}{\mathrm{vol}\,\Omega} \epsilon_v(x)^{5/4}\right)^{3/5},$$

where K is a universal constant, provided a certain conjecture of Lieb and Thirring is satisfied.

References

[1] S. Aubry, The Devil's staircase transformation in incommensurable lattices, Lecture Notes in Math., pp. 221–245 in *The Riemann problem, complete integrability and arithmetic applications*, D. Chudnovsky and G. Chudnovsky, eds., Lecture Notes in Math. No. 925 (Springer, Berlin, 1982).

[2] P. Bak, Chaotic behavior and incommensurate phases in the anisotropic Ising model with competing interactions, Phys. Rev. Letters 46 (1981) 791–794.

[3] H. Brolin, Invariant sets under iteration of rational functions. Arkiv för Mat. 6 (1965) 103–144.

[4] A.J. Chorin, The evolution of a turbulent vortex, Commun. Math. Phys. 83 (1982) 517–535.

[5] J. Curry, On computing the entropy of the Hénon attractor, preprint.

[6] U. Frisch and R. Morf, Intermittency in nonlinear dynamics and singularities at complex times, Phys. Rev. A 23 (1981) 2673–2705.

[7] U. Frisch, P.-L. Sulem and M. Nelkin, A simple dynamical model of intermittent fully developed turbulence, J. Fluid Mech. 87 (1978) 719–736.

[8] A. Janner and T. Janssen, Symmetry of incommensurate crystal phases in the superspace group approach, Match 10 (1981) 5–26.

[9] T. Janssen and J.A. Tjon, Bifurcations with nonuniversal exponents in a lattice model, preprint.

[10] J.L. Kaplan and J.A. Yorke. Chaotic behavior of multidimensional difference equations, in *Functional Differential equations and approximation of fixed points*, Lecture notes in Math. no. 730 (Springer, New York, 1979) pp. 228–237.

[11] E. Lieb and W. Thirring, Inequalities for the moments of the eigenvalues of the Schrödinger equation and their relation to Sobolev inequalities, in Studies in Mathematical Physics: Essays in Honor of Valentine Bargman, E. Lieb, B. Simon, and A.S. Wightman, eds. (Princeton Univ. Press, Princeton, NJ, 1976), pp. 269–303.

[12] B. Mandelbrot. Fractals: forms, chance, and dimension (W.H. Freeman, San Francisco, 1977).

[13] S. Newhouse, D. Ruelle and F. Takens, Occurrence of Strange Axiom A attractors near quasiperiodic flows on T^m, $m \geq 3$, Commun. Math. Phys. 64 (1978) 35–40.

[14] L. Onsager, Statistical hydrodynamics, Suppl. Nuovo Cim. 6 (1949) 279–287.

[15] H. Poincaré, Théorie des tourbillons (Georges Carré, Paris, 1893).
[16] Y. Pomeau and P. Manneville, Intermittent transition to turbulence in dissipative dynamical systems, Commun. Math. Phys. 74 (1980) 189–197.
[17] D. Ruelle, What are the measures describing turbulence? Progr. Theor. Phys. Suppl. 64 (1978) 339–345.
[18] D. Ruelle, An inequality for the entropy of differentiable maps, Bol. Soc. Bras. Mat. 9 (1978) 83–87.
[19] D. Ruelle, Repellers for real analytic maps, Ergod. Th. and Dynam. Syst., To appear.
[20] D. Ruelle, Do turbulent crystals exist? Physica 113A (1982) 619–623.
[21] D. Ruelle and F. Takens, On the nature of turbulence, Commun. Math. Phys. 20 (1971) 167–192; 23 (1971) 343–344.
[22] D.A. Russel, J.D. Hanson and E. Ott, Dimension of strange attractors, Phys. Rev. Letters 45 (1980) 1175–1178.
[23] I. Shimada, Gibbsian Distribution on the Lorenz Attractor, Progr. Theor. Phys. 62 (1979) 61–69.
[24] Ya. G. Sinai, On the entropy per particle for the dynamical system of hard spheres, preprint.
[25] L.-S. Young, Dimension, entropy and characteristic exponents, preprint.

CHAPTER 2

EXPERIMENTAL OBSERVATIONS
OF CHAOTIC SYSTEMS

OSCILLATIONS AND CHAOS IN CHEMICAL SYSTEMS

Irving R. EPSTEIN
Department of Chemistry, Brandeis University, Waltham, MA 02254, USA

The chemical community today views chemical chaos much as it did chemical oscillation 20 to 30 years ago. There are a number of "enlightened" students of and believers in the phenomenon, but the vast majority of chemists are either ignorant of or skeptical about the possibility of genuine chaos in a well-controlled chemical system.

Major developments in the understanding of periodic chemical oscillation, including the recent systematic design of a new family of oscillators, are reviewed. The question of how closely linked periodic and chaotic behavior are in chemical systems is considered briefly, and implications for the design of chaotic systems are noted.

Two systems are considered in some detail. The first, earlier reported as a photochemical oscillator and/or chaotic reaction, is shown to be a system of hydrodynamic rather than chemical interest. The second, one of the newly designed family of chlorite oscillators, exhibits many of the features found in experiments on chaos in the Belousov–Zhabotinskii reaction. It illustrates the close relation between complex periodic oscillations and chaos. Both examples serve to point out the origins of the present controversy over the existence of chemical chaos.

1. Introduction: periodic chemical oscillators

Although chaotic behavior has been thoroughly studied and broadly accepted in physics and mathematics, for many chemists reports of such phenomena might bring to mind Shakespeare [1] rather than science:

"There are more things in heaven and earth, Horatio,
Than are dreamt of in your philosophy"

The vast majority of chemists have never heard of chaos, and of those that have, many, perhaps a majority are extremely skeptical of its validity as a phenomenon intrinsic to any *chemical* system. In a sense, chemical chaos has a public relations problem. To understand this problem, it is useful to review briefly some of the history of periodic chemical oscillation, since it shows a number of parallels and relations with chemical chaos, both in its scientific and its "public relations" aspects.

The first observations of chemically driven periodic oscillation were probably those of the prehistoric men and women who first noted the rhythmic beating of their hearts or the coincidence of menstrual periods with the phases of the moon. Several temporally and spatially periodic phenomena were reported in heterogeneous chemical systems [2–5] in the nineteenth and early twentieth centuries, but these were agreed to be strongly or even totally derived from the physical processes, such as electrodeposition or precipitation, occurring in those systems. The first single, homogeneous oscillating chemical reaction was discovered accidentally by Bray [6] in 1921.

The Bray reaction, which involves hydrogen peroxide and iodate in acidic aqueous solution, was largely ignored by chemists for nearly half a century. Bray had considerable difficulty in finding students to work on it. Of those chemists who became aware of the Bray reaction, many attempted to show that the oscillations resulted not from the chemistry but from impurities or dust particles [7]. Some saw chemical oscillation as a potential violation of the second law of thermodynamics – a sort of chemical perpetual motion machine. Richard Noyes, one of the pioneers in the study of chemical oscillation, summarizes the situation:

"In 1921, Bray reported that the iodate–hydrogen peroxide system exhibited homogeneous chemical oscillations and ascribed the observations to a complex autocatalytic mechanism. However, as late as 1968, efforts were made to dismiss his claims as artifacts..." [8]

The modern era of oscillating chemical reactions begins in 1958 with the serendipitous discovery by Belousov [9] of oscillations in the color of a mixture of citric and sulfuric acids, bromate and a cerium salt. Belousov's work, perhaps because of skepticism among chemists, was published in an obscure Soviet medical journal. However, by this time Prigogine and his school [10] had shown that far from equilibrium a wealth of nonlinear phenomena could occur. Chemical oscillation is not necessarily a violation of thermodynamics, only of reversible, equilibrium thermodynamics. This increasing comprehension of the significance of chemical oscillation led others, notably Zhabotinskii [11], to pursue the study of Belousov's reaction in depth.

By 1972, Field, Körös and Noyes [12] had proposed a detailed mechanism – i.e. a set of some 20 elementary steps or molecular encounters – which they claimed gave an explanation of the observed oscillations in the system, by then known as the Belousov–Zhabotinskii or BZ reaction. A detailed computer simulation [13] soon verified the success of that mechanism.

The discovery of the BZ reaction was followed by the development of a number of variants of that system [14–15] in which one or more components are deleted or replaced by chemically related species or even by the constituents of the Bray reaction [16]. A large array of biologically derived oscillators, generally cell-free extracts from such starting materials as yeast or beef heart, were found and were shown to derive from a fundamental biochemical oscillator, the process of glycolysis [17].

By 1980, a relatively large number of homogeneous systems were known to exhibit periodic chemical oscillation. All had been discovered by one of three techniques:
1) Accident
2) Variation of known oscillators
3) Extraction from biological systems.

Unfortunately for the development of a general theory of chemical oscillation, the first method had yielded only two examples and was too unreliable to depend upon as a source of others. The second approach does not produce oscillators of sufficient novelty to provide new insights, and biological systems tend to be too complex to lend themselves to generalization.

In late 1980 there occurred a major breakthrough – the first successful systematic design of a new chemical oscillator by our group at Brandeis [18]. The key elements in that design were three in number. First, the system was held far from equilibrium by the use of a continuous flow stirred tank reactor (CSTR). Then, an autocatalytic reaction was selected in order to provide the appropriate enhancement of reaction rate by a product. Finally, the system was probed in the CSTR until conditions resulting in bistability were found. Introduction of an appropriate feedback species in accordance with the cross-shaped phase diagram model of Boissonade and De Kepper [19] led to oscillations. The resulting behavior of that chlorite–iodate–arsenite reaction is shown in fig. 1. Note the relative simplicity of the waveforms, typical of periodic chemical oscillators, and the possibility of measuring two (or more) oscillatory responses simultaneously.

With the first deliberate construction of a chemical oscillator, the field began to move rapidly. The first chlorite oscillator has been developed into a family of nearly 20 such systems [20]. "Missing links" to the BZ reaction have been found [21]. The pieces are beginning to fit together, and, as shown in fig. 2, a tentative taxonomy of chemical oscillators is beginning to emerge. The increased activity in the field as well as the growing awareness among chemists of the principles of irreversible

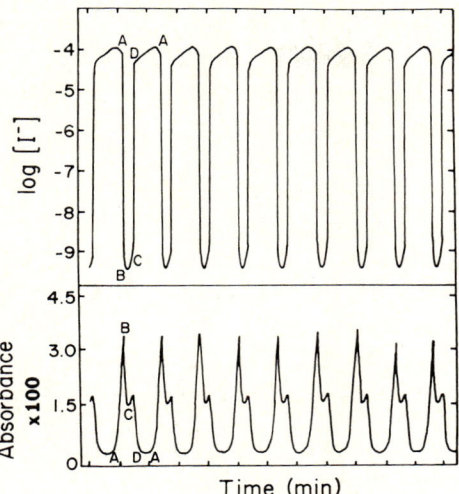

Fig. 1. Oscillations of the iodide concentration and the absorption per cm path length at 460 nm (proportional to iodine concentration) for a reaction in a stirred tank reactor with $[KIO_3]_0 = 24 \times 10^{-3}$ M, $[As_2O_3]_0 = 2 \times 10^{-3}$ M, $[NaClO_2]_0 = 2 \times 10^{-3}$ M, $[Na_2SO_4]_0 = 0.1$ M, $[H_2SO_4]_0 = 0.01$ M, residence time 400 s and $T = 25°C$. Concentrations are given in the reactor after mixing, but before any reaction takes place.

thermodynamics have combined to bring the study of periodic chemical oscillation into the mainstream of chemistry.

2. Chemical chaos

As the number of chemists studying oscillating reactions and the number of these oscillators have grown, the field has branched out toward related, but more exotic phenomena – spatial structures [22–23], coupled oscillators [24], and chemical chaos. Chaos in chemistry today is a bit like periodic oscillation was some 30 years ago – a subject of considerable controversy. Recalling Noyes' disapproving comments above on the treatment of the Bray reaction, it is of interest to note a more recent statement of his on chaos.

"The modeling computations indicate the range of chaotic behavior is orders of magnitude narrower than that observed experimentally...".

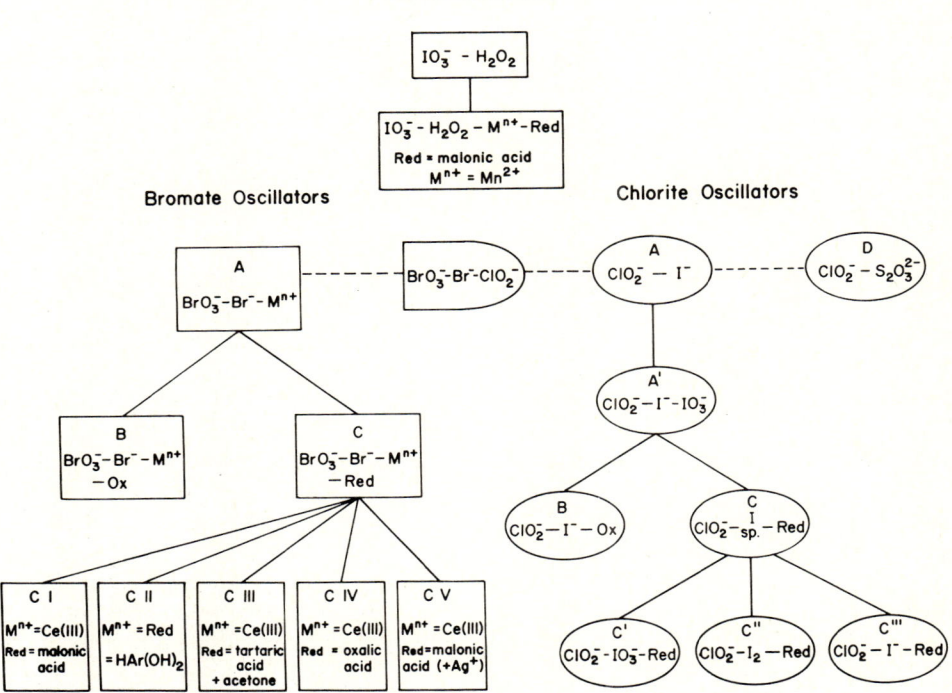

Fig. 2. Schematic classification of bromate and chlorite oscillators. I sp. denotes an iodine containing species (I^-, I_2 or IO_3^-) Ox = oxidizing agent. Red = reducing agent.

The resolution of the discrepancy is not clear. Perhaps a substitution of continuous for peristaltic pumping will so narrow the range of chaotic behavior that it can be explained by the ultimately uncontrollable fluctuations inherent in any experimental system.... The jury is still out" [25].

The shoe is now on the other foot. The defender of chemical periodicity is now raising the possibility that chemical chaos is an experimental artifact. The experiments referred to above are those of Hudson et al. [26] on the BZ system. Similar, but more elegant (and perhaps more convincing) experiments have been done on that system by groups in Austin [29] and in Bordeaux [30] and are discussed elsewhere in this collection. The reported examples of chemical chaos are summarized in table I. We turn now to the last two entries in that table, which give some insight into the origins of the controversy among chemists over the nature of chemical chaos.

Table I
Examples of chemical chaos

System	Ref.	Comments
Mathematical Models	27–28	Chemical relevance of differential equations questionable
BZ reaction	26, 29–31	Most extensive experiments, model calculation by Turner
Peroxidase catalyzed oxidation of NADH	32	Biological system limited data reported
Photochemical systems	33–36	See text
Chlorite–thiosulfate	37	See text

3. "Photochemical chaos"

The chemical literature contains at least four reports [33–36] of periodic and aperiodic oscillation in photochemical reactions and a fifth has recently been discovered in our own laboratories [38]. These systems are listed in table II. Theoretical work by Nitzan and Ross [39] suggests that illuminated chemical systems may exhibit multiple stationary states and, by implication, oscillatory behavior as well.

Table II
"Photochemical oscillators"

System	Ref.
$(CH_3)_2CO$ in CH_3CN	33
(naphthyridine) in C_6H_{12}	34
$C_{14}H_{10}$ or $C_{14}H_8(CH_3)_2$ in CH_3Cl	35
Rhodamine B in $C_2H_4Cl_2$	36
$(CH_3CO)_2 + O_2$ in several solvents	38

The experiments in which this behavior is observed show a number of features in common. An organic species is dissolved in an organic solvent. The dissolved species, when illuminated at an appropriate excitation wavelength, emits light of a longer wavelength; it fluoresces or phosphoresces. When a cell containing the solution is irradiated at the excitation wavelength with constant intensity, a time-varying, sometimes periodic intensity of emission is observed. A typical cell geometry is shown in fig. 3a. As emphasized in fig. 3b, the cell may be thought of as a pseudo-flow reactor, with the illuminated portion (generally 10–20% of the total solution volume) playing the role of the reactor, and the dark portion and the light source serving as the reservoirs. It does not appear unreasonable, then, that such a system might undergo (photo)chemical oscillation.

On closer observation, however, these systems are found to have several disturbing characteristics. The oscillatory behavior, particularly the periodic oscillation, is exceedingly irreproducible. On stirring the system, the phenomenon disappears and only constant or smoothly varying emission intensities may be observed. Large effects, possibly beyond what may be attributable to the chemistry, occur on changing the solvent. The behavior is extremely sensitive to the geometry of the cell.

The most startling observation was obtained when we removed the top of the monitoring instrument and observed the cell directly. In the biacetyl $((CH_3CO)_2)$ system, the emitted light is an intense green, visible to the naked eye. Instead of seeing the illuminated portion of the cell flashing alter-

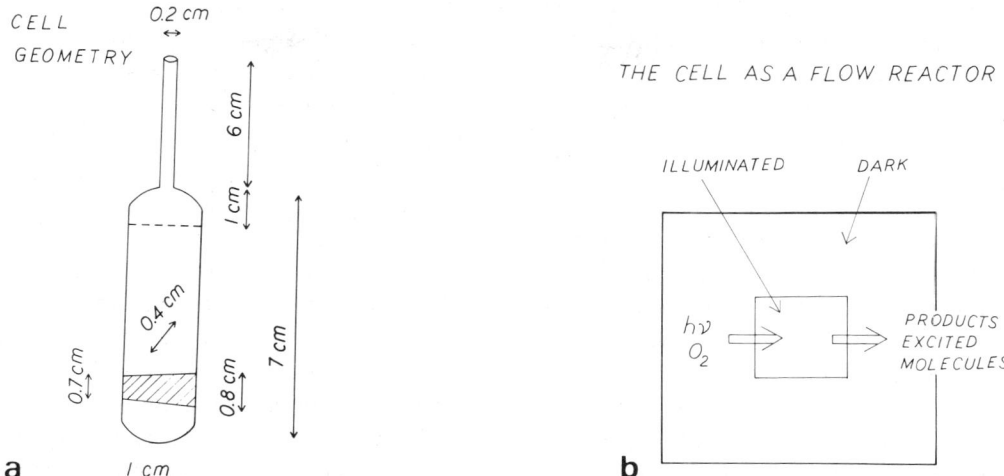

Fig. 3. a) Cell geometry for experiments on photochemical systems. Hatched area is illuminated by the incident beam. b) Schematic view of the cell illustrating how it may be viewed as a flow reactor.

nately between green and dark, we observed bands of green moving across the cell (fig. 4), apparently in and out of the region sampled by the detector in the instrument. The solution was undergoing not temporal, but spatial oscillation! While this discovery in retrospect explains most of the anomalies mentioned above, it requires careful investigation and analysis.

It is most instructive to attempt to estimate the Rayleigh number for a system of this type. The excitation lamp produces a certain amount of heat, and the heating effect is considerably larger at the bottom than at the top of the cell. Typical values of the viscosity, density and thermal expansion coefficient of organic liquids are available in the literature. Although we have been able to make only a rather crude measurement of the temperature gradient in the cell, we estimate it to be

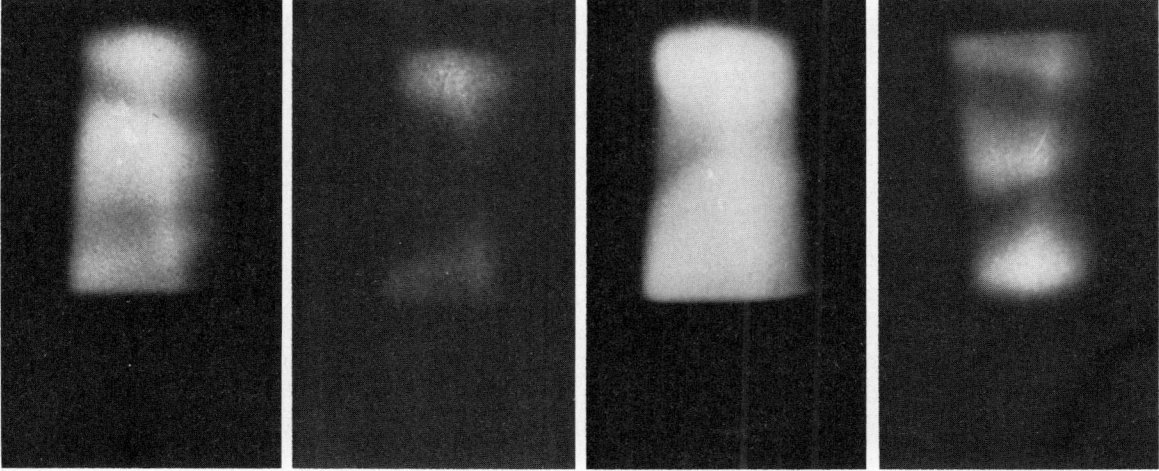

Fig. 4. Photographs of the illuminated region of the cell shown in fig. 3 containing biacetyl and oxygen dissolved in acetonitrile and illuminated at 415 nm. Light colored regions show emission by the solution at 515 nm (green phosphorescence).

≥ 0.01°/cm. The *lowest* plausible value for the Rayleigh number in a typical "photochemical oscillator" of this sort is found to be ≈ 2000. The system is unstable to convective motion of the liquid!

We now attribute the observed periodic oscillation and chaos not to the dynamics of the photochemistry, but to the convective instability which is monitored by the photochemical emission. The system is not a photochemical oscillator, but a Rayleigh–Bénard system with chemical reaction [40] as probe. To test this hypothesis we have gently heated the cell both from above and below. Heating from above stops the intensity fluctuations in all systems tested, while heating from below tends, though not always reproducibly, to start or enhance these variations.

It thus appears that "photochemical" oscillation and chaos, at least of the type reported thus far, is an artifact of the temperature gradient induced by the excitation light source. It is to just such uncontrolled or unnoticed experimental factors that many chemists attribute *all* reports of chemical chaos.

4. The chlorite–thiosulfate reaction

Until 1982, the BZ reaction was the only well characterized example of chemical chaos. Several carefully designed and analyzed experiments [26, 29, 30] as well as a model calculation by Turner [29] confirm the existence of chaos intrinsic to the dynamics of this reaction. However, as noted above, a measure of skepticism prevailed among chemists, in part perhaps because of the phenomenon being confined to a single reaction.

In a study [41] of the first iodine-free chlorite oscillator, we noted the presence of complex periodic oscillations. A more thorough investigation [37] reveals the existence, as a function of reactor residence time, of a remarkably rich set of complex periodic modes varying from pure large amplitude oscillations (L) to one large and one small peak per cycle (L + S), to as many as sixteen small for each large peak (L + 16S), until pure small amplitude oscillations (S) are obtained. This behavior is summarized in fig. 5. In a narrow but reproducible range of residence time between each pair of periodic (L + nS, L + (n + 1)S) modes, we observe a region of chaos in which L + nS and L + (n + 1)S groups alternate, apparently at random. Some chaotic behavior of this type is shown in fig. 6. Phase diagrams, which define the regions of periodic and chaotic behavior as functions not only of residence time, but of input chlorite and thiosulfate concentration as well, have been determined and are shown in figs. 7 and 8.

While the discovery of a totally new chaotic chemical system would seem to argue overwhelmingly for the validity of the phenomenon, the

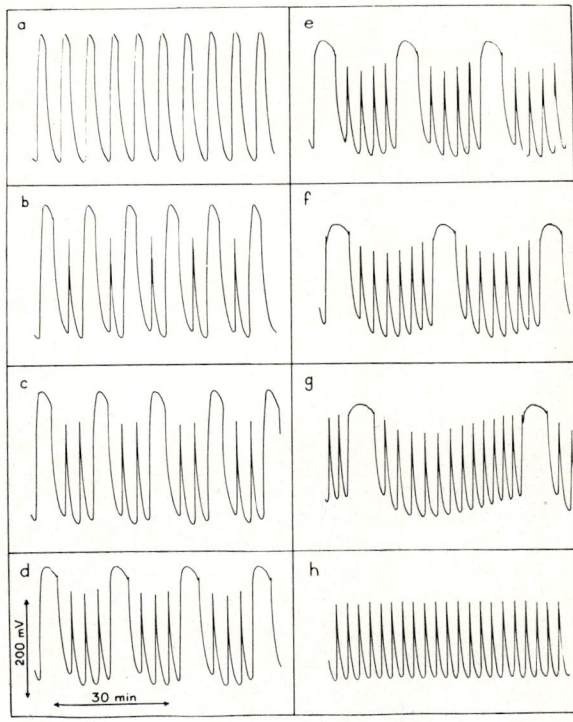

Fig. 5. Complex periodic oscillations with $[ClO_2^-]_0 = 5 \times 10^{-4}$ M, $[S_2O_3^{2-}]_0 = 3 \times 10^{-4}$ M, pH 4, $T = 25.0°C$. a) L, $\tau = 5.9$ min; b) L + S, $\tau = 9.5$ min; c) L + 2S, $\tau = 10.8$ min; d) L + 3S, $\tau = 13.5$ min; e) L + 4S, $\tau = 15.8$ min; f) L + 6S, $\tau = 20.6$ min; g) L + 12S, $\tau = 26.3$ min; h) S, $\tau = 47.3$ min. τ is residence time in the reactor. L + nS signifies one large and n small amplitude oscillations per cycle.

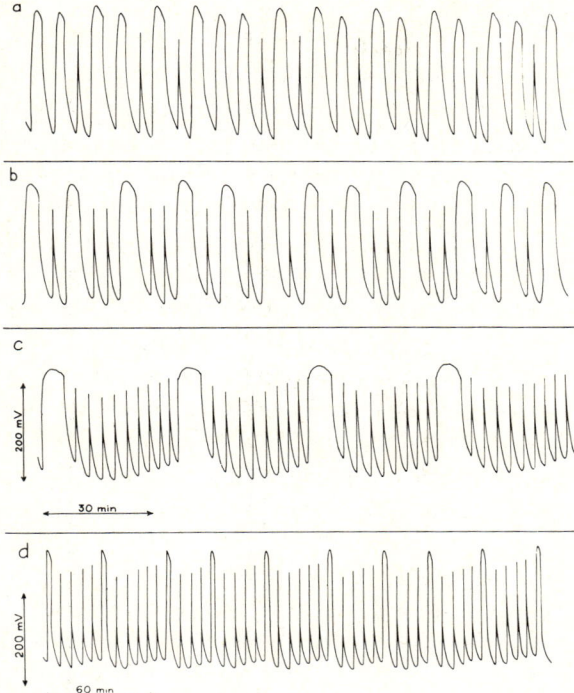

Fig. 6. Aperiodic oscillations with input concentrations and temperature as in fig. 1. a) (L, L + S), $\tau = 6.8$ min; b) (L + S, L + 2S), $\tau = 10.5$ min; c) (L + 8S, L + 9S), $\tau = 23.6$ min; d) $[S_2O_3^{2-}]_0 = 4 \times 10^{-4}$ M and other constraints unchanged: (L + 3S, L + 4S, L + 5S), $\tau = 11.42$ min.

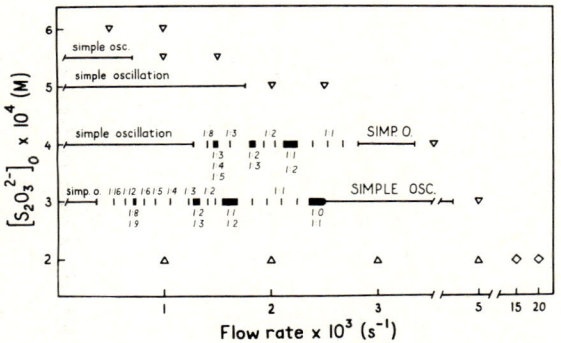

Fig. 7. A section of the phase diagram of the chlorite-thiosulfate system in the $[S_2O_3^{2-}]_0 - k_0$ plane with $[ClO_2^-]_0 = 5 \times 10^{-4}$ M, pH 4, $T = 25°C$. "SIMPLE" and "simple" denote pure L and pure S oscillations, respectively. Symbols: Vertical segments (|) show flow rate at which periodic L + nS pattern appears with the ratio above the segments; ■, chaotic region with mixed pattern of L + nS, L + (n + 1)S, (L + (n + 2)S); ▽, low potential steady state (SS I); △, high potential steady state (SS II); ◇, bistability (SS I/SS II).

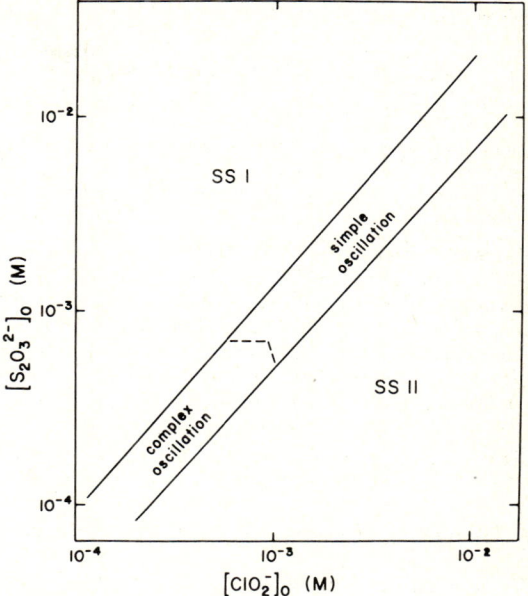

Fig. 8. A section of the phase diagram in the $[S_2O_3^{2-}]_0 - [ClO_2^-]_0$ plane. The two solid lines enclose the oscillatory region. The dashed line separates regions of simple and complex oscillation. Fixed constraints: $k_0 = 2 \times 10^{-3}$ s^{-1}, pH 4, $T = 25.0°C$.

evidence is not totally unambiguous. Under optimal conditions, the chaotic ranges are narrow and reproducible. However, if older, less rigid tubing is used in the peristaltic pump, making possible larger fluctuations in the pumping rate, the width of the chaotic region increases. Perhaps even more disturbingly, as shown in table III, as the temperature is increased by about 4°C the system passes through the entire series of periodic and chaotic modes from pure L to pure S.

The above observations make it possible to entertain what might be termed the "flip–flop" interpretation of chemical chaos. That is, there may exist two (or more) different periodic states very close together in terms of some system parameter such as flow rate or temperature. Random fluctuations in this parameter, uncontrolled by the experimentalist, then knock the system back and forth between these neighboring states, producing an apparently chaotic, but neither intrinsic nor deterministic behavior. Such an interpretation is difficult, though perhaps not impossible to rule out.

The direct approach to distinguishing between intrinsic and flip–flop chaos would be to repeat the same experiment several times and observe whether the sequence of $L + nS$ and $L + (n + 1)S$ cycles was the same in all trials. Deterministic chaos, one would think, should yield reproducible results, while flip–flop chaos should not. Unfortunately, however, the sensitivity of intrinsically chaotic systems to their initial conditions (divergence of neighboring trajectories) implies that a true repetition of an experiment will be impossible to achieve, and variation in the observed sequence of cycles should appear in either case. The inevitable small differences in experimental conditions must give rise to variations in the behavior observed in "identical" experiments on the same inherently chaotic system. We have a sort of dynamic catch-22.

Table III
Modes of oscillation observed as a function of temperature[a]

T (°C)	Mode of oscillation[b]
22.0	L
22.8	L + S
24.0	L + S
24.4	L + S, L + 2S
25.0	L + 2S
25.2	L + 2S
25.4	L + 2S, L + 3S
25.6	L + 3S
25.8	L + 3S, L + 4S
26.1	L + 4S, L + 5S
26.6	S

[a] $[ClO_2^-]_0 = 5 \times 10^{-4}$ M, $[S_2O_3^{2-}]_0 = 4 \times 10^{-4}$ M, pH 4, $k_0 = 2.05 \times 10^{-3}$ s^{-1}.
[b] A single letter (L or S) signifies simple periodic oscillation; $L + nS$ signifies complex periodic oscillation; $L + nS$, $L + (n + 1)S$ signifies aperiodic oscillation in which the two different types of cycles are mixed in a random fashion.

5. Conclusion

The weight of evidence in favor of chaos intrinsic to the dynamics of at least some chemical systems appears to be growing. However, physicists and mathematicians should not expect to create converts among chemists simply by showing them a few Poincaré maps, however beautiful. The arguments involved are "strange" and mathematical, and the chemist will respond (quite correctly) that plotting his data in a more sophisticated way cannot produce any structure that was not already present in his chart recordings, though he will admit that the right plot can make the structure more easily visible.

As chemists become more aware of nonlinear dynamics, just as they earlier absorbed irreversible thermodynamics, chaos is likely to become a more accepted and more thoroughly studied phenomenon. At present, most chemists are not even aware of the notion of chaos, and many who are see it as a case of mathematicians and physicists trying to shoehorn chemistry into a model which it doesn't really fit.

The eventual acceptance of chaos as a phenomenon of genuine interest to chemists will depend upon the answers to some, perhaps all of the following questions:

1) Is it possible to distinguish experimentally between intrinsic and "flip–flop" chaos?

While the difficulties of a direct approach to this question were alluded to above, recent observations of the folding of the attractor [42] and of a portion of the U-sequence [43] in the BZ reaction may be the beginnings of an unambiguously positive answer.

2) Can realistic models of chemical chaos be produced?

The systems of equations proposed by Rössler [27], for example, while displaying a variety of mathematically interesting behavior, are of questionable chemical relevance. Turner's calculation [29] on a four-parameter model of the BZ reaction is considerably more convincing, but similar results need to be obtained with a more detailed model, and the effects of numerical error [25] must be probed more thoroughly.

3) Are all chemical oscillators potentially chaotic?

The two most thoroughly studied classes of oscillators, bromate and chlorite, have each yielded an example of chaos. It is an open question whether the complexity of dynamics necessary to generate oscillation will always be sufficient, for some appropriate set of parameters, to produce chaos as well. A deductively rather than an inductively derived answer to this question would be highly desirable and extremely significant.

4) Is there an algorithm for the deliberate design of chaotic chemical systems?

An affirmative answer to question 3 above would provide most of the answer to this question. One could then systematically build a chemical oscillator and search near the boundaries of the oscillatory region in parameter space, which is where chaos has been observed in the BZ and chlorite-thiosulfate reactions.

5) Are there batch (i.e., closed) chaotic chemical systems, and, if so, can they generate spatial chaos?

This question needs to be framed more precisely, since oscillatory behavior in batch systems is inherently aperiodic; the system must ultimately reach equilibrium. However, it may be possible to distinguish the nearly periodic batch oscillation most familiar to chemists from genuinely chaotic batch behavior. The waves and spatial structures generated by batch oscillators [22–23] hold considerable interest for chemists and biologists. The variety of spatial phenomena which might correspond to batch chaos, and their possible chemical and biological implications, are a fascinating matter for speculation.

Acknowledgements

Our work on periodic oscillation has been carried out by a number of able coworkers, most notably Kenneth Kustin, Patrick de Kepper, Miklós Orbán, Christopher Dateo and Mohamed Alamgir. The discovery of chaos in the chlorite-thiosulfate system was made in conjunction with Dr. Orbán, while the photochemical systems have been studied together with Colin Steel, Oscar Vades-Aguilera and Meredith Morgan. Both studies have benefited from discussions with Harry Swinney. This work was supported by Grant No. CHE 7905911 from the National Science Foundation.

References

[1] W. Shakespeare, Hamlet, Act I, Scene 5.
[2] R.E. Liesegang, Z. Phys. Chem. 52 (1905) 185.
[3] J.S. Morgan, J. Chem. Soc. 43 (1921) 1262.
[4] W. Ostwald, Z. Phys. Chem. 35 (1900) 33.
[5] A.T. Fechner, Schweigg. J. Chem. Phys. 53 (1828) 141.
[6] W.C. Bray, J. Am. Chem. Soc. 43 (1921) 1262.
[7] F.O. Rice and O.M. Reiff, J. Phys. Chem. 31 (1927) 1352.
[8] R.M. Noyes and R.J. Field, Ann. Rev. Phys. Chem. 25 (1974) 95.
[9] B.P. Belousov, Sb. Ref. Radiats. Med. 1958 (Megdiz, Moscow, 1959), p. 145.
[10] I. Prigogine and R. Balescu, Bull. Cl. Soc. Acad. Roy. Belg. 41 (1955) 912.
[11] A.M. Zhabotinskii, Dokl. Nauk SSSR 157 (1964) 392.
[12] R.J. Field, E. Körös and R.M. Noyes, J. Am. Chem. Soc. 94 (1972) 8649.
[13] D. Edelson, R.J. Field and R.M. Noyes, Int. J. Chem. Kinet. 7 (1975) 417.
[14] P.G. Bowers, K.E. Caldwell and D.F. Prendergast, J. Phys. Chem. 76 (1972) 2185.
[15] M. Orbán and E. Körös, J. Phys. Chem. 82 (1978) 1692.
[16] T.S. Briggs and W.C. Rauscher, J. Chem. Educ. 50 (1973) 496.
[17] A. Ghosh and B. Chance, Biochem. Biophys. Res. Commun. 16 (1964) 174.
[18] P. De Kepper, I.R. Epstein and K. Kustin, J. Am. Chem. Soc. 103 (1981) 2133.
[19] J. Boissonade and P. De Kepper, J. Phys. Chem. 84 (1980) 501.
[20] M. Orbán, C. Dateo, P. De Kepper and I.R. Epstein, J. Am. Chem. Soc., 104 (1982) 5911.
[21] M. Orbán, P. De Kepper and I.R. Epstein, J. Am. Chem. Soc. 104 (1982) 2657.
[22] A.T. Winfree, Science 175 (1972) 634.
[23] P. De Kepper, I.R. Epstein, K. Kustin and M. Orbán, J. Phys. Chem. 86 (1982) 170.
[24] K. Bar-Eli and W. Geiseler, J. Phys. Chem. 85 (1981) 3461.
[25] N. Ganapathisubramanian and R.M. Noyes, J. Chem. Phys. 76 (1982) 1770.

[26] R.A. Schmitz, K.R. Graziani and J.L. Hudson, J. Chem. Phys. 67 (1977) 3040.
[27] O.E. Rössler, Z. Naturforsch. 31a (1976) 259, 1664.
[28] A.S. Pikovsky, Phys. Lett. 85A (1981) 13.
[29] J.S. Turner, J.-C. Roux, W.D. McCormick and H.L. Swinney, Phys. Lett. 85A (1981) 9.
[30] C. Vidal, J.-C. Roux, S. Bachelart and A. Rossi, Ann. N.Y. Acad. Sci. 357 (1980) 377.
[31] J. Maselko, Chem. Phys. 51 (1980) 473.
[32] L.F. Olsen and H. Degn, Nature 267 (1977) 177.
[33] T.L. Nemzek and J.E. Guillet, J. Am. Chem. Soc. 98 (1976) 1032.
[34] I. Yamazaki, M. Fujita and H. Baba, Photochem. Photobiol. 23 (1976) 69.
[35] R.J. Bose, J. Ross and M.S. Wrighton, J. Am. Chem. Soc. 99 (1977) 6119.
[36] R.W. Bigelow, J. Phys. Chem. 81 (1977) 88.
[37] M. Orbán and I.R. Epstein, J. Phys. Chem. 86 (1982) 3907.
[38] I.R. Epstein, M. Morgan, C. Steel and O. Valdes-Aguilera, J. Phys. Chem., Submitted.
[39] A. Nitzan and J. Ross, J. Chem. Phys. 59 (1973) 241.
[40] D. Gutkowicz-Krusin and J. Ross, J. Chem. Phys. 72 (1980) 3577, 3588.
[41] M. Orbán, P. De Kepper and I.R. Epstein, J. Phys. Chem. 86 (1982) 431.
[42] J.C. Roux and H.L. Swinney, in Nonlinear Phenomena in Chemical Dynamics, C. Vidal and A. Pacault, eds. (Springer, Berlin, 1981) p. 38.
[43] R. Simoyi, A. Wolf and H.L. Swinney, Phys. Rev. Lett. 49 (1982) 245.

EXPERIMENTAL STUDIES OF BIFURCATIONS LEADING TO CHAOS IN THE BELOUSOF–ZHABOTINSKY REACTION

J.-C. ROUX[†]

Department of Physics, The University of Texas at Austin, Austin, Texas, 78712, USA

A review is presented of the experimental evidences of chemical chaos in the Belousof–Zhabotinsky reaction, with special emphasis on the bifurcations leading to the chaotic dynamics. It is shown that a simple 1D map can account for most of the experimental observations.

1. Introduction

We shall present here a general overview of the various non-linear behaviors observed experimentally in the B–Z reaction, stressing mainly chaotic dynamics rather than the regularly oscillating regimes. Furthermore, we propose a unified description of the dynamics based mainly on 1-D maps. No reference will be systematically made to the results of simulations of chemical models yielding chaotic states. Those interested in this subject should see refs. 1–4.

The B–Z reaction consists in the oxidation of malonic acid in acidic media by bromate in the presence of a catalytic metallic ion such as cerium III or manganese II. The oxidation is performed in a well-stirred and well-thermostated flow reactor and the chemical reaction is monitored by measuring the variations of the concentration of some intermediate species: e.g. CeIV is measured by its optical density or the bromide ion is measured by a selective electrode. Table I reports the experimental conditions used in the four more complete experimental studies of chaos in the B–Z reaction. Among the control parameters that appear table I the flow rate is the easiest to vary, and, for this reason it was used in all these studies as a bifurcation parameter.

Another reason for this choice is that in the limit of zero flow rate the system approaches thermodynamic equilibrium; an increase of the flow rate will drive the system out of equilibrium, thus allowing for the appearance of non-linear behavior (oscillations...). For ease of the comparison between different data the flow rate is often expressed by the residence time τ in the reactor which takes into account the volume of the reactor:

τ = volume/flux .

However, the flow rate is not directly analogous to other bifurcation parameters such as Rayleigh or Reynold number simply because, in the limit of flow rates going to infinity, the residence time in the reactor is too small for the chemicals to have time to react with each other; then there is no chemical reaction. We must emphasize that the B–Z reaction does not reach oscillating states for arbitrary values of these control parameters in table I. They have to be chosen carefully, thus defining in the space of control parameters a volume (V_p) inside of which the reaction is oscillating (regularly or not) and outside of which it is time independent (18).

2. Regularly oscillating states

For values of the control parameter belonging to a subvolume v_p of V_p one may observe three

[†] Permanent address: Centre de Recherche Paul Pascal, Universite de Bordeaux I, Domaine Universitaire, 33405 Talence Cedex, France.

Table I
Values of the control parameter used in experiments on chaos in the B.Z. reaction. (The chemical concentrations are given relative to the reference concentration (6): malonic acid: 0.3 mole/l, bromate: 0.14 mole/l, cerous sulfate: 0.001 mole/l, sulfuric acid: 0.2 mole/l).

	Malonic acid	Bromate	Cerous sulfate	Sulfuric acid	Residence time (h)	Temperature (°C)	Relaxation oscillations period (min)
Virginia [5–7]	1	1	1	1	0.08–0.15	25	9
Texas [8–11]	0.83	1	0.83	1	0.5–3.0	28.3	1.5
Bordeaux I [12–15]	0.26	0.26	0.25	7.2	0.06–0.21	39	0.4
Bordeaux II [16–17]	0.19	0.013	0.58	7.2	0.09–1.7	39	2.8

different oscillating states depending on the value of the flow rate; see fig. 1 [8, 16]. Starting at near zero flow rate and increasing it, we observe after what is certainly a Hopf bifurcation, small amplitude quasi sinusoidal oscillations. This limit cycle will be called A in the following. If we increase the flow rate further, the system may develop a new oscillating behavior which could be described as large amplitude relaxation oscillations. In this case the peak in the power spectra can be as high as 60 to 70 db. above the instrumental background noise level depending on the quality of the probe (the best results have been obtained when measuring the concentrations of the cerium IV by spectrophotometric method). The period of relaxation oscillations is roughly twice the period of the small amplitude oscillations described previously. We shall call this cycle B in the following.

At high flow rate we may observe a new kind of small amplitude oscillations that we call type C. The mean concentration of the cerium IV for these oscillations is much larger than for the type A oscillations. We are thus led to the conclusion that the two quasi sinusoidal oscillations are different in nature.

This succession of states, summarized in fig. 1 is

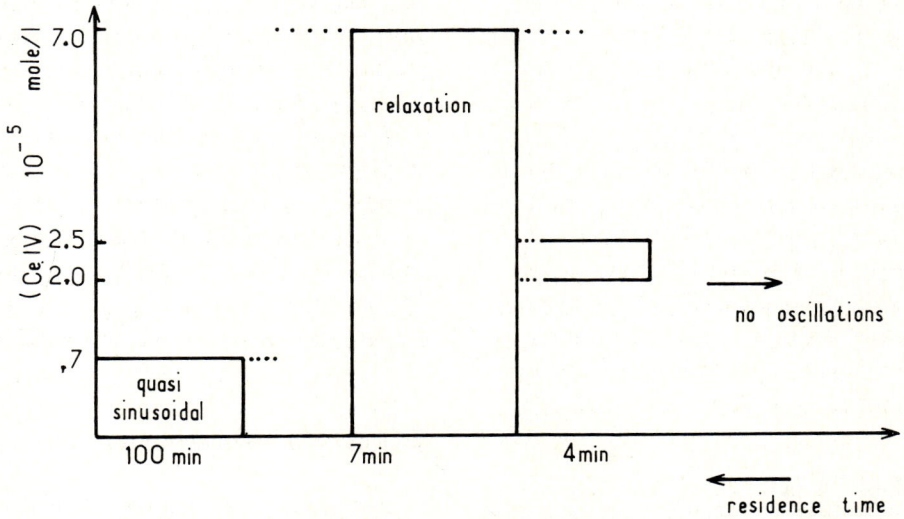

Fig. 1. Amplitude of the successive simple oscillating states as a function of flow rate (from ref. 16).

a very general result that can be observed for a wide range of values of concentrations and temperature. However, for some range of the experimental control parameters one may observe only part of this sequence.

The chaotic states observed up to now in this system have always been found between state A and B or between state B and C.

3. Chaos between low amplitude oscillations and relaxation oscillations at low flow rate

Evidence for intermittency illustrated in fig. 2 strongly suggests that the small amplitude oscillations disappear by a tangent bifurcation. The intermittency observed at high and low acidity will now be discussed in terms of a proposed 1-D map.

3.1. *Intermittency at high acidity* (experiment 4, table I–Bordeaux)

Above a critical value of the flow rate the small amplitude oscillations are no longer stable; the trajectories, fig. 2, never reach the limit cycle A but, instead, make from time to time an excursion of large amplitude. We must note that neither the times between each large excursion nor their amplitudes are constant. This is clearly Pomeau–Manneville type 1 intermittency [19–20]; "laminar" periods in the vicinity of the previous stable limit cycle (A) are interrupted by a visit to a strange attractor. The one-dimensional map constructed from the amplitudes of the oscillations in the "laminar" regime clearly shows (fig. 2) the situation which is expected to occur after such a tangent bifurcation. Thus we are driven to the conclusion that the increase of flow rate has moved the curve upward producing the disappearance of the two fixed points A' and E' (see fig. 3). Nonperiodic behavior is insured by the visit to the strange attractor between points D' and F'.

Following a proposal by D. Ruelle [21] we will now show that a global 1D-map consisting of a periodic wave tilted on its horizontal axis can describe the observed transition to intermittency (see fig. 3),

$$x_{n+1} = ax_n + cP(x_n - x_0),$$

where $P(x)$ is a periodic function. (The map proposed by D. Ruelle had $P(x) = \sin(x)$.) The very steep slope of the curve in the vicinity of the unstable fixed point B' explains why the visit in the strange attractor is very short and actually corresponds to only one point between two "laminar" regimes.

We will show that, if the map parameters a and c are monotone increasing functions of flow rate, then the proposed map can provide a complete qualitative description of the dynamical behavior observed in experiments on the B–Z reaction.

3.2. *Intermittency at low acidity* (experiment 2, table I–Texas)

The situation depicted in fig. 4 [22] shows that, after a critical value of the flow rate the small amplitude oscillations are, from time to time, interrupted by long excursion into a chaotic regime. As a first approximation this can still be interpreted by intermittency, the consequence of a tangent bifurcation. However, (i) the durations of the laminar periods and of the chaotic ones are orders of magnitude larger than in the corresponding Bordeaux experiments and (ii) no evolution of the shape or period of the small amplitude oscillations was noticed. This situation, still relevant to a tangent bifurcation indicates that we are now close to a case of bistability. The periodic wave proposed in fig. 5 illustrates this tentative explanation: the basin of attraction of the chaotic attractor and of the stable fixed point barely communicate (solid line in fig. 4) and we can imagine that small fluctuations might drive the system for one basin of attraction to the other. Experiments are in progress in Texas to verify this point.

Fig. 5 deserves a more careful analysis: we can see that if we move the curve upward again

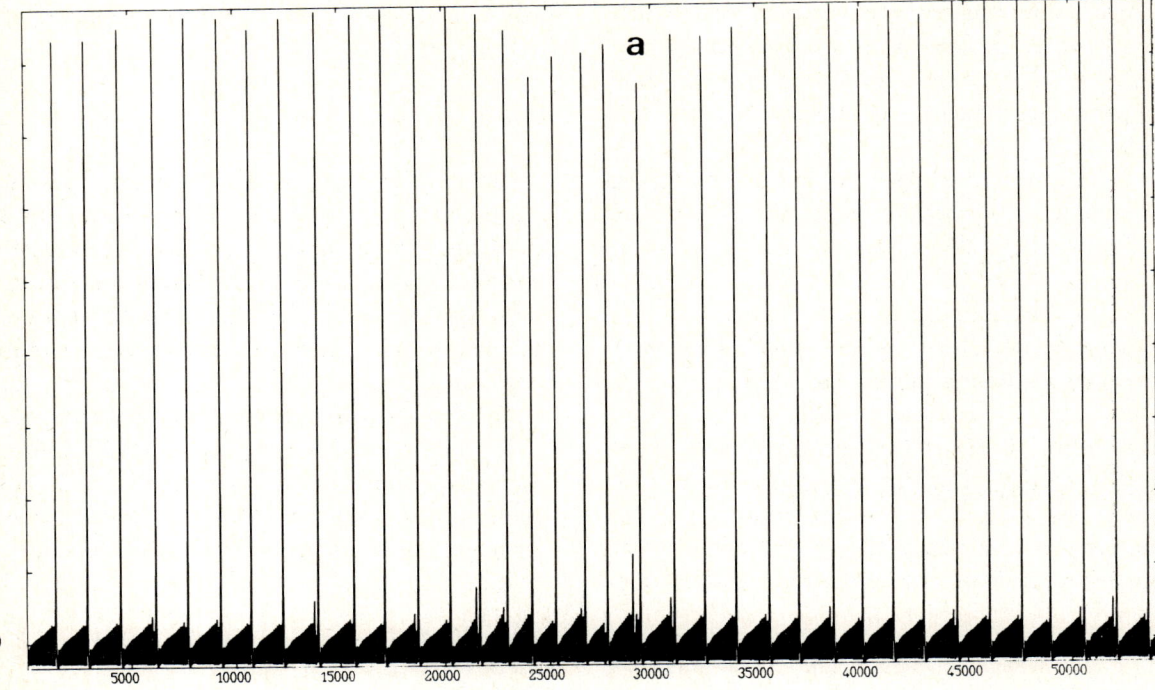

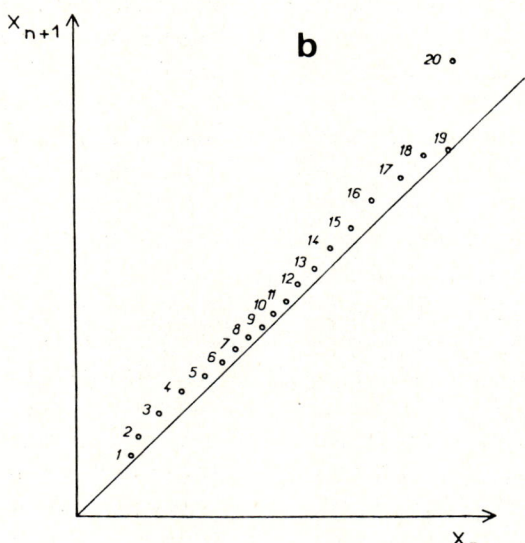

Fig. 2. (a) Time series in the intermittent regime (b) Next amplitude map for the small oscillations in the "laminar" regime. (from ref. 17).

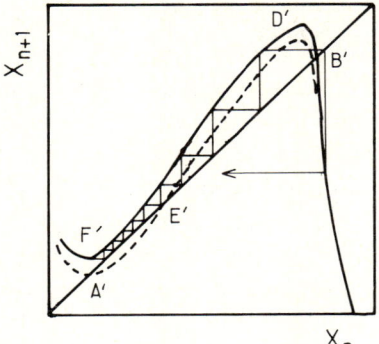

Fig. 3. 1D-map describing the intermittency observed (see text). The dashed line presents one stable fixed point A' (corresponding to limit cycle A) and two unstable fixed points E' and B'. The full line represents the intermittent regime after an increase in flow rate had moved the curve upward.

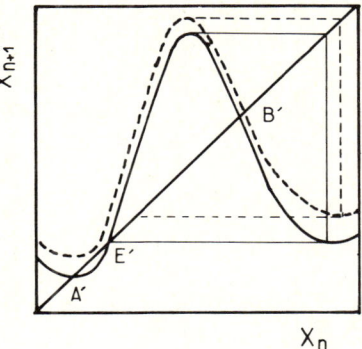

Fig. 5. 1D-map (again built from a periodic wave, but with an axis with a positive slope) corresponding to the situation shown fig. 4. The basins of attraction of the strange attractor and of the stable fixed point A' (solid line) are not completely separated.

(increasing the flow rate) we might observe reinjections in the lower right part of the curve where the slope of the tangent at the curve is zero. This situation allows, as noted by Pikovsky [23, 24], for complicated oscillating regimes. Thus moving the curve upward will result in a succession of periodic and chaotic regimes exactly in the same way as the Pikovsky map does. Actually, [8], we do observe in these experiments after the regime of fig. 4 a succession of alternating chaotic and periodic re-

* A periodic regime noted P_n^m is a regular succession of n large oscillations followed by m small ones.

gimes with periodic regimes P_1^6, P_1^4, P_1^3, P_1^2, P_1^1*. The chaotic states observed in between these periodic states are always "stochastic" [30] mixtures of the preceding and following regularly oscillating states.

4. Bifurcation leading to the large amplitude relaxation oscillations

If we trust the 1D-map of fig. 3 or 5 the net result of increasing the flow rate is to move the periodic wave upward. So doing we will reach a situation where the slope of the tangent at the map at point

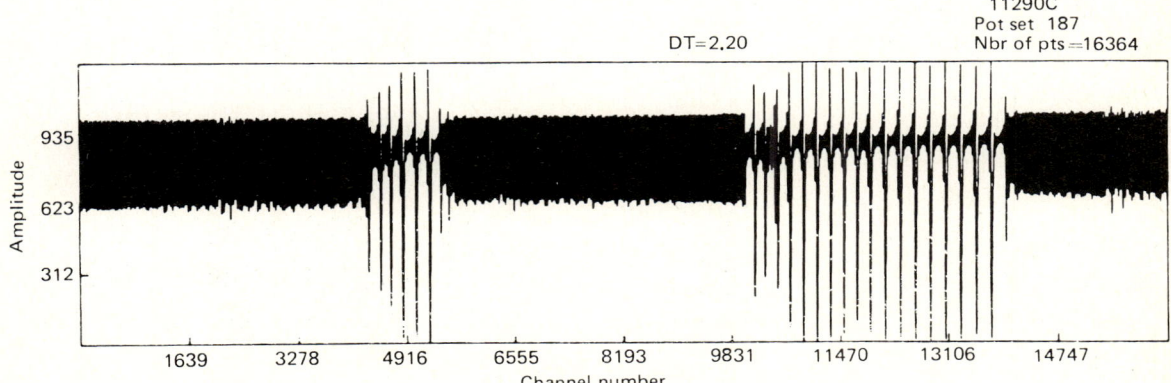

Fig. 4. Intermittency observed at low acidity. (from ref. 22).

B' will become equal to -1. At this point the "universal" behavior of one humped 1D-map may be observed and is actually in the experiments in Texas. Starting from the stable relaxation oscillations and going toward low flow rate, we were able to sort out three stages of the period-doubling sequence (periods 2, 4 and 8) and some evidence of other 2^n regimes. In the chaotic regime that follows this sequence Simonyi, Wolf and Swinney [11] have been able to show the occurrence of windows of period 3, 6, ... appearing in the order predicted by the theory.

Thus the 1D-map proposed is able to describe qualitatively the sequence of dynamical states observed in the experiments in the low flow rate range. We start from a tangent bifurcation, go through a sequence of periodic and chaotic regimes, and finally reach a new limit cycle (B') after a reverse pitchfork bifurcation.

5. Bifurcations leading to the disappearance of the large amplitude relaxation oscillations

If we continue to move upward the 1D-map of figs. 3 or 5, the next bifurcation after the pitchfork bifurcation for this map would be a tangent bifurcation.

5.1. High acidity (experiment 3, table I–Bordeaux–)

The situation is depicted in fig. 6 [15], indeed, the data can again be interpreted in terms of a tangent bifurcation. There is a strong similarity to the behavior observed at low flow rate–low acidity in the sense that we cannot discern any evolution of the shape or period of the oscillations in the laminar regime (the large amplitude oscillations in this case). The analogy of the two situations is reinforced by the further evolutions of the dynamics which are strictly parallel in these two flow rate ranges. We also observe at high flow rate a succession of complicated periodic regimes separated by chaotic states which are hybrids of the preceding and following periodic states. However, for the concentrations studied we reach a steady state rather than a stable oscillating state at the end of the sequence. Thus we cannot expect that our simple 1D-map to be of great help in describing the situation after this last tangent bifurcation.

Another argument in support of the existence a tangent bifurcation in this region is the following. Tresser et al. [25] have demonstrated that a tangent bifurcation for a map topologically equivalent to our periodic wave can produce the three different behaviors shown in fig. 7, thus predicting the occurrence of a bistability. Indeed recent results

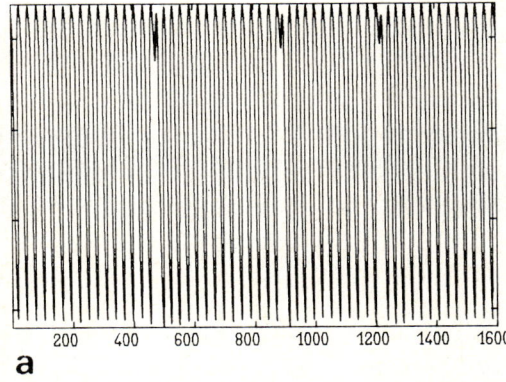
a

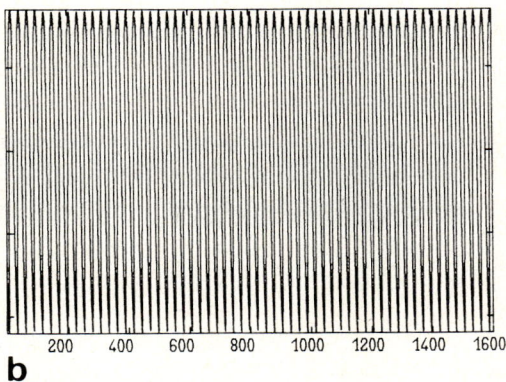
b

Fig. 6. Large amplitude stable relaxation oscillations obtained at high acidity (a) give way, after a critical flow rate value, to the regime (b).

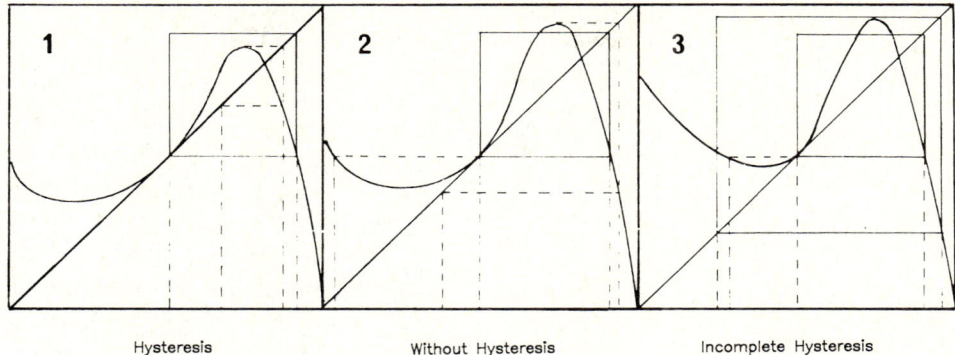

Fig. 7. Type of transitions after a tangent bifurcation (from ref. 25).

[26] (experimental conditions close to case 2, table I) do show the existence of hysteresis between the large and small amplitude oscillations.

We can see in fig. 8 that the state of the system for a value of the flow rate between point L and H depends on the history.

5.2. At low acidity (experiment 1, table I, –Virginia–)

Hudson and coworkers [5–6] carefully studied the behavior of this system in the high flow rate range. They were the first to observe the sequence of chaotic and periodic states between the relaxation oscillations and the small amplitude ones reached at very high flow rate. This again supports the idea that the limit cycle B disappears by a tangent bifurcation but, to our knowledge the nature of the bifurcation has not been determined in these experiments.

Thus, as a partial conclusion at this point we emphasize the similarity of the dynamical behaviors observed at low and high flow rate. This supports the idea that the 1D-map could be similar in the two cases.

6. Bifurcation on a torus

Among the well-recognized routes [27] leading to chaos we described the occurrence of two of them

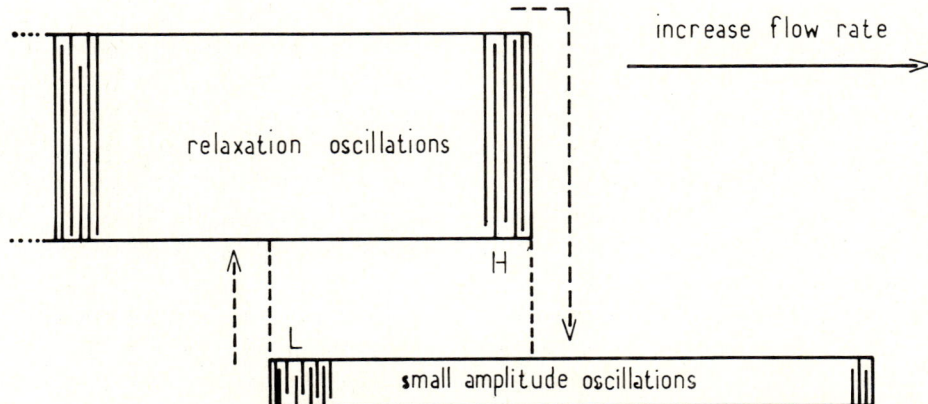

Fig. 8. Bistability observed at the disappearance of the large amplitude oscillations. (The exact nature of the small amplitude oscillations is difficult to know precisely because of the experimental noise).

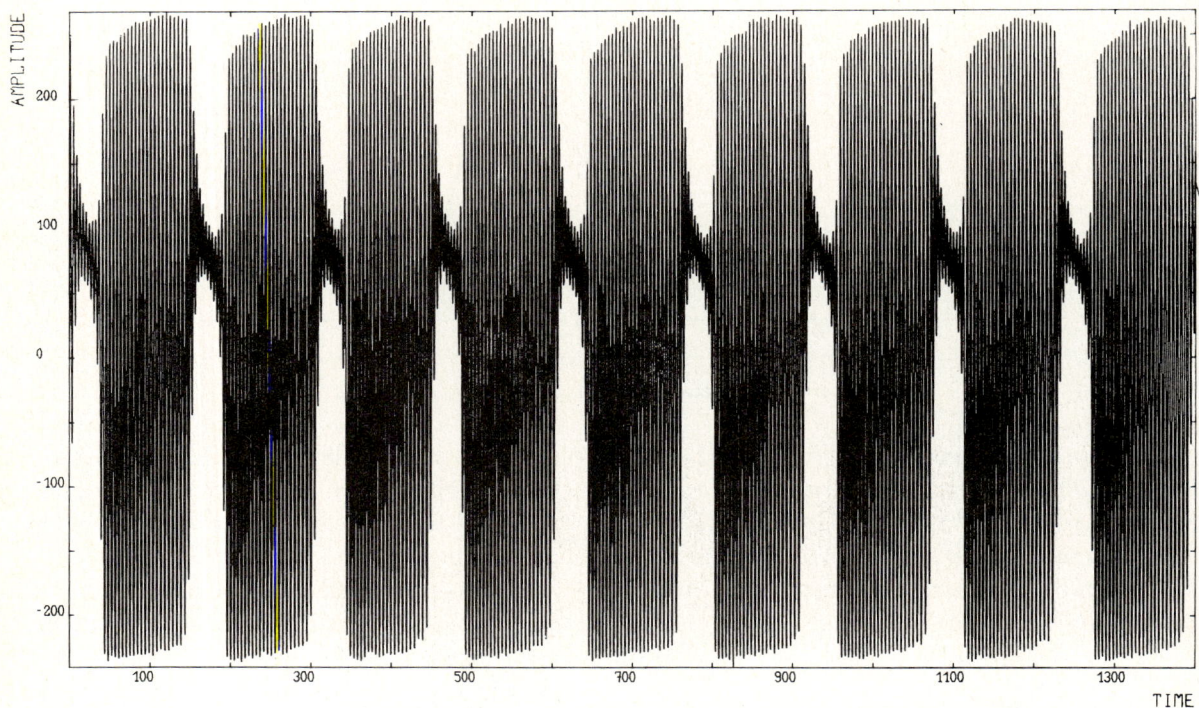

Fig. 9. Time series of a quasi-periodic regime.

in the B–Z, reaction, namely, the Feigenbaum cascade and the tangent bifurcation.

In a recent experiment [26] with a set of chemical concentrations slightly different from the set for which we have observed the bistability we observe an abrupt transition from the large amplitude oscillations to the regime depicted in figs. 9, 10 and 11; this regime clearly looks nearly quasiperiodic. The power spectrum shows two frequencies with a ratio of 40 and the reconstructed attractor looks like a torus as shown schematically in fig. 12. The Poincaré sections of this torus perpendicular to the plane of fig. 11 and going through the dashed lines on this figure are shown on the same figure. The sections are closed curved as expected for a torus.

The evolution of the shape of the Poincaré sections from intersection 2 to intersection 8 clearly shows the appearance of a folding followed by a stretching. This will result in some mixing of the trajectories, mixing which can be traced in every Poincaré section as a back and forth movement of the consecutive intersections in the part of the map where the folding occurs. The first return map will be in this case a map of the circle onto itself and is obtained by plotting the angular coordinate of one intersection as a function of the angular coordinate of the preceding one. With the center of the curve chosen as the center of gravity we obtained the 1D-map shown fig. 13. The mixing process which would have appeared as a hump in the region 1 of the map is undetectable and results simply in a thickening of the curve produced by the back and forth movement previously noted. The part of the 1D-map corresponding to the second area of accumulation of points is cleaner and, again, looks like a case of intermittency. When we increase further the flow rate the number of points in this area increases at the expense of the number of points in the region 1. This means that the number of "turns" inside the hole in the sphere increases while the number of "turns" outside decreases.

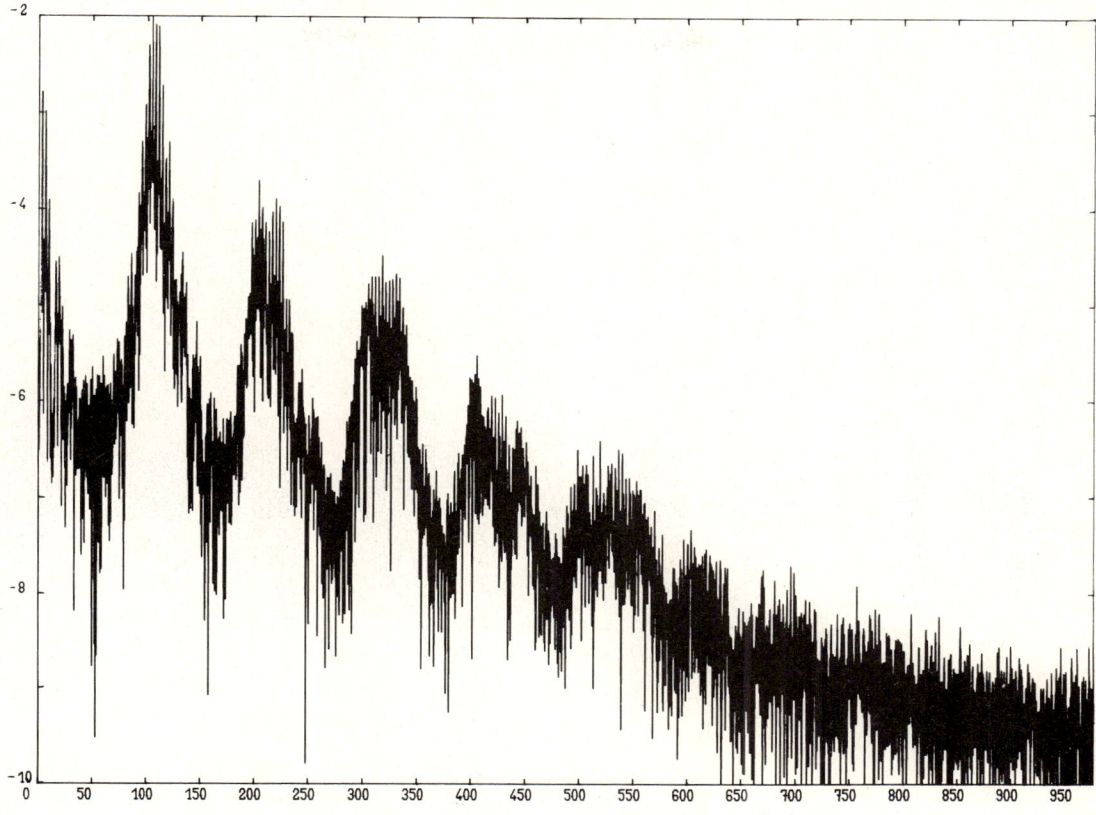

Fig. 10. Power spectrum of the regime shown fig. 9.

Guckenheimer [28] has suggested that such a torus is to be expected when a tangent bifurcation appears in the vicinity of a steady state [29]. More experimental work is needed to confirm this hypothesis.

7. Discussion

The main criticism addressed to reports of chemical chaos such as the ones we have presented concerns the true "chemical" nature of the chaos observed. Two different interpretations of the observations have been proposed

(i) The so-called chaos can always be described as a mixture of the preceding and of the following periodic states. This "mixture" may come simply from drifts of the flow rate that inevitably occur with the peristaltic pumps used.

(ii) Peristaltic pumps always produce some kind of pulsing in the flow rate and the chaos could appear because of the coupling of this oscillator with the chemical one.

The second argument can be ruled out because some experiments were done by Hudson's group with gravity feeds and the chaotic states were still observed.

The first argument is stronger, the stability of the flow rate being actually the weakest part of the experimental systems. However, during experiments in Bordeaux precise flowmeters were used and all records for which a significant change in the fluxes was observed were discarded. With these flowmeters we were unable to notice any fluctuations in the flow rate during a time interval corresponding to the supposed shift from one periodic state to the other. Furthermore, it is

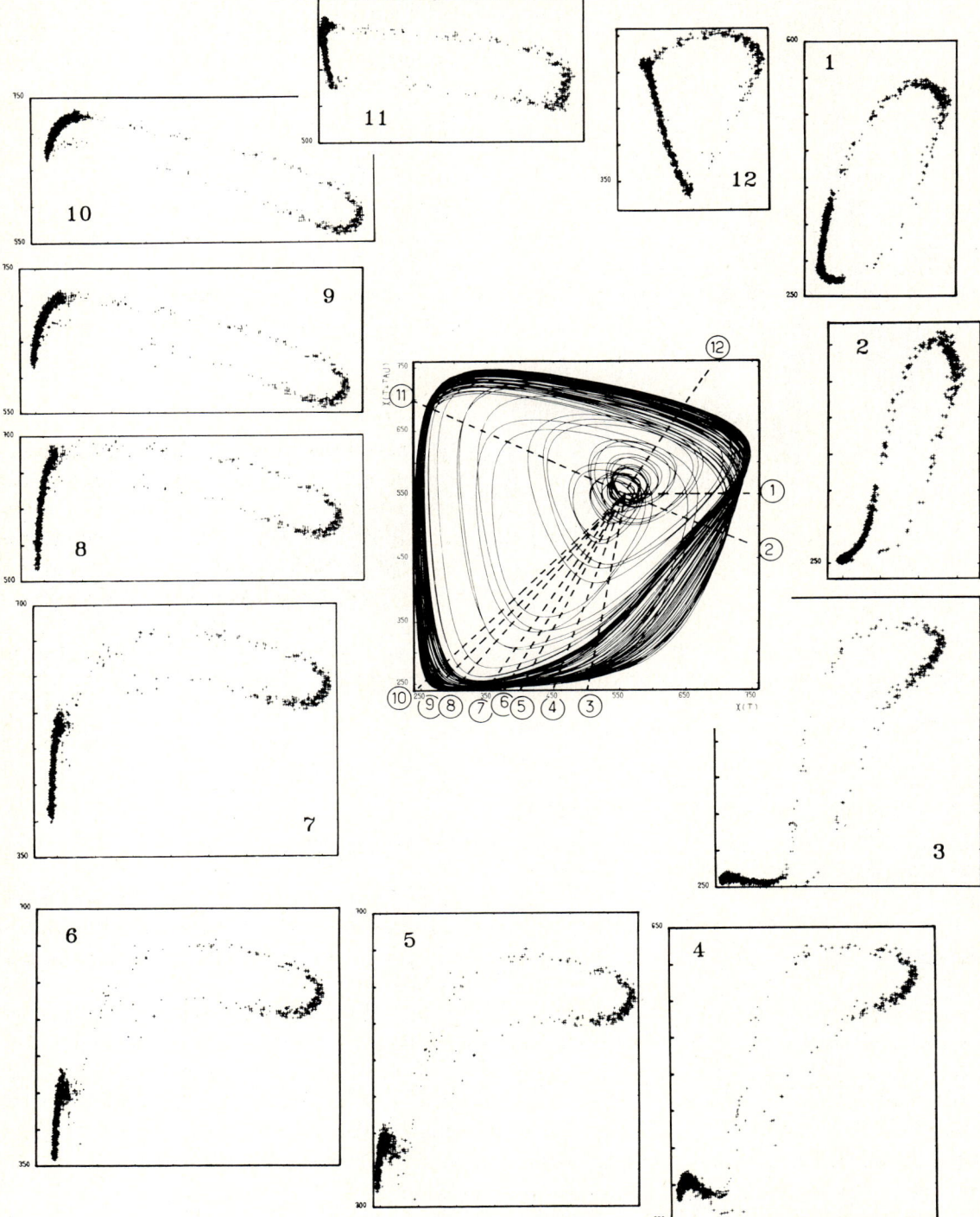

Fig. 11. Attractor of the quasi periodic regime reconstructed by Ruelle's method [16]. Cross sections through the dashed lines 1 to 12 are shown around the attractor.

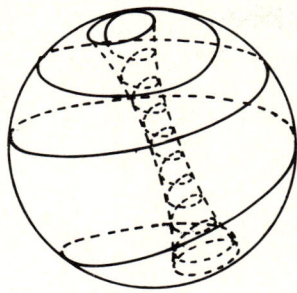

Fig. 12. Schematic representation of the attractor of fig. 11 (proposed by C. Lobry). (See for comparison fig. 8 of ref. 29).

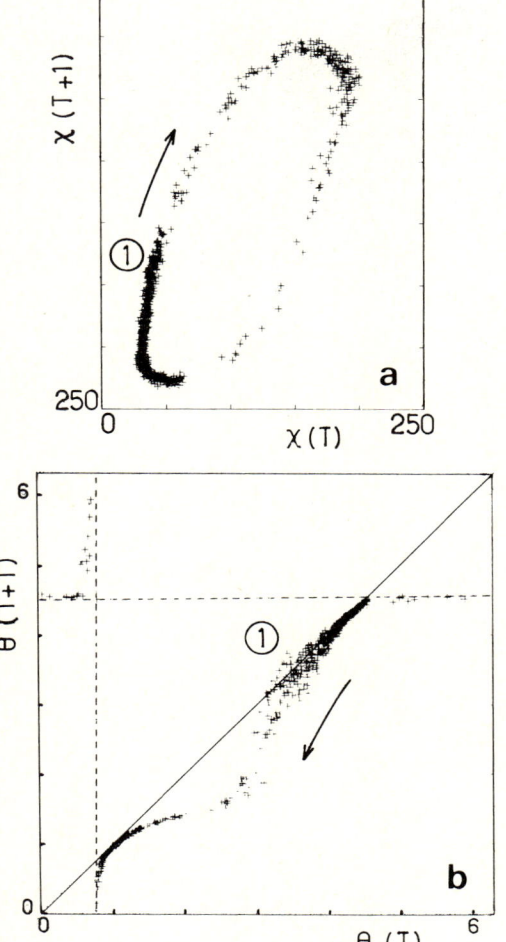

Fig. 13. 1D-map built from section 1 of fig. 11. θ_n is the angle, measured at the center of gravity, of the direction of one point with reference to the horizontal line.

difficult to imagine stochastic fluctuations able to yield chaotic states presenting the specific features reported, namely the period doubling sequence, the universal (U) sequence, a well defined 1D-map, a tangent bifurcation, etc. . . .

In our opinion the main interest of these studies in chemical chaos is the confirmation of "universal behavior" in the chaotic regimes and in the bifurcations leading to it. In this respect the exact nature of the origin of the chaotic behavior is of less importance. We do believe that the body of evidence for deterministic chaos is quite strong.

8. Conclusion

We have shown that the B–Z reaction displays the three well-recognized routes leading to chaos. Experiments have clearly established two of them, the tangent bifurcation and the pitchfork bifurcation, while the third, bifurcation to quasiperiodic behavior and then chaos needs further confirmation.

A simple 1D-map similar to that proposed by Ruelle appears to give, at least qualitatively, a fairly good description of most of the dynamical behaviors observed and is certainly worthy of a more detailed analysis.

Acknowledgements

This presentation greatly benefited from many stimulating discussions with H.L. Swinney and with people working in this field in Bordeaux and in Austin and with A. Arneodo and R. Lozi of the University of Nice and C. Lobry of the University of Bordeaux. This research was supported by C.N.R.S., National Science Fondation Grant CME 79-23627, and Robert A. Welch Foundation Grant F-805.

References

[1] J.J. Tyson, J. Math. Biol. 5 (1978) 351.
[2] J.S. Turner, in Self-Organisation and Dissipative Struc-

ture, W.C. Schieve and P. Allen, eds. (Univ. of Texas Press, 1982) p. 41.
[3] C. Lobry and R. Lozy, in Nonlinear Phenomena in Chemical Dynamics, A. Pacault and C. Vidal, eds. (Springer, Berlin, 1981) p. 67.
[4] K. Tomita and I. Tsuda, Phys. Lett. 71A (1979) 489; Prog. Theor. Phys. 64 (1980) 1138.
[5] R.A. Schmitz, K.R. Graziani and J.L. Hudson, J. Chem. Phys. 71 (1977) 3040.
[6] J.L. Hudson, M. Hart and D. Marinko, J. Chem. Phys. 71 (1979) 1601.
[7] J.L. Hudson, J.C. Mankin, J. Chem. Phys. 74 (1981) 6171.
[8] J.C. Roux, J.S. Turner, W.D. McCormick and H.L. Swinney, in Non-Linear Problems: Present and Future, A.R. Bischop, ed. (North-Holland, Amsterdam, 1982) p. 409.
[9] J.S. Turner, J.-C. Roux, W.D. McCormick and H.L. Swinney, Phys. Lett. 85A (1981) 9.
[10] J.-C. Roux and H.L. Swinney, in Non-Linear Phenomena in Chemical Dynamics, A. Pacault and C. Vidal, eds. (Springer, Berlin, 1981) p. 38.
[11] R. Simoyi, A. Wolf and H.L. Swinney, Phys. Rev. Lett. 48 (1982) 245.
[12] C. Vidal, J.-C. Roux, S. Bachelart, and A. Rossi, C.R. Acad. Sci. Paris 289C (1979) 73.
[13] C. Vidal, J.-C. Roux, A. Rossi and S. Bachelart, Ann. N.Y. Acad. Sci. 357 (1980) 377.
[14] J.-C. Roux, A. Rossi, S. Bachelart and C. Vidal, Phys. Lett. 77A (1980) 391.
[15] C. Vidal, S. Bachelart and A. Rossi, J. Phys. Paris 43 (1982) 7.
[16] J.-C. Roux, A. Rossi, S. Bachelart and C. Vidal, Physica 2D (1981) 395.
[17] J.-C. Roux, A. Rossi, S. Bachelart and C. Vidal, J. Phys. Lett. 42 (1981) L271.
[18] P. de Kepper, A. Pacault and A. Rossi, C. R. Acad. Sci. (Paris) 283C (1976) 375.
[19] Y. Pomeau and P. Manneville, Physica 1D (1980) 395.
[20] Y. Pomeau and P. Manneville, Comm. Math. Phys. 74 (1980) 189.
[21] D. Ruelle in Non-Linear Phenomena in Chemical Dynamics, A. Pacault and C. Vidal, eds. (Springer, Berlin, 1981).
[22] H.L. Swinney and J.-C. Roux, to be published.
[23] A.S. Pikovsky, Phys. Lett. 85A (1981) 13.
[24] A.S. Pikovsky and M.I. Rabinovitch, Physica 2D (1981) 8.
[25] C. Tresser, P. Coullet and A. Arneodo, J. Phys. Lett. 41 (1980) L243.
[26] J.-C. Roux and A. Rossi, to be published.
[27] J.P. Eckmann, Rev. Mod. Phys. 53 (1981) 643.
[28] J. Guckenheimer, Personal communication, see also re 29.
[29] J. Guckenheimer, in Dynamical Systems and Turbuler Warwick 1980, Lecture notes in Mathematics ₹ (Springer, Berlin, 1981) p. 99.
[30] H.L. Swinney, this volume.

PHASE SPACE ANALYSIS OF CONVECTION IN A ^{3}He–SUPERFLUID ^{4}He SOLUTION†

Hans HAUCKE* and Yoshiteru MAENO*
Los Alamos National Laboratory, Box 1663, MS M764, Los Alamos, New Mexico 87545, USA

Observations have been made on thermal convection below 1 K in a dilute solution of ^{3}He in superfluid ^{4}He contained in a cylindrical cell of aspect ratio $\Gamma = 1.20$. Complicated oscillatory phenomena were observed with a high degree of reproducibility using two temperature sensors. Phase space analysis suggests a description in terms of strange attractor dynamics.

We present some preliminary results for the time-dependent behavior of a convecting dilute solution of ^{3}He in superfluid ^{4}He. A new oscillatory state has been observed, with complex behavior which we interpret in terms of strange attractor concepts. In previous work [1] using similar apparatus, we reported the steady-state behavior of this system [2].

Our convection cell was filled with a solution of 1.6 mole% ^{3}He in superfluid ^{4}He and operated at 0.7 K. This solution is believed to be effectively a one-component fluid with a calculated Prandtl number of about 0.05 and a negative thermal expansion coefficient. The cell was constructed with top and bottom plates of copper and an annular spacer fitting between the plates made of a graphite-resin material of high thermal resistivity [3]. The inner wall of this removable spacer, together with the top and bottom plates, defined the cell geometry. This spacer fitted snugly into a thin-walled stainless steel tube which served to confine the fluid. The cell height was 0.80 cm, and for the data presented here the fluid cavity was cylindrical with aspect ratio (radius/height) $\Gamma = 1.20$. A small copper piece was inserted into a hole in the center of the top plate and thermally isolated from the rest of the top plate by a 0.05 mm thick Mylar sheet. This piece, which we call the "probe" had a diameter of 0.30 cm and an area comprising 2.5% of the total exposed area of the top plate. Two differential thermocouples were used to infer the fluid motion; one was attached between the top and bottom plates (top–bottom TC) and the other between the probe and the top plate (probe TC). Each thermcouple was connected to a SQUID ammeter.

The bottom plate was maintained at a fixed temperature of 0.700 K by means of a ^{3}He refrigerator and a resistance heater. The power applied to the top plate heater ($\dot{Q}_{TOP}$) was kept constant while the top–bottom and probe TC outputs were recorded. We use the time-averaged temperature difference across the cell (ΔT) as a measure of the strength of the drive. We define ΔT_c as the value of ΔT at which convection starts. Above ΔT_c two separate steady states exist [4], characterized by different relations between ΔT and $\dot{Q}_{TOP}$. Both states have time-dependent regimes; it was more difficult to prepare the time-dependent state with higher conductance ($\dot{Q}_{TOP}/\Delta T$). As we increased $\dot{Q}_{TOP}$ slowly, only the state with higher conductance exhibited a reproducible sequence of oscillations having low dimensional characteristics. The sequence included regimes of large and small amplitude oscillations with very similar frequencies, followed by a regime in which both the large and small amplitudes appeared in an alternating manner.

We concentrate here on the value $\Delta T/\Delta T_c = 4.15$, at which aperiodic character is evident but the

---

† Work performed under the auspices of the U.S. Department of Energy.
* Permanent address: Department of Physics, University of California, San Diego, La Jolla, CA 92093, USA

structure of the attractor is still relatively simple. Fig. 1 shows the time series for both thermocouple sensors. The oscillation amplitudes of ca. 1 mK p–p may be compared to $\Delta T = 28.0$ mK. The comparable size of the two oscillations suggests a coherent structure extending throughout the cell rather than a localized effect. In fig. 2 we show the power spectral density of the probe TC output. Here and in all succeeding analyses we filtered the TC outputs with a two-pole filter cutting off at 0.1 Hz. For all computer-generated figures (figs. 2, 4, 5, and 6) the sampling period was 0.7 s. Other data taken at smaller $\Delta T/\Delta T_c$ show a signal with a comparable amplitude but a noise floor below 20 dB. Thus, the spectrum in fig. 2 suggests that there is appreciable broadband power not caused by noise in the system parameters (e.g., the bottom plate temperature) or the TC sensor noise. However, with the present

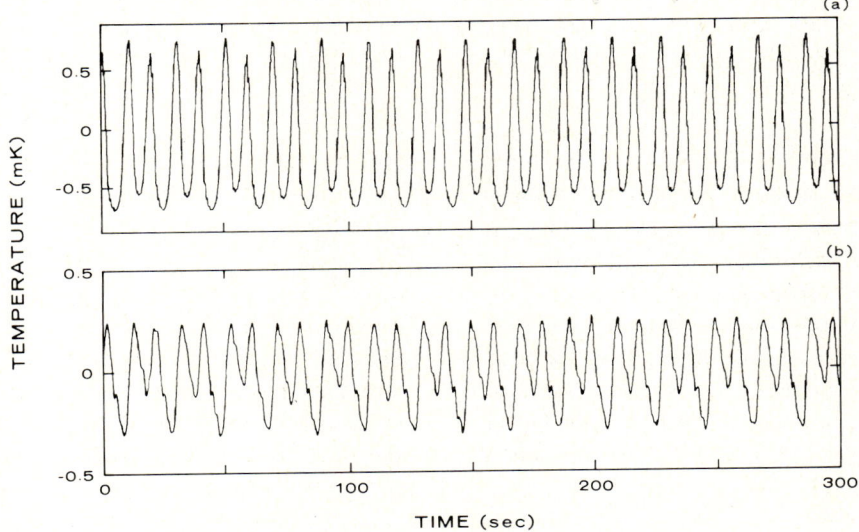

Fig. 1. Time series at $\Delta T/\Delta T_c = 4.15$: (a) the probe thermocouple and (b) the top–bottom thermocouple. The origins are chosen arbitrarily.

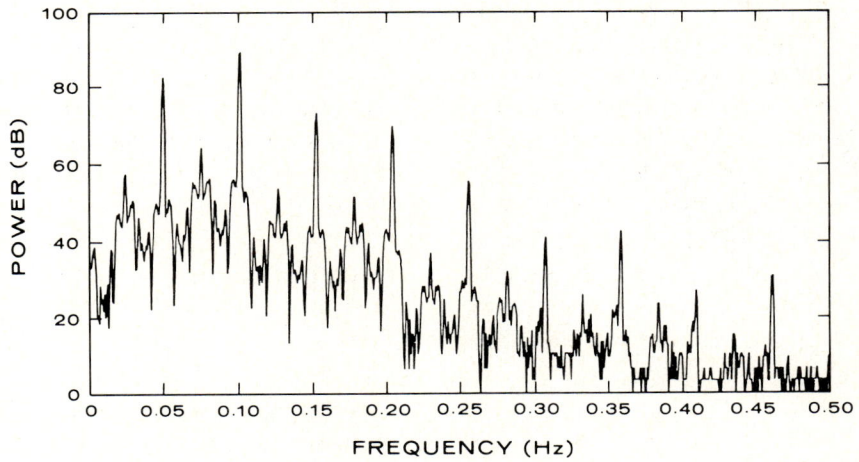

Fig. 2. Power spectral density of the probe thermocouple output at $\Delta T/\Delta T_c = 4.15$. A 2048-point time series was used with a Blackman window [5].

frequency resolution, which is typified by the width of the sharp peaks in fig. 2, it is unfortunately not possible unambiguously to distinguish multiple peaks from broadband noise.

We employed two methods of projecting the system trajectory onto a two-dimensional phase space: in fig. 3 we use top–bottom and probe TC outputs, and in fig. 4 we use the probe TC output alone by means of delay coordinates [6]. These figures suggest the trajectory lies on a sheet-like attractor and that a three-dimensional phase space suffices to closely approximate the observed dynamics. Using three delay coordinates generated from the probe TC output and taking a Poincaré section (fig. 5), we confirm this. Next, we parameterize the Poincaré section by associating each point in the section with its value T_n on the horizontal axis. The subscript n labels the points in the section in order of increasing time. Fig. 6a is the return map generated by plotting T_{n+1} against T_n. Fig. 6b is a return map for a larger drive of $\Delta T/\Delta T_c = 4.32$. Although T_{n+1} does not appear in these figures to be uniquely determined by T_n, there

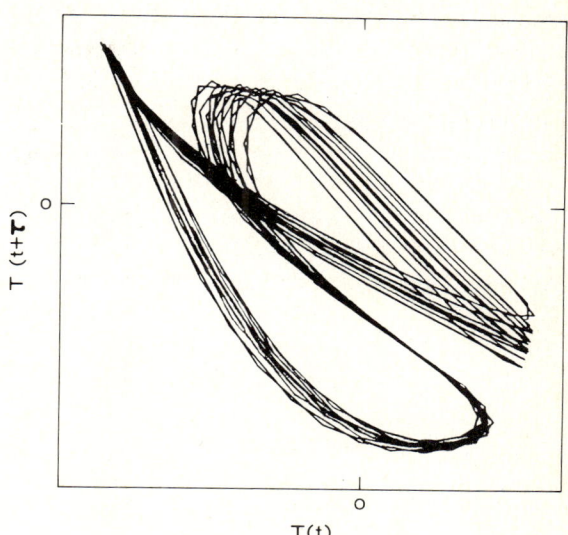

Fig. 4. Delay map at $\Delta T/\Delta T_c = 4.15$. The probe thermocouple output $T(t)$ is plotted against $T(t+\tau)$. $\tau = 4.9$ s is seven times the sampling period. 500 data points from the time series are used.

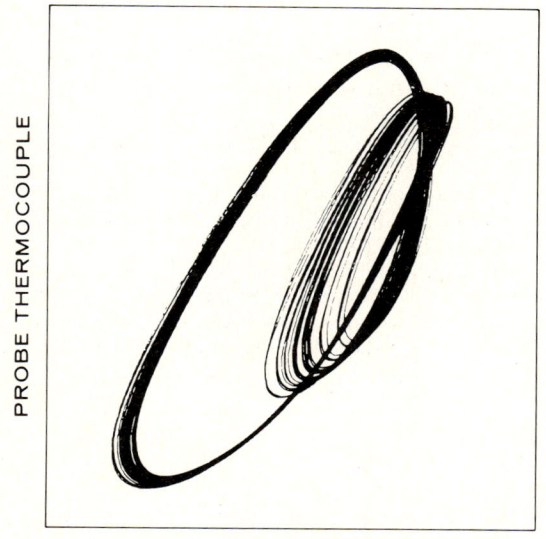

Fig. 3. Two-dimensional phase space plot at $\Delta T/\Delta T_c = 4.15$. The probe and top–bottom thermocouple outputs are plotted on the vertical and horizontal axes respectively.

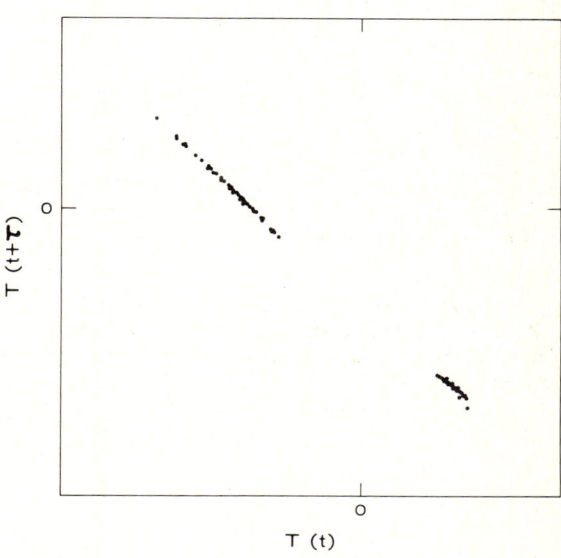

Fig. 5. Poincaré section of the three-dimensional delay map with axes $T(t)$, $T(t+\tau)$ and $T(t+2\tau)$ at $\Delta T/\Delta T_c = 4.15$. Points are shown whenever the trajectory crosses the plane $T(t+2\tau) = 0$ with increasing $T(t+2\tau)$. A 2100 point time series is used to obtain 149 points on the Poincaré section.

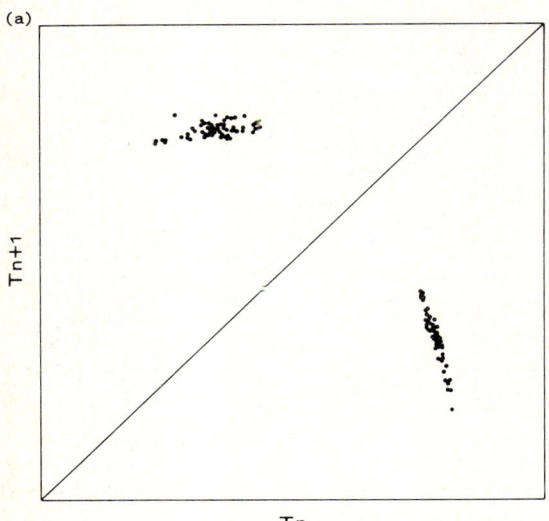

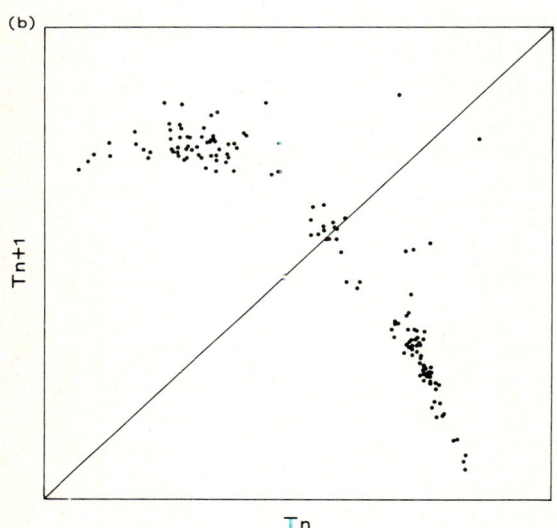

Fig. 6. Return maps (a) at $\Delta T/\Delta T_c = 4.15$ and (b) at $\Delta T/\Delta T_c = 4.32$. The value of $T(t)$ on the Poincaré section for point n is plotted against its succeeding value. 2100 point time series are used.

is clearly a correlation. The two point clusters in fig. 6a become spread out along a curve in fig. 6b.

Although it is not certain at this point whether the oscillations observed are a quality of low-Prandtl-number, one-component fluids or a result of superfluid or multi-component fluid characteristics, the interesting dynamics and good signal-to-noise ratio found in this system make further study promising. We are continuing our efforts toward a full characterization of the evolution of the attractor with increasing $\Delta T/\Delta T_c$.

Acknowledgement

We would like to thank our advisor J. Wheatley for numerous helpful comments and J.D. Farmer for valuable conversations.

References

[1] P. Warkentin, H. Haucke and J. Wheatley, Phys. Rev. Lett. 45 (1980) 918.
[2] V. Steinberg, Phys. Rev. A24 (1981) 975. A. Fetter, Physica 107B (1981) 149.
[3] Vespel SP-22 polymide resin, Du Pont Company.
[4] H. Haucke, Y. Maeno, P. Warkentin and J. Wheatley, J. Low Temp. Phys. 44 (1981) 505.
[5] F. Harris, Proc. IEEE 66 (1978) 51.
[6] J.D. Farmer, J. Hart and P. Weidman, Phys. Lett. 91A (1982) 22. N. Packard, J. Crutchfield, J.D. Farmer and R. Shaw, Phys. Rev. Lett. 45 (1980) 712.

TWO-PARAMETER STUDY OF THE ROUTES TO CHAOS

A. LIBCHABER, S. FAUVE and C. LAROCHE

Ecole Normale Supérieure, Laboratoire de Physique, 24 rue Lhomond, 75005 Paris, France

We study the routes to chaos for a Rayleigh–Bénard experiment in mercury, as a function of two parameters, the Rayleigh number (R) and the Chandrasekhar number (Q). For low Q the main route is a period-doubling cascade of bifurcations occurring at low R. For higher values of Q, two routes are observed, one related to a soft mode instability for moderate R, and a second one related to a three oscillators state, occurring at higher Rayleigh number values.

1. Introduction

This paper is related to the well-known Rayleigh–Bénard experiment, and will focus on the various bifurcations leading to weak turbulence in a low Prandtl number fluid ($P = 0.025$) mercury [1, 2]. The original aspect of it is the study in a two-dimensional parameter space, which has useful consequences. As mercury is an electrically conducting fluid, we apply an horizontal magnetic field to the horizontal fluid layer heated from below. We then study the routes to chaos, for increasing values of this parameter B_0.

As mercury is a low Prandtl number fluid, the first instability which appears, as the Rayleigh number is increased beyond convection, is a time dependent one, the oscillatory instability [3]. It consists of a transverse oscillation of the convective rolls pattern, which propagates along the rolls axis. This is the dynamic mode from which various routes to chaos will bifurcate.

Let us now present the effect of an horizontal magnetic field. In the convective state an electric current J will be induced. This in turn will add a pondermotive force, $J \times B_0$, to the Navier–Stokes transport equation. Also from Maxwell equations we can write the equation for the magnetic induction [4],

$$\frac{\partial \boldsymbol{B}}{\partial t} = \boldsymbol{V} \times (\boldsymbol{V} \times \boldsymbol{B}) + \frac{1}{\mu\sigma}\nabla^2 \boldsymbol{B} \qquad (1)$$

(σ electrical conductivity, v fluid velocity field). Let us define the magnetic diffusivity,

$$v_m = \frac{1}{\mu\sigma}.$$

Given the poor conductivity of the fluid, the magnetic viscosity is large and eq. (1) becomes

$$\frac{\partial B}{\partial t} = v_m \nabla^2 B. \qquad (2)$$

The magnetic diffusion time is very short for mercury when compared to the viscous diffusion time and the heat diffusion time ($d^2/v_m \approx 10^{-3}$ s whereas $d^2/K \approx 10$ s for $d = 7$ mm, where d is the layer depth). This implies that the magnetic lines of force will not be frozen in with the fluid velocity field. The fluid can undergo internal deformation during a time in which the spontaneous decay of the electromagnetic field is very fast. In short we don't have to worry about Alfvén modes [4].

Coming back to the Navier–Stokes transport equation the ponderomotive force gives to the fluid an added stiffness which is equivalent to an increase of the effective eddy viscosity. The importance of this effect is related to another dimensionless number, the Chandrasekhar number [3] Q ($Q = \sigma B_0^2 d^2 / \rho v$ where v is the kinematic viscosity and ρ the fluid density). As shown by Busse and Clever [5] an increase of Q shifts the onset of the oscillatory instability to a higher Rayleigh number. It will also induce an extra damping to the mode.

We can now briefly describe our experimental results. For low Q values, the whole domain, from the oscillatory instability to the onset of weak turbulence, lies in a low Rayleigh number range, i.e. small nonlinearities. There we find that mode locking and period-doubling bifurcations to chaos are prevalent. For higher Q values we test larger nonlinearities. We observe then only quasi-periodic behavior, without mode locking. The roads to chaos are then of two types. For moderate Q values a mode softening occurs, for the low frequency mode. At high Q values we observe a transition from a two mode quasi periodic state to a three-mode quasi-periodic state. But concomittant with the onset of the third mode noise abrupty sets in, with a well-defined spectrum: the noise amplitude decreases exponentially with frequency. In conclusion, what the magnetic field does to the experiment is to shift the range of Rayleigh numbers to higher and higher values, allowing to test higher nonlinearities. We get then distinct routes to chaos.

Let us now briefly describe another important effect of the magnetic field B_0, pointed out by Chandrasekhar [4] and equivalent to the Proudman–Taylor-one for fluid rotation. In the presence of a magnetic field any velocity gradient along the field direction will be strongly reduced. In particular convective rolls perpendicular to the magnetic field direction will be inhibited at high enough field (about IKG in our experiment) and convective rolls with axis parallel to the field direction will build up. This effect was experimentally shown by Lehnert and Little [6] and we use this phenomena to align rolls in large aspect ratio cells. An equivalent point of view is to notice that a transverse field induces closed pattern of large eddy currents. Looking at convection, the potential energy released by the buoyancy force will now have to balance not only viscous dissipation but also joule heating. Consequently, at high enough field, convection will be inhibited for this orientation of the rolls.

2. Experimental set-up and procedure

We have used in this experiment two cells of aspect ratio $\Gamma = 4$ ($7 \times 7 \times 28$ mm) and $\Gamma = 6$ ($6 \times 6 \times 36$ mm) of parallelepiped shape. The lateral boundaries are made of plexyglass. The top and bottom plates consists of copper blocks, well regulated. The whole cell is placed in a vacuum chamber. All our measurements are related to the temperature field. Small NTC thermistors (negative temperature coefficient) are placed in holes drilled in the copper blocks, and in a position tangent to the mercury layer.

The experimental procedure goes as follows. We increase the Rayleigh number above the convective state. We then apply a large magnetic field to align the rolls. We then reduce the magnetic field to the Q value we want and we start the experiment by increasing very slowly the Rayleigh number up to the first time-dependent instability. Let us note that we have verified the effect of orientation of the rolls in a large magnetic field. For this purpose we use a set-up where the top boundary is a sapphire plate, under which a thin layer of cholesteric liquid crystal visualizes the temperature profile of the convective state [1].

3. The oscillatory instability and its magnetic field dependence

As shown by Busse [3], the oscillatory instability corresponds to the amplification of three dimensional disturbances by the two-dimensional flow, and it is a time-dependent mode. Its onset value corresponds to the inertial term of the Navier–Stokes equation, $(\mathbf{v} \cdot \nabla)\mathbf{v}$ exceeding the diffusion term $\nu \Delta \mathbf{v}$. The onset is thus related to a critical Reynolds number [7] for the convective flow, $Re = vd/\nu$. So whereas the onset of convection is related to a critical Rayleigh number, the onset of the oscillatory instability is associated with a critical Reynolds number. Applying those ideas to a convective cell one finds for the onset of oscillations [7],

$$(R - R_c)^{1/2} \approx \alpha P,$$

where P is the Prandtl number and α the wavenumber of the convective structure. The period of the oscillatory instability is given by the characteristic convective circulation time, so one gets easily for the frequency

$$f \approx K\alpha(R - R_c)^{1/2} d^{-2},$$

where K is the heat diffusivity.

Going now to the real world, i.e. a cell with $\Gamma = 4$ we find experimentally that the general evolution follows the theory. The results are shown on fig. 1 ($Q = 0$). But there is a strong discrepancy at onset. There one finds that the oscillatory frequency decreases first as one increases the Rayleigh number. This may be related to the presence of the side walls which break the symmetry of the four rolls state. The two rolls tangent to the lateral boundaries have a lower convective velocity than the two central ones, thus a lower Reynolds num-

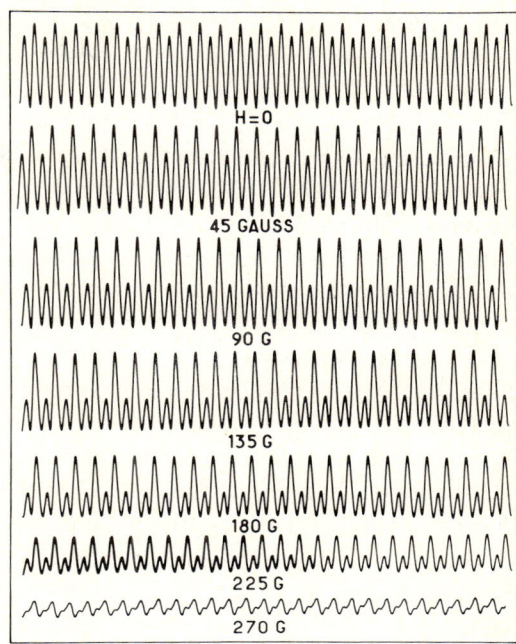

Fig. 2. Time recording of the oscillatory instability and its magnetic field dependence. Note that frequency f and $f/2$ are always present, but with varying relative amplitude.

ber. The central rolls will be destabilized first, entraining the lateral one, and a collective behavior will occur only at a higher Re number. This may also explain the line shape of the signal, in fig. 2 ($H = 0$). We find there that not only do we have a frequency f but also $f/2$. It may be related to the existence of two wavenumbers for the convective structure, with a ratio close to two.

3.1. The magnetic field dependence

A magnetic field parallel to the rolls axis gives to the fluid an added stiffness [4] which is equivalent to an increase of the effective eddy viscosity, inhibiting velocity variations along the magnetic field direction. When this added viscosity becomes larger than the kinematic viscosity, the new onset of the oscillatory instability corresponds to a balance [1] between the inertial term $(v \cdot V)v$ and this magnetic term $(\sigma B^2/\rho)\Delta^{-1}\delta^2_{yy}v$ (where y is the magnetic field direction), which thus defines a new Reynolds number, Re $= vd/vQ$, strongly reduced.

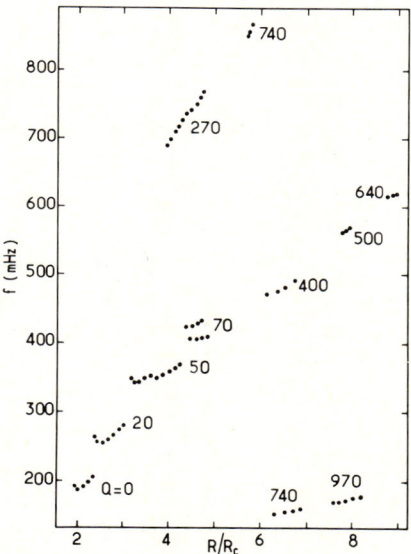

Fig. 1. Aspect ratio $\Gamma = 4$. Frequency dependence of the oscillatory instability as a function of the reduced Rayleigh number, for increasing values of Q.

Fig. 1 shows the evolution of the oscillatory instability for increasing Q values. Up to $Q = 200$ only one mode is present and the onset of oscillation is shifted towards higher Rayleigh numbers, as Q increases. For $Q > 200$ two and then three modes can be destabilized, depending on initial conditions. This indicates the presence of a new instability which could be the skewed varicose [8] one, but we shall come back to this point later.

Fig. 2 shows, for a given heat flux, the evolution of the temperature signal as a function of the applied field H [1, 2]. Above 90 G, one observes a strong damping of the mode and an increase in its frequency, both effects related to be inhibition of the oscillatory instability field [1].

Let us now turn to the study of the routes to chaos which will be of a different nature, as Q increases.

4. Low Q regime, the period-doubling cascade and the subharmonic route

For low Q values the often observed route to chaos is the period-doubling cascade [9], and all of the observed routes follow subharmonic bifurcations and lock in regimes. The period doubling scenario is by now well documented [10, 2]. It is customary to compare the experimental results to a one-dimensional mapping. In such a model, the mapping of the interval as a dynamical system, there is a strict hierarchy in the order of appearance of bifurcations as a function of the control parameter R. As R increases a cascade of period-doubling bifurcations evolves to its limit R_∞. Beyond this value a chaotic state sets in, within which laminar windows appear with periods of various length.

But one-dimensional mappings relate to experiments where the modes are highly damped (infinite contraction). As previously indicated, in our experiment, the damping depends on the Chandrasekhar number. Thus a more realistic mapping is a two dimensional one with two parameters, a constraint and a contraction rate. The Hénon mapping [11] is a good example of it and was chosen to analyze our previous experiments in liquid helium [12]. In such mappings several competing scenarios are possible. They differ only in that the period-doubling cascade can be interrupted, the next bifurcation being a period multiplication by integers other than two, period three having the largest basin of attraction. Let us note that if we vary adiabatically the Rayleigh number one stays on the continuous branches of the bifurcation tree, i.e. the period-doubling cascade evolves to its limit and this is what experiment I is about. In experiment II with the same cell and apparently very close operating conditions, we show that period three appears after the first period doubling bifurcation. In experiments I and II a small magnetic field was applied ($Q = 22$) i.e. moderate damping of the modes. Experiment III relates to a larger cell ($\Gamma = 6$) where we could follow a period-doubling scheme at a much higher magnetic field ($Q = 260$) and there the whole scenario of a one-dimensional map appeared. There were no competing scenarios, and in the chaotic state a number of laminar windows were observed, with the correct order of appearance.

It thus seems that the general analogy between the experiment and the Hénon mapping is supported. Experiments I and III correspond respectively to moderate and large damping of the modes. Experiment II where the cascade is interrupted may correspond to non-adiabatic change of the Rayleigh number.

4.1. Experiment I ($\Gamma = 4$, $Q = 22$, $B_0 = 270$ G)

The scenario described here corresponds to the observation of the period doubling cascade, which develops up to $f/32$, well resolved as far as the onset values up to $f/16$. In fig. 3 the temperature recordings are presented showing the bifurcations to period 4, 8 and 16 for $R/R_c = 3.52$, 3.62 and 3.65. Their respective Fourier spectrum are presented in figs. 4b, c, d for values of R/R_c close to the preceeding ones.

If we compute the Feigenbaum number δ for the

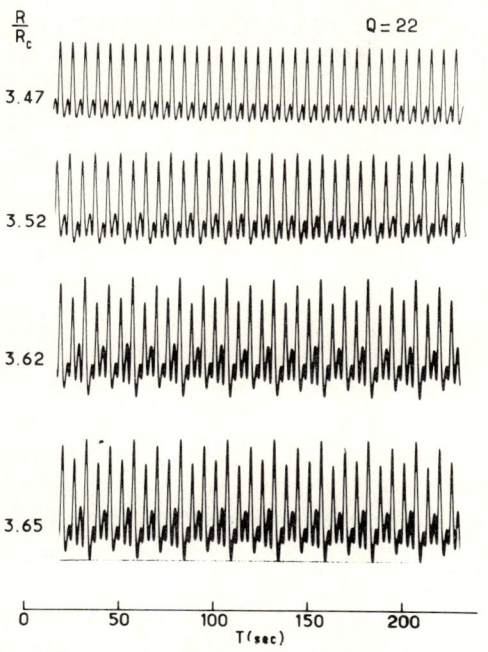

Fig. 3. The period doubling cascade ($\Gamma = 4$, $Q = 22$).

last three bifurcations, we get

$$\delta = \frac{R_8 - R_4}{R_{16} - R_8} = 4.4 \pm 0.1.$$

This is to be compared with the theoretical asymptotic value [9] $\delta = 4.669\ldots$

We can also compute the ratio of the successive subharmonics amplitude called μ. This ratio is measured directly on the temperature recordings. We show on fig. 5 an enlargement of the temperature signal after the $f/16$ bifurcation and in fig. 6 an enlargement of its Fourier spectrum. The last value of μ measured is $\mu \approx 5$ to be compared with theoretical values between 4.58 and 6.5 (the first one is given by Nauenberg and Rudnick [13], the second one by Feigenbaum [9]). This measurement of μ led to some confusion in the past, which can be understood if we look at the Fourier spectrum in figs. 4d and 6. It is clear there that the odd

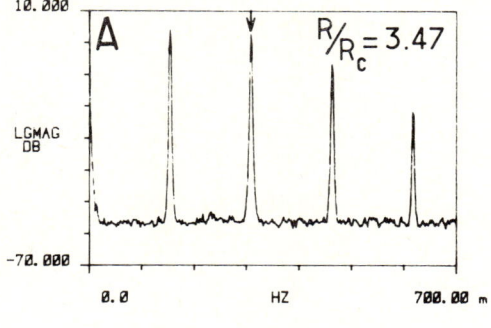

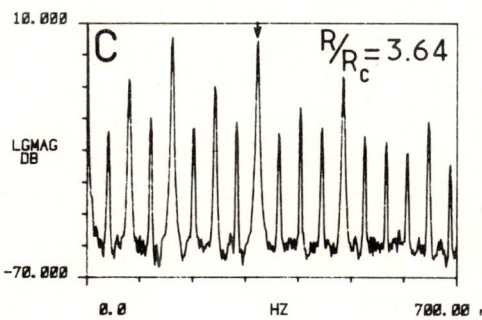

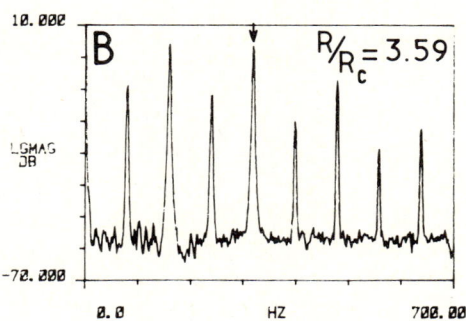

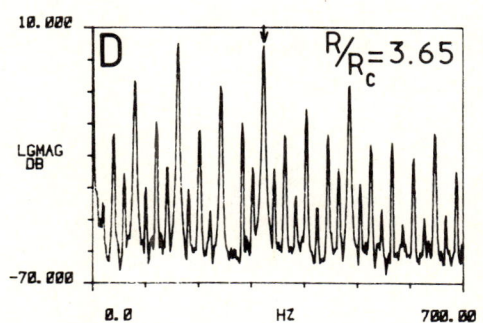

Fig. 4. Fourier spectrum for the period-doubling cascade.

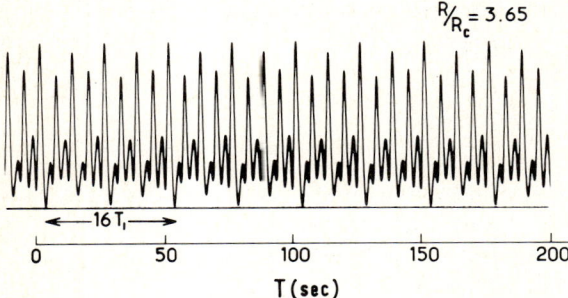

Fig. 5. The developed cascade, time recording.

harmonics of $f/16$ have an amplitude which is modulated and depends on the order of the harmonics. It is thus not clear how to calculate the ratio from the Fourier spectrum. The μ value given is derived from the direct measurement of the signal amplitude and not its Fourier transform.

4.2. Experiment II ($\Gamma = 4, Q = 22, B_0 = 270$ G)

This scenario illustrates the competition between several subharmonic bifurcations in a two-dimensional mapping. Going back to the Hénon mapping [12], one finds that period 3 (which has the largest basin of attraction) and period 2 are present for similar values of the parameter a, when the area contraction rate is not too large ($-0.3 < b < -0.1$). In the experiment, parameter a corresponds to the Rayleigh number and parameter b to the damping of the oscillatory mode.

Experiment I showed that the period-doubling cascade can be pushed far in the bifurcation

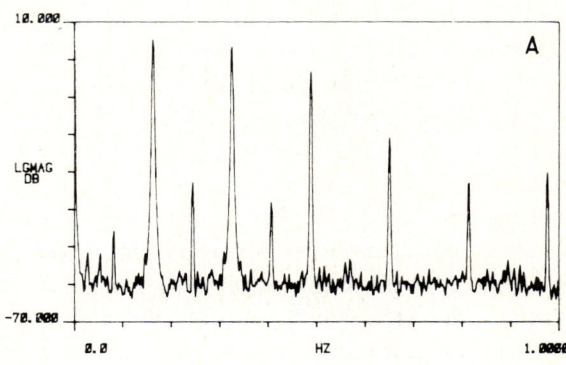

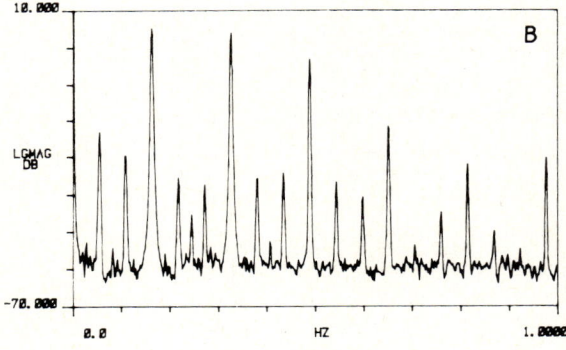

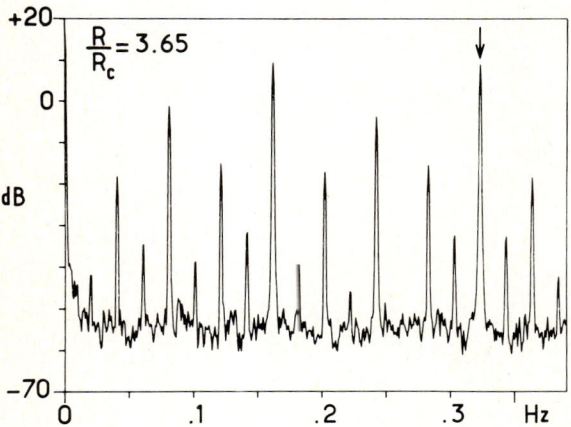

Fig. 6. The developed cascade Fourier spectrum. The arrow indicates the main frequency.

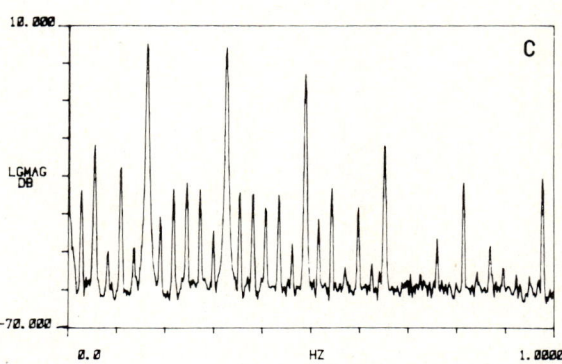

Fig. 7. Competition between period 2 and period 3. A) onset of period 2 bifurcation; B) coexistence of period 2 and period 3, intermittently; C) beginning of a period-doubling cascade from period 3. A. B. and C correspond to increased R.

scheme. In experiment II we use the same cell, the same Q value. Nevertheless one finds a somewhat different scenario. In fig. 7a the onset of a period-doubling bifurcation is clearly observed and corresponds to fig. 4b of the Feigenbaum route. But, then, for a small change of the Rayleigh number one finds that period 3 develops. Somehow we fell on the basin of attraction of period 3 (a tangent bifurcation). Finally, for a larger value of R, on leaves period 3 by a period-doubling bifurcation, the beginning of it being shown in fig. 7c.

What conclusions can be drawn from experiments I and II? For finite dissipation (small Q) we have a competition between the period 2 series of pitchfork bifurcations, and tangent bifurcations. If we could vary the Rayleigh number in an adiabatic fashion, once the pitchfork bifurcation starts one would stay on it (experiment I). But this is difficult to achieve, and one often falls on a scenario like experiment II. Furthermore, from our experiments, it seems that once a period 3 appears one always gets it. If we reduce the Rayleigh number and then increase it again, the first bifurcation encountered is period 3, as if the corresponding initial conditions have a long lifetime.

In order to achieve the conditions of a one-dimensional mapping, a larger mode damping is necessary, and this is shown in experiment III where $Q = 260$.

4.3. Experiment III ($\Gamma = 6$, $Q = 260$, $B_0 = 1100$ G)

We use a cell of large aspect ratio ($\Gamma = 6$) and a higher value of the Chandrasekhar parameter ($Q = 260$). With a larger aspect ratio the lateral walls have a reduced effect on the onset of the instabilities, and the whole scenario occurs for $R/R_c < 3$, even with a magnetic field present. This field plays a major role in orienting the convective rolls; it acts as an ordering field and allows a period-doubling cascade to appear even with six convective rolls present. Without it the cell would have a somewhat disordered structure which leads to a low frequency noise present for those Rayleigh number values [14, 7].

In fig. 8 we show the frequency dependence of the oscillatory instability and the onset values of the bifurcations. For this small range of R values the oscillatory instability frequency varies linearly with R/R_c, showing that we keep the same dynam-

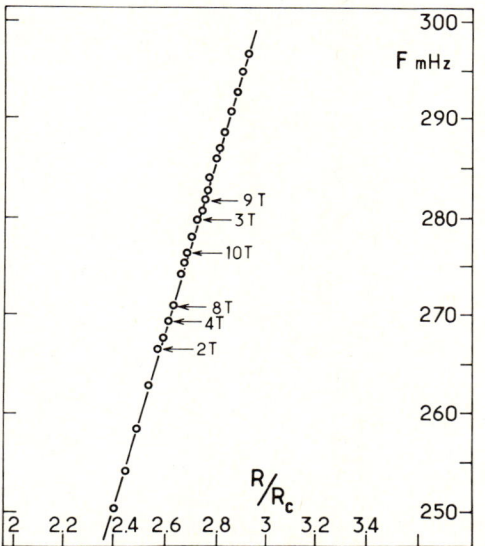

Fig. 8. Aspect ratio $\Gamma = 6$, $Q = 260$. Evolution of the oscillatory instability frequency. Arrows indicate the onset of bifurcations with various periods.

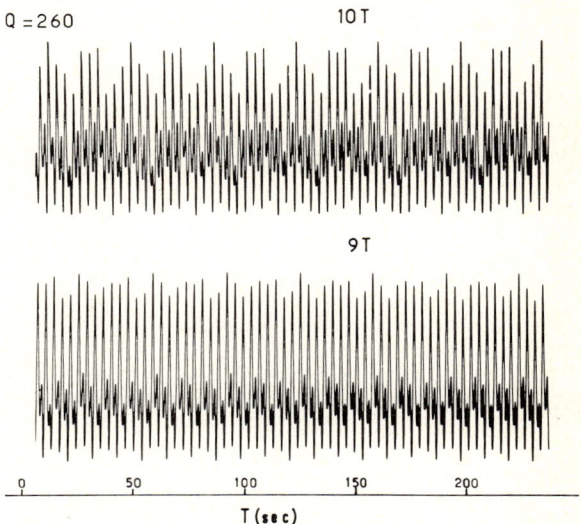

Fig. 9. Time recording for period 9 T and 10 T.

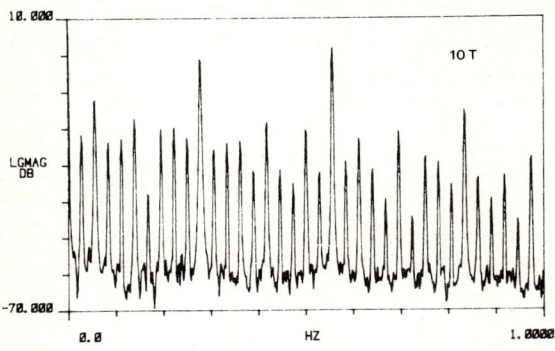

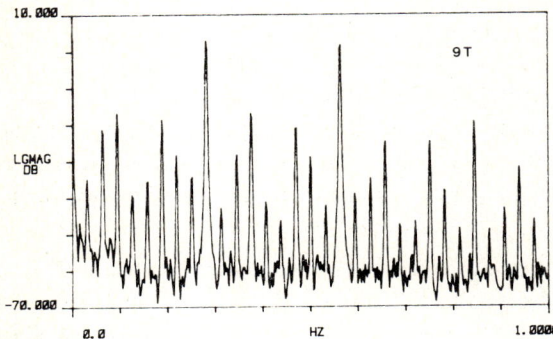

Fig. 10. Fourier spectrum for period 9 T and 10 T.

ical mode. The period-doubling bifurcation is very close to the one described in experiment I. Once in the turbulent side we observe the reverse bifurcation sequence [15] as in our previous work in liquid helium [7], but also some numbers of the Universal sequence of Metropolis et al. [16]. This means that in the turbulent regime we observe three "laminar" flows with periods 10 T, 3 T and 9 T, respectively, in the right order for a U sequence [17]. This is the first observation of this phenomena in a R.B. experiment.

Fig. 9 shows the local temperature oscillations corresponding to periods 10 T and 9 T and fig. 10 their Fourier spectrum. The range in Rayleigh number for period 3 T was large enough, so that we could observe intermittency in entering period 3 T and the period-doubling cascade [18], leaving it. For example we show in fig. 11 a spectrum for the reverse bifurcation sequence which terminates period 3 T. There, period 3 T and 3×2 T appear as sharp peaks, while period 3×4 T and 3×8 T are noisy.

This experiment shows that the whole picture of one-dimensional map applies as a possible scenario

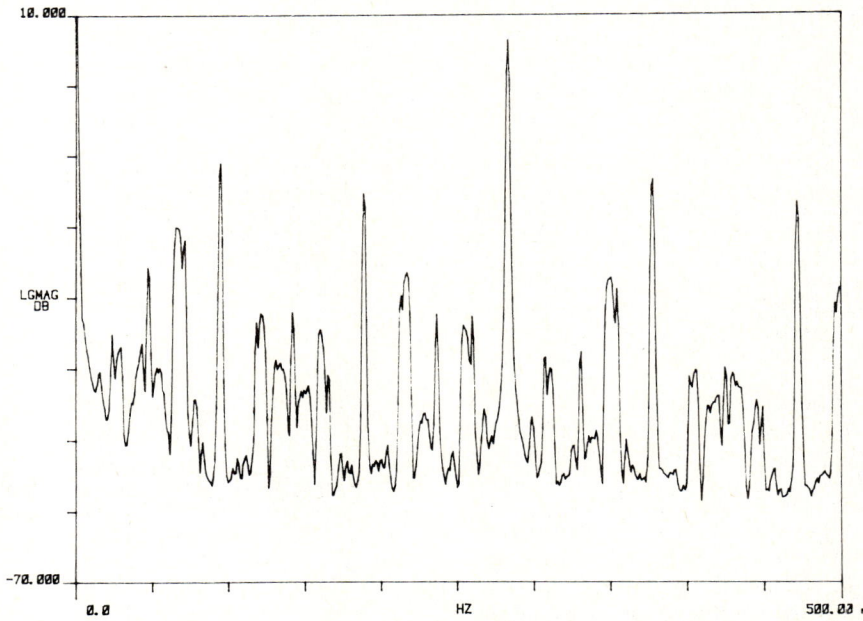

Fig. 11. Fourier spectrum of the reversed cascade, starting from period 3 T.

in a R.B. experiment, even in a cell with a large aspect ratio.

5. Moderate Q regime, quasiperiodicity and mode softening ($\Gamma = 4$, $Q = 400$, $B_0 = 1150$ G)

Let us come back to the first cell with aspect ratio $\Gamma = 4$ and increase the magnetic field up to $Q = 400$. We find another route to turbulence, in no way related to period doubling. Fig. 12 shows the evolution of the temperature signals. Given our Q value, the range of Rayleigh number where the new scenario evolves, is higher $6.69 < R/R_c < 6.81$. For $R/R_c = 6.69$ we find as previously the oscillator with frequency f and $f/2$. For $R/R_c = 6.7$ there is an Hopf bifurcation and the state becomes quasiperiodic with a very low frequency oscillator as the second mode. As the Rayleigh number increases the period of this mode keeps increasing (frequency decrease) from $R/R_c = 6.70$ to $R/R_c = 6.80$. Finally for $R/R_c = 6.81$ the state becomes chaotic. Within our measuring precision, in all this evolution we follow a quasiperiodic state (two torus in phase space).

Fig. 13 shows 3 typical Fourier spectrum, starting from the periodic state (f and $f/2$) then, in the middle spectrum, the low frequency mode its harmonics and its modulation of the high frequency mode. Finally, the turbulent state is reached for $R/R_c = 6.81$ with a very low frequency broad band noise. Let us remark that in the turbulent regime $f/6$ is also noticeable (division by 3 of $f/2$). A final experimental observation is that it all looks like a competition in amplitude between the mode of frequency f and its subharmonic $f/2$.

Let us try to understand this new route and for that matter come back to fig. 1. Two experimental observations show that for $Q = 400$ we leave the simple picture of convection with cylindrical rolls. First 3 modes exist for the oscillatory instability and also the onset value of the mode does not follow any more the theory [19]. We conjecture that such behavior is associated with the presence of a stationary instability which breaks the symmetry of the convective structure, and could be of the skewed varicose type [8]. Indeed, increasing the magnetic field, is similar to increasing the fluid Prandtl number, for it introduces a large magnetic viscosity. We know that, as the Prandtl number increases, the skewed varicose instability is the first one encountered. Unfortunately there is no theory to support this conjecture in the magnetic field case.

But now this new road to turbulence can be explained as resulting from the interaction between the oscillatory instability and the stationary instability. The very low frequency mode comes from the competition between those two modes [20, 21], and the mode softening relates to the presence of an homoclinic orbit [20]. In simple terms let us say that the coupling between an oscillating mode and a stationary one may lead to an oscillation of the amplitude of the oscillatory one, the period of which can go towards infinity, as the interaction increases.

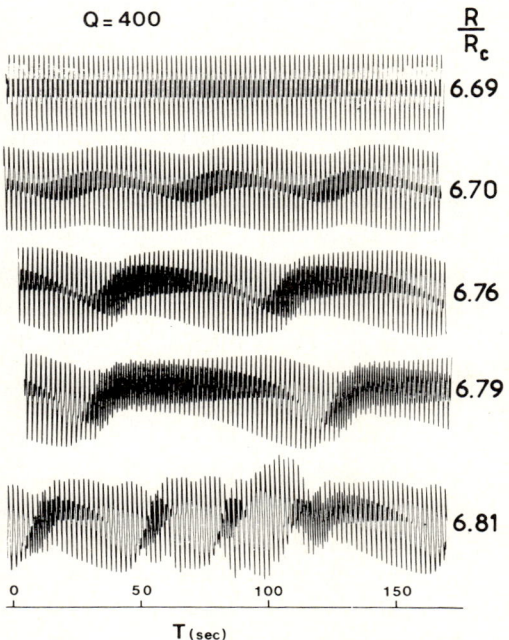

Fig. 12. Aspect ratio $\Gamma = 4$, $Q = 400$. Time recording of the soft mode instability. Note the increase in period of the low frequency modulation. For $R/R_c = 6.81$ the state is chaotic.

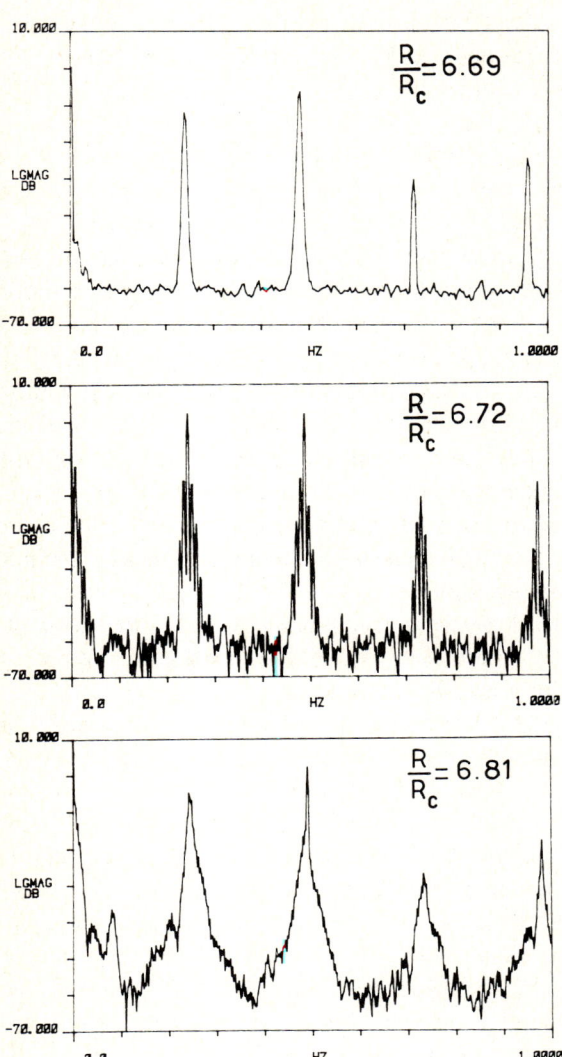

Fig. 13. Fourier spectra for the soft mode instability ($\Gamma = 4$, $Q = 400$). $R/R_c = 6.69$ before the Hopf bifurcation; $R/R_c = 6.72$ after the bifurcation; $R/R_c = 6.81$ chaotic regime.

6. Large Q regime, a scenario with three oscillators ($\Gamma = 4$, $Q = 675$, $B_0 = 1500$ G)

Increasing again the magnetic field ($B_0 = 1500$ G) a new scenario appear, for higher Rayleigh numbers. For $R/R_c = 8.91$ we reach a quasi-periodic state with frequencies f_1 and f_2 shown on

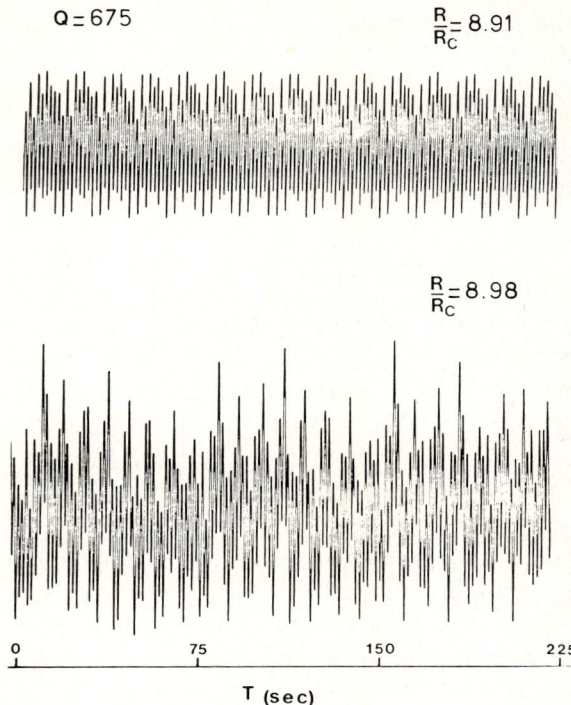

Fig. 14. Aspect ratio $\Gamma = 4$, $Q = 675$. Time recording for the Ruelle–Takens-like scenario. $R/R_c = 8.91$ quasi periodic state; $R/R_c = 8.98$ chaotic state.

fig. 15, the direct recording being presented in fig. 14. In the spectrum all the peaks can be analyzed as combination of f_1 and f_2. Let us note again that for large Rayleigh numbers the states are always quasiperiodic, and that lock in and period doubling behavior is typical of small nonlinearities.

Abruptly the regime becomes chaotic for $R/R_c = 8.98$ as shown clearly in the time recording (fig. 14). If we now study the Fourier spectrum one makes the following observations: the peaks at frequency f_1 and f_2 and their combination are still present, but a new oscillator of frequency f_3 appears, which was absent for $R/R_c = 8.91$ (the position of oscillator f_3 is indicated by a vertical line). Also an exponential noise appears concomitant with the oscillator f_3. The noise amplitude follows a law $A \approx \exp{-f/f_0}$ where $f_0 \approx f_3$.

In this scenario one bifurcates from a quasi-periodic state to a chaotic state with three oscil-

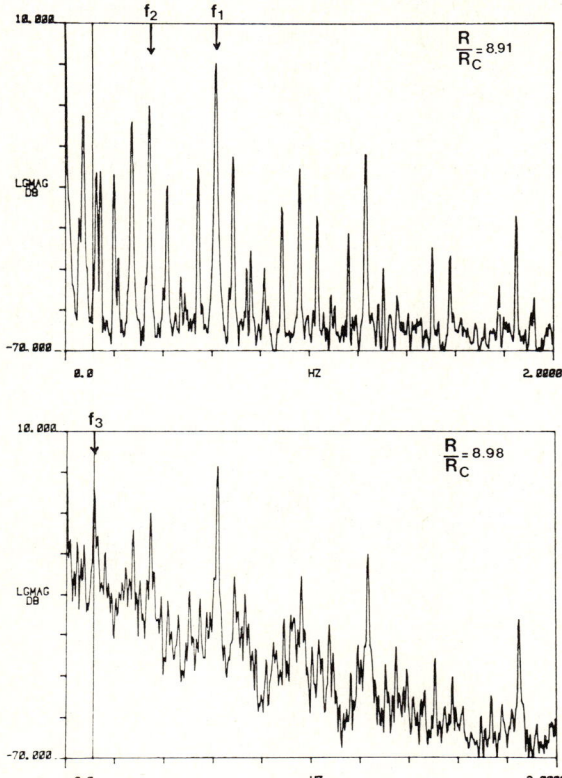

Fig. 15. Fourier spectra corresponding to the Ruelle–Takens-like scenario. $R/R_c = 8.91$ quasi periodic state with frequencies f_1 and f_2 (vertical line indicate the position of future oscillator f_3); $R/R_c = 8.98$. Oscillator f_3 is present. Exponential noise is present (in a log–linear plot a constant slope in the recording).

intermittent bursts for the high frequency response of an experiment.

7. Conclusion

The main results of this experiment are:

The magnetic field acts as an ordering field for the convective rolls, which leads to simple dynamics even in a large aspect ratio cell, $\Gamma = 6$.

Increasing the magnetic field, one can test regions of higher Rayleigh number, i.e. larger nonlinearities.

Mode lock in and subharmonic bifurcations are associated with low Rayleigh numbers, small nonlinearities.

For high Rayleigh numbers, quasiperiodic states are prevalent.

A soft mode transition appears as the result of a competition between a stationary and an oscillating instability.

For very high Rayleigh numbers, a Ruelle–Takens-like scenario exists.

lators and an exponentialy decaying noise. This scenario has some similarity with the Ruelle Takens [22] conjecture, which reads as follows [23]: "If the steady state solution of a system has lost its stability through a Hopf bifurcation three times in succession, and the three newly created modes are essentially independent then a strange attractor may occur".

This is not the first observation of the scenario with three oscillators [24]. But in our case the third oscillator bifurcation is associated with the presence of noise, which decays exponentialy. This exponential decay of noise in the chaotic regime has been recently discussed [25] in connection with

References

[1] S. Fauve, C. Laroche and A. Libchaber, J. Physique Lett. 42 (1981) L 455.
[2] A. Libchaber, C. Laroche and S. Fauve, J. Physique Lett. 43 (1982) L211.
[3] F.H. Busse, J. Fluid Mech. 52 (1972) 97.
[4] S. Chandrasekhar, Hydrodynamic and hydromagnetic stability (Oxford Univ. Press, Oxford, 1961).
[5] F.H. Busse and R.M. Clever, 'On the stability of convection rolls in the presence of a vertical magnetic field", preprint.
[6] B. Lehnert and N.C. Little, Tellus 9 (1957) 97.
[7] A. Libchaber and J. Maurer, Nonlinear Phenomena at Phase Transition and Instabilities, T. Riste, ed. (Plenum, New York, 1981), pp. 259–286.
[8] F.H. Busse and R.M. Clever, J. Fluid Mech. 91 (1979) 319.
[9] M.J. Feigenbaum, Phys. Lett. 75A (1979) 375. P. Collet and J.P. Eckmann, Iterated Maps on the Interval as Dynamical Systems (Birkhaüser, Boston, 1980).
[10] A. Libchaber and J. Maurer, J. Physique 41 (1980) 51. M. Giglio, S. Muzzati and U. Perini, Phys. Rev. Lett. 47 (1981) 243.
[11] M. Hénon, Comm. Math. Phys. 50 (1976) 69. S.D. Feit, Comm. Math. Phys. 61 (1978) 249.

[12] A. Arneodo, P. Coullet, C. Tresser, A. Libchaber, J. Maurer and D. d'Humiéres, On the observation of an uncompleted cascade in a R.B. experiment", Physica D, in press.
[13] M. Nauenberg and J. Rudnick, Phys. Rev. B24 (1981) 493.
[14] G. Ahlers and R.W. Walden, Phys. Rev. Lett. 44 (1980) 445.
[15] E.N. Lorenz, Ann. N.Y. Acad. Sci. 357 (1980) 282.
[16] N. Metropolis, M.L. Stein and P.R. Stein, J. Comb. Theory A15, (1973) 25
[17] For similar observations in chemical reactions: R.H. Simoyi, A. Wolf and H.L. Swinney, One-dimensional dynamics in a multicomponent chemical reaction", preprint.
[18] D.J. Scalapino, J.E. Hirsch and B.A. Huberman, Melting, Localization, and Chaos, Kalia and Vashishta, eds. (North-Holland, Amsterdam 1982).
[19] S. Fauve and A. Libchaber, to be published..
[20] W.F. Langford, A. Arneodo, P. Coullet, C. Tresser and J. Coste, Phys. Lett. 78A (1980) 11.
[21] P. Holmes, Ann. N.Y. Acad. Sci. 357 (1980) 473.
[22] D. Ruelle and F. Takens, Comm. Math. Phys. 20 (1971) 167.
[23] J.P. Eckmann, Rev. Mod. Phys. 53 (1981) 643.
[24] J.P. Gollub and S.V. Benson, J. Fluid Mech. 100 (1980) 449. J. Maurer and A. Libchaber, J. Phys. Lett. 41 (1980) L515.
[25] B. Malraison and P. Atten, preprint. U. Frisch and R. Morf, Phys. Rev. A23 (1981) 2673.

BIFURCATION AND THE UNIVERSAL SEQUENCE FOR FIRST-SOUND SUBHARMONIC GENERATION IN SUPERFLUID HELIUM-4

Charles W. SMITH and Manu J. TEJWANI
Department of Physics and Astronomy, University of Maine, Orono, Maine 04469, USA

Measurements show that below the superfluid transition, the generation of first-sound subharmonics in the low-megahertz range quantatively follows the Feigenbaum universal convergence for period doubling and qualitatively includes frequencies which correspond to the universal sequence. In addition, by using ion-trapping techniques the physical nature of the onset of the first bifurcation is identified as the threshold for the generation of quantum vortex line, not the threshold for the production of macroscopic classical turbulence, i.e., acoustic cavitation.

Recent experimental studies of driven nonlinear systems have included Rayleigh–Bérnard convection [1], couette flow [2], optically bistable laser cavities [3], acoustic cavitation noise [4], charge-density waves [5], pinning dynamics of dislocation lines [6], chemical reactions [7], and nonlinear discrete electronic circuits [8]. These systems generate output signals rich in spectral detail. Generally one is interested in how the frequency content of the output signals varies as some driving parameter is changed. Typically, driven nonlinear systems exhibit (1) harmonic generation, (2) subharmonic generation (which displays onset thresholds but may or may not show period doubling), (3) ultraharmonic generation (harmonics of the subharmonics), and at a sufficiently large value of the driving parameter, (4) a transition to a noisy, chaotic or turbulent regime.

The recent theoretical work by Feigenbaum [9] concerning certain universal properties in period-doubling bifurcations of iterated one-dimensional maps has catalyzed a search for analogous universal behaviors in experimental systems. He has shown that for a transition to chaotic behavior via a sequence of period-doubling bifurcations, an ordered set of values of the driving parameter, λ_n, for which bifurcations occur, converges to a universal number δ, where $\delta = (\lambda_{n+1} - \lambda_n)/(\lambda_{n+2} - \lambda_{n+1}) = 4.669\ldots$. In addition, the ratio of the amplitudes of the Fourier components of adjacent fully developed bifurcated subharmonics scales by $\mu = 6.57$ or $10 \log_{10} \mu = 8.2$ dB, again a universal number. As a result of this expected commonality, a rather large literature is emerging which describes many nonlinear systems exhibiting subharmonic routes to chaotic behavior. However, it must be emphasized that to date only three types of experiments quantitatively show the Feigenbeum period-doubling bifurcation universalities. These are Rayleigh–Bénard experiments, chemical reactions of the Belousov–Zhabotinski type, and nonlinear discrete electronic circuits.

This paper presents the results of measurements of the finite-amplitude first-sound response of liquid helium-4, a system known to exhibit an acoustic subharmonic spectrum [10]. Above the superfluid transition, $T_\lambda = 2.17$ K, liquid helium behaves macroscopically as a classical liquid. One observes a very rich nonlinear response, dominated by vapor bubble dynamics of the type recently carefully documented for the case of water [4]. This includes a subharmonic route to chaos (acoustic cavitation), but does not show bifurcation of the Feigenbaum type. However, below the superfluid transition where the existence of conventional vapor bubbles is excluded by quantum-fluid properties, we observed not only a sharper subharmonic spectral response but a well-defined bifurcation sequence which quantitatively exhibits the Feigenbaum convergence ratio, δ.

The experiment consists of radiating liquid helium with first sound and measuring the frequency response of the transport of energy through the liquid. A pair of piezoelectric (PZT4) thickness mode transducers are positioned parallel and centered on a common axis in an open geometry in the liquid helium, that is, not in a closed resonant cavity. One transducer, employed as a first-sound source, is driven at its electromechanical resonant frequency, f_0. (For the data presented herein, $f_0 = 2.6883$ MHz.) The other transducer, carefully chosen to have a relatively flat response in the frequency range of interest, receives the signal, which is frequency analyzed (a power spectral density is obtained) and digitally signal averaged. At low acoustic-pressure amplitudes harmonic generation, typical of a driven nonlinear dynamic system, is observed. We emphasize that the nonlinearity in this case is the response of the liquid helium to finite-amplitude first sound. It is simple to confirm, as we have done, that the sound source and analysis instrumentation do not constitute a nonlinear electronic network in the range of the drive voltages and frequencies employed.

The first subharmonic, $f_0/2$, appears for a drive voltage on the ultrasonic transducer which corresponds to a maximum sound displacement velocity of 0.15 cm/s. As the sound intensity is increased, other subharmonics as well as ultraharmonics appear. Fig. 1 shows 256 digitally signal averaged power spectra at 1.60 K. It should be explicitly mentioned that because the driving parameter (sound-pressure amplitude) is swept through its full range of values each sound cycle, the power spectrum shows the system's full range of nonlinear responses up to that amplitude. Therefore, several subharmonics, ultraharmonics and harmonics are observed in a given power spectrum. Fig. 2 shows the peak amplitude in decibels, corresponding to the power contained in each of the first three components of the first subharmonic-

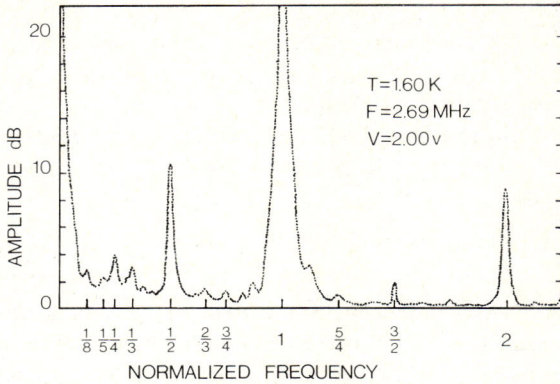

Fig. 1. The acoustic response of superfluid helium-4 (256 digitally signal-averaged spectra) clearly exhibiting harmonics, subharmonics, and ultraharmonics. These spectra were recorded at a sound level 6.25 times greater than the threshold for the production of the first subharmonic and approximately 1/15 the sound level for the onset of acoustic cavitation.

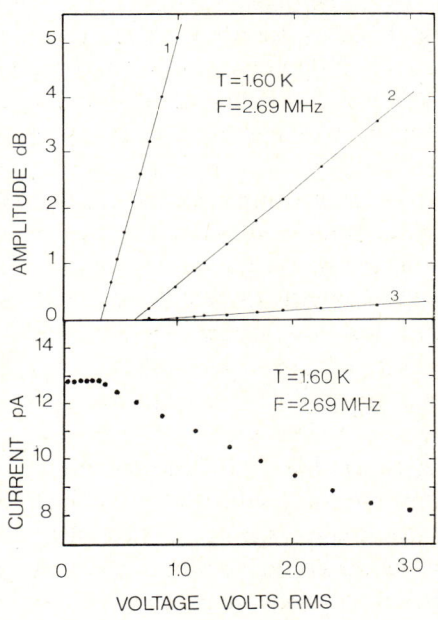

Fig. 2. The upper graph shows the peak amplitude in decibels corresponding to the power in each of the first three components of the first bifurcation sequence as a function of the rms voltage on the drive transducer (proportional to the sound-pressure amplitude). Linear regression analysis yields $\lambda_1 = 0.32$, $\lambda_2 = 0.61$, and $\lambda_3 = 0.67$. The lower curve shows the onset of ion trapping due to vortex line generation occurring at the same drive voltage as the onset of the first bifurcation.

bifurcation sequence f_0/n, where $n = 2, 4,$ and 8. The convergence ratio, δ, was calculated using linear regression analysis to estimate the onset threshold for each subharmonic. Four independent sets of data at 1.60 K similar to those presented in fig. 2 yield $\delta = 4.83 \pm 0.6$ in quantitative agreement with the predicted value. Determinations at two other temperatures and at $f_0 = 9.8751$ MHz fall within the above range. Attempts to measure the rescaling factor μ for fully developed adjacent members of the bifurcation sequence have met with less success. Analysis limitations arising from the combined effects of the $f_0/8$ peak falling on a low-frequency noise shoulder and of a tendency for the fully developed $f_0/4$ peak to phase lock to nearby subharmonics permit us only to place a range of 6–10 dB on the scaling parameter. However, this result is not in conflict with the predicted value of 8.2 dB.

Above the superfluid transition one still observes subharmonics. The threshold for the generation of $f_0/2$ is larger. Bifurcation of the Feigenbaum type is not present: $f_0/4$ is difficult to generate and all peaks appear broader and set upon a noisy background.

Using a gated tritium source, an appropriate drift space, and a guarded collector, one can produce a current of negative ions (electron microbubbles) with which to probe the sound field. If quantum vortex line is present, ions can be trapped on the line and the current to the collector will decrease [11]. We observe that as the sound-pressure amplitude is increased from zero, no ion trapping occurs until the threshold for the production of the first subharmonic-bifurcation sequence is exceeded; see fig. 2. This effect is large and has the unmistakable lifetime edge and polarity dependence for ion trapping on vortex line. In this manner, for the case of acoustic subharmonic bifurcation in superfluid helium, we identify the physical aspect associated with the universal convergence as the threshold for the production of quantum turbulence (vortex-line generation), not the threshold for the production of classical turbulence by sound (acoustic cavitation), which lies 2 orders of magnitude higher in sound-pressure amplitude.

There are several other features in our acoustic data which show systematic behaviors. (1) For values of the driving voltage greater than that corresponding to the bifurcation accumulation point (about 0.69 rms volts) subharmonics appear which correspond to members of the universal sequence [12]. The qualitative aspects of these states are not difficult to observe. Periods 10, 6, 5, 3 appear in the power spectra, in that order, as the sound intensity is increased. Period 10, the first to appear, is easiest to observe by noting its ultraharmonics (especially 9/10 and 11/10) as side bands on the driving frequency. Period 6 is next but difficult to observe since its signal is easily confused with period 8 from the bifurcation sequence. Period 5 and 3 appear at slightly higher sound intensity although they are both difficult to distinguish from the period 10 and 6 signals which preceed them and one must carefully monitor asymmetries between the even and odd ultraharmonics of periods 10 and 6 to confirm their presence. (2) At a fixed sound intensity, sufficient to produce a spectrum rich in subharmonics, a slight departure from thermal equilibrium, for example a temperature increase of 10 mK for 2 s, will result in slow amplitude oscillations of the subharmonic peaks. Peaks of smaller amplitude may momentarily grow at the expense of larger peaks. Some subharmonics may temporarily disappear. Such transients last a few minutes before the system settles down. (3) At sound intensities well above the accumulation point occasionally neighboring subharmonic peaks (particularly $f_0/4$ with $f_0/3$ or with $f_0/5$) move toward each other, merge into a band, and then move back to their correct positions. This banding is relatively unstable and can easily be disturbed by a very small change in temperature, sound amplitude, or sound frequency.

We conclude that liquid helium-4 presents an interesting system in which to study the production of acoustic subharmonics in that below the superfluid transition a bifurcation sequence of the Feigenbaum type is quantitatively observed and above the superfluid transition it is not. Below the superfluid transition and at sufficiently high sound intensity, frequencies appear in the power spectrum which are members of the universal sequence. Furthermore, below the superfluid transition, we are able to identify the physical nature of the onset of the first subharmonic-bifurcation sequence not to be the threshold for the acoustic production of macroscopic classical turbulence (acoustic cavitation) but to be the threshold for the ultrasonic generation of microscopic quantum turbulence (vortex-line generation).

Acknowledgements

We would like to acknowledge the helpful participation of James A. Rooney and Dale A. Farris in this study. This work was supported in part by the U.S. Air Force Office of Scientific Research, Grant No. 76-3113 and the National Science Foundation, Grant No. DMR-8005358.

References

[1] A. Libchaber and J. Maurer, J. Phys. (Paris), Colloq. 41 (1980) C3-51.
J.P. Gollub and S.V. Benson, J. Fluid Mech. 100 (1980) 449.
M. Giglio, S. Musazzi, and U. Perini, Phys. Rev. Lett. 47 (1981) 243.
[2] P.R. Fenstermacher, H.L. Swinney and J.P. Gollub, J. Fluid Mech. 94 (1979) 103.
[3] K. Ikeda, H. Daido and O. Akimoto, Phys. Rev. Lett. 45 (1980) 709.
H.M. Gibbs, F.A. Hopf, D.L. Kaplan and R.L. Shoemaker, Phys. Rev. Lett. 46 (1981) 474.
[4] Werner Lauterborn and Eckehart Cramer, Phys. Rev. Lett. 47 (1981) 1445.
[5] B.A. Huberman and J.P. Crutchfield, Phys. Rev. Lett. 43 (1979) 1743.
[6] C. Herring and B.A. Huberman, Appl. Phys. Lett. 36 (1980) 975.
[7] J.S. Turner, J.C. Roux, W.D. McCormick and H.L. Swinney, Phys. Lett. 85A (1981) 9. J.C. Roux and H.L. Swinney, in *Nonlinear Phenomena in Chemical Dynamics*, C. Vidal and A. Pacault, eds. (Springer, Berlin, Heidelberg, New York, 1981) p. 38.
[8] Paul S. Linsay, Phys. Rev. Lett. 47 (1981) 1349.
James Testa, Jóse Pérez and Carson Jeffries, Phys. Rev. Lett. 48 (1982) 714.
[9] M.J. Feigenbaum, J. Stat. Phys. 19 (1978) 25; 21 (1979) 665; Phys. Lett. 74A (1979) 375; Commun. Math. Phys. 77 (1980) 65.
[10] R.F. Carey, J.A. Rooney, and C.W. Smith, J. Acoust. Soc. Am. 66 (1979) 1801.
[11] R.J. Donnelly, Experimental Superfluidity (Univ. of Chicago Press, Chicago, 1967), chap. 6.
[12] N. Metropolis, M.L. Stein and P.R. Stein, J. Comb. Theory A15 (1973) 25.

BIFURCATION AND CHAOS IN A PERIODICALLY STIMULATED CARDIAC OSCILLATOR

Leon GLASS, Michael R. GUEVARA and Alvin SHRIER
Department of Physiology, McGill University, 3655 Drummond Street, Montreal, Quebec, Canada H3G 1Y6

and

Rafael PEREZ
Instituto de Fisica, Universidad Nacional de Mexico, Mexico City, Mexico

Periodic stimulation of an aggregate of spontaneously beating cultured cardiac cells displays phase locking, period-doubling bifurcations and aperiodic "chaotic" dynamics at different values of the stimulation parameters. This behavior is analyzed by considering an experimentally determined one-dimensional Poincaré or first return map. A simplified version of the experimentally determined Poincaré map is proposed, and several features of the bifurcations of this map are described.

1. Introduction

Cardiac dysrhythmias are abnormal cardiac rhythms that often occur in diseased hearts. In what follows, we analyze biological and theoretical models for the generation of cardiac dysrhythmias that are associated with a lack of synchronization between autonomous pacemaker sites in the heart. The theoretical techniques are also applicable to a broader range of problems dealing with the synchronization of nonlinear oscillators to a periodic input.

In the normal human heart, the primary pacemaking site is located in the sinoatrial node (SAN). From the SAN the cardiac impulse spreads sequentially through the atrial musculature, the atrioventricular node (AVN), and specialized conducting tissues to the ventricles [1]. Electrical excitation of the atria is followed 0.08–0.12 sec later by excitation of the ventricles. The electrochemical events responsible for the spread of excitation can be monitored noninvasively by the electrocardiogram (ECG) which is a record of the potential differences between different points on the surface of the body. On the ECG, the P wave is associated with the excitation of the atria, the QRS complex with excitation of the ventricles, and the T wave with recovery of the excitability of the ventricles. Normally, excitation of cardiac tissue is associated with mechanical contraction. Fig. 1A shows a normal ECG. In a class of disorders called the atrioventricular (AV) heart blocks, there are abnormalities in the relative timing of the atrial and ventricular contractions. In first degree AV block the delay between the atrial and ventricular contractions is elevated above its normal maximal value. In second degree AV block, atrial contractions are not always followed by ventricular contractions. The second degree AV blocks are often periodic and are characterized by a ratio which gives the relative frequencies of atrial to ventricular contractions. Fig. 1B shows an example in which one cycle of 3:2 AV block (i.e. 3 atrial to 2 ventricular contractions) is followed by five cycles of 2:1 AV block. As well, second degree AV block can show extremely irregular or fluctuating rhythms (fig. 1C). In third degree AV block, there is an apparent lack of correlation between the atrial and ventricular rhythms, with separate pacemakers for each rhythm (fig. 1D). Finally, electro-

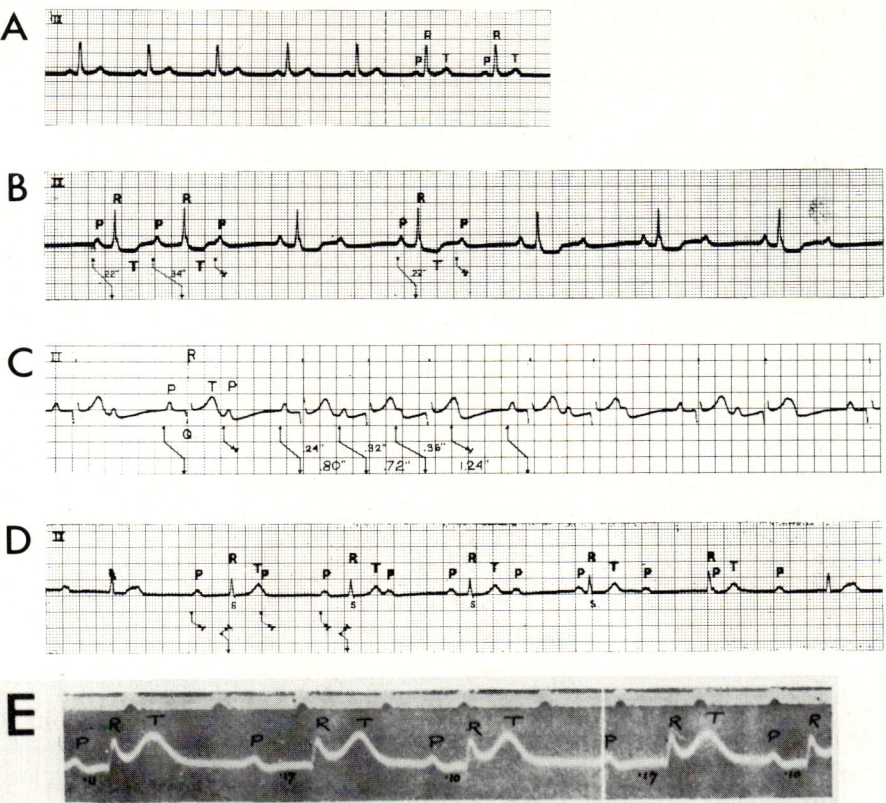

Fig. 1. Electrocardiograms illustrating normal and pathological cardiac rhythms. (A) Normal sinus rhythm. Each P wave is followed a fixed time later (the PR interval) by a QRS complex (1:1 ratio in atrioventricular conduction). (B) Second degree AV block with one 3:2 Wenckebach cycle followed by five consecutive 2:1 cycles. In the Wenckebach phenomenon, the PR interval following a dropped ventricular beat gradually increases until a P wave is not followed by a QRS complex (i.e. a ventricular beat is dropped). (C) Second degree AV block showing segments with 2:1, 4:3, and 3:2 conduction ratios. (D) Third degree heart block, Note the dissociation of the atrial and ventricular rhythms. (E) Sinus rhythm with alternating PR intervals (2:2 conduction). The laddergrams below the tracings of (B)–(D) schematically show the path of impulse formation and conduction. The small dots in these diagrams show sites of impulse generation, arrows show the spread of excitation, while short dark bars show regions of block of conduction. The sinoatrial node is at the top of the laddergram, the ventricles at the bottom. The time between the heavier vertical lines on the electrocardiographic paper is 0.20 sec, while the voltage difference between adjacent heavier horizontal lines is 0.5 mV. Time intervals in seconds are marked on some panels. (A)–(D) show human electrocardiograms reproduced with permission from [1], while (E) is from a cat reproduced with permission from [2].

cardiograms with alternating PR intervals (2:2 and 4:2 AV block) have been observed in animals [2] and in man [3, 4]. Fig. 1E shows an instance of 2:2 AV block published in 1910.

A fundamental hypothesis underlying our work is that changes in the cardiac rhythm can be associated with bifurcations in the qualitative dynamics of mathematical models describing generation and conduction of the cardiac impulse. The equations describing the electrical activity of the heart are complex and presumably vary from individual to individual. However, the qualitative features of the dynamics of these equations for the heart, and in particular their bifurcation structure, may well be similar in different individuals. Indeed, cardiac dysrhythmias have been classified by non-mathematicians (i.e. cardiologists) on the basis of their qualitative dynamics [1].

An early theoretical study of cardiac dysrhythmias was made by van der Pol and van der Mark in 1928 [5]. They showed that when three nonlinear oscillators (representing pacemaker sites of the heart) are coupled together, many different rhythms similar to the AV blocks can be observed as the relative frequencies and coupling parameters of the oscillators are varied. This pioneering work has since been extended by other researchers using analogue and digital models [6, 7]. As well, physiological studies have provided additional evidence that the canine heart contains a subsidiary pacemaker situated just below the AVN [8, 9]. However, most cardiac electrophysiologists ascribe the AV heart blocks to blocked conduction in the region of AVN [1], where there is assumed to be impulse conduction but no impulse generation. Mathematical models for this situation show correspondences with the AV heart blocks but do not show patterns with alternating PR intervals [10–12]. Our emphasis in the current work is on the effects of periodic stimulation of an autonomous cardiac oscillator. This situation may be applicable to the AV heart blocks, and also is relevant to a wide range of other situations, such as the artificial pacing of the heart and dysrhythmias such as parasystole.

Analysis of the effect of periodic forcing on relaxation oscillations of the van der Pol type have had a major impact on mathematics. Early studies demonstrated bistability (in which one of two different stable oscillating patterns is possible, depending on the initial conditions) as well as aperiodic dynamics [13, 14]. This observation of aperiodic dynamics in the 1940s played a role in the subsequent formulation of the horseshoe map by Smale (see the account in [15]). More recent studies of periodic forcing of nonlinear oscillators have shown that multistability and aperiodic dynamics can be accounted for by consideration of 1-dimensional maps of the circle into itself [16–18]. It has also been shown that such maps can display aperiodic dynamics that result from a cascade of period-doubling bifurcations [19–23].

Recent experimental studies on hydrodynamic [24–26], chemical [27, 28] and electronic [29–31] systems have shown the presence of complex dynamic behavior such as quasiperiodicity, period-doubling bifurcations, intermittency, and chaos. There is great interest in analyzing in detail the "universal" features of the transitions from periodic to chaotic dynamics [32–36].

In our research we have tried to develop biological models for cardiac dysrhythmias and to analyze experimentally observed dynamics. In section 2 we show that, under certain assumptions, the periodic stimulation of a cardiac oscillator by brief current pulses can be reduced to the analysis of a 1-dimensional map. In section 3 we consider the effects of periodic stimulation of spontaneously beating aggregates of cells obtained from embryonic chick heart [37]. In this work a current pulse generator is analogous to the SAN, whereas the heart cell aggregate is analogous to the subsidiary pacemaker located below the AVN. As the period of the stimulation changes, complex bifurcations including period-doubling bifurcations can be observed. In section 4 we investigate a 1-dimensional map that has been proposed as a simple model for periodic forcing of nonlinear oscillators. This map depends on 2 parameters corresponding to the strength and frequency of the periodic forcing.

2. Pulsatile stimulation of a cardiac oscillator: theory

In order to analyze the bifurcations of periodically stimulated cardiac oscillations mathematically, a number of assumptions are needed. In this section the main assumptions are explicitly stated. As well, the resulting properties of the dynamics are briefly discussed. Similar approaches have been previously employed [38–40]. More details on the mathematical aspects can be found in [41–43].

Assumption (*i*). A cardiac oscillator under normal conditions can be described by a system of ordinary differential equations with a single unstable steady state and displaying an asymptotically stable limit cycle oscillation which is globally attracting except for a set of singular points of measure zero.

Assume the limit cycle has period T_0 and that we start with initial conditions $x(t=0) = x_0$, with x_0 being an arbitrary point on the limit cycle. Set the phase of x_0 to be zero. Then the phase of the point $x(t)$ is defined to be t/T_0 (mod 1). Thus, a phase ϕ ($0 \leq \phi < 1$) can be assigned to every point on the limit cycle. The eventual phase of points in the basin of attraction of the cycle can now be defined. Let $x(t=0)$ and $x'(t=0)$ be the initial conditions of a point on the cycle and a point not on the cycle respectively, and $x(t)$, $x'(t)$, be the coordinates of the trajectories at time t. If $\lim_{t\to\infty} d(x(t), x'(t)) = 0$, where d is the Euclidean distance, then the eventual phase of $x'(t=0)$ is the same as the phase of $x(t=0)$. A locus of points all with the same eventual phase is called an isochron. If the equations describing the system are of dimension 2, then there must be at least one singular point at which the eventual phase is not defined (fig. 2). If the system of equations is of order greater than 2, then there must exist a set of dimension ≥ 1 consisting of points where the eventual phase is not defined [41–43].

Consider the effect of delivering a stimulus starting at a time when the oscillator is at some (old) phase ϕ ($0 \leq \phi < 1$). In general, at the end of the perturbation, the orbit will lie on a different isochron of (new) phase θ. Then

$$\theta = g(\phi), \quad (1)$$

where the function g depends on the stimulus amplitude and duration. The function g is often called the phase transition curve (PTC) [42]. Note that there can also be a stimulus which will perturb the oscillator to a phaseless point.

Assumption (ii). Following a short perturbation, the time course of the return to the limit cycle is much shorter than the spontaneous period of oscillation or the time between periodic pulses.

This assumption allows one to measure the PTC for the cardiac preparation and use the experimentally measured PTC to compute the effects of periodic stimulation. Assume that one of the variables in the system is experimentally measurable. Define a reference or marker event to be a particular point on the waveform (e.g. the maximum value attained). The time between two consecutive marker events of the spontaneous cycle is the period T_0 of the oscillation. Fig. 3A shows the situation schematically. A stimulus (which we assume is a delta function) is delivered at a time δ following the preceding spontaneous event. The duration T of the perturbed cycle is the time from the event preceding the stimulus to the event immediately subsequent to the stimulus. Note that due to fast relaxation back to the limit cycle, the post-stimulus cycles are approximately of duration T_0. Thus, essentially all of the eventual phase shift occurs within the perturbed cycle. Therefore

$$T/T_0 = 1 + \phi - \theta, \quad (2)$$

where $\phi = \delta/T_0$ (fig. 3B). Since T can be experimentally measured for a given ϕ, and T_0 is known, the PTC can be determined (fig. 3C).

Now consider the effects of periodic stimulation with a time τ between successive stimuli. Calling ϕ_i (mod 1) the phase of the oscillator immediately before the ith stimulus we obtain

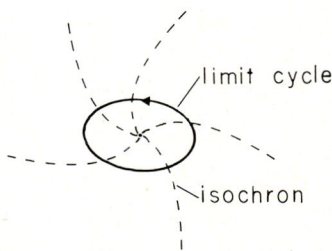

Fig. 2. Phase space of a 2-dimensional nonlinear oscillator. The solid oriented closed curve is the limit cycle, while the dashed curves represent isochrons. The point in the center is an equilibrium point, and therefore is phaseless and does not lie on any isochron.

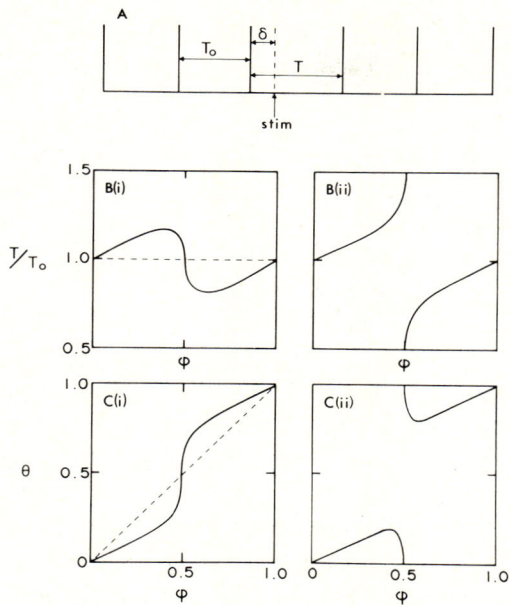

Fig. 3. (A) Schematic diagrams illustrating the effect of perturbation of a limit cycle oscillator by a single impulse. The vertical lines represent the occurrence of a marker event (see text). The spontaneous time between marker events (the period of the limit cycle) is denoted by T_0. A stimulus is delivered at a time δ after a marker event, causing a change in the cycle length from T_0 to T. The durations of the cycles following the perturbed cycle are all very close to T_0. (B) The normalized perturbed interbeat interval T/T_0 plotted as a function of the old phase $\phi = \delta/T_0$. (C) The new phase-old phase curve (phase transition curve or PTC) computed from (B) using eq. (2). In part (i) of (B) and (C), the stimulus is weak enough to produce type 1 phase resetting, while in (ii) it is sufficiently strong to result in type 0 behavior. The curves of (B) and (C) are taken from a very simple limit cycle model [19].

$$\phi_{i+1} = f(\phi_i) = g(\phi_i) + \tau/T_0, \qquad (3)$$

where $g(\phi)$ is the experimentally determined PTC for that stimulus strength and $g(\phi + j) = g(\phi) + j$ for j an integer. Eq. (3) represents a first return or Poincaré map. Starting from an initial phase ϕ_0, one can iterate eq. (3) to generate a sequence ϕ_0, $\phi_1 = f(\phi_0), \ldots, \phi_N = f(\phi_{N-1}) = f^N(\phi_0)$. If

$$\begin{aligned}\phi_N^*(\bmod 1) &= \phi_0^*(\bmod 1),\\ \phi_i^*(\bmod 1) &\neq \phi_0^*(\bmod 1), \quad 1 \leq i < N,\end{aligned} \qquad (4)$$

then ϕ_0^* is said to be a fixed point of period N and $\phi_0^*, \phi_1^*, \ldots, \phi_{N-1}^*$ is a cycle of period N. The cycle is stable if

$$\left|\left(\frac{\partial f^N}{\partial \phi_i}\right)_{\phi_i = \phi_0^*}\right| = \prod_{j=0}^{N-1}\left|\left(\frac{\partial f}{\partial \phi_i}\right)_{\phi_i = \phi_j^*}\right| < 1. \qquad (5)$$

If an extremum of the function f is a point on a cycle, the slope computed from eq. (5) equals zero, and the cycle is called a superstable cycle. A stable cycle of period N corresponds to stable $N:M$ phase locking with

$$M = \sum_{i=1}^{N}[g(\phi_i^*) - \phi_i^* + \tau/T_0]. \qquad (6)$$

The rotation number is defined as

$$\rho = \lim_{i \to \infty}\frac{\phi_i - \phi_0}{i}. \qquad (7)$$

For stable $N:M$ phase locking the rotation number is rational, $\rho = M/N$. In section 3 we use eq. (3) to predict the properties of the periodically stimulated cardiac preparation. The observation of good agreement between the theoretical predictions and experimental observations provides a posteriori justification for assumption (ii). However, a careful experimental and theoretical analysis of the consequences of the breakdown of assumption (ii) has not yet been performed.

Assumption (iii). The topological characteristics of the PTC change in stereotyped ways as the stimulus strength increases.

As a consequence of assumption (i) the PTC is a continuous map of the unit circle into itself $g: S^1 \to S^1$. The topological degree (or winding number) of g is the number of times θ traverses the unit circle as ϕ traverses the unit circle once. At zero perturbation strength $\theta = \phi$ so that by continuity the PTC is a monotonic map of degree 1 for sufficiently small perturbation (fig. 3C(i)). As the stimulus strength increases from zero, our experimental studies seem to indicate that the PTC remains of degree 1, but undergoes a transition from a monotonic to a non-monotonic function. Winfree [43] has given evidence that at high stimulus strength the PTCs for neural and cardiac oscillators can be of degree zero and therefore non-monotonic (fig. 3C(ii)). We conjecture that as stimulus strength increases the PTC for biological oscillators will, provided assumption (i) is satisfied, in general undergo the sequence of transitions

degree 1 (monotonic)→degree 1 (nonmonotonic)→degree 0.

We have previously considered the case of a model oscillator in which, as the stimulus strength is increased, the PTC undergoes a direct transition from degree 1 (monotonic) to degree 0 [19]. In section 4 we consider the properties of a simple model of phase locking in which the PTC undergoes a transition from degree 1 (monotonic) to degree 1 (non-monotonic).

3. Pulsatile stimulation of a cardiac oscillator: experiment [37]

As an application of the theory presented in section 2, we consider the effects of single and periodic stimulation of spontaneously beating aggregates of cells taken from the ventricles of 7-day-old embryonic chick heart. Spheroidal aggregates (100–200 μm in diameter) of electrically coupled cells were maintained in tissue culture [44, 45]. Each aggregate beats spontaneously with a period between 0.4 and 1.3 sec. Cells within a single aggregate are very nearly isopotential [46]. The voltage difference across the cell membrane is measured using a glass microelectrode filled with KCl, inserted into one cell of a beating aggregate. Current pulses are delivered through the same microelectrode.

In Fig. 4 are shown the effects of brief injection of current at different phases ($\phi = \delta/T_0$) of the cardiac cycle (for two different current strengths). Note that at the higher current strength there is a

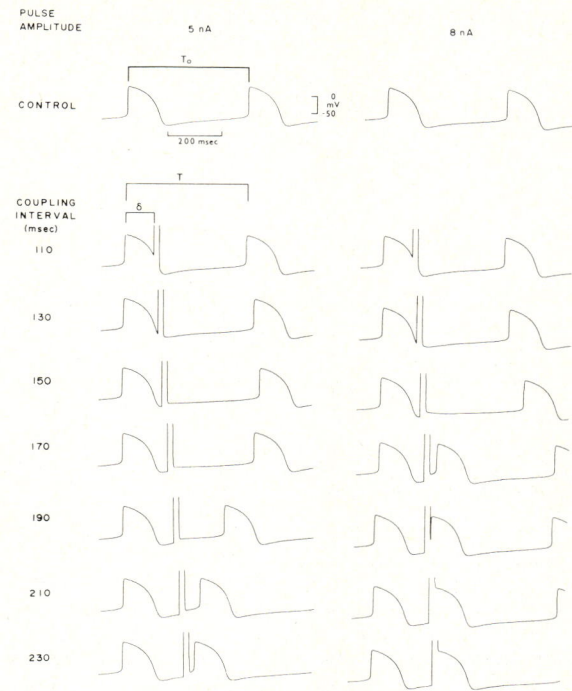

Fig. 4. Transmembrane voltage recorded from an aggregate. The uppermost tracings each show one cycle of spontaneous activity. The left and right panels show the effects of delivering 5 nA and 8 nA constant current pulses (20 msec duration) at different coupling intervals. The coupling interval δ is the time to the beginning of the stimulus from the upstroke of the immediately preceding action potential. The perturbed interbeat interval is denoted by T. Note that as the stimulus intensity is increased from 5 nA to 8 nA, the transition from prolongation to shortening of the perturbed cycle becomes more abrupt and occurs at a shorter coupling interval. The interbeat interval of the cycle following the perturbed cycle is almost equal to the control interbeat interval in the three traces of the right-hand panel that show such cycles.

very rapid change in the perturbed interbeat interval T for stimuli delivered at times from 150 ms to 170 ms into the cycle. One expects, based on theoretical considerations that the plot of T versus ϕ can be discontinuous for sufficiently high current strengths [42]. From experiments, it is difficult (and perhaps impossible) to establish whether such curves are indeed discontinuous.

Using the T/T_0 data obtained from the single pulse perturbation experiments (fig. 5C) and eq. (2), it is possible to construct the return map given

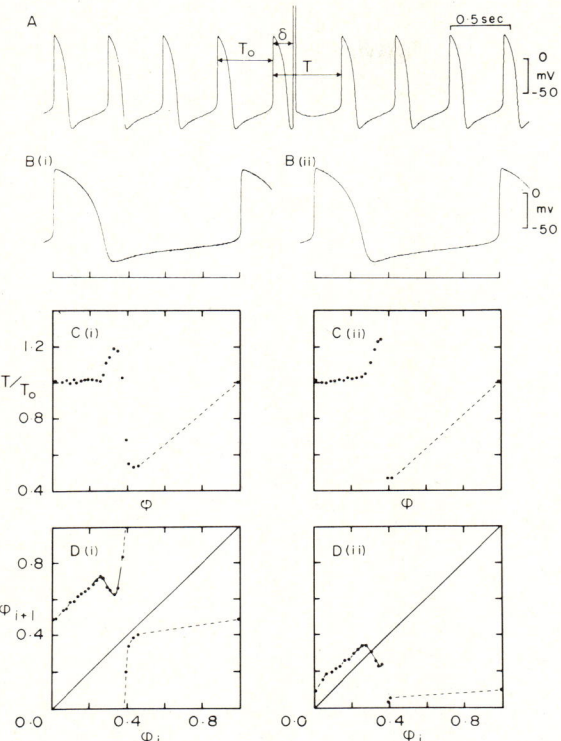

Fig. 5. (A) Transmembrane potential from an aggregate as a function of time, showing spontaneous electrical activity and the effect of a 20 msec, 9 nA depolarizing pulse delivered at an interval of 160 msec following the action potential upstroke. In (B to D), parts (ii) show results from this aggregate (aggregate 2), while parts (i) are from aggregate 1, taken from a different culture. (B) Membrane voltage as a function of phase ϕ, $0 \leq \phi < 1$. (C) Phase-resetting data, showing the normalized length T/T_0 of the perturbed cycle as a function of ϕ. (i) Pulse duration 40 msec, pulse amplitude 5 nA; (ii) pulse duration 20 msec, pulse amplitude 9 nA. For approximately $0.4 < \phi < 1.0$ the action potential upstroke occurs during the stimulus artifact and hence the perturbed cycle length cannot be exactly determined. The dashed line represents a linear interpolation that approximates the data. During collection of these data, the average control interbeat intervals (± 1 standard deviation) were (i) $T_0 = 515 \pm 5.7$ msec and (ii) $T_0 = 434 \pm 5.5$ msec. (D) Poincaré maps computed from eq. (3) and the data in fig. 5C; (i) $\tau = 250$ msec, (ii) $\tau = 480$ msec. The dashed line represents a linear interpolation used in iterating the Poincaré map; the solid line through the data points is a quartic fit for $0.22 < \phi_i < 0.37$. Reprinted from [37] with permission.

by eq. (3). Examples of return maps from two different aggregates are shown in fig. 5D. Using eqs. (3)–(5), the experimentally derived return maps can be iterated to compute the response to periodic stimulation. The numerically predicted phase locking regions (fig. 6B) are compared with experimental observations (fig. 6A). In fig. 6C are shown the stable phases in the period-doubling zone.

Representative traces of microelectrode recordings at different stimulation frequencies are shown in fig. 7. Regular and irregular dynamics theoretically predicted in fig. 6 were observed. Notice that in fig. 7C(i) (aggregate 1) there is a spontaneous change from 1:1 to 2:2 phase locking at a stimulation period of 550 ms. Stimulation of the second aggregate at a period of 490 ms produced both 4:4 and irregular patterns (fig. 7C(ii) and fig. 7C(iii) respectively). Computations show that 2:2 and 4:4 phase-locking as well as chaotic dynamics are predicted in the range 460–490 ms for this aggregate (fig. 6C).

There are several reasons to expect discrepancies between the theoretically predicted and experimentally observed dynamics. First, assumption (ii) in section 2 is not strictly satisfied since there can be slight changes in the rate of the preparation for several beats following the injection of a single stimulus. The following two additional physiological considerations are also important:

i) *There is biological "noise" intrinsic to the experimental preparation as well as environmental "noise" which is not accounted for in the theory.* Although the fluctuations in interbeat interval in the absence of stimulation are comparatively small, some of the experimentally observed irregular patterns may not be due to intrinsic aperiodicity in the dynamical system itself, but rather to the effects of biological "noise" generated within the preparation or environmental "noise" arising from fluctuating ambient conditions. In addition, some of the irregularity in deterministically aperiodic patterns (such as that shown in fig. 7C (iii) which is ascribed to "chaotic" dynamics arising out of a cascade of period-doubling bifurcations) may well be accounted for by "noise". Additional analysis is needed to determine the effect of "noise" on

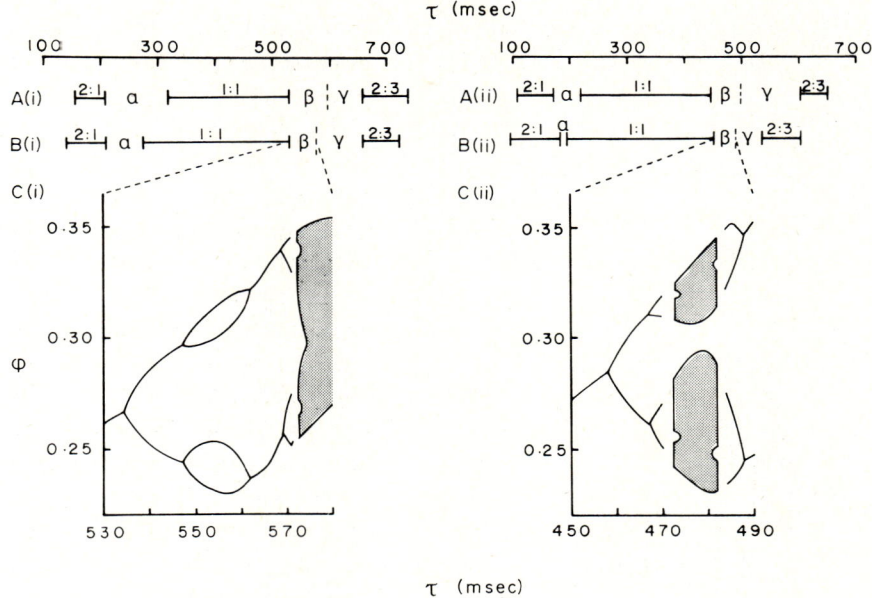

Fig. 6. Experimentally determined and theoretically computed responses to periodic stimulation of period τ with the same pulse durations and amplitudes as in fig. 5C. Parts (i) refer to aggregate 1, parts (ii) to aggregate 2. (A) Experimentally determined dynamics: there are three major phase-locking regions (2:1, 1:1, 2:3) and three zones of complicated dynamics labelled α, β and γ. (B) Theoretically predicted dynamics; note agreement with (A). (C) Theoretically predicted dynamics in zone β: curves give phase or phases in the cycle at which the stimuli fall during 1:1, 2:2 and 4:4 locking; stippled regions show the range of phases in which the stimulus falls during irregular dynamics. Reprinted from [37] with permission.

"chaos" and thus to determine the extent to which "noise" is implicated in generating the irregular dynamics experimentally observed.

ii) *Prolonged periodic stimulation of the aggregate changes the properties of the aggregate.* Following a long period of periodic stimulation with 1:1 synchronization at frequencies higher than the intrinsic aggregate beating frequency, the cessation of stimulation leads to a transient slowing of the beat rate below the control value ("overdrive suppression"). Conversely, after a period of "underdrive", "underdrive acceleration" occurs during the post stimulation recovery period [47]. The spontaneous transition from 1:1 to 2:2 phase locking observed in fig. 7C(i) during stimulation at a fixed frequency could arise from an increase in the intrinsic frequency of the aggregate secondary to underdrive.

Despite these considerations, there is good agreement between theory and experiment. However, probably as a consequence of "noise", we have not been able to observe phase locking patterns which are theoretically computed to extend over small regions of parameter space [48–50]. Consequently, experimental observation of the Feigenbaum constant, such as has been performed in periodically forced electronic oscillators [30, 31] and in experiments on turbulence [25] will be exceedingly difficult in this and other biological preparations. On the other hand, biological preparations may be ideal systems to analyze the effects of "noise" in systems which would be predicted to be deterministically "chaotic" in the absence of noise [51].

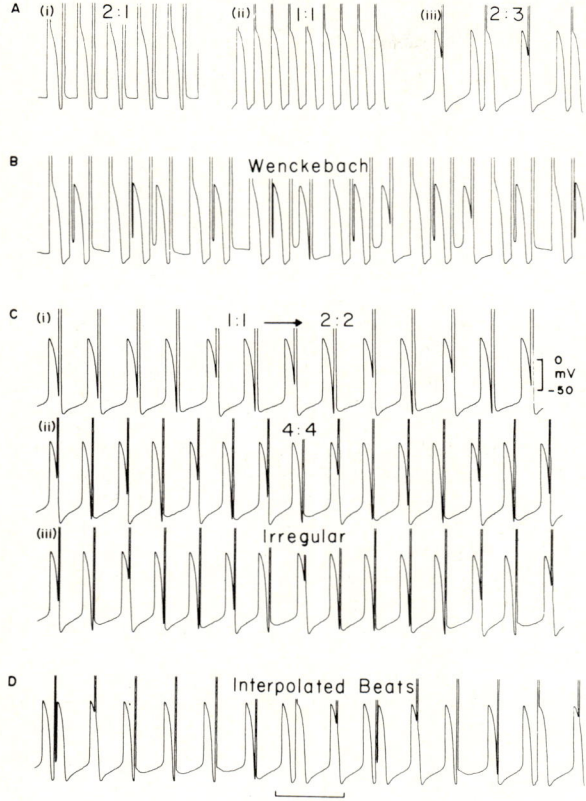

Fig. 7. Representative transmembrane recordings from both aggregates showing the effects of periodic stimulation with the same pulse durations and amplitudes as in fig. 5C. (A) Stable phase-locked patterns: (i) 2:1 (aggregate 1, $\tau = 210$ ms); (ii) 1:1 (aggregate 2, $\tau = 240$ ms); (iii) 2:3 (aggregate 2, $\tau = 600$ ms). (B) Dynamics in zone α: irregular dynamics displaying the Wenckebach phenomenon (aggregate 1, $\tau = 280$ ms). (C) Dynamics in zone β: (i) 1:1 phase locking spontaneously changing to 2:2 phase locking (aggregate 1, $\tau = 550$ ms). During 2:2 phase locking there are two distinct phases of the cycle at which the stimuli fall. (ii) 4:4 phase locking (aggregate, 2, $\tau = 490$ ms). There are four distinct phases of the cycle at which the stimuli fall. (iii) Irregular dynamics with one action potential in each stimulus cycle (aggregate 2, $\tau = 490$ ms). There is a narrow range of phases in which the stimuli fall. (D) Dynamics in zone γ: irregular dynamics displaying escape or interpolated beats (aggregate 2, $\tau = 560$ ms). Reprinted from [37] with permission.

4. A model map [22, 23]

Numerical computation of phase locking using the experimentally determined Poincaré map has shown complex bifurcations. In this section we discuss the possibility that a large topological class of maps might display the same bifurcation structure as the experimentally determined maps. Such behavior may be anticipated since the bifurcations for the one parameter map of the interval into itself with a single maximum are largely independent of the functional form of the map [32, 33].

Consider the map given in eq. (3) where $g(\phi)$ is a continuous function defined on the real line and $\deg[g(\phi) \pmod 1] = 1$. Assume there is a single maximum at $\phi_{\max}$ and a single minimum at $\phi_{\min}$ in the interval [0, 1] (fig. 5D(i)). Let $\phi_i = H_i(\tau)$ where the $H_i(\tau)$ are functions found by iterating eq. (3) from $\phi_0 = \phi_{\max}$. For j an integer, $g(x+j) = g(x) + j$, and

$$H_N(j) - H_N(j-1) = N. \tag{8}$$

There will be a superstable cycle for each value of τ for which $\phi_0 \pmod 1 = H_N(\tau) \pmod 1$. Since $H_N(\tau) \pmod 1$ equals any fixed value between 0 and 1 at least N times as τ varies from $j-1$ to j, there will be a minimum of N superstable cycles associated with N different rotation numbers occurring at N distinct values of τ. The iterates of the minimum also give rise to superstable cycles. Consequently, *there will be at least two values of τ at which there exist superstable cycles for each rational rotation number* [22]. This fixed point theorem does not depend on the functional form of g.

The 1-dimensional, two parameter map

$$\phi_{i+1} = f(\phi_i) = \phi_i + b \sin 2\pi\phi_i + \tau, \tag{9}$$

where b is a real number has recently been discussed by several workers as a model for periodically forced nonlinear oscillators [22, 23, 35, 36, 40]. For $b < 1/2\pi$, $f(\phi_i)$ is a monotonic function and it is known from early studies that the rotation number is a monotonic function of τ and that only periodic and quasiperiodic dynamics exist [52]. In addition, recent studies have described the application of renormalization methods to analyze the dynamics for $0 < b \leqslant 1/2\pi$ [35, 36]. We primarily consider the dynamics for $b > 1/2\pi$ [20–23]. In this region $f(\phi_i)$ is not monotonic.

For eq. (9) it is straightforward to show that there exists a stable fixed point of period 1 at

$$\phi^* = \frac{1}{2\pi} \sin^{-1}\left(\frac{1-\tau}{b}\right) \quad (10)$$

(corresponding to 1:1 phase locking) for

$$(1-\tau)^2 < b^2 < (1-\tau)^2 + \pi^{-2}. \quad (11)$$

The stable period 1 orbit appears via a tangent bifurcation $(\partial f/\partial \phi)_{\phi^*} = 1$ at $b = 1 - \tau$ and loses its stability via a period-doubling bifurcation $(\partial f/\partial \phi)_{\phi^*} = -1$ at $b = |(1-\tau)^2 + \pi^{-2}|^{1/2}$. In addition, for $\tau = 1$ it is easy to compute that there is a further bifurcation to two stable period 2 orbits with rotation number 1 (2:2 phase locking) at $b = 0.5$ and a bifurcation to two stable period 4 orbits (4:4 phase locking) at $b = (0.25 + 1/2\pi^2)^{1/2}$. For $\tau = 1.5$ there is a period-doubling bifurcation from a period 2 cycle (2:3 phase locking) to a period 4 cycle (4:6 phase locking) at $b = \pi^{-1} 2^{-1/2}$.

Further results on the boundaries of the phase locking zones were obtained by numerical analysis. The Poincaré map in eq. (9) was numerically iterated from different ϕ_0 at many points in the (b, τ) parameter space. The results are shown in fig. 8. For many regions in the parameter space there is bistability in that one of two stable cycles is asymptotically reached depending on the initial condition ϕ_0. In addition, evidence for chaotic dynamics (a positive Lyapunov number) was found [23].

The 1:1 phase locking region arises by a tangent bifurcation along one boundary and is lost via a period doubling bifurcation along the other boundary. Consequently, since the slope of the period 1 cycle is a continuous function of b and τ, there must be a locus of points with period 1 along which the slope is equal to zero. Along the locus, the maximum (or minimum) of the Poincaré map will be on the cycle and the cycle is called a superstable cycle. This will occur for

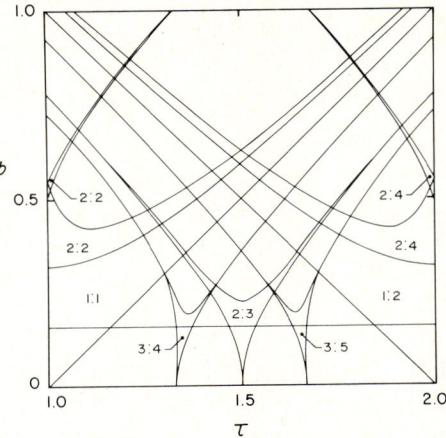

Fig. 8. Locally stable phase locking regions for eq. (9). The line at $b = 1/2\pi$ separates the regions in which eq. (9) is a monotonic function on the unit circle ($b < 1/2\pi$) and a nonmonotonic function ($b > 1/2\pi$). The widths of some of the phase locking zones (e.g. 3:4, 2:3) become so narrow as b increases that the boundaries have been collapsed into a single line in the drafting of the figure. In the non-labelled regions are phase-locked, quasiperiodic and chaotic dynamics. Slightly modified from [23] with permission.

$$(1-\tau)^2 + 1/4\pi^2 = b^2, \quad (12)$$

where for $\tau < 1$ the points defined by eq. (12) are derived from the maximum and for $\tau > 1$ the points defined by eq. (12) are derived from the minimum.

The loci of the superstable cycles are called the skeleton [22]. In fig. 9 are shown the boundaries of the $N:M$ phase locking zones ($1 \leq N \leq 5$) for $0 < b < 1/2\pi$ and the associated skeleton for $b > 1/2\pi$. Fig. 10 shows the skeleton derived from the maximum $\phi_{\max}$ for $N \leq 3$.

A branch of the skeleton extending downwards to $b = 1/2\pi$ is called a primary branch, and all other branches are called secondary. (This terminology is due to S. Shenker who has independently observed the following properties). Since no two branches of the skeleton derived from the maximum can intersect, and the skeleton of the 1:1 phase locking asymptotically approaches the line $b = 1 - \tau$ the slope of all branches of the skeleton derived from the maximum must be asymptotically

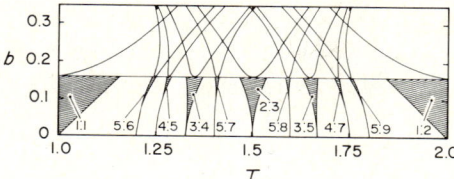

Fig. 9. The $N:M$ phase locking regions for $1 \leq N \leq 5$ for $b < 1/2\pi$ and the associated superstable cycles for $b > 1/2\pi$ for eq. (9). From [22] with permission.

equal to -1 for sufficiently large values of b. Further, all primary branches derived from the maximum must maintain the ordering given by monotonically increasing rotation numbers for fixed b as τ increases (fig. 10). Finally, in fig. 11, we show the skeleton of the phase locking zones up to period 4 with $\rho = 1$. Limited numerical analysis indicates a topologically equivalent skeleton appears in other V-shaped regions formed by the primary branches of the skeleton for each rational rotation number.

On the basis of these numerical and analytical results we have proposed the following structure for the skeleton and phase locking zones of eq. (9) [22]. Each zone of stable phase locking for $b < 1/2\pi$ extends through $b = 1/2\pi$ and then splits into two branches (figs. 8 and 9). In the V-shaped region of the extensions of each "Arnold tongue"

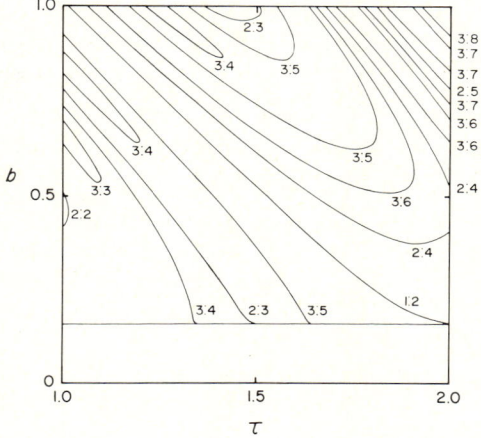

Fig. 10. The period N superstable cycles for $1 \leq N \leq 3$ derived from ϕ_{max} for eq. (9). There is a symmetrically located set of superstable cycles found from iteration of ϕ_{min}.

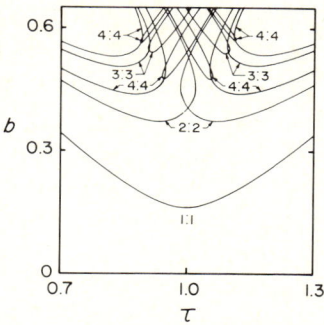

Fig. 11. Superstable cycles associated with $N:N$ phase locking patterns, $1 \leq N \leq 4$ (rotation number $\rho = 1$). From [22] with permission.

are period-doubling bifurcations. The skeleton of the phase locking zones in each one of these V-shaped regions is topologically equivalent to the skeleton for $\rho = 1$ (fig. 11), but with a different rotation number.

There is considerable interest in the transition from quasiperiodic to chaotic dynamics in physical systems and mathematical models [35, 36, 53]. In the current context one question is, "What happens to the quasiperiodic orbits (i.e. those with irrational rotation number) known to exist when $b < 1/2\pi$ as b passes through the value $b = 1/2\pi$?" For a map of the circle into itself, for a set of parameters for which there is bistability with two different rotation numbers, there will be initial conditions which give all intermediate rotation numbers [17]. A consequence of the bistability of the sine map for $b > 1/2\pi$ is that for $b > 1/2\pi$ orbits with a given irrational rotation number will be present in a wedge shaped region whose tip is on the line $b = 1/2\pi$. It should be possible to describe the aperiodic orbits associated with an irrational rotation number using techniques developed in symbolic dynamics and kneading theory [16, 17]. However, if there is bistability it is still not known if almost all points will attract to one of the two stable periodic orbits. Finally, this discussion has been concerned with analysis of 1-dimensional maps. A careful analysis of the breakdown of quasiperiodic dynamics in 2-dimensional maps has shown an overlapping of Arnold tongues similar to that shown in figs. 8 and 9 [53].

5. Conclusions

We have shown that the effects of periodic pulsatile stimulation of a cardiac oscillator can be analyzed by consideration of a 1-dimensional map which is obtained from experimental measurements. The dynamics in response to periodic stimulation are predicted by iterating this experimentally derived map and bear a close correspondence to the experimentally observed dynamics. In particular, stable phase locking, period-doubling bifurcations and aperiodic dynamics are all theoretically computed and experimentally observed. The experimentally observed dynamics show patterns similar to many commonly observed cardiac dysrhythmias. Furthermore, this work gives a novel perspective to some of the aperiodic cardiac dynamics clinically observed as well as uncommon phase locked rhythms such as 2:2 and 4:2 AV block. We have also analyzed the dynamics for a simple Poincaré map (the sine map in eq. (9)) and discussed period-doubling bifurcations, bistability and chaos in this model system.

Experimentally observed transitions from periodic to chaotic dynamics in physical systems can often be accounted for, at least qualitatively, by the bifurcations in simple 1- and 2-dimensional maps under parametric changes [26–29]. Unfortunately, in this work, it is not yet possible to compute from first principles the underlying maps which give a phenomenological correspondence with theory. Consequently, the study of periodic forcing of oscillations by short pulsatile inputs has advantages over other experimental techniques and may be useful in other physical systems. There are many similarities between the experimentally measured 1-dimensional map for the cardiac oscillator and the maps derived from other experimental systems [26–29]. Clearly, further work is needed to carefully probe experimentally observed dynamics in physical and biological systems and to clarify the bifurcation structure of maps which display quasi-periodic, periodic and aperiodic dynamics.

Analysis of the mathematics describing cardiac dysrhythmias has potentially rich rewards in the diagnosis and treatment of heart disease. We hope that our work will help stimulate interest in these problems by mathematicians and physicists.

Acknowledgments

This research has been supported by grants from the Natural Sciences and Engineering Research Council of Canada and the Canadian Heart Foundation. MRG is a recipient of a predoctoral traineeship from the Canadian Heart Foundation. We thank John Guckenheimer, James Sethna and Scott Shenker for helpful conversations. The manuscript was partially written while LG was in residence at the Aspen Center for Physics.

References

[1] S. Mangiola and M.C. Riota, Cardiac Arrhythmias: Practical ECG Interpretation, 2nd Edition (J.B. Lippincott, Philadelphia, 1982).
[2] T. Lewis and G.C. Mathison, Heart 2 (1910) 47.
[3] M. Segers, Arch. Mal. Coeur 44 (1951) 525.
[4] R. Langendorff, Am. Heart J. 55 (1958) 181.
[5] B. van der Pol and J. van der Mark, Phil. Mag. 6 (1928) 763.
[6] D.A. Sideris and S.D. Moulopoulos, J. Electrocardiol. 10 (1977) 51.
[7] C.R. Katholi, F. Urthaler, J. Macy Jr. and T.N. James, Comp. Biomed. Res. 10 (1977) 529.
[8] F.A. Roberge, R.A. Nadeau and T.N. James, Cardiovasc. Res. 2 (1968) 19.
[9] T.N. James, J.H. Isobe and F. Urthaler, Circ. Res. 45 (1979) 108.
[10] M.B. Berklinblit, in I.M. Gelfand, V.S. Garfinkel, S.V. Fomin and M.L. Tsetlin (eds.), Models of the Structural-Functional Organization of Certain Biological Systems (MIT Press, Cambridge, 1971) p. 155.
[11] H.D. Landahl and D. Griffeath, Bull. Math. Biophys. 33 (1971) 27.
[12] J.P. Keener, J. Math. Biol. 12 (1981) 589.
[13] M.L. Cartwright and J.E. Littlewood, J. Lond. Math. Soc. 20 (1945) 180.
[14] N. Levinson, Ann. of Math. 50 (1949) 127.
[15] S. Smale, The Mathematics of Time (Springer, New York, 1980) p. 147.
[16] M. Levi, Memoirs Amer. Math. Soc. 32, Number 244 (1981).
[17] J. Guckenheimer, in Dynamical Systems (Birkhauser, Boston, 1980) p. 115.

[18] P. Coullet, C. Tresser and A. Arneodo, Phys. Lett. 77A (1980) 327.
[19] M.R. Guevara and L. Glass, J. Math. Biol. 14 (1982) 1.
[20] T. Geisel and J. Nierwetberg, Phys. Rev. Lett. 48 (1982) 7.
[21] M. Schell, S. Fraser and R. Kapral, Phys. Rev. A 26 (1982) 504.
[22] L. Glass and R. Perez, Phys. Rev. Lett. 48 (1982) 1772.
[23] R. Perez and L. Glass, Phys. Lett. 90A (1982) 441.
[24] P.R. Fenstermacher, H.L. Swinney, S.V. Benson and J.P. Gollub, Ann. N.Y. Acad. Sci. 316 (1979) 652.
[25] A. Libchaber, C. Laroche and S. Fauve, J. Physique Lett. 43 (1982) L-211.
[26] A. Arneodo, P. Coullet, C. Tresser, A. Libchaber, J. Maurer and D. d'Humières, in press (1982).
[27] A.S. Pikovsky, Phys. Lett. 85A (1981) 13.
[28] R. Simoyi, A. Wolf and H. Swinney, Phys. Rev. Lett. 49 (1982) 245.
[29] J.P. Gollub, E.J. Romer and J.E. Socolar, J. Stat. Phys. 23 (1980) 321.
[30] P.S. Linsay, Phys. Rev. Lett. 47 (1981) 714.
[31] J. Testa, J. Perez and C. Jeffries, Phys. Rev. Lett. 48 (1982) 714.
[32] R.M. May, Nature (London) 261 (1976) 459.
[33] M.J. Feigenbaum, J. Stat. Phys. 19 (1978) 25.
[34] J.-P. Eckmann, Rev. Mod. Phys. 53 (1981) 643.
[35] D. Rand, S. Ostlund, J. Sethna and E.D. Siggia, Phys. Rev. Lett. 49 (1982) 132.
[36] M.J. Feigenbaum, L. P. Kadanoff and S.J. Shenker, in press (1982).
[37] M.R. Guevara, L. Glass and A. Shrier, Science 214 (1981) 1350.
[38] D.H. Perkel, J.H. Schulman, T.H. Bullock, G.P. Moore and J.P. Segundo, Science 145 (1964) 61.
[39] S.W. Scott, Ph.D. Thesis, S.U.N.Y. (Buffalo), (1979).
[40] G.M. Zaslavsky, Phys. Lett. 69A (1978) 145.
[41] M. Kawato and R. Suzuki, Biol. Cybernetics 30 (1978) 241.
[42] M. Kawato, J. Math. Biol. 12 (1981) 13.
[43] A.T. Winfree, The Geometry of Biological Time (Springer, New York, 1980).
[44] R.L. DeHaan, Dev. Biol. 23 (1970) 226.
[45] R.L. DeHaan and L.J. DeFelice, Theor. Chem. 4 (1978) 181.
[46] L.J. DeFelice and R.L. DeHaan, Proc. IEEE 65 (1977) 796.
[47] D.L. Ypey, W.P.M. VanMeerwijk and R.L. DeHaan, in L.N. Bouman and H.J. Jongsma (eds.), Cardiac Rate and Rhythm (Martinus Nijhoff, The Hague, 1982) p. 363.
[48] J.P. Crutchfield and B.A. Huberman, Phys. Lett. A77 (1980) 407.
[49] L. Glass, C. Graves, G.A. Petrillo and M.C. Mackey, J. theor. Biol. 86 (1980) 455.
[50] R. Guttman, L. Feldman and E. Jakobsson, J. Membr. Biol. 56 (1980) 9.
[51] J. Guckenheimer, Nature 298 (1982) 358. See also J. Crutchfield and N. Packard, this volume.
[52] V.I. Arnold, Translations A.M.S. 2nd Series, 46 (1965) 213.
[53] D.G. Aronson, M.A. Chory, G.R. Hall and R.P. McGehee, Commun. Math. Phys. 83 (1982) 303.

CHAPTER 3

MATHEMATICAL PROPERTIES
AND MODEL SYSTEMS

PERSISTENT PROPERTIES OF BIFURCATIONS

John GUCKENHEIMER*

Division of Natural Sciences, University of California, Santa Cruz, CA, USA

This paper reviews results about bifurcations which occur in families of differential equations. Persistent properties are defined to be those which remain when the family of equations is perturbed. We provide a list of such properties which is relevant for numerical studies of dynamical systems.

1. Introduction

This lecture is a short survey about bifurcations of ordinary differential equations. Since 1975 there has been considerable success in using qualitative results about bifurcation theory to interpret a range of observations involving transitions of dynamical behavior in experimental systems [4]. These observations have been organized around the theme of exploring the "routes to turbulence" where turbulence is interpreted as meaning aperiodic flow [3, 12]. The analogy is drawn between experimental transitions which occur as physical parameter(s) are varied and the bifurcations in a family of ordinary differential equations. There are some physical systems, seemingly characterized by those which can express only few degrees of freedom, for which this analogy appears to be well founded. In this lecture we review the different bifurcation phenomena of ordinary differential equations which are persistent (a term defined below) and therefore can be expected to be observable in physical systems.

The basic object of study here is a system of ordinary differential equations

$$\dot{x} = f_\mu(x), \qquad (1)$$

with $x \in \mathbb{R}^n$ and $\mu \in \mathbb{R}^k$ a (multidimensional) parameter. I shall assume without further comment that f_μ is smooth (C^∞) and depends smoothly on μ. The flow $\Phi_\mu : \mathbb{R}^n \times \mathbb{R} \to \mathbb{R}^n$ is defined by the requirement that the curves $t \to \Phi_\mu(x, t)$ be the solutions of (1) with initial conditions x. For technical simplicity,

I also assume that there is a large domain $D \subset \mathbb{R}^n$ such that all flow trajectories enter D, crossing its boundary with non-zero velocity.

Dynamical systems theory [21] considers qualitative properties of the system (1) and its associated flows, concentrating on "typical" systems and "typical" features of these. The determination of a precise meaning of "typical" is a subtle one that requires careful consideration. The most stringent definitions are based upon the concepts of structural stability and topological equivalence. Two flows are *topologically equivalent* if there is a continuous change of coordinates (*homeomorphism*) which sends trajectories of one flow to trajectories of the other. A system of equations is *structurally stable* if all C^1 perturbations of the system have flows which are topologically equivalent to one another. When families such as (1) are involved, there are several choices for the definition of structural stability of a family [15, 23]. A *bifurcation* is a transition from one topological equivalence class to another within a family. None of these definitions is entirely satisfactory for classification problems of bifurcations or for the full range of applications because important cases appear pathological. Nonetheless, the description of "typical" bifurcations has been one of the most useful aspects of dynamical systems theory for applications. Therefore, it is worthwhile to formulate definitions that focus upon the issues of central interest in the applications. An example here illustrates these difficulties and motivates the definition which follows.

Example [5]. Assume that the system (1) has $n = 3$

---

*Research partially supported by the National Science Foundation.

and $k = 1$ and that there is a *periodic orbit* γ for the flow Φ_{μ_0}. The flow near γ is studied by passing a surface Σ transversally through γ and introducing the return map $\theta : \Sigma \to \Sigma$ defined by setting $\theta(y)$ to be the first intersection of the trajectory starting at y with Σ. If Σ intersects γ just once, the map θ has a fixed point p at its intersection with γ and the Jacobian derivative $D\theta(p)$ gives information about the flow near γ. If all eigenvalues of $D\theta(p)$ have absolute value smaller than one, then γ is attracting: there is a neighborhood of γ in which all trajectories approach γ as $t \to \infty$.

Now letting μ increase from μ_0, γ and θ vary smoothly with μ provided that $D\theta(p)$ does not have an eigenvalue one. Assume that a pair of complex eigenvalues of $D\theta(p)$ cross the unit circle at λ, $\bar{\lambda}$ when $\mu = \mu_1$. This implies that the stability of γ changes and a bifurcation occurs, called a *Hopf* bifurcation for the periodic orbit. If certain inequalities on the Taylor series of θ at p are satisfied when $\mu = \mu_1$, then the flows Φ_μ for $\mu > \mu_1$ have attracting two-dimensional tori T_μ whose diameter in directions transverse to γ is of order $|\mu - \mu_1|^{1/2}$.

Each invariant torus T_μ contains periodic or quasiperiodic orbits. The analysis of how many periodic orbits occur for which parameter values is delicate, and the numbers can be changed by perturbation. The system (1) can always be perturbed to a family which contains flows from a different set of equivalence classes than the original family. Nonetheless, the fact that (1) has attracting invariant tori is significant for applications while the differences in the topologically inequivalent perturbations are too subtle to have much relevance. Power spectra of experimental observations often show the sharp, discrete frequencies which one expects to be associated with Hopf bifurcations of periodic orbits.

Definition. Let X be a space of families of dynamical systems defined by equations (1) and let P be a dynamical property of a family of flows $\Phi_\mu \in X$. Then P is *persistent* if sufficiently small perturbations of Φ_μ in the space X all have the property P.

This definition should be compared with the "stability dogma" which asserts that only structurally stable systems and families are physically meaningful. The stability dogma is justified by noting that one never expects complete exact knowledge of the set of equations (1) appropriate to a particular physical system; therefore, the system should have qualitative properties that do not change with perturbation. The definition of persistence allows one to replace the stability dogma by the following principle: *only persistent properties of families of flows have physical significance.* In stating this principle, one should clearly recognize the necessity of establishing which space X is appropriate for the problem being considered. For example, any symmetries of the physical system being modelled should be embodied in the choice of X.

We now list important persistent properties of flows and their bifurcations.

1) *Invariant manifolds* [11]:

Theorem. Assume that the system (1) has an equilibrium point $(x_0, \mu_0) \in \mathbb{R}^n \times \mathbb{R}^k$; i.e. $f_{\mu_0}(x_0) = 0$. Divide the spectrum of the linearization $Df_{\mu_0}(x_0)$ into parts σ_s, σ_c, and σ_u characterized by negative, zero, and positive real parts, respectively. Let E^s, E^c, and E^u be the (generalized) eigenspaces of σ_s, σ_c, and σ_u. Then there exist invariant manifolds W^s, W^c and W^u for the flow Φ_{μ_0} with tangent spaces E^s, E^c, and E^u. The *stable* and *unstable manifolds* W^s and W^u are C^∞ and unique while the *center manifold* W^c need not be unique and may be only finitely differentiable.

Remarks.

(i) There is a comparable invariant manifold theorem for fixed points of discrete systems such as the return maps of periodic orbits. The spectrum of the linearization at the fixed points is divided into σ_s, σ_c, and σ_u with σ_s lying inside the unit circle, σ_c lying on the unit circle, and σ_u lying outside the unit circle.

(ii) The center manifold allows the "reduction" of a bifurcation problem for an equilibrium or periodic orbit to one in which the linearization has all its eigenvalues with neutral stability. The variation of the system with respect to its parameters can be included in the reduction by treating the parameters as part of system (1) with the inclusion of the equations $\dot{\mu} = 0$. This increases the dimension of the center manifold by k, so that the enlarged center manifold contains all of the interesting dynamics which occur entirely within small neighborhoods of the bifurcating equilibrium or periodic orbit.

(iii) There is a center manifold theorem for systems with an infinite number of degrees of freedom which allows the reduction of bifurcation problems for these systems to finite-dimensional ones [13].

(iv) Stable and unstable manifolds of a periodic orbit γ have dimensions which allow them to intersect transversally if σ_c is empty. Smale's homoclinic theorem [20] states that the existence of a transversal intersection of $W^u(\gamma)$ with $W^s(\gamma)$ (off γ) implies that the flow has a hyperbolic invariant set Λ [21]. The set Λ produced by Smale's theorem contains aperiodic trajectories as well as periodic trajectories of arbitrarily long periods. In applications, one is frequently interested in knowing whether an invariant set is attracting, but this is a much more difficult issue to resolve. For geometric reasons, it is evident that many limit sets which appear to be attracting cannot be uniformly hyperbolic (axiom A [21]). One of the major unresolved issues for the theory is to gain a better understanding of the limit sets of examples like the Hénon map [8].

2. Simple bifurcations of equilibria

Assume that the system (1) has an equilibrium point p_μ for μ varying in some region of the parameter space. If the Jacobian matrix $Df_\mu(p_\mu)$ has no eigenvalues with real part zero, then p_μ is *hyperbolic* or *elementary* and the geometry of the flows near p_μ are determined by the dimensions of their stable and unstable manifolds. A *simple* bifurcation of the equilibrium occurs when there is a simple zero eigenvalue or a simple pair of pure imaginary eigenvalues. Persistent properties of simple bifurcations are determined by inequalities imposed upon the Taylor expansions of (1).

Theorem. Let X be one of the following spaces of one parameter families of vector fields on the line: (i) general systems (1), (ii) systems (1) which satisfy $f_\mu(0) = 0$ for all μ, (iii) systems which satisfy $f_\mu(-x) = -x$. If the system (1) has an equilibrium at $(0, \mu_0)$, then within the spaces of families (i)–(iii), the following bifurcation properties are persistent:

(i) If $f''_{\mu_0}(0) \neq 0$ and $\partial f(0, \eta_0)/\partial \mu \neq 0$, then the equilibria of (1) form a smooth curve tangent to the line $\mu = \mu_0$ with non-zero curvature at $(0, \mu_0)$. There are $\delta, \epsilon > 0$ such that for $0 < |\mu - \mu_0| < \delta$, then there are either zero or two equilibria of f_μ. If there are two, then one is attracting and one is repelling.

(ii) If $f''_{\mu_0}(0) \neq 0$ and $\partial^2 f(0, \mu_0)/\partial x \partial \mu \neq 0$, then the equilibria of (1) near $(0, \mu_0)$ form two curves which intersect transversally, one of which is given by $x = 0$. The stability of the equilibria along each curve changes at $(0, \mu_0)$.

(iii) If $f'''_{\mu_0}(0) \neq 0$ and $\partial^2 f(0, \mu_0)/\partial x \partial \mu \neq 0$, then the equilibria of (1) near $(0, \mu_0)$ form two curves, one of which is given by $x = 0$ and one of which is tangent to the line $\mu = \mu_0$ with non-zero curvature. The stability of solutions along the curve $x = 0$ changes while the stability of solutions along the curve tangent to $\mu = \mu_0$ does not change.

Remarks.
(i) The three cases are called the *saddle-node*, *transcritical*, and *pitchfork* bifurcations. Transcritical bifurcations arise in homogeneous problems with *trivial* solutions while pitchfork bifurcations occur in problems with a reflectional symmetry.

(ii) The simplest systems with the simple bifurcations described above are

$$\dot{x} = \mu - x^2,$$
$$\dot{x} = \mu x - x^2, \qquad (2)$$
$$\dot{x} = \mu x - x^3,$$

respectively.

Theorem (13). Suppose system (1) is one parameter family in the plane. If p is an equilibrium at $\mu = \mu_0$ for which $Df_{\mu_0}(p)$ has pure imaginary eigenvalues, then there is a smooth change of coordinates which gives (1) the form

$$\begin{aligned}\dot{x} &= -(w + c(x^2 + y^2))y + (a(\mu - \mu_0) \\ &\quad + b(x^2 + y^2))x + R_1, \\ \dot{y} &= (w + c(x^2 + y^2))x + (a(\mu - \mu_0) \\ &\quad + b(x^2 + y^2))y + R_2,\end{aligned} \qquad (3)$$

where the remainder terms tend to zero faster than $(x^2 + y^2)^{3/2}$ or $\mu(x^2 + y^2)^{1/2}$. If $a \neq 0$ and $b \neq 0$, then the following properties of (3) are persistent. There is a curve of equilibria along which the stability changes between attractor and repellor at $\mu = \mu_0$. There is a smooth surface consisting of periodic orbits which has tangent plane $\mu = \mu_0$ and non-zero curvature at $(0, \mu_0)$.

Remark. This case is called *Hopf* bifurcation.

3. Simple bifurcation of periodic orbits

There is a theory of simple bifurcations for periodic orbits analogous to that for equilibria. The results are naturally expressed in terms of fixed points for discrete systems by using cross-sections to the orbits and return maps. For simple bifurcations of a family of maps $F_\mu : \mathbb{R}^n \to \mathbb{R}^n$ at a fixed point p, there are three cases to consider: $DF_\mu(p)$ can have eigenvalues $+1$, -1, or a pair of complex eigenvalues $\lambda, \bar{\lambda}$ with $|\lambda| = 1$. The cases in which $\lambda^3 = 1$ or $\lambda^4 = 1$ are special [1].

Theorem. Let $F_\mu : \mathbb{R}^1 \to \mathbb{R}^1$ be a one-parameter family of maps. If (p, μ_0) is a fixed point at which $F'_{\mu_0}(p) = 1$, $F''_{\mu_0}(p) \neq 0$, and $\partial F(p, \mu_0)/\partial \mu \neq 0$, then the fixed points of F form a smooth curve tangent to the line $\mu = \mu_0$ at the point (p, μ_0). If (p, μ_0) is a fixed point with $F'_{\mu_0}(p) = -1$, $(F_{\mu_0} \circ F_{\mu_0})'''(p) \neq 0$, and $\partial^2(F \circ F)(p, \mu_0)/\partial \mu \partial x \neq 0$, then there is a smooth curve of fixed points whose stability changes at (p, μ_0) and a smooth curve of trajectories of period two passing through (p, μ_0) with non-zero curvature.

Theorem [13]. Let $F_\mu : \mathbb{R}^2 \to \mathbb{R}^2$ be a one-parameter family of maps of the plane which has a fixed point p when $\mu = \mu_0$ at which $DF_{\mu_0}(p)$ has a pair of complex eigenvalues $\lambda, \bar{\lambda}$ with $|\lambda| = 1$. If $\lambda^3 \neq 1$ and $\lambda^4 = 1$, then there is a smooth change of coorinates which gives the mappings F_μ Taylor expansion

$$F_\mu(z) = [(\lambda + ic|z|^2) + (a\mu + b|z|^2)]z + R, \qquad (4)$$

where $z = x + iy$, R is a remainder term which tends to zero faster than $|z|^3$ and $\mu|z|$, and a, b, c are real constants. If a and b are both non-zero, then there is a family of invariant curves of F_μ, each homeomorphic to a circle, which forms a surface passing through (p, μ_0) with tangent plane $\mu = \mu_0$ and non-zero curvature. If $c \neq 0$, then the rotation number of F_μ on this family of invariant is a (weakly) monotone function of μ.

Remarks.
(i) The case of eigenvalue $+1$ is called a *saddle-node* or *tangent* bifurcation. Pomeau and Manneville [16] have described the relationship between this bifurcation and certain types of intermittency observed in physical systems.

(ii) The case of eigenvalue -1 is called a *flip*, *period-doubling*, or *subharmonic* bifurcation. There are one-parameter families of maps which display sequences of period-doubling bifurcations as a persistent route to aperiodic behavior. There are *universal* features of these cascades related to scaling properties that are independent of the particular example in which they occur [2].

(iii) The case of pure imaginary eigenvalues is the *Hopf bifurcation* for periodic orbits. The behavior of rotation numbers is discussed more fully below. The invariant curves of the theorem typically are not C^∞ when they contain periodic orbits, but the smoothness increases unboundedly as $\mu \to \mu_0$.

4. Periodicity and quasiperiodicity

Hopf bifurcation for periodic orbits demonstrates that flows on invariant two-dimensional tori occur naturally as "attractors" in dynamical systems. It is therefore important to have a good understanding of the typical behavior of a family of flows on a torus. The concept of *rotation number* or *winding number* plays a prominent role: flows on a two-dimensional torus have periodic orbits if and only if the rotation number is rational. Otherwise all trajectories are quasiperiodic and dense in the torus.

Denote $p(\mu)$ the function which describes the rotation number in terms of the parameter in a one-parameter family of flows on the torus. A highly sophisticated analysis (involving the work of Kolmogorov, Arnold, Moser and Herman [9]) is required to obtain a thorough understanding of the function $p(\mu)$ in a typical one parameter family. The results of this analysis are the following:

(i) $p(\mu)$ is a continuous function;

(ii) it is highly exceptional that $p(\mu)$ assumes a rational value at an isolated point of μ. Typically, $p(\mu)$ assumes each rational value on a closed interval;

(iii) the set of μ for which $p(\mu)$ is rational usually contains an open-dense set of μ;

(iv) if $p(\mu)$ is not constant, then the set of μ for which $p(\mu)$ is irrational has positive Lebesgue measure [10].

There is a paradox here involving properties (iii) and (iv): the topological concept of "generic" (properties true in a countable intersection of open–dense sets) clashes with the measure theoretic concept of "full measure" (properties true except on sets of measure zero). One is left with the question of whether one should expect to observe flows with only rational rotation numbers in experiments since they are generic even though they lack full measure.

There is one cautionary note about the structure of $p(\mu)$ described above. If the flows have a circular symmetry, then the entrainment phenomenon (ii) will not occur. Historically, this remark is relevant to early experimental attempts to test the theories of Ruelle and Takens [18] since Taylor–Couette flow between rotating cylinders does have such a symmetry and attempts to detect entrainment there were unsuccessful. Rand [17] has given an analysis of Hopf bifurcation for periodic orbits in the presence of circular symmetry.

5. Homoclinic trajectories

Homoclinic trajectories are asymptotic to an equilibrium or to a periodic orbit as $t \to \pm \infty$. As we remarked in (i), transversal homoclinic orbits are contained in invariant sets with aperiodic trajectories. There are also persistent bifurcation phenomena associated with homoclinic trajectories of different types.

Theorem [22]. There are one-parameter families of flows Φ_μ in the plane for which the following property is persistent. There is a saddle point equilibrium of Φ_μ, which has a homoclinic orbit. This homoclinic orbit lies in the closure of a continuous family of periodic orbits whose period is unbounded as $t \to \infty$.

Theorem (Silinikov [19]). Consider the class of three-dimensional flows Φ which have an equilibrium p at which there are complex eigenvalues $\lambda_1, \bar{\lambda}_1$ and real eigenvalue λ_2 such that $0 < -\operatorname{Re} \lambda_1 < \lambda_2$ and such that p has a homoclinic trajectory. Within the class of flows satisfying these properties there is an open dense set of flows which have transversal homoclinic orbits.

Theorem (Newhouse [14]). Let $F_\mu : \mathbb{R}^2 \to \mathbb{R}^2$ be a one-parameter family of maps having saddle fixed points p_μ such that when $\mu = \mu_0$, $W^s(p_\mu)$ and $W^u(p_\mu)$ have a point of tangency at which the curvatures of $W^s(p_\mu)$ and $W^u(p_\mu)$ differ. If this intersection of $W^s(p_{\mu_0})$ with $W^u(p_{\mu_0})$ is a point of transverse intersection between $\cup_\mu W^s(p_\mu)$ and $\cup_\mu W^u(p_\mu)$ and if $|\det Df_\mu(p_\mu)| < 1$, then there are parameter values μ near μ_0 for which Df_μ has an infinite number of attracting periodic orbits.

6. Multiple bifurcations

There are many persistent properties associated with bifurcations in families containing multiple parameters. These have been analyzed systematically for bifurcations of equilibria in two parameter families [7], but there is a scope for much more work in this area. A general principle is that phenomena which occur globally in families with fewer parameters can occur with small amplitude in families with more parameters.

References

[1] V.I. Arnold, Loss of stability of self oscillations close to resonance and versal deformations of equivariant vector fields, Funct. Anal. Appl. 11 (1977) 85–92.
[2] M.J. Feigenbaum, Quantitative universality for a class of nonlinear transformations, J. Stat. Phys. 19 (1978) 25–52.
[3] J.P. Gollub and S.V. Benson, Many routes to turbulent convection, J. Fluid Mech. 100 (1980) 449–470.
[4] J.P. Gollub and H. Swinney, Onset of Turbulence in Rotating Fluid, Physical Review Letters 35 (1975) 927–930.
[5] J. Guckenheimer, One-parameter families of vector fields on two-manifolds: another nondensity theorem in Dynamical Systems, M.M. Peixoto, ed (Academic Press, New York, 1973) pp. 111–128.
[6] J. Guckenheimer, Lectures on Bifurcation Theory, Dynamical Systems, CIME Lectures 1978 (Birkhauser, 1980). Boston, Basel, Stuttgard.
[7] J. Guckenheimer and P. Holmes, Nonlinear Oscillations, Dynamical Systems, and Bifurcation Theory, (Springer, Berlin, 1983).
[8] M. Hénon, A two-dimensional mapping with a strange attractor, Comm. Math. Phys. 50 (1976) 69–78.
[9] M. Herman, Mesure de Lebesgue et nombre de rotation, Lecture Notes in Math. 597 (1977) 271–293.
[10] M. Herman, Sur la conjugaison differentiable des diffeomorphismes du circle a des rotations, Publ. I.H.E.S. 49 (1979) 5–234.
[11] M. Hirsch, C. Pugh and M. Shub, Invariant Manifolds, Lecture Notes in Mathematics 583 (Springer, Berlin, 1977).
[12] A. Libchaber and J. Maurer, A Rayleigh–Bénard Experiment: helium in a small box, Geilo NATO School, April 1981.
[13] J. Marsden and M. McCracken, The Hopf bifurcation and its applications (Springer, Berlin, 1976).
[14] S.E. Newhouse, The abundance of wild hyperbolic sets and non-smooth stable sets for diffeomorphisms, Publ. I.H.E.S. 50 (1979) 101–151.
[15] S.E. Newhouse, J. Palis, and F. Takens, Stable Arcs of Diffeomorphisms, Bull. Am. Math. Soc. 82 (1976) 499–502.
[16] Y. Pomeau and P. Manneville, Intermittent Transition to Turbulence in Dissipative Dynamical Systems, Comm. Math. Phys. 75 (1980) 189–197.
[17] D. Rand, Dynamics and symmetry, Predictions for modulated waves in rotating fluids, Arch. Rat. Mech. Anal., to appear.
[18] D. Ruelle and F. Takens, On the nature of turbulence, Comm. Math. Phys. 20 (1971) 167–192.
[19] L.P. Sil'nikov, A contribution to the problem of the structure of an extended neighborhood of a structurally stable equilibrium of saddle-focus type, Math. USSR Sb. 10 (1970) 91–102.
[20] S. Smale, Diffeomorphisms with many periodic points, in Differential and Combinatorial Topology (Princeton Univ. Press, Princeton, 1965) pp. 63–80.
[21] S. Smale, Differentiable dynamical systems, Bull. Am. Math. Soc. 73 (1967) 747–817.
[22] J. Sotomayor, Bifurcations of vector fields on two-dimensional manifolds, Publ. I.H.E.S. 43 (1973) 1–46.
[23] F. Takens, Singularities of Vector Fields, Publ. I.H.E.S. 43 (1973) 47–100.

ON THE ATTRACTING SET FOR DUFFING'S EQUATION

II. A GEOMETRICAL MODEL FOR MODERATE FORCE AND DAMPING

Philip HOLMES and David WHITLEY

Department of Theoretical and Applied Mechanics and Center for Applied Mathematics, Cornell University, Ithaca, NY 14853, USA

After a brief review of some earlier work on Duffing's equation in the small force and damping regions, we use the results of numerical integrations to construct a geometrically defined Poincaré map which captures the qualitative features of the attracting set of larger force and damping levels. This map has a (small) constant Jacobian determinant and can be regarded as a perturbation of a non-invertible one-dimensional map. We give a partial analysis of the map and pose some important open questions regarding perturbations of one-dimensional maps and the creation of "strange attractors" during bifurcation to horseshoes.

1. Introduction

In this paper we develop a geometrical model for the Poincaré map associated with Duffing's equation:

$$\ddot{x} + \delta\dot{x} - x + x^3 = \gamma \cos \omega t, \tag{1.1}$$

for moderate force level γ and dissipation δ. The excitation frequency does not play a prominent role in our analysis and henceforth we will assume $\omega = 1$. For information on applications and related work on the Duffing equation, and particularly for global analytical results in the small γ, δ regime, we refer the reader to the companion review paper by Holmes and Whitley [17], and to Holmes [15], Moon and Holmes [32], Holmes and Marsden [16] and Greenspan and Holmes [8, 9]. We remark that the methods developed in this paper can be applied to many problems in non-linear oscillations. For a rather different example (with non-constant Jacobian) see Levi's [24] work on the van der Pol equation.

The Poincaré map of (1.1) is a crucial tool in our analysis. Letting $u = x$, $v = \dot{x}$ and $\theta = t$, we convert (1.1) into an autonomous system with 2π-periodic phase space $(u, v; \theta) \in \mathbb{R}^2 \times S^1$:

$$\dot{u} = v,$$
$$\dot{v} = u - u^3 - \delta v + \gamma \cos \theta, \tag{1.2}$$
$$\dot{\theta} = 1.$$

We next take a cross section $\Sigma : \{(u, v, \theta) \mid \theta = 0\}$ and define the time 2π first return or *Poincaré map*:

$$P_\gamma : U \to \Sigma; \quad U \subseteq \Sigma \tag{1.3}$$

induced by the flow $\phi_t : \mathbb{R}^2 \times S^1 \to \mathbb{R}^2 \times S^1$ of (1.2). Letting $\phi_t(u, v, \theta)$ denote the solution based at (u, v, θ) at $t = 0$, we have

$$P_\gamma(u, v) = \pi \cdot \phi_{2\pi}(u, v, 0), \tag{1.4}$$

where π denotes projection onto the first factor. It is easily proved (Holmes [15], Holmes and Whitley [17]) that all solutions of (1.2) are bounded for all t, provided $\delta > 0$. Hence P_γ is globally defined.

Moreover, it is easy to show that P_γ preserves orientation (or solution curves of (1.2) would cross) and that it contracts areas uniformly at the rate $e^{-2\pi\delta}$, for

$$DP_\gamma = \exp(2\pi Df), \tag{1.5}$$

and

$$\det(DP_\gamma) = \exp(2\pi\, \mathrm{Tr}(Df)). \tag{1.6}$$

Here $\mathrm{Tr}\, Df = \partial \dot{u}/\partial u + \partial \dot{v}/\partial v + \partial \dot{\theta}/\partial \theta = -\delta$ (from (1.2)) measures the volume contraction of the flow. Of course, P_γ cannot be computed without solving the equation, and hence is only accessible via perturbation methods, as in the literature cited above, or numerical integrations and inspired guesswork, as in the present paper.

In what follows we shall regard the dissipation δ as a fixed positive constant and vary γ. Thus we denote the explicit parameter dependence of the Poincaré map: P_γ. Since the cross section Σ is isomorphic to the plane, we shall be dealing with families of orientation preserving diffeomorphisms of the plane with constant Jacobian.

In section 2 we outline previous (perturbation) results on the Duffing equation in the small γ, δ regime. In section 3 we give some numerical results for moderate γ, δ due to Ueda [40]. Then, in section 4 we develop a geometrical model for the Poincaré map and partially analyze it in section 5. The desire to complete this analysis leads us, in sections 6–7, to consider families of two-dimensional diffeomorphisms with small Jacobian which are close to one dimensional maps of the form

$$y \to f_\mu(y), \tag{1.7}$$

where $f_\mu: I \to R$ is defined on the unit interval and has a single critical point and negative Schwarzian derivative. Such one-dimensional maps are now quite well understood and, under suitable restrictions it is known that the set of parameter (μ) values for which (1.7) has a strange attractor is of positive measure (Jakobsen [20]). The one-dimensional theory carries over in a limited way to high iterates of two-dimensional maps such as the Duffing Poincaré map, and we discuss some positive results in this area and point out some important differences and problems.

The space available here prevents us from giving a full background to the differentiable dynamical and global analytical methods used in this paper. For general background we recommend the books by Chillingworth [3], the notes of Newhouse [35] and Bowen [2] and the forthcoming text by Guckenheimer and Holmes [12].

In particular, some familiarity with the horseshoe construction due to Smale [37, 38], cf. Moser [33], Newhouse [35], will be assumed in this paper. A nice elementary discussion of the horseshoe can be found in Chillingworth [3]. We feel that a proper understanding of this example is an essential prerequisite to attempts to understand chaos and strange attractors in dynamical systems.

The reader may also find it helpful to consult a number of earlier papers on the Duffing equation and variants of it, especially Holmes [15], Moon and Holmes [32], and Greenspan and Holmes [8, 9]. Andronov, Vitt end Khaiken [1] continues to provide the best background in nonlinear oscillations and planar systems.

2. The attracting set for small γ and δ

In Holmes and Whitley [17] (cf. Holmes [15]) it is shown that an open disc $D \subset \mathbb{R}^2$ can be chosen such that $\overline{P_\gamma(D)} \subset D$ (overbar denotes closure). We can therefore define the attracting set A_γ as

$$A_\gamma = \overline{\bigcap_{n \geq 0} P_\gamma^n(D)}. \tag{2.1}$$

Since $\det(DP_\gamma) = \mathrm{e}^{-2\pi\delta} < 1$, and the boundary of D is a simple closed curve, it follows that A_γ is a compact, invariant connected set with empty interior.

We suppose that δ is fixed positive, and sufficiently small, for validity of perturbations. Then, the work of Holmes [15], Greenspan and Holmes [8] and Holmes and Whitley [17], using Melnikov's [28] method; establishes the following:

Theorem 2.1. If γ and δ are sufficiently small and $\gamma > (4/3\sqrt{2}\pi) \cosh(\pi/2)\delta \approx 0.753\delta$, then
(i) $A_\gamma = \overline{W^u(p_{2m+1})}$, where $W^u(p_{2m+1})$ is the un-

stable manifold of a subharmonic ($2m + 1$-periodic cycle) of P_γ.

(ii) A_γ contains invariant hyperbolic sets Λ^N (Smale horseshoes) on which some iterate P_γ^N of P_γ is conjugate to a shift on two symbols.

(iii) A_γ contains finitely many stable periodic orbits, and hence is not indecomposable. The number of stable periodic orbits grows without bound as $\gamma, \delta \to 0$.

For pictures of this attracting set, see Greenspan and Holmes [8] and Holmes and Whitley [17].

The geometrical structure of A_γ is such that it contains orbits which circulate arbitrarily many times to the left or right in any sequence, but all such chaotic orbits are, alas, unstable and one expects to see stable subharmonic behavior after a period of transient chaos. Numerical integrations of (1.1) for small γ and δ support this conclusion. Thus A_γ *is not a strange attractor*, although the high period stable subharmonics may well be unobservable even on the most expensive computers.

There is no generally accepted definition of a strange attractor, except in the case of one-dimensional maps (cf. section 6). In this paper we therefore adopt the following:

Definition. A strange attractor is a closed, invariant attracting set which contains a dense orbit and hence is indecomposable, and in which solutions exhibit sensitive dependence on initial conditions: that is, almost all orbits in the attractor separate locally exponentially fast.

We will see that finding a dense orbit, or equivalently, getting rid of stable periodic orbits, is the hard part. We have seen that such stable orbits *always* exist for small γ and $\delta > 0$. We now will see what happens to them for larger γ and δ.

3. Numerical work

In this section we review some numerical results. We first point out that, for zero and small force γ, with $\delta > 0$ fixed and moderate, the Poincaré map P_γ is particularly simple. In fig. 1 we show results due to Ueda [40], who plotted segments of the stable and unstable manifolds of the saddle point $p_1 \approx (0, 0)$ which represents a continuation for $\gamma \neq 0$ of the original saddle at $(u, v) = (0, 0)$ of the unforced Duffing equation. For low γ (fig. 1a), the manifolds do not intersect and almost all solutions approach either one or other of the sinks near $(u, v) = (\pm 1, 0)$. Here it is clear that $A_\gamma = \overline{W^u(p_1)}$ and A_γ contains two sinks. As γ is increased, however, the stable and unstable manifolds approach, touch, and thereafter intersect transversely

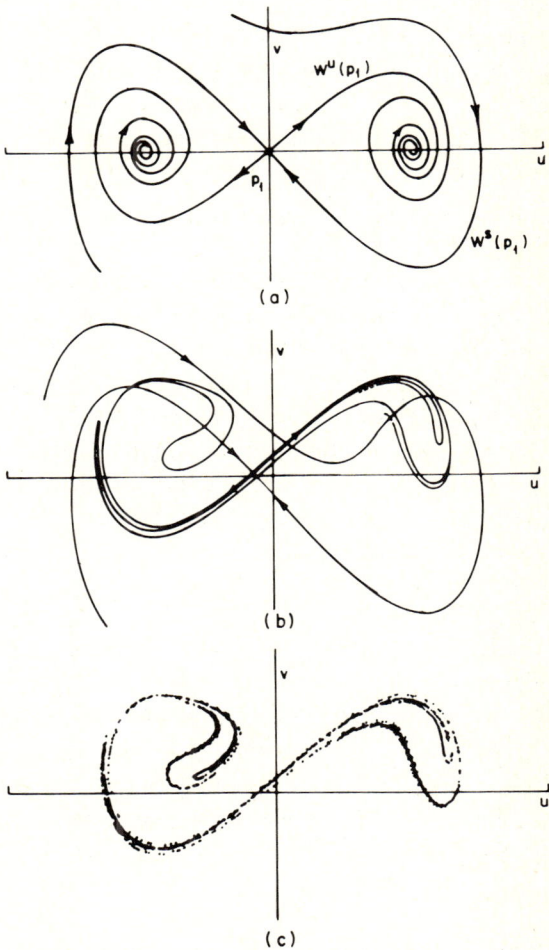

Fig. 1. Poincaré maps for Duffing's equation, $\delta = 0.25$, $\omega = 1.0$. (a), (b) invariant manifolds of p_1. (c) A single orbit.

in a (pair of) homoclinic orbits (fig. 1b). This leads to the horseshoes and 'unstable chaos' of theorem 2.1. In fig. 1c we show a typical single orbit $\{P^n(x)\}_{n=50}^{1000}$ for the same parameter values as 1b. The starting transient has been deleted. We note that this orbit appears to lie on a set equal to the closure of the unstable manifold $\overline{W^u(p_1)}$. This set is locally the product of a curve and a Cantor set (cf. Hénon [14])*. The set of parameter values for which such apparently chaotic motions are observed is relatively large, but careful work has revealed thin bands within it, in which stable motions of periods 3, 4, 5 and even 7 exist. However these orbits have different structures from the stable subharmonics found analytically for small γ and δ (Greenspan and Holmes [7]). The map developed below will share this type of behavior.

4. A geometrically defined attracting set

We now construct a geometrical attractor on the basis of the numerical work outlined above. A careful examination of the structure of manifolds of fig. 1b reveals the topological structure 'straightened out' and sketched in fig. 2a. Symmetry implies that the same structure exists in the left-hand half of the plane also. The transverse homoclinic points labelled q, r, s, t are mapped under P_γ to q^+, r^+, s^+, t^+ and under P_γ^{-1} to q^-, r^-, s^-, t^-. The point p_1 is the (fixed) saddle point. In fig. 2b we show three regions bounded by pieces of these invariant manifolds and their images under P_γ.

Those familiar with Smale's construction (Smale [37, 38], Moser [33], Chillingworth [3]) will clearly recognize a horseshoe in fig. 2b in the rectangle p_1qrq^+ and its preimage $p_1q^-r^-q$, but we still have not constructed an attracting set, since the image of the vertically shaded region falls outside the

---

* With some reservations: there must be points where the unstable manifold 'turns back on itself', almost everywhere in the attracting set, cf. Zeeman [42].

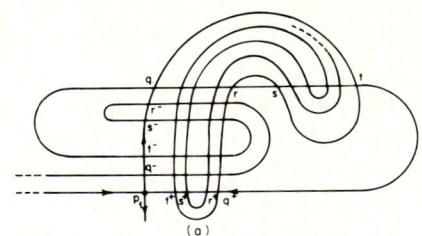

(a) The topological structure of the manifolds.

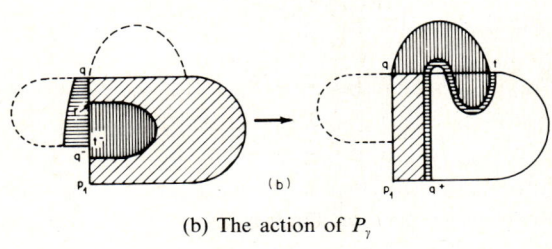

(b) The action of P_γ

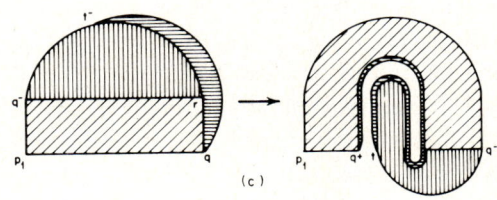

(c) An equivalent map

Fig. 2. Towards a geometric model of the attracting set (not to scale).

original three regions chosen. To construct a trapping region analogous to D of section 2, we enlarge our basic domain and once more 'straighten out' (cf. fig. 2c), so that the set $p_1q^-r^-q$ becomes a 'horizontal' rectangle. Taking a symmetric structure in the negative half-plane, we finally obtain the diffeomorphism of a (closed) disc D into its interior shown in fig. 3. To distinguish this from the true (incalculable) Poincaré map, we denote this geometrically defined map G_γ.

We now state various properties which we shall assume for the map G_γ:

(A1) The map has constant Jacobian determinant $\alpha(=e^{-2\pi\delta}) < 1$.

(A2) The eleven horizontal strips H_1–H_{11} are mapped into the corresponding vertical strips V_2, V_4, V_6, V_8, V_{10}, the 'buttons' V_1, V_{11}, and the arches A_3, A_5, A_7, A_9.

(A3) The map is linear hyperbolic (or at least its

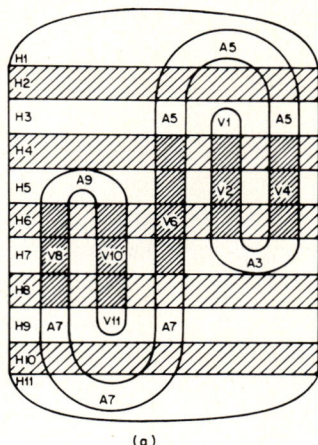

(a) The map G_γ

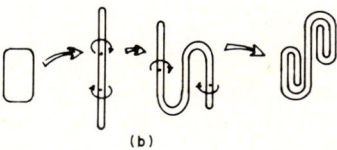

(b) Cartoon of the action of G_γ

Fig. 3. The geometrical attracting set.

expansion and contractions are nicely bounded, cf. Moser [33]) on the strips $H_2, H_4, H_6, H_8, H_{10}$.

We believe that careful numerical estimates on the Duffing equation would reveal that, for P_γ, regions and their images could be chosen essentially consistent with our model G_γ. Although P_γ would not, of course, be piecewise linear, it still seems likely that hyperbolicity estimates would be obtainable. We also note that, from Ueda's computations, we can deduce that an increase in the force level γ has the effect of increasing the hyperbolic expansion and contraction rates, so that the arches A_3 and A_7 are pulled down and A_5 and A_9 pulled up. Such parameter dependence will be important later.

We now turn to a partial analysis of the invariant attracting set for this map, which we denote

$$\Lambda_\gamma = \overline{\bigcap_{n=0}^{\infty} G_\gamma^n(D)}.$$

5. Shoes and sinks for G_γ

We first show explicitly that the invariant set Λ_γ of G_γ contains horseshoes. In particular, the three horizontal strips H_4, H_6 and H_8 form the basis of a Markov Partition (Bowen [2], Newhouse [35], Guckenheimer and Holmes [12]) and, since

$$G_\gamma(H_4) \cap H_4, H_6 \neq \emptyset,$$
$$G_\gamma(H_6) \cap H_4, H_6, H_8 \neq \emptyset,$$
$$G_\gamma(H_8) \cap H_6, H_8 \neq \emptyset,$$

we have the associated *transition matrix*

$$A = (a_{ij}) = \begin{matrix} & (4) & (6) & (8) \\ & \begin{bmatrix} 1 & 1 & 0 \\ 1 & 1 & 1 \\ 0 & 1 & 1 \end{bmatrix} & \begin{matrix} (4) \\ (6), \\ (8) \end{matrix} \end{matrix} \quad (5.1)$$

where $a_{ij} = 1$ if an orbit passing from i to j exists and $a_{ij} = 0$ if $G_\gamma(H_i) \cap H_j = \emptyset$. (In fact we need more than just non-empty intersections: the images of the strips H_j must "overlap" the H_j suitably, cf. Bowen [2].) Thus the invariant set Λ_γ of G_γ contains a subshift on the three symbols 4, 6, 8; the combinations 48 and 84 being disallowed. The general theory of symbolic dynamics permits us to conclude that, to any bi-infinite sequence containing the symbols 4, 6, 8, but not the pairs 48 or 84, there corresponds precisely one point of Λ_γ. Moreover, the orbit of this point visits the strips H_4, H_6 and H_8 in the order prescribed by the occurrence of symbols in the sequence. That is, if $a(x) = \{a_j\}_{j=-\infty}^{\infty}$ is the symbol sequence for the point x, that $a_j = 4$, 6 or 8 depending on whether $G_\gamma^j(x) \in H_4, H_6$ or H_8, respectively.

Thus the symbolic dynamics shows that there are three fixed points ...444.444...; ... ...666.666... and ...888.888...; two orbits of period two (...46.46...; ...68.68...) and so forth. In general G_γ^k has $N = \text{Tr}(A^k)$ fixed points. All the periodic orbits are of saddle type, since on $H_4 \cup H_6 \cup H_8$, G_γ expands vertically and contracts horizontally. We remark that the 'central' fixed point ...666.666... is just the original saddle of Duffing's equation near $(u, v) = (0, 0)$. In addition to this countable set of periodic orbits, there is an uncountable set of orbits which are not asymptotically periodic and which correspond to symbol sequences with positive or negative (infinite) nonperiodic tails. Thus, chaotic orbits visiting the left (V_8) and the right (V_4) halves of the disc in *any* prespecified order exist. This is in essence part of the result we had in theorem 2.1. except that the numerical evidence has permitted us to conclude that such orbits exist for the map G_γ itself- and not merely for some (perhaps high) iterate.

Examination of fig. 3 reveals that regions H_4, H_6 and H_8 are the only ones intersected by their images ($G(H_j)$), and so to construct more of the invariant set Λ_γ we must go on to look at G_γ^2, G_γ^3, etc. We will now sketch a little of this analysis. First consider G_γ^2 (H_2) (or, equivalently, by the symmetry, $G_\gamma^2(H_{10})$). In fig. 4 we show the two successive images, indicating that there are points of H_2 which return to H_2 after two iterates. Once more a Markov partition can be chosen, based on the horizontal substrips H_{21}, $H_{22} \subset H_2$ with images $V_{21} = G_\gamma^2(H_{21})$, $V_{22} = G_\gamma^2(H_{22}) \subset H_{22}$, and we deduce that $G^2|_{H_2}$ has a shift on two symbols. However, the nonlinearity of the map of H_5 needs to be specified more precisely before we can conclude that the invariant set is hyperbolic.

In a similar fashion we can find horseshoe-like (hyperbolic) subshifts of finite type for higher iterates of G_γ, by restricting our attention to orbits which do not contain points near the peaks of the arches A_3, A_5, A_7 or A_9. For the specific example G_γ of fig. 3, G_γ^5 turns out to be the lowest iterate for which orbits can pass near these peaks and return to their starting points. We now discuss the implications of this.

We consider an orbit passing through the following regions: $..H_9 \rightarrow H_5 \rightarrow H_3 \rightarrow H_7 \rightarrow H_{11} \rightarrow H_9 \rightarrow ...$ (It has a symmetric partner $H_3 \rightarrow H_7 \rightarrow H_9 \rightarrow H_5$ $H_1 \rightarrow H_3 \rightarrow ...$). A happy morning's drawing should convince the reader that a thin horizontal substrip $H \subset H_{11}$ can be chosen such that $G_\gamma^5(H)$ lies in H_{11} as shown in fig. 5 for three force levels γ. (Recall that, as γ is increased, the images $A_9 = G_\gamma(H_9)$ and $A_3 = G_\gamma(H_3)$ move up and down respectively). Thus, as γ varies, we create a full shift on two symbols – a horseshoe – for G_γ^5. As Newhouse noted [34, 35], in this situation we necessarily create stable sinks (of period 5), cf. fig. 5b.

This construction, and analogues of it, show how stable periodic orbits can "suddenly spring out of chaos" as a parameter is varied in the Duffing equation. However, we note that motions such as these are not directly related to the weakly dissipative "regular" subharmonics found in earlier analyses (section 2 and Greenspan and Holmes [8], Holmes and Whitley [17]). In that case we note that we had to take γ and δ smaller the higher the order of the subharmonic, so that averaging remained valid. In fact a careful examination of the order of the terms involved shows that in studying

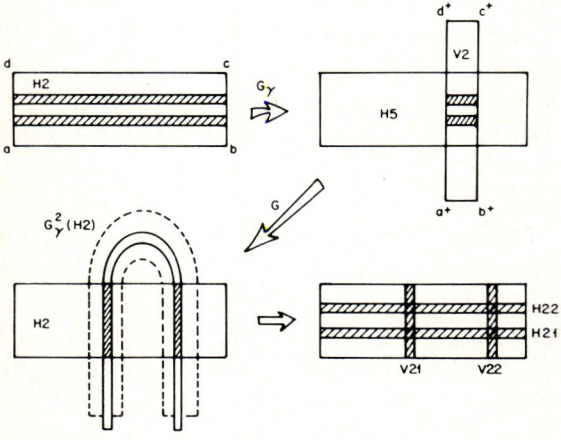

Fig. 4. $G_\gamma^2|_{H_2}$, showing an orbit with sequence ...2525....

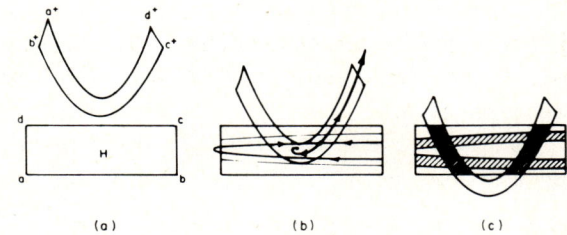

Fig. 5. $G_\gamma^5|_{H_{11}}$ The creation of a 5-shoe $a, b, c, d \rightarrow a^+, b^+, c^+, d^+$: (a) γ small (b) γ medium (c) γ high.

P_γ^N by perturbation methods we must take $\det(P_\gamma^N) \approx \exp(-2\pi N^6 \exp(-4\pi N)) \to 1$ as $N \to \infty$. In studying the geometrically defined map G_γ, in contrast, we fix $\det(G_\gamma) = \alpha$ and thus $\det(G_\gamma^N) = \alpha^N \to 0$ as $N \to \infty$. For the particular values of Ueda's computations, on which our geometrical construction is based we have

$$\det DP_\gamma = \exp\left(-\frac{2\pi \times 0.25}{1}\right) \approx 0.208, \quad (5.2)$$

and

$$\det(DP_\gamma^5) \approx 0.00039. \quad (5.3)$$

The two analyses can therefore be expected to yield different (and complementary) results.

Examination of higher iterates G_γ^N show that the qualitative picture of fig. 5 occurs over and over again in various regions of the disc D as γ is varied. For example, the reader might like to imagine γ decreased so that the peaks of the arches A_5 and A_7 of fig. 3 lie in regions H_2 and H_{10} respectively. In this case, stable orbits of period three can be constructed which pass through regions $H_2 \to H_4 \to H_5$ and $H_{10} \to H_8 \to H_7$, respectively. Such orbits correspond to numerically obtained period three subharmonics (Ueda [40]).

We end this section by noting that it follows from our geometrical construction that the attracting set Λ_γ is the closure of the unstable manifold of the stable point ...666.666... in $H_6 \cap V_6$. This follows in much the same way as the usual horseshoe construction, where it is shown that the invariant set lies in the closure of any of the unstable manifolds of saddle points within it (cf. Holmes and Whitley [18]).

Thus we have a similar structure to that proven to exist for the Duffing equation in theorem 2.1, with the added advantage that the features of iterates of G_γ relevant to the creation of periodic orbits and horseshoes can be understood by the analysis of small perturbations of one-dimensional maps. Before going on to sketch such analyses, we outline how an iterate G_γ^N such as that of fig. 5 can be regarded as a small perturbation of such a noninvertible map.

Consider the family of maps

$$F_{(\epsilon,\gamma)}(x, y) = (y, -\epsilon x + f_\gamma(y)), \quad (5.4)$$

where $f_\gamma: I \to \mathbb{R}$ is a one-dimensional map with a single critical point. A more precise formulation is provided in section 6. For $\epsilon > 0$ (5.4) is a diffeomorphism while for $\epsilon = 0$ it is non-invertible, the image of all points in $\mathbb{R}^2$ lying on the curve $y = f_\gamma(x)$. If we pick the family f_γ appropriately, then $F_{\epsilon,\gamma}$ behaves as shown in fig. 5, and as $\epsilon \to 0$ the image of the rectangle $abcd$ shrinks in area to a (parabolic) arch, the points a^+, b^+ and c^+, d^+ coalescing. Thus we expect the features of $F_{\epsilon,\gamma}$, $\epsilon > 0$ small, to be well approximated in some sense by those of $F_{0,\gamma}$ and hence of the one dimensional map f_γ. In the final sections we will review some of the many results on maps of the interval and describe recent work on perturbations of such maps in the form of diffeomorphisms with small Jacobian determinant, like (5.4). This will enable us to provide a partial description of the bifurcation sequence in which sinks, saddles and ultimately horseshoes are created for iterates G_γ^N as γ increases.

6. Maps of the interval

In this section we describe and illustrate some of the many results on one-parameter families of maps of the interval. We will work with the class $\mathscr{C}$ of C^3 maps $f: I \to I$ of a closed interval $I = [a, b] \subset \mathbb{R}$, containing the origin, which satisfy the following properties:

(1) $f(a) = f(b) = a$;

(2) f has a single critical point, which we assume to lie at the origin $x = 0$, and $f''(0) < 0$ so that this critical point is a maximum; and

(3) the Schwarzian derivative,

$$Sf(x) = \frac{f'''(x)}{f'(x)} - \frac{3}{2}\left[\frac{f''(x)}{f'(x)}\right]^2, \quad (6.1)$$

is negative on $I \setminus \{0\}$.

Condition (3) is included since for maps f with negative Schwarzian derivative a result due to Singer [36] (see also Misiurewicz [30]) tells us that if p is a stable periodic point of f, then there is either a critical point of f or an endpoint of I whose ω-limit set is the orbit of p. If we further assume within the class $\mathscr{C}$ that, either the fixed point $x = a$ at the left-hand end of the interval is unstable, or that $\omega(0) = \{a\}$ then our maps will have at most one stable periodic orbit. In this case Singer's theorem implies that, if the orbit of the critical point does not tend to a stable periodic orbit, then f has no stable periodic orbit.

Since the appearance of Li and Yorke's [25] 'period 3 implies chaos' result, itself a corollary of an earlier theorem of Sarkovskii (see Stefan [39]), there has been much interest in the dynamics of one-dimensional maps. For maps in $\mathscr{C}$ we now have a topological classification (Guckenheimer [11]) and a decomposition of the nonwandering set (Jonker and Rand [21], van Strien [41]), both largely based on the unpublished 'kneading theory' of Milnor and Thurston [29]. Apart from these topological results, Feigenbaum [6] has discovered universal metric properties, and Jakobsen [19, 20] has important theorems on the existence of invariant measures and resulting chaotic dynamics in families of maps. For more general introductory material we refer the reader to the monograph by Collet and Eckmann [4].

Here our main concern is to outline some of the properties of one-parameter families of maps in $\mathscr{C}$. We consider families f_γ, depending continuously on a real parameter γ, which are *full* in the sense that there are parameter values $\gamma_1^s < \gamma_1^h$ so that:

(a) $f_\gamma \in \mathscr{C}$ for $\gamma \in [\gamma_1^s, \gamma_1^h]$;

(b) the nonwandering set $\Omega_{\gamma_1^s}$ of $f_{\gamma_1^s}$ consists of a single fixed point;

(c) for $\gamma > \gamma_1^h$, f_γ satisfies conditions (1)–(3) above and maps I onto (but not into) itself; in which case one may show (cf. van Strien [41]) that $f_\gamma | \Omega_\gamma$ is conjugate to a full (one-sided) shift on two symbols. Ω_γ is a one-dimensional version of the standard Smale horseshoe.

In general the interval I (i.e. its endpoints) will vary with the parameter as in the archetypal example

$$x \to \gamma - x^2; \qquad \gamma_1^s = -\tfrac{1}{4}, \quad \gamma_1^h = 2. \qquad (6.2)$$

Graphs of a typical family f_γ for various values of γ are shown below.

We will now describe the major features of the bifurcation set in the parameter range $\gamma_1^s \leq \gamma \leq \gamma_1^h$. We assume that, as in the quadratic family (6.2), f_γ has a fold or saddle-node bifurcation at $\gamma = \gamma_1^s$ which creates two fixed points, one stable and one unstable, which exist for $\gamma > \gamma_1^s$, and that γ_1^h is the least γ-value such that $f_\gamma(0) = b$. Let $\gamma_1^{ss} = \sup\{\gamma \, | \, f_\gamma(0) \leq 0\}$ and $\gamma_2^s = \sup\{\gamma \, | \, f_\gamma(p) = p$ and $f'(p) = -1\}$ where $p = p_\gamma$ is the unique positive fixed point of f_γ. Clearly $\gamma_1^{ss} \in (\gamma_1^s, \gamma_1^h)$ and $\gamma_2^s \in (\gamma_1^{ss}, \gamma_1^h)$.

Now one can find parameters $\gamma > \gamma_2^s$ so that $f_\gamma^2(0) > 0$ and, since $f_{\gamma_1^h}^2(0) = a$, it follows by continuity that there is at least one γ with $f_\gamma^2(0) = p' \in (a, 0)$, where $f_\gamma(p') = f_\gamma(p)$. Let γ_2^h be the infimum of such γ's. Then $\gamma_1^s < \gamma_2^s < \gamma_2^h < \gamma_1^h$ and for $\gamma \in [\gamma_2^s, \gamma_2^h]$ f_γ^2 maps the interval $[p', p]$ into itself. In fact f_γ maps $[p', p]$ onto $[p, f_\gamma(0)]$ and $[p, f_\gamma(0)]$ into $[p', p]$ so that, since all points in $I \setminus \{a, b\}$ eventually map into $[p', p]$, all periodic points of f_γ, $\gamma \in [\gamma_2^s, \gamma_2^h]$, except the fixed points, have even periods. Moreover, for this parameter range, $f_\gamma^2 | [p', p]$ is a full family and behaves just like $f_{\gamma|_I}$ 'in miniature'. Repeating this argument inductively we have (cf. Collet, Eckmann and Lanford [5]):

Proposition 6.1. There are two sequences of parameters $\gamma_{2^n}^s$, $\gamma_{2^n}^h$ with $\gamma_1^s < \gamma_2^s < \gamma_4^s < \cdots < \gamma_{2^n}^s < \cdots < \gamma_{2^n}^h < \cdots < \gamma_4^h < \gamma_2^h < \gamma_1^h$ so that, for $n \geq 1$:

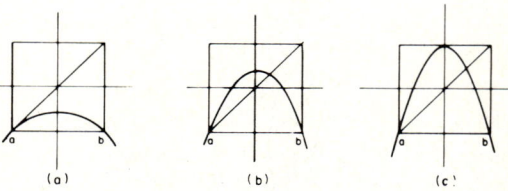

Fig. 6. Three graphs of f_γ.

(1) $f^s_{\gamma 2^n}$ has a point of period 2^{n-1} with eigenvalue $\lambda = -1$;

(2) $f^{2^{n+1}}_{\gamma^h_{2^n}}(0)$ is an unstable point of period 2^{n-1};

(3) If $\gamma \in [\gamma^s_{2^n}, \gamma^h_{2^n}]$ there are 2^n closed subintervals $J_1, \ldots, J_{2^n}$ with $0 \in J, f^{2^n}_\gamma | J_1 \in \mathscr{C}$, and the nonwandering set $\Omega(f_\gamma)$ is $\Omega(f_\gamma | \cup^{2^n}_{i=1} J_i)$ together with an unstable orbit of period 2^k, one for each $k = 0, 1, \ldots, 2^{n-2}$, plus the fixed point a.

Thus we see that the bifurcation set is as indicated in fig. 7. There is a fold bifurcation at γ^s_1 creating one stable and one unstable fixed point. The stable fixed point loses stability at γ^s_2 where a stable orbit of period 2 is created through a flip bifurcation, and this process repeats as γ increases: at $\gamma^s_{2^n}$ an orbit of period 2^{n-1} flips to period 2^n. At $\gamma = \gamma^h_{2^n}$, $f^{2^n}_\gamma$ maps the subinterval J_1 exactly onto itself with a single fold at the origin, and $f_{\gamma^h_{2^n}}$ is conjugate to the piecewise linear map

$$x \to s - 1 - s|x|; \quad s = 2^{1/2^n}. \tag{6.3}$$

For a class of maps including the quadratic family (6.2) Jakobsen [19] has shown that the $f_{\gamma^h_{2^n}}$'s have absolutely continuous, ergodic invariant measures supported on the subintervals J_i.

In the interval $[\gamma^s_{2^n}, \gamma^h_{2^n}]$ all periodic orbits except those involved in the initial period-doubling sequence have periods which are multiples of 2^n. For families with a quadratic maximum satisfying a certain transversality condition, of which (6.2) is an example, (Lanford [22]), the sequences $\gamma^s_{2^n}$ and $\gamma^h_{2^n}$ converge, from below and above respectively, on a parameter γ^F_1. Furthermore, if $l_n = \gamma^h_{2^n} - \gamma^s_{2^n}$ denotes the length of the nth nested 'box' in the parameter space, then these lengths converge to zero at Feigenbaum's [6] universal rate:

$$\lim_{n \to \infty} \frac{l_n - l_{n+1}}{l_{n+1} - l_{n+2}} = 4.669 \ldots, \tag{6.4}$$

independently of the particular family.

This view of the bifurcation set shows clearly the central role of the parameter value γ^F_1. For $\gamma < \gamma^F_1$ the nonwandering set consists of finitely many periodic orbits each with period a power of 2, while when γ crosses γ^F_1 there is an explosion into a region of complicated, *although not necessarily always chaotic*, dynamics. The map $f_{\gamma^F_1}$ itself, which could be thought of as an 'organising centre' for the whole family, falls into the class of maps with zero topological entropy studied by Misiurewicz [30]. Its nonwandering set consists of the fixed point a in the boundary of I, one (unstable) orbit of period 2^k for each $k \geq 0$ and an invariant minimal set homeomorphic to a Cantor set. Almost all points in I are attracted to the Cantor set, but do not display sensitive dependence upon initial conditions.

Embedded in the large box $[\gamma^s_1, \gamma^h_1]$ between each pair $\gamma^h_{2^{n+1}}, \gamma^h_{2^n}$ are other 'boxes' corresponding to periods of the form $k \cdot 2^n$, in which $f^{k \cdot 2^n}_\gamma$ is a full family on some subinterval of I. The internal structure of these boxes is exactly the same as the outer one. (The embedded box idea was first stressed by Mira, cf. Gumowski and Mira [13].) For example, in (γ^h_2, γ^h_1) there is a 3-box $[\gamma^s_3, \gamma^h_3]$. At γ^s_3, f_γ has a saddle-node of period 3 and at $\gamma^h_3 f^3_\gamma$ maps a subinterval exactly onto itself with a single fold. There are sequences $\gamma^s_{3 \cdot 2^n}$ and $\gamma^h_{3 \cdot 2^n}$ for which at $\gamma^s_{3 \cdot 2^n}$, f_γ has a flip bifurcation from period $3 \cdot 2^{n-1}$ to period $3 \cdot 2^n$ and $f^{3 \cdot (2^{n+1})}_{\gamma^h_{3 \cdot 2^n}}(0)$ is a point of period $3 \cdot 2^{n-1}$, exactly analogous to the behavior of f_γ for $\gamma \in [\gamma^s_1, \gamma^h_1]$.

The position of the interior boxes is restricted by the form of the periods allowed in each box $[\gamma^s_{2^n}, \gamma^h_{2^n}]$, and by Šarkovskii's theorem on the ordering of orbits (Stefan [39], cf. Guckenheimer [10]). Also, we know that boxes corresponding to periods $k \cdot 2^n$, k odd, accumulate on $\gamma^h_{2^{n+1}}$ from above. This is because, for $\gamma > \gamma^h_{2^{n+1}}$ the periodic orbit which at $\gamma^h_{2^{n+1}}$ contains the point $f^{2^{n+1}+1}_{\gamma^h_{2^{n+1}}}(0)$ is a snap-back repellor for f^{2^n} (Marotto [26], Marotto

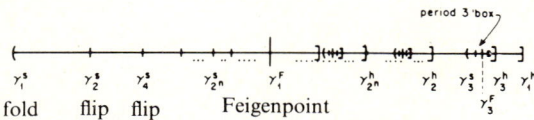

Fig. 7. The bifurcation set of f_γ.

and Rogers [27]). This means that, for $\gamma > \gamma_{2^n+1}^h$, f^{2^n} has periodic points of all periods greater than some period N. In particular f^{2^n} has periodic points with odd periods, i.e. f has periodic points with periods of the form odd $\times 2^n$. These periodic orbits do not exist when $\gamma < \gamma_{2^n+1}^h$, since then all periods have the form 2^k or $k \cdot 2^{n+1}$. Thus $\gamma_{2^n+1}^h$ is an accumulation point of boxes with associated periods odd $\times 2^n$.

The parameter γ_1^F is then seen to be an accumulation point of accumulation points and it is γ_1^F which divides the parameter space into those maps with simple dynamics ($\gamma < \gamma_1^F$) from those which possibly have chaotic behavior ($\gamma > \gamma_1^F$). We emphasize however that not all f_γ with $\gamma > \gamma_1^F$ are chaotic. For many parameter values (most likely for an open, dense set, though this is not proven) f_γ has a stable periodic orbit and in this case an open, dense set of points in I, with full Lebesgue measure, are attracted to the stable orbit (van Strien [41]). There may be complicated transient behavior but we do not consider this to be 'chaos'.

Even if the map has no stable periodic orbit it may not display the sensitive dependence on initial conditions (Guckenheimer [11]) that we feel a 'strange' attractor should possess. (This is the case, for example, when $\gamma = \gamma_1^F$). However, Jakobsen [20] has shown that in the quadratic family (6.2) there is a set Γ of parameter values of positive Lebesgue measure for which f_γ has an absolutely continuous, ergodic invariant measure. These maps exhibit sensitive dependence, have indecomposable attractors, and may be considered truly chaotic. This set Γ contains the countable set $\{\gamma_{k \cdot 2^n}^h\}_{k,n=0}^\infty$ for which $f^m(0)$ falls on an unstable periodic point, and also an uncountable set (Misiurewicz [30, 31]) for which $f^m(0)$ falls on an unstable Cantor set whose closure does not contain 0. Both these sets are of zero measure, and Jakobsen's set involves maps for which the orbit of critical point returns arbitrarily closely to the critical point.

7. Diffeomorphisms of $\mathbb{R}^2$ near maps of the interval

In this penultimate section we consider diffeomorphisms with small Jacobian determinant of the form

$$F_{(\epsilon,\gamma)}(x,y) = (y, -\epsilon x + f_\gamma(y)), \quad \epsilon \ll 1, \qquad (7.1)$$

where f_γ is a one-dimensional map of the type discussed in the previous section. When $\epsilon = 0$, $F_{(0,\gamma)}$ maps $\mathbb{R}^2$ onto the parabola $y = f_\gamma(x)$ which is the graph of f_γ, and the resulting dynamics on the invariant parabola are identical to those of f_γ. We describe two theorems of van Strien [41] which indicate both the similarities and important differences in the case when $\epsilon \neq 0$. Our proof of the second theorem differs from van Strien's and is, we feel, more direct. We note that the diffeomorphism studied by Hénon [14] can be put in the form (7.1).

First we show that for some open sets of γ-values the behavior of $F_{(0,\gamma)}$ persists for $F_{(\epsilon,\gamma)}$, ϵ small. Let $\tilde{P}$ be the interior of the set of parameter values for which f_γ has an attracting periodic orbit, and let P be a component of $\tilde{P}$. For $\delta > 0$, define $P(\delta) = \{\gamma \mid (\gamma - \delta, \gamma + \delta) \in P\}$ so that $P(\delta)$ is a slightly smaller set than P.

Two one-parameter families of maps ϕ_γ, ψ_γ of $\mathbb{R}^2$ are Ω-semiconjugate on Q if there is a homeomorphism $\rho: Q \to Q$ and a family of continuous surjections $h_\gamma: \Omega(\phi_\gamma) \to \Omega(\psi_\gamma)$ so that $\psi_{\rho(\gamma)} \circ h_\gamma = h_\gamma \circ \phi_\gamma$ for $\gamma \in Q$, and so that h_γ is a bijection from Per(ϕ_γ), the set of periodic points of ϕ_γ, to Per(ψ_γ). We say that ϕ_γ and ψ_γ are Ω-conjugate on Q if h_γ is a homeomorphism.

We will assume that the family f_γ has generic flip bifurcations, i.e. if f_{γ^*} has a periodic point p with period k and eigenvalue -1 then $d(df_\gamma^k(p)/dx)/d\mu \big|_{\gamma=\gamma^*} \neq 0$. We remark that the second condition required for a generic flip, $d^3(f_{\gamma^*}^{2k}(p))/dx^3 \neq 0$, is automatically satisfied since our maps have negative Schwarzian derivative.

Van Strien proves the following:

Theorem 7.1. For $\epsilon \geq 0$ there is a $\delta(\epsilon)$ depending continuously on ϵ, with $\delta(0) = 0$, so that

(1) $\forall \epsilon > 0$, $F_{(\epsilon,\gamma)}$ is Ω-semi conjugate to $F_{(0,\gamma)}$ on $P(\delta(\epsilon))$;

(2) If $\epsilon_1, \epsilon_2 \neq 0$ and $\epsilon = \max\{\epsilon_1, \epsilon_2\}$ then $F_{(\epsilon_1,\gamma)}$ and $F_{(\epsilon_2,\gamma)}$ are Ω-conjugate on $P(\delta(\epsilon))$.

Essentially this says that for γ values such that f_γ is in a period doubling sequence, the behavior of $F_{(0,\gamma)}$ persists for $F_{(\epsilon,\gamma)}$, ϵ small, on a possible smaller γ-range. It follows from the implicit function theorem that the bifurcation points in the period doubling sequences of f_γ extend to sets of 'parallel' curves in the ϵ–γ plane which are bifurcation curves for period doubling sequences of $F_{(\epsilon,\gamma)}$. Lanford [23] has remarked that this result can be made uniform in ϵ by using the universality ideas carefully as one approaches γ_1^F. In fact for fixed $\epsilon < 1$ the flip bifurcations accumulate at the same asymptotic rate $\delta = 4.669\ldots$ as in the one-dimensional case.

At other points, however, the dynamics of $F_{(\epsilon,\gamma)}$ are radically different from those of $F_{(0,\gamma)}$:

Theorem 7.2. Suppose there is a parameter γ_* so that

(1) For γ near γ_*, f has a periodic point $p(\gamma)$ of period k with $|f_\gamma^k(p(\gamma))| > 1$;

(2) there is an integer n with $f_{\gamma_*}^n(0) = p(\gamma_*)$,

(3) $\dfrac{\partial}{\partial \mu}(f_\mu^n(c) - p(\gamma))\big|_{\gamma = \gamma_*} \neq 0$.

Then there is $\epsilon_0 > 0$ so that for each ϵ with $0 < |\epsilon| < \epsilon_0$

(a) For γ near γ_*, $F_{(\epsilon,\gamma)}$ has a periodic saddle $p(\epsilon, \gamma)$ of period k with $p(0, \gamma) = p(\gamma)$;

(b) there is a continuous curve $\gamma(\epsilon)$, with $\gamma(0) = \gamma_*$, so that the stable and unstable manifolds of $p(\epsilon, \gamma(\epsilon))$ have a tangency.

We outline the proof of this result (cf. van Strien [41]) which, together with the results of Newhouse [34, 35], implies that close to maps $F_{(\epsilon,\gamma)}$ satisfying the conditions of the theorem there are diffeomorphisms of $\mathbb{R}^2$ which have infinitely many sinks. For example, we can take $\gamma_* = \gamma_{k \cdot 2^n}^h$; $k, n \in \mathbb{Z}$. In contrast, the maps $F_{(0,\gamma)}$ can, as we remarked in section 6, have at most one sink. More details will appear in Holmes and Whitley [18].

Conclusion (a) of the theorem is an immediate consequence of the implicit function theorem. For (b), we define $g_\epsilon : I \to \mathbb{R}^2$ by $g_\epsilon(\gamma) = F_{(\epsilon,\gamma)}^n(0, f_\gamma(0))$ and let $M_\epsilon = \{(\gamma, W^s(p(\epsilon, \gamma))\} \subset \mathbb{R}^3$. Hypothesis (3) says that g_0 is transverse to M at some point $(\gamma_*(\epsilon), p(\epsilon, \gamma_*)))$ near $(\gamma_*, p(0, \gamma_*))$. We then find small open balls $B_r(g_\epsilon(\gamma_1))$ and $B_r(g_\epsilon(\gamma_2))$, $\gamma_1 < \gamma_* < \gamma_2$, whose radii are independent of ϵ, and which lie on opposite sides of $W^s(p(\epsilon, \gamma))$ for all sufficiently small ϵ.

Since $F_{(0,\gamma_*)}^n(0, f_{\gamma_*}(0)) = p(0, \gamma_*)$, for any $\delta > 0$ we can choose ϵ small enough so that $B_\delta(0, f_\gamma(0))$ contains part of the unstable manifold $W^u(p(\epsilon, \gamma))$ for $\gamma \in (\gamma_1, \gamma_2)$. The continuity of F^{-1} ensures that δ may be chosen so that

$$F^n(B_\delta(0, f_\gamma(0))) \subset B_r(g_\epsilon(\gamma)).$$

Then for $\gamma \in (\gamma_1, \gamma_2)$, $B_r(g_\epsilon(\gamma))$ contains a piece of $W^u(p(\epsilon, \gamma))$. Varying γ from γ_1 to γ_2 this ball crosses $W^s(p(\epsilon, \gamma))$ and so must part of the unstable manifold.

As an illustration of the theorem consider the case $\gamma = \gamma_2^h$ where $f_\gamma^3(0)$ is the unstable fixed point in the interior of I. Conditions (1) and (2) of (9.2) are satisfied and (3) is easily checked in any given family (e.g. (8.1)). At a point of tangency (the 'first') of the stable and unstable manifolds of $p(\epsilon, \gamma(\epsilon))$ these manifolds are arranged as shown in fig. 8.

Note that there will be more than one tangency for $F_{(\epsilon,\gamma)}$ as γ passes through γ_2^h with ϵ small, in fact there will be infinitely many. This follows because when $\epsilon = 0$ the unstable manifold of p collapses onto the parabola $y = f_\gamma(x)$ and the points at the ends of the loops in $W^u(p)$ (such as s, t in fig. 8) will coincide. Thus at $\epsilon = 0$ there are in effect countably many (coincident) tangencies occurring between $W^u(p)$ and $W^s(p)$. When $\epsilon \neq 0$ these tangencies separate into fans of curves through $\epsilon = 0$, $\gamma = \gamma_*$, each curve representing a tangency. More-

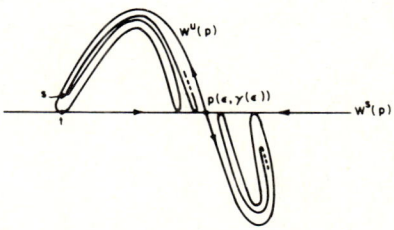

Fig. 8. Homoclinic Tangency.

over, since higher order invariant hyperbolic sets Λ^j created in preceding fans already exist, each fan contains a further *uncountable* set of curves radiating from γ_* on which we have tangencies between invariant manifolds $W^u(x)$, $W^s(y)$ of arbitrary pairs of points x, $y \in \Lambda^j$. This leads to the picture of the ϵ–γ plane of fig. 9. There are further "parallel" bands of curves representing period doubling sequences along with fans of homoclinic tangencies. Each tangency is preceded and followed by sequences of saddle-node and period doubling bifurcations (Gavrilov and Silnikov [7]). However, it appears that all these periodic points have their counterparts in the one-dimensional map and this lends weight to the conjecture of van Strien [41] that there are no more periodic orbits for $F_{(\epsilon,\gamma)}$, $\epsilon \neq 0$ than there are for $F_{(0,\gamma)}$; they are simply created in a different order.

These results show that, while the bifurcation sequence during the creation of the horseshoe for a strongly dissipative two-dimensional map is grossly the same as that for the one-dimensional family sketched in fig. 7, the details differ dramatically. Specifically, for an orbit of period $\leq N$ (fixed), we can choose ϵ sufficiently small to guarantee the 'correct' (i.e. Sarkovskii's) order of bifurcation, but for any $\epsilon > 0$ we can find orbits of sufficiently high period M which come 'out of order' and such that more than one stable orbit can coexist.

8. Conclusions and some open problems

We leave the reader to draw his own conclusions on precisely how the results of sections 6–7 should be applied to G_γ, and end with a conjecture based on the following ideas. Recall that, for f_γ, the points $\gamma^h_{k \cdot 2^n}$ correspond to strange attractors. In contrast, in the fans above these points for $\epsilon \neq 0$ we expect to find residual subsets of open sets of parameter values for which $F_{(\epsilon,\gamma)}$ has countably many sinks (Newhouse [34, 35]). However, the 'windows' in which these sinks are stable are extremly short and it is still possible and even likely, that there is a set of parameter values of positive measure for which $F_{(\epsilon,\gamma)}$ has no stable sinks. We conjecture that this is the case, but note that, as in Jakobsen's work, a proof (or disproof) of this conjecture will require careful estimates of the derivatives along orbits which continually reenter a small neighborhood of the 'bends' in the manifolds of fig. 8.

Conjecture 8.1. For ϵ sufficiently small there is a set $\Gamma_\epsilon \in (\gamma^F_1, \gamma^h_1)$ of positive Lebesgue measure such that for $\gamma \in \Gamma_\epsilon$, $F_{(\epsilon,\gamma)}$ has a strange attractor.

To cut our possible losses, we remark that, even if this conjecture is untrue, and all the Jakobsen points vanish when $\epsilon \neq 0$, the stable periodic orbits remaining are of such high periods that they are effectively unobservable and one would see a pseudo-strange attractor.

Acknowledgements

This work was partially supported by NSF grant MEA-8017570. The first author would like to thank the organizing committee of 'Order in Chaos' for inviting him to speak at that meeting and thus providing a stimulus for 1) some of this work and 2) writing it up.

References

[1] A.A. Andronov, E.A. Vitt and S.E. Khaiken, Theory of Oscillators, (Pergamon, Oxford, 1966).
[2] R. Bowen, 'On Axiom A Diffeomorphisms', Amer. Math. Soc. Regional Conference Series in Math #35 (1978).
[3] D.R.J. Chillingworth, Differential Topology with a View to Applications (Pitman, London, 1976).

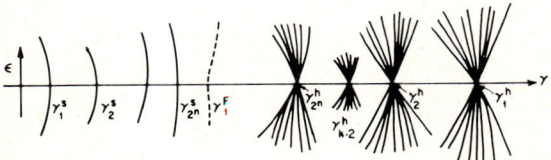

Fig. 9. The bifurcation set of $F_{(\epsilon,\gamma)}$.

[4] P. Collet and J.P. Eckmann, Iterated Maps on the Interval as Dynamical Systems (Birkhauser, Boston, 1980).
[5] P. Collet, J.P. Eckmann and O.E. Lanford, Universal Properties of Maps on an Interval, Comm. Math. Phys. 76 (1980) 211–254.
[6] M. Feigenbaum, Qualitative Universality for a Class of Nonlinear Transformations., J. Stat. Phys. 19 (1978) 25–52.
[7] N.K. Gavrilov and L.P. Silnikov, On Three-Dimensional Dynamical Systems Close to Systems with a Structurally Unstable Homoclinic Curve I and II, Math. USSR Sbornik. 88 (1972) 467–485; 90 (1973) 139–156.
[8] B.D. Greenspan and P.J. Holmes, Homoclinic Orbits, Subharmonics and Global Bifurcations in Forced Oscillations. To appear in *Nonlinear Dynamics and Turbulence*, G. Barenblatt, G. Iooss and D.D. Joseph, eds. (Pitman, London, 1983).
[9] B.D. Greenspan and P.J. Holmes, Repeated Resonance and Homoclinic Bifurcation in a Periodically Forced Family of Oscillators (submitted for publication).
[10] J. Guckenheimer, On the Bifurcation of Maps of the Interval. Invent. Math. 39 (1977) 165–178.
[11] J. Guckenheimer, Sensitive Dependence on Initial Conditions for One-Dimensional Maps. Comm. Math. Phys. 70 (1979) 133–160.
[12] J. Guckenheimer and P.J. Holmes, Nonlinear Oscillations, Dynamical Systems and Bifurcations of Vectorfields (Springer Verlag, New York in press).
[13] I. Gumowski and C. Mira, Recurrences and Discrete Dynamical Systems, Springer Lecture Notes in Math #809 (1980).
[14] M. Hénon, A Two-Dimensional Mapping with a Strange Attractor, Comm. Math. Phys. 50 (1978) 69–77.
[15] P.J. Holmes, A Nonlinear Oscillator with a Strange Attractor, Phil. Trans. Roy. Soc. A292 (1979) 419–448.
[16] P.J. Holmes and J.E. Marsden, A Partial Differential Equation with Infinitely Many Periodic Orbits: Chaotic Oscillations of a Forced Beam, Archive for Rational Mechanics and Analysis 76 (1981) 135–166.
[17] P. Holmes and D.C. Whitley, On the Attracting Set for Duffing's Equation, I: Analytical Methods for Small Force and Damping, Proc. University of Houston 'Year of Concentration in Partial Differential Equations and Dynamical Systems', to appear.
[18] P. Holmes and D.C. Whitley, Bifurcations of One and Two Dimensional Maps (submitted for publication).
[19] M.V. Jakobson, Topological and Metric Properties of One-Dimensional Endomorphisms, Soviet Math. Dokl. 19 (1978) 1452–1456.
[20] M.V. Jakobson, Absolutely Continuous Invariant Measures for One Parameter Families of One-Dimensional Maps, Comm. Math. Phys. 81 (1981) 39–88.
[21] L. Jonker and D.A. Rand, Bifurcations in One Dimension. I. The Non-Wandering Set. Invent. Math. 62 (1981) 347–365. II. A Versal Model for Bifurcations. Invent. Math. 63 (1981) 1–15.
[22] O.E. Lanford, A Computer Assisted Proof of the Feigenbaum Conjectures, Bull. Amer. Math. Soc. 6 (1982) 427–434.
[23] O.E. Lanford, personal communication and this volume (1983).
[24] M. Levi, Qualitative Analysis of Periodically Forced Relaxation Oscillations, Memoirs. Amer. Math. Soc. 32 (1981) No. 244.
[25] T.Y. Li and J. Yorke, Period Three Implies Chaos, Amer. Math. Monthly 82 (1978) 985–992.
[26] F. Marotto, Snap-Back Repellers Imply Chaos in $\mathbb{R}^n$, J. Math. Anal. Appl. 63 (1978) 199–223.
[27] F. Marotto and T. Rogers (in preparation).
[28] V.K. Melnikov, On the Stability of the Center for Periodic Perturbations. Trans. Moscow Math. Soc. 12 (1963) 1–57.
[29] J. Milnor and W. Thurston, On Iterated Maps of the Interval I and II, Princeton preprint (1977).
[30] M. Misiurewicz, Structure of Mappings of an Interval with Zero Entropy, Publ. I.H.E.S. 53 (1981) 5–16.
[31] M. Misiurewicz, Absolutely Continuous Measures for Certain Maps of the Interval, Publ. IHES 53 (1981) 17–52.
[32] F.C. Moon and P.J. Holmes, A Magnetoelastic Strange Attractor, J. Sound and Vibration 65 (1979) 275–296.
[33] J. Moser, Stable and Random Motions in Dynamical Systems (Princeton Univ. Press, Princeton, 1973).
[34] S.E. Newhouse, The Abundance of Wild Hyperbolic Sets and Non-smooth Stable Sets for Diffeomorphisms, Publ. I.H.E.S. 50 (1979) 101–151.
[35] S.E. Newhouse, 'Lectures on Dynamical Systems', in Dynamical Systems, C.I.M.E. Lectures Bressanone, Italy, June 1978, Progress in Mathematics #8, (Birkhauser, Boston, 1980).
[36] D. Singer, Stable Orbits and Bifurcations of Maps of the interval, SIAM J. Appl. Math. 35 (1978) 260–267.
[37] S. Smale, Diffeomorphisms with Many Periodic Points, in 'Differential and Combinatorial Topology, S.S. Cairns ed. (Princeton Univ. Press, Princeton, 1963) pp. 63–80.
[38] S. Smale, Differentiable Dynamical Systems, Bull. Amer. Math. Soc. 73 (1967) 747–817.
[39] P. Stefan, A Theorem of Sarkovskii on the Existence of Periodic Orbits of Continuous Endomorphisms of the Real Line, Comm. Math. Phys. 54 (1977) 237–248.
[40] Y. Ueda, Personal communication; also see Explosion of Strange Attractors Exhibited by Duffings Equation, Ann. N.Y. Acad. Sci. 357 (1981) 422.
[41] S. Van Strien, On the Bifurcations Creating Horseshoes, in Springer Lecture Notes in Math 898, D.A. Rand and L.S. Young, eds. (Springer, New York, 1981).
[42] E.C. Zeeman, Bifurcation, Catastrophe, and Turbulence, Preprint, Mathematics Institute, University of Warwick, U.K. (1980).

PERIOD DOUBLING IN ONE AND SEVERAL DIMENSIONS

Oscar E. LANFORD III*†

Institute for Mathematics and its Applications, University of Minnesota, Minneapolis MN 55455, USA

Feigenbaum cascade—infinite sequences of successive period doublings–form a route from periodic to aperiodic behavior of dynamical systems. These sequences of bifurcations exhibit some striking universal features. The simplest of these features to formulate concerns the rate of accumulation of the bifurcations: If μ_n denotes the parameter value at which the nth doubling occurs, then, asymptotically,

$$\mu_n = \mu_\infty - c(4.6692\ldots)^{-n} + \text{"higher order terms"}.$$

The rate $4.6692\ldots$ appears to be universal, i.e., it shows up in many apparently unrelated systems such as
—one-dimensional non-invertible mappings, such as the one-parameter family $x \to 1 - \mu x^2$ on $[-1, 1]$, where $0 < \mu < 2$;
—dissipative (volume-decreasing) invertible mappings such as the Hénon system (see below);
—dissipative differential equations, such as the Lorenz system and the five-component truncation of the two-dimensional Navier–Stokes equations studied by Franceschini et al.

The main point we want to make here is that, despite their apparent diversity, these are really all instances of *precisely the same* mathematical phenomenon, and can be understood relatively easily once one has understood period doubling for one-dimensional mappings. (There is, on the other hand, another period doubling process, occurring for area-preserving mappings of the plane, which, although analogous to dissipative period doubling, seems to be an independent mathematical phenomenon. See Eckmann, Koch, and Wittwer [1] and the references cited therein.)

To undergo dissipative period doubling, a family of mappings–or, more generally a restriction of some iterate of the mappings–must have a characteristic behavior illustrated by the Hénon family

$$(x, y) \to (1 - \mu x^2 + cy, -cx),$$

with c small. (We are restricting ourselves here, for definiteness, to the orientation-preserving case, and have called the parameters μ and $-c^2$ instead of the more traditional a and b. We think of the bifurcations as occurring as we vary μ with c held fixed.) These mappings can be visualized as acting by:
1) contracting vertically:

$$(x, y) \to (x, cy);$$

2) bending and stretching by an x-dependent vertical shift:

$$\to (x, 1 - \mu x^2 + cy);$$

3) rotating a quarter-turn clockwise:

$$\to (1 - \mu x^2 + cy, -x);$$

4) contracting again vertically:

$$\to (1 - \mu x^2 + cy, -cx).$$

The general feature we want to emphasize is that the mapping contracts its multi-dimensional domain to an almost one-dimensional one, then folds that approximately one-dimensional set back into the original domain. Furthermore, the region of strong

---

*Current address: IHES, 91440 Bures-sur-Yvette, France..
†Work supported in part by NSF grant MCS81-07086AO1

folding–a vertical strip about the x-axis in the above example–is mapped away from itself and into a region of gentle folding. Finally, it is characteristic of mappings undergoing period doubling that the strong-folding region, although mapped away from itself by a first application of the mapping, is sent back into itself by a second.

The key to understanding repeated period doubling is the introduction of a *renormalization* or *doubling* operator $\mathcal{T}$ which carries a mapping F to one obtained by
– composing F with itself;
– restricting to an appropriate subdomain;
– making a change of coordinates to magnify the subdomain up to the original domain.

Roughly speaking, applying $\mathcal{T}$ divides the periods of all cycles by two but preserves their stability properties.

The idea now is to apply the renormalization group program to $\mathcal{T}$. To account for the observed universality, what one needs to show is that $\mathcal{T}$ has a fixed point and that, in the neighborhood of the fixed point, $\mathcal{T}$ is expanding in one direction and contracting in all others (i.e., that the linearization of $\mathcal{T}$ at the fixed point has a single simple eigenvalue with modulus greater than one and that the remainder of its spectrum is strictly inside the unit circle.) These facts have been established for one-dimensional mappings [2]; the proof rests on complicated numerical estimates verified (rigorously) by computer. Up to now, no one has succeeded in giving a conceptual proof.

By contrast, the theory of multi-dimensional period doubling can be reduced to the one-dimensional theory by a relatively simple conceptual argument. The argument goes roughly as follows: The space of one-dimensional mappings may be imbedded in the space of multi-dimensional mappings by associating with the one-dimensional mapping f the multi-dimensional mapping

$F_0: (x, y) \to (f(x), 0).$

(Here, y may have any number of components.) Such as F_0 is of course not invertible, but an arbitrarily small perturbation on F_0 can give an invertible mapping; the Hénon mapping with c small is an example. We can think of the space of F_0's as a surface M_0 in the space of all F's. What is now done is to construct a multi-dimensional doubling operator which
1) maps M_0 into itself;
2) agrees with the ordinary one-dimensional doubling operator on M_0;
3) is contractive in the directions transverse to M_0, i.e., when applied to an F near but not on M_0, gives a new mapping which is still closer to M_0.

In order to get 3) to hold, it is necessary to choose the change of variables in the construction of the doubling operator with some care.

A multi-dimensional doubling operator satisfying 1)–3) has as a fixed point the mapping

$(x, y) \to (g(x), 0),$

where g is the fixed point for the one-dimensional operator. Contractivity in directions transverse to M_0 guarantees that allowing the operator to act on mappings which are not strictly one-dimensional does not introduce any new expanding directions.

An analysis similar to the one described above was first given by Collet, Eckmann, and Koch [3]. In precisely this form, it is unpublished work of the author.

References

[1] J.-P. Eckmann, H. Koch and P. Wittwer, A computer-assisted proof of universality for area-preserving maps, Université de Genèva preprint UGVA-DPT 1981/04-345, to appear in Memoirs A.M.S.
[2] O.E. Lanford, A computer-assisted proof of the Feigenbaum conjectures, Bull. A.M.S. (New Series) 6 (1982) 427–434.
[3] P. Collet, J.-P. Eckmann, and H. Koch, Period-doubling bifurcations for families of maps on $\mathbb{R}^n$, J. Stat. Phys. 25 (1981) 1–14.

THE REAL AND COMPLEX LORENZ EQUATIONS AND THEIR RELEVANCE TO PHYSICAL SYSTEMS

A.C. FOWLER
Department of Mathematics, Massachusetts Institute of Technology, Cambridge, MA 02139, USA

J.D. GIBBON*
Center for Nonlinear Studies, Los Alamos National Laboratory, Los Alamos, NM 87545, USA

and

M.J. McGUINNESS
Department of Applied Mathematics, California Institute of Technology, Pasadena, CA 91125, USA

We summarize some recently obtained results on real and complex Lorenz equations and discuss their possible significance in relation to real fluid dynamical processes.

1. Why Lorenz?

Two seminal papers (Lorenz [41], Ruelle and Takens [65]) in the recent development of the study of aperiodic motions in dynamical systems both proceeded from a motivation suggested by the phenomenon of turbulence in fluid mechanics. Indeed, 'turbulence' is often referred to in recent dynamical studies, and presumably, much of the continuing interest in such chaotic and aperiodic trajectories in dynamical systems stems from the hope that new and exciting results will provide an impetus and a methodology for understanding old and long-standing problems in the theory of turbulence.

To some extent, it is clear that this has already happened: the phenomena of period doubling (May [47], Feigenbaum [14] and Collet and Eckmann [7]), intermittency (in the sense of Manneville and Pomeau [43]), and hysteresis involving multiple stable states, have been widely observed in difference equations (above references; also Tresser et al. [69]), differential equations (Robbins [61], Fraser and Kapral [22], Franceschini and Tebaldi [21]), differential-delay equations (May [48], Hopf et al. [34]); they are also constituent parts of various experimentally observed routes to turbulence in convective and Taylor–Couette flows, as reviewed by Ott [51], Eckmann [13], and Swinney (this volume) and reported by many authors, for example Libchaber and Maurer [38, 39], Libchaber et al. [40], Maurer and Libchaber [45, 46], Fenstermacher et al. [15], Di Prima and Swinney [10], Gollub and Benson [28]. Other 'real-life' systems exhibiting the same sort of chaotic behaviour are Nicholson's blowfly populations (Gurney et al. [29]), white blood-cell populations and the respiratory system (Mackey and Glass [42]), chemical oscillations in the Belousov reaction (Rössler and Wegmann [64]) and the irregular reversals of the earth's magnetic field (Robbins [59, 60]), and at least some of these can (at least in principle) be realistically modelled by a set of ordinary differential equations.

However, the wealth of phenomena and of sys-

---

* On leave of absence May–August 1982. Permanent address: Department of Mathematics, Imperial College, Queen's Gate, London SW7 2BZ, UK.

tems to which they apply (both model and experimental) obscures the fact that certain dominant and enduring concerns of fluid dynamicists, that is, the study of transition and turbulence in shear flows (flat plate, pipe flow, Couette flow, Poiseuille flow) and in convection *in an unbounded layer*, have yielded relatively little to these new techniques; consequently, turbulence of these types has received scant attention in the dynamical systems literature. Other experiments (Taylor–Couette flow, rotating annulus) yield results which can be *interpreted* by dynamical systems results, but not *predicted*. Our purpose in this paper, therefore, is to outline some possible ways in which the study of simple dynamical systems, and in particular the Lorenz equations (and a complex generalisation thereof) may have some bearing on certain situations in fluid experiments where some kind of 'turbulence' is observed.

Although the equations which bear his name were 'derived' (via truncation) from a model of two-dimensional convection, Lorenz was not interested in convection per se, but in the possible deterministic aperiodicity of atmospheric motions. It is thus remarkable that in the study of idealised models of baroclinic instability which are designed to simulate certain features of the atmospheric circulation, the Lorenz equations have reappeared, but now being *derived* in a formal manner by the method of multiple scales in situations where the basic flow is marginally unstable.

Baroclinic instability in the atmosphere occurs when an equilibrium state can be maintained in which surfaces of constant density are not parallel to surfaces of constant gravitational potential. Some potential energy is available to be converted into kinetic energy of fluid motions. If small disturbances can grow at the expense of available potential energy, the fluid is said to be baroclinically unstable. The simplest theoretical and experimental model which displays baroclinic instability is the so-called two-layer model which consists of two immiscible fluids, the lighter overlying the heavier and in relative motion, which are rapidly rotated about their vertical axis. Instability occurs at some critical shear in *both* the inviscid and the viscous cases although the neutral curves differ in each case. For the inviscid case, Pedlosky [53, 55] calculated the amplitude equations in the neighbourhood of the critical point and found that the amplitude of a disturbance $A(X, T)$ modulating a carrier wave was governed by the equations

$$\left(\frac{\partial}{\partial T} + c_1 \frac{\partial}{\partial X}\right)\left(\frac{\partial}{\partial T} + c_2 \frac{\partial}{\partial X}\right) A = \alpha A - \beta AB, \quad (1a)$$

$$\left(\frac{\partial}{\partial T} + c_2 \frac{\partial}{\partial X}\right) B = \left(\frac{\partial}{\partial T} + c_1 \frac{\partial}{\partial X}\right) |A|^2. \quad (1b)$$

These equations were subsequently shown to be a completely integrable system by Gibbon et al. [23] by use of the inverse scattering method. It is easily shown that the sine–Gordon equation is embedded in eqs. (1a) and (1b) when the amplitude function A has no phase variation. Define

$$A = (2\beta)^{-1/2}\left(\frac{\partial}{\partial T} + c_1 \frac{\partial}{\partial X}\right) \Phi, \quad B = \alpha(1 - \cos \Phi)/\beta \quad (1c)$$

and (1a, b) reduces to

$$\left(\frac{\partial}{\partial T} + c_1 \frac{\partial}{\partial X}\right)\left(\frac{\partial}{\partial T} + c_2 \frac{\partial}{\partial X}\right) \Phi = \alpha \sin \Phi. \quad (1d)$$

Consequently, a form of pendulum behaviour is contained in this model. Similar results have been found by Moroz and Brindley [49] using a continuously stratified model (see Drazin [11, 12]). Gibbon and McGuinness [25] have shown that (1a) and (1b) are typical for a certain class of dispersive unstable systems in general, one further example being the self-induced transparency equations of optical pulse propagation.

The connection with the Lorenz model comes when one wants to consider a weakly viscous model instead of the purely inviscid case. This limit has been studied by various authors, the first being Pedlosky [54, 56] followed by Hart [32], although at the time it was not recognised that Lorenz type equations had been produced. This was later realised by Pedlosky and Frenzen [57] and Gibbon and McGuinness [24]. Again there is a connection with optics, since it was Haken [31] who showed that the laser equations (which can be expressed in the form of eqs. (1a) and (1b) when damping is excluded) can be transformed into the Lorenz equations when spatial scales are excluded. How the Lorenz equations occur in all these examples as a class (see Gibbon and McGuinness [24, 26] can most easily be described intuitively by considering eqs. (1a) and (1b), in the spatially independent case when weak viscosity (i.e. damping) is added. This is done by the phenomenological introduction of weak viscosity (specifically viscous term $\sim$ amplitude perturbation). This is a distinguished limit which links a purely dispersive to a purely dissipative instability. It can be seen intuitively that the class of examples governed by (1a, b) in the purely dispersive limit, become

$$\frac{d^2A}{dT^2} + \Delta_1 \frac{dA}{dT} = \alpha A - \beta AB,$$

$$\frac{dB}{dT} + \Delta_2 B = \frac{d}{dT}(|A|^2) + \Delta_3 |A|^2,$$
(2)

when weakly viscous terms are added. Δ_1 and α are complex numbers if weak dispersive effects are also added (e.g. weak beta plane in the two layer model). A similar form of eqs. (2) has been found by Moroz and Brindley [49] for the continuously stratified model of baroclinic instability.

Under the transformations

$$t = \Omega T, \quad \Omega = \text{Re}(\Delta_1) - \tfrac{1}{2}\Delta_3,$$
$$X = (2\beta)^{1/2} \Omega^{-1} A, \quad (3)$$
$$Z = 2\beta \Omega^{-1} \Delta_3^{-1},$$

eqs. (2) can be rewritten as (Fowler et al. [20], Gibbon and McGuinness [26], Gibbon [27])

$$\dot{X} = -\sigma X + \sigma Y,$$
$$\dot{Y} = (r - Z)X - aY, \quad (4)$$
$$\dot{Z} = \tfrac{1}{2}(XY^* + X^*Y) - bZ,$$

where $r = r_1 + ir_2$, $a = 1 - ie$, $\dot{X} \equiv dX/dt$, * denotes complex conjugate, and

$$\sigma = \Delta_3/2\Omega, \quad b = \Delta_2/\Omega, \quad r_1 = 1 + 2\,\text{Re}(\alpha)/\Delta_3\Omega,$$
$$r_2 = [2\,\text{Im}(\alpha) + \Delta_3\,\text{Im}(\Delta_1)]/\Omega\Delta_3, \quad e = -\text{Im}(\Delta_1)/\Omega.$$
(5)

For example, a weak β-effect in the two-layer model produces complex Δ_1 and α, and so complex r and a, real b. In such circumstances, we call (4) the complex Lorenz equations. If no weak dispersive effects are present, $r = r_1$ is real, $a = 1$, and we regain the (real) Lorenz equations. For higher dimensional models (e.g. the baroclinic models with more than one cross-stream mode), one can obtain an infinite set of Lorenz-type equations (Brindley and Moroz [5], Hart [32], Booty et al. [4]).

Thus we see that, at least in some systems, particularly the baroclinic models, one can really expect the behaviour of the Lorenz equations to have some bearing on the system. In the following two sections, we will discuss some recent results on the behaviour of real and complex Lorenz equations, and their possible implication both for baroclinic and other models.

2. 'Real' behaviour

The behaviour of the real Lorenz equations is well known. As the bifurcation parameter r is increased (dissipation is decreased), the origin bifurcates to two non-trivial steady states, which undergo subcritical Hopf bifurcation at a critical value r_c. Above r_c, one can see a bewildering variety of motions, including alternating régimes of chaotic and periodic orbits separated by intermittent and period-doubling transitions. In addition (Fowler and McGuinness [18]), hysteresis occurs between periodic and chaotic/periodic solutions. All these phenomena, and a rationale which helps an understanding of them, are detailed in the recent book by Colin Sparrow [67]. Geometrically, one can understand the existence of a strange invariant set by the occurrence at some value $r = r_h$ of a homoclinic orbit (that is, an orbit which begins and ends at an equilibrium point of the equations). As r passes through r_h, a homoclinic 'explosion' takes place which 'produces' an infinite number of periodic and aperiodic orbits. These orbits can only disappear in further homoclinic orbits, period-doubling cascades (reversed) or Hopf bifurcations. Topological considerations imply that there will be many period-doubling windows whose 'function' is to remove orbits: this suggests that it might be useful in any system to concentrate more on the homoclinic events, rather than on the myriad period-doubling events which these sprawn, except insofar as the observation of a period-doubling cascade may indicate the existence somewhere in parameter space of homoclinicity: of course, these remarks may need adjustment in higher dimensions.

Obviously, the weakly nonlinear baroclinic models contain homoclinicity. This also occurs (heteroclinicity, in fact) at a codimension two bifurcation point in double diffusion (Knobloch and Proctor [36], Da Costa et al. [8]) where one might be able to prove the existence of 'chaos' using essentially a theorem of Shilnikov (Arneodo et al. [1]). It is an interesting point that the Lorenz attractor which obtains as a result of the homoclinicity is parametrically unrelated to the sub-critical Hopf bifurcation at r_c, which need not even exist at all. Thus, one should view homoclinic bifurcation as a new and separate entity, which must be analysed in its own right. This observation *might* be interesting in view of the long-standing effort to connect the onset of turbulence in plane Poiseuille flow at Re $\approx$ 1000 with the sub-critical Hopf bifurcation at Re = 5772 (e.g. Stewartson and Stuart [68]) by means of a threshold amplitude, although it must be admitted that it is becoming clear that such a threshold does play a part, by virtue of its long (viscous) relaxation time (Orszag and Patera [50]). Similar, less justifiable, efforts for pipe flow (Davey and Nguyen [9]) might be restored using the concept of bifurcation from infinity (Rosenblat and Davis [63]): however, see Patera and Orszag [52].

It is more obviously likely (or plausible) that convection may contain homoclinicity of some kind, particularly in view of the experimental observations of period-doubling already cited. In fact, it is the homoclinicity in the Lorenz equations (a primitive 'model' of convection) which is responsible for the non-monotone difference map relating successive maxima of Z. One might ask whether there is any way to analytically tackle the homoclinicity in these equations, and thereby 'predict' the chaotic behaviour by producing a Lorenz map analytically. To do this, one needs some help, and it turns out this is available in the form of a region in parameter space where r and σ are both large (and comparable). In this case the solutions behave relaxationally, as shown in fig. 1, consisting (for X and Y) of a series of spikes interspersed with flat 'quiet' periods. In the limit $r/\sigma \sim \mathcal{O}(1)$, $b/\sigma \ll 1$, one can use asymptotic techniques to 'solve' the equations, and produce a difference

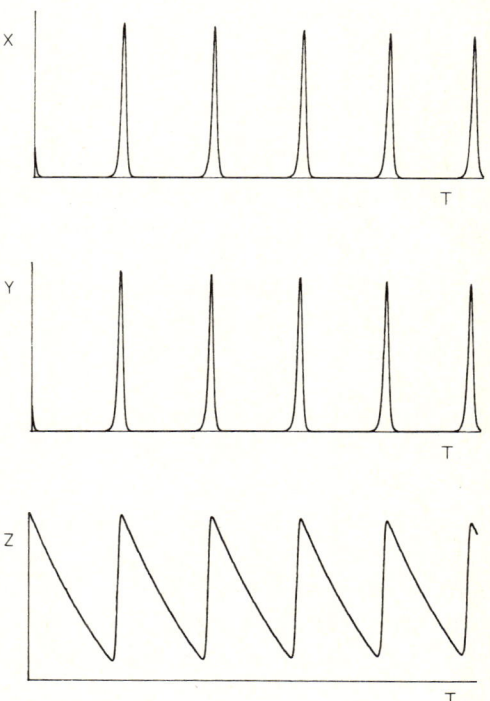

Fig. 1. Relaxational behaviour of X, Y and Z versus time in the real Lorenz equations at parameter values $r = 240$, $\sigma = 300$, $b = 1$.

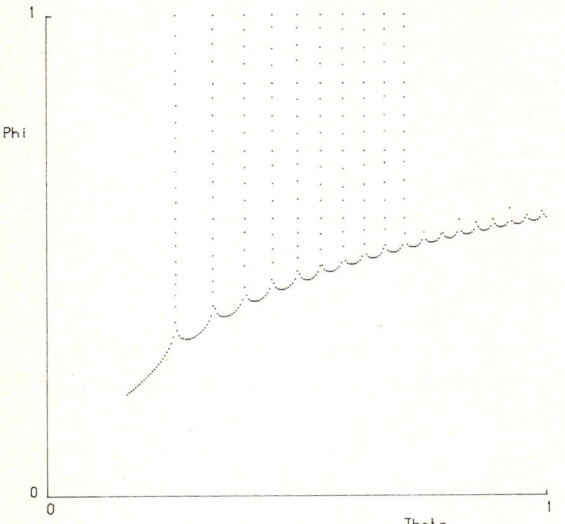

Fig. 2. Difference map for high r and σ ($r = 160$, $\sigma = 100$, $b = 1$) relating successive maxima of Z in the real Lorenz equations. Here the nth maximum is $Z \approx r[1 + \theta]$; the $(n+1)$th is $Z \approx r[1 + \phi]$ and the curve is plotted from the analytic results of Fowler and McGuinness [17]: the first ten cusps are indicated.

map, such as shown in fig. 2 (Fowler and McGuinness [17]). Using this analytic map, we have been able to predict quantitatively the onset of period doubling, and hysteresis (or intermittency) with a fair measure of sucess.

What we wish to draw attention to here, however, is the special pulse-like nature of the solutions. Such pulses occur in many other kinds of system (logistic delay equation, geodynamo, spruce budworm population, jökulhlaups): see Fowler [16] and discussion therein. More germanely, turbulence in shear flows is well known to be intermittent or 'bursty' in character (Maslowe [44]), and an attempt has been made to formalise these observations in an analysis which incorporates two different time scales (Landahl [37]): we recall also the slow viscous time scale of two-dimensional relaxation and the fast advective time scale of three-dimensional instability (Orszag and Patera [50]).

Virtually the only quasi-analytic theory of convective turbulence at high Rayleigh number is the 'bubble' model of Howard [35], incidentally recently much espoused by geophysicists, which has certain experimental support (e.g. Tritton et al. [72]), and seems never to have been experimentally contradicted. In this model, long quiescent conductive phases in which thermal boundary layers grow into an isothermal fluid are interspersed by short-lived, violent convective overturns in which the unstable boundary layers are eradicated. This model requires Ra (Rayleigh number) and σ (Prandtl number) to be large for its validity. The interesting thing is that if we interpret X, Y and Z in the Lorenz equations in terms of their original physical variables (Lorenz [41]), we find that Z is the vertical departure from the conductive temperature profile, Y is the horizontal departure (so producing buoyancy), and X is essentially the velocity. At high Ra (r) and σ, we find relaxational motions in the Lorenz equations, in which there are long 'quiet' phases, in which $X \approx 0$ (almost no motion) and Z decays exponentially (thermal boundary layers grow), interspersed with rapid pulses in which Z jumps up, and X becomes significant (convective overturn). In other words, we *see* Howard's postulated P.D.E. solution in the Lorenz truncation! This obviously suggests a means whereby one might explicitly 'solve' the full convection equations: suffice it to say that this problem has not been satisfactorily resolved as yet. We might also mention that the slow phase is essentially a linear problem, recalling the fast linear three-dimensional instability of Orszag and Patera [50].

What we are tentatively suggesting is that it may be possible to gain a small analytic foothold on the problem of convective turbulence at high Ra by using direct asymptotic methods on the full equations (and implicitly hoping these equations admit homoclinic solutions), rather than trying to get there through a series of transitions, which apparently precludes any moderately simple explicit computation. One also retains fully nonlinear dynamics in the solutions, and there is no limitation as to being close to any curve in parameter space (in fact, as Ra and $\sigma \to \infty$, one is really in a neighborhood of infinity, quite a large area). Such an idea is apparently novel in convection (though many authors have studied steady solutions when Ra→∞, e.g. Roberts [62]), but has been touched on in shear flow, where the asymptotic nature of flow disturbances as R (Reynolds number)→∞ has long been a consideration in both linear and non-linear studies (Maslowe [44], Benney and Bergeron [3]), as well as in experiments, where the threshold nature of transition has been recognised since the original work of Reynolds [58].

3. 'Complex' behaviour

When r and a are complex, the set of equations (4) exhibits a notably distinct sequence of bifurcations to those of the real Lorenz equations. Recalling $r = r_1 + ir_2$, $a = 1 - ie$, we find that for $r_1 > r_{1c} = 1 + (e + r_2)(e - \sigma r_2)/(\sigma + 1)^2$, the origin is oscillatorily unstable, and a Hopf bifurcation to a limit cycle occurs. Moreover, this limit cycle takes the *exact* form

$$X = A\,e^{i\omega t}, \quad Y = \left[1 + \frac{i\omega}{\sigma}\right] A\,e^{i\omega t}, \quad Z = |A|^2/b, \quad (6)$$

where

$$\omega = \frac{\sigma(e + r_2)}{\sigma + 1}, \quad |A|^2 = b(r_1 - r_{1c}). \quad (7)$$

To find an exact limit cycle in a nonlinear system is remarkable, and worthy of some comment. First, we might hope that a corresponding exact limit cycle might exist in the full equations: for example, in the underlying vorticity equations of the two-layer and continuously stratified models, although this might be modulationally unstable in much the same way as the Benjamin–Feir instability [2] occurs. However, such exact solutions have yet to be found. More realistically, the exact limit cycle in (4) is really due to the fact that when written as a fifth order differential system, the system can be put in the form

$$\dot{x} = f(x), \quad (8)$$

where f is rotationally invariant, that is, there is a rotation matrix $R(\alpha)$ representing a rotation of the x-axes through an angle α, such that

$$Rf(x) = f(Rx), \quad (9)$$

and in terms of R, the exact limit cycle has the form

$$x = R(\omega t)x_0, \quad (10)$$

where x_0 is constant. Such a solution is possible, using $R(\alpha)R(\beta) = R(\alpha + \beta)$, if x_0 and ω satisfy the nonlinear algebraic eigenvalue problem

$$f(x_0) = \omega R'(0)x_0. \quad (11)$$

The rotational symmetry of (9) suggests an analogy with the rotational invariance of the baroclinic models: thus one might hope that (for example) the steady wave régime in rotating annulus experiments (Hide and Mason [33]) might be susceptible to a representation of the form (10), and indeed one might hope for similar results for other rotationally invariant systems, e.g. Taylor vortex flow.

For the complex Lorenz equations, one can do explicit stability analysis of the limit cycle, and we find that a Hopf-like bifurcation to an invariant 2-torus occurs at a value $r_1 = r'_{1c} > r_{1c}$, and the form of the corresponding doubly periodic motion is (in terms of (8))

$$x = R(\tilde{\omega}t)u(\tilde{\Omega}t), \quad (12)$$

where u is periodic, and when $|r_1 - r'_{1c}| \sim \epsilon^2$,

$$\tilde{\omega} = \omega + \mathcal{O}(\epsilon^2),$$
$$\tilde{\Omega} = \Omega + \mathcal{O}(\epsilon^2), \quad (13)$$
$$u = x_0 + \epsilon\{u_\Omega e^{i\tilde{\Omega}t} + (*)\} + \mathcal{O}(\epsilon^2);$$

here Ω is the marginally stable frequency of the perturbation to the limit cycle. Eq. (13) is obtained by using the method of multiple scales, but one may ask whether the special form of (12) extends beyond the limited range of (13). It turns out (Fowler and McGuinness [19]) that one can do explicit analysis at $r_1 \gg 1$ (following Robbins [61], Shimizu [66], Sparrow [67]) which confirms (12) in this limit. Also, spectral analyses at intermediate values all support this form of the torus. Thus, the rotational invariance again gives a special form to the doubly periodic motion, and one might ask if such a form is of relevance to certain rotationally invariant experimental systems.

The bifurcation to a 2-torus in the complex Lorenz equations is numerically found to be subcritical, as its real counterpart is also. (See fig. 3 for a stability map in $r_1 - r_2$ space). As r_2 is reduced, one observes (numerically) a period-doubling cascade analogous to that in the real Lorenz equations (Robbins [61]): this is easily seen in phase plots of $|X|^2$ versus $|Y|^2$ for example (which removes the precessing frequency $\tilde{\Omega}$); we do not know if this cascade is preceded by frequency locking (e.g. Maurer and Libchaber [45], but conjecture that it is the precessing limit cycle u which period-doubles, and we also conjecture that the presence of a period-doubling torus is due to the occurrence of a homoclinic torus at some point in parameter space: this may be suspected on analytic grounds at high r_1 (Fowler and McGuinness [19]), and probably also at high r_1 and σ, although we have not done the corresponding analysis in this case. See also fig. 8 of Fowler et al. [20], where the chaotic torus gets very close to the origin.

Lastly, some comments on scales. In our discussion of the complex Lorenz equations, we have deliberately omitted inclusion of long space derivatives $\partial/\partial X$, as in (1). Apart from simplicity, there is one good reason for this. In many of the real laboratory systems to which we have referred, the geometry is essentially *finite* in the direction of wave propagation. Thus in the rotating annulus experiments, there is really no X scale near criticality, and purely time-dependent motions should ensue (e.g. amplitude vacillation). It is possible that as the supercriticality (Δ, say) and the wave-number increase, there comes a point where (numerically) the circumference is of a length $X \sim \mathcal{O}(1)$. Then the appropriate spatial derivatives might enter, with a requirement of periodicity. It is also plausible (but a guess) that when Δ thus increases, the first appearance of X dependence is as a bifurcating travelling wave (wave-number vacillation?), as is well known to occur in reaction-diffusion systems as the domain size increases.

Part of the achievement of Pedlosky was to show that the dissipation parameter r ($= E^{1/2}/\epsilon$ in the baroclinic problems, where E is the Ekman number and ϵ is the Rossby number) defines a distinguished limit when $r \sim |\Delta|^{1/2}$. Thus when $r \gg |\Delta|^{1/2}$, viscosity dominates, and amplitude equilibration occurs; when $r \ll |\Delta|^{1/2}$, one obtains conservative (e.g. sine-Gordon) equations whose amplitude depends entirely on initial conditions, and thus experimentally will be very susceptible to noise. If $r \sim |\Delta|^{1/2}$, one regains the rich structure of the Lorenz equations. When $r \ll |\Delta|^{1/2}$, however, conservative waves (e.g. solitons) will evolve over a long viscous time scale, no matter how small r is: in a sense, the viscous time scale is the true scale over which transients die out, and the 'state' of the system should be under-

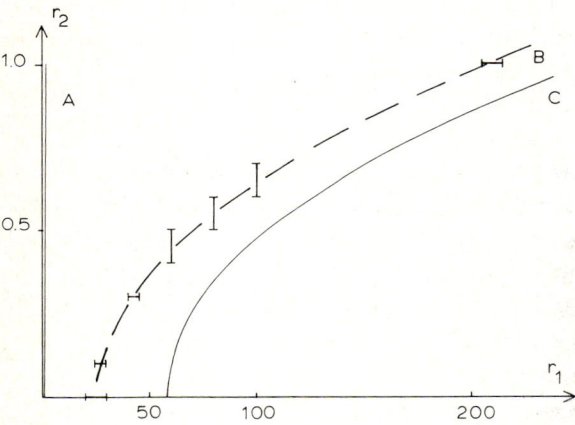

Fig. 3. Parameter space diagram in the r_2 versus r_1 plane depicting regions of stability. Between the curves A and C the limit cycle is linearly stable ($r_1 < r'_{1c}$). The finite amplitude 2-torus is numerically observed below the approximate dashed curve B and loses stability by period doubling below C as r_2 is reduced. The 'Hopf' bifurcation of the limit cycle to a 2-torus on C is sub-critical, the (unstable) torus existing for r_2 above C. Here $\sigma = 2$, $b = 0.8$ and $e = 3r_2$.

stood in these terms. Even if one waits a viscous time, however, it seems likely that substantial noise could drown the signal. In this sense, conservative equations such as sine-Gordon are not the whole story, particularly for closed systems (e.g. rotating annuli): it is well understood, for example, in convection experiments, that one must wait $\gtrsim$ a conductive time scale for patterns to evolve to a steady state (e.g. Busse [6]).

4. Concluding remarks

Fluid dynamicists tend to be of the opinion that studies of chaos in simple dynamical systems are not of particular use in the 'real' problems of shear flow and convection (and perhaps other 'nasty' examples), although they have undoubtedly given understanding to a wide variety of other 'weakly' turbulent phenomena in fluids, chemical reactions, population dynamics, etc. Our aim here has been to give some possible reasons why the study of even one simple set of equations may help understanding of physical systems, both by direct application, and by indirect analogy. If any of these ideas eventually bear fruit, then the effort will have been worthwhile.

References

[1] A. Arneodo., P. Coullet and C. Tresser, J. Stat. Phys. 27 (1982) 171.
[2] T.B. Benjamin and J.E. Feir, J. Fluid Mech. 27 (1967) 417.
[3] D.J. Benney and R. F. Bergeron, Stud. Appl. Math. 48 (1969) 181.
[4] M. Booty, J.D. Gibbon and A.C. Fowler, Phys. Lett. 87A (1982) 261.
[5] J. Brindley and I.M. Moroz, Phys. Lett. 77A (1980) 441.
[6] F.H. Busse, in Hydrodynamic Instabilities and the Transition to Turbulence, H.L. Swinney and J.P. Gollub, eds. (Springer, Berlin, 1981), p. 97.
[7] P. Collet and J.-P. Eckmann, Iterated Maps on the Interval as Dynamical Systems (Birkhäuser, Boston, 1980).
[8] L.N. Da Costa, E. Knobloch and N.O. Weiss, J. Fluid Mech. 109 (1981) 25.
[9] A. Davey and H.P.F. Nguyen, J. Fluid Mech. 45 (1971) 701.
[10] R.C. Di Prima and H.L. Swinney, in Hydrodynamic Instabilities and the Transition to Turbulence, H.L. Swinney and J.P. Gollub, eds., (Springer, Berlin, 1981) p. 139.
[11] P.G. Drazin, Quart J. R. Met. Soc. 96 (1970) 667.
[12] P.G. Drazin, J. Fluid Mech. 55 (1972) 577.
[13] J-P. Eckmann, Rev. Mod. Phys. 53 (1981) 643.
[14] M.J. Feigenbaum, J. Stat. Phys. 19 (1978) 25.
[15] P.R. Fenstermacher, H.L. Swinney and J.P. Gollub, J. Fluid Mech. 94 (1979) 103.
[16] A.C. Fowler, IMA J. Appl. Math. 28 (1982) 41.
[17] A.C. Fowler and M.J. McGuinness, A description of the Lorenz attractor at high Prandtl number, Physica 5D (1982) 149.
[18] A.C. Fowler and M.J. McGuinness, Hysteresis, period doubling and intermittency at high Prandtl number in the Lorenz equations. Stud. Appl. Math., in press (1983).
[19] A.C. Fowler and M.J. McGuinness, A nonlinear torus at high r_1 in the complex Lorenz equations. Submitted to SIAM J. Appl. Math. (1983).
[20] A.C. Fowler, J.D. Gibbon and M.J. McGuinness, Physica 4D (1982) 139.
[21] V. Francheschini and C. Tebaldi, J. Stat. Phys. 21 (1979) 707.
[22] S. Fraser and R. Kapral, Phys. Rev. 23A (1981) 3303.
[23] J.D. Gibbon, I.N. James and I.M. Moroz, Proc. R. Soc. Lond. A 367 (1979) 219.
[24] J.D. Gibbon and M.J. McGuinness, Phys. Lett. 77A (1980) 295.
[25] J.D. Gibbon and M.J. McGuinness, Proc. R. Soc. Lond. A 377 (1981) 185.
[26] J.D. Gibbon and M.J. McGuinness, The real and complex Lorenz equations in rotating fluids and lasers, Physica 5D (1982) 108.
[27] J.D. Gibbon, "The Complex Lorenz Equations", in Chaos and Order in Nature, H. Haken ed. (Springer, Berlin, 1981) (note; some of the figures in this article have been mixed up with those of other articles).
[28] J.P. Gollub and S.V. Benson, J. Fluid Mech. 100 (1980) 449.
[29] W.S.C. Gurney, S.P. Blythe and R.M. Nisbet, Nature 287 (1980) 17.
[30] H. Haken, Encyclopedia of Physics; Light and Matter, vol. 25/2c (Springer, Berlin, 1970).
[31] H. Haken, Phys. Lett. 53A (1973) 77.
[32] J.E. Hart, J. Atmos. Sci. 30 (1973) 1017.
[33] R. Hide and P.J. Mason, Adv. Phys. 24 (1975) 47.
[34] F.A. Hopf, D.L. Kaplan, H.M. Gibbs and R.L. Shoemaker, Phys. Rev. A25 (1982) 2172.
[35] L.N. Howard, Int. Cong. Appl. Mech. Munich 1964, H. Görtler, ed. (Springer, Berlin, 1966), p. 1109.
[36] E. Knobloch and M.R.E. Proctor, J. Fluid Mech. 108 (1981) 291.
[37] M.T. Landahl, SIAM J. Appl. Math. 28 (1975) 735.
[38] A. Libchaber and J. Maurer, J. Phys. Lett. 39 (1978) L369.

[39] A. Libchaber and J. Maurer, J. Phys. 41 (1980) C3, 51.
[40] A. Libchaber, C. Laroche and S. Fauve, J. Phys. Lett. 43 (1982) L211.
[41] E.N. Lorenz, J. Atmos. Sci. 20 (1963) 130.
[42] M.C. Mackey and L. Glass, Science 197 (1977) 287.
[43] P. Manneville and Y. Pomeau, Physica 1D (1980) 219.
[44] S. Maslowe, in Hydrodynamic Instabilities and the Transition to Turbulence, H.L. Swinney and J.P. Gollub, eds. (Springer, Berlin, 1981) p. 181.
[45] J. Maurer and A. Libchaber, J. Phys. Lett. 40 (1979) L419.
[46] J. Maurer and A. Libchaber, J. Phys. Lett 41 (1980) L515.
[47] R.M. May, Nature 261 (1976) 459.
[48] R.M. May, Ann. N.Y. Acad. Sci. 357 (1980) 267.
[49] I.M. Moroz and J. Brindley, Proc. R. Soc. Lond. A 377 (1981) 379.
[50] S.A. Orszag and A.T. Patera, Phys. Rev. Lett. 45 (1980) 989.
[51] E. Ott, Rev. Mod. Phys. 53 (1981) 655.
[52] A.T. Patera and S.A. Orszag, J. Fluid Mech. 112 (1981) 467.
[53] J. Pedlosky, J. Atmos. Sci. 27 (1970) 15.
[54] J. Pedlosky, J. Atmos. Sci. 28 (1971) 587.
[55] J. Pedlosky, J. Atmos. Sci. 29 (1972) 680.
[56] J. Pedlosky, J. Atmos. Sci. 38 (1981) 717.
[57] J. Pedlosky and C. Frenzen, J. Atmos. Sci. 37 (1980) 1177.
[58] O. Reynolds, Phil. Trans. Roy. Soc. 174 (1983) 935.
[59] K.A. Robbins, Proc. Nat. Acad. Sci. USA 73 (1976) 4297.
[60] K.A. Robbins, Math. Proc. Camb. Phil. Soc. 82 (1977) 309.
[61] K.A. Robbins, SIAM J. Appl. Math. 36 (1979) 457.
[62] G.O. Roberts, Geophys. Astrophys. Fluid Dyn. 12 (1979) 235.
[63] S. Rosenblat and S.H. Davis, SIAM J. Appl. Math. 37 (1979) 1.
[64] O.E. Rössler and K. Wegmann, Nature 271 (1978) 89.
[65] D. Ruelle and F. Takens, Commun. Math. Phys. 20 (1971) 167.
[66] T. Shimizu, Physica 97A (1979) 383.
[67] C. Sparrow, The Lorenz Equations: Bifurcations, Chaos, and Strange Attractors. Lecture Notes in Applied Mathematics, Vol. 41 (Springer, Berlin, 1982).
[68] K. Stewartson and J.T. Stuart, J. Fluid Mech. 48 (1971) 529.
[69] C. Tresser P. Coullet and A. Arneodo, J. Phys. Lett. 41 (1980) L243.
[70] D.J. Tritton, D.M. Rayburn and M.A. Forrest, in Mechanisms of Continental Drift and Plate Tectonics, P.A. Davies and S.K. Runcorn, eds. (Academic Press, New York, 1980), p. 267.

TRANSITIONS TO CHAOS IN THE GINZBURG–LANDAU EQUATION

H.T. MOON, P. HUERRE and L.G. REDEKOPP
Department of Aerospace Engineering, University of Southern California, Los Angeles, California 90089-1454, USA

The amplitude evolution of instability waves in many dissipative systems is described close to criticality, by the Ginzburg–Landau partial differential equation. A numerical study of the long-time behavior of amplitude-modulated waves governed by this equation allows the identification of two distinct routes of the Ruelle–Takens–Newhouse type as the modulation wavenumber is decreased. The first route involves a sequence of bifurcations from a limit cycle to a two-torus to a three-torus and to a turbulent régime, the last stage being preceded by frequency locking. The turbulent régime is itself followed by a new two-torus. In the second route, this two-torus exhibits a single subharmonic bifurcation which immediately results in transition to chaos. A description of the various possible dynamical states is tentatively given in the plane of the two control parameters c_d and c_n.

1. Introduction

The notion that transition to chaos could arise from purely deterministic evolution models, has been made strikingly clear by numerous investigations of systems of nonlinear ordinary differential equations with a few degrees of freedom. In the pioneering work of Lorenz [1], the Oberbeck–Boussinesq equations, which govern Rayleigh–Bénard convection in a fluid layer heated from below, are drastically truncated to obtain a system of three nonlinear equations. In this case, the flow may undergo chaotic transitions, if it is assumed that only three spatial Fourier modes are adequate to describe the motion. Higher order truncation models of the Rayleigh–Bénard problem also lead to turbulence [2–4], but via a different sequence of bifurcations. Thus, in contrast with "fully developed" fluid turbulence which is traditionally thought to involve a very large number of modes, deterministic chaos already arises in systems of only very few (at least three) degrees of freedom. However, in order to test the relevance of these new concepts to real transition processes in fluids flows, it is advisable to choose evolution models which potentially may display an infinite number of excited degrees of freedom, i.e., partial differential equations. It is also essential that those partial differential equations rationally describe unstable flows in at least some well-defined region of control parameter space. The Ginzburg–Landau equation fulfills both of these criteria. The general context in which it arises is described below.

The linear stability of fluid systems is usually characterized by a dispersion relation

$$D(\omega, k; r) = 0 \qquad (1)$$

linking the frequency ω to the wavenumber k, the control parameter r typically representing a Reynolds number. In many situations, there exists a stability boundary or neutral curve $r_n(k)$ (see fig. 1) separating stable ($\omega_i < 0$) and unstable ($\omega_i > 0$) régimes of the (k, r) plane. The minimum of the neutral curve (k_c, r_c) corresponds to the critical condition above which the fluid becomes linearly unstable. For instabilities occurring in *dissipative systems*, one usually has

$$D_\omega(\omega_c, k_c; r_c) \neq 0, \qquad (2)$$

so that only one branch $\omega(k; r)$ crosses the real ω-axis into the half-plane $\omega_i > 0$ as r becomes larger than r_c. By making use of singular perturbation methods [5, 6], it is then possible to study

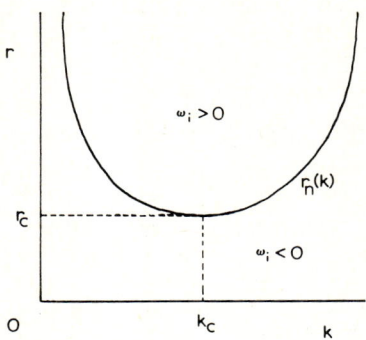

Fig. 1. Sketch of stability boundary in the plane of the wavenumber k and the control parameter r. The region $\omega_i > 0$ is unstable. The region $\omega_i < 0$ is stable.

the weakly nonlinear development of a disturbance of the form $\epsilon A(X, T) e^{i(k_c x - \omega_c t)}$ in the neighborhood of (k_c, r_c). When condition (2) is met and $\epsilon \ll 1$, the complex amplitude $A(X, T)$ satisfies the Ginzburg–Landau equation

$$i\partial A/\partial T + \tfrac{1}{2} \partial^2 \omega/\partial k^2 \big|_c \partial^2 A/\partial X^2$$
$$= \partial \omega/\partial r \big|_c \Delta r A + a|A|^2 A, \tag{3}$$

where $X = \epsilon(x - \partial \omega/\partial k \big|_c t)$ and $T = \epsilon^2 t$ are slow space and time scales, and the level of supercriticality is normalized as $\Delta r = (r - r_c)/\epsilon^2 = \mathcal{O}(1)$. (The linear operator in eq. (3) reflects the nature of the Taylor series expansion

$$\omega = \omega_c + \partial \omega/\partial k \big|_c (k - k_c) + \tfrac{1}{2} \partial^2 \omega/\partial k^2 \big|_c (k - k_c)^2$$
$$+ \partial \omega/\partial r \big|_c (r - r_c) + \cdots, \tag{4}$$

representing the dispersion relation in the vicinity of the critical point. Note that the group velocity $\partial \omega/\partial k \big|_c$ is real but that $\partial^2 \omega/\partial k^2 \big|_c$, $\partial \omega/\partial r \big|_c$, and a are in general complex. We also remark that the evolution of A is viewed from a frame of reference moving at the group velocity $\partial \omega/\partial k \big|_c$. If A, X and T are suitably rescaled, eq. (3) reduces to the following normalised form:

$$\partial A/\partial t = \text{sgn } \Delta r A + (1 + ic_d) \partial^2 A/\partial x^2$$
$$+ \text{sgn } a_i (1 + ic_n)|A|^2 A, \tag{5}$$

where c_d and c_n are real parameters. For simplicity we have used the symbols x and t for space and time although these slow variables bear no relationship with the fast scales appearing in the phase of the instability wave. From this brief discussion, the Ginzburg–Landau equation is seen to govern the finite-amplitude evolution of wave packet envelopes for small subcriticality or supercriticality, provided that conditions (2) are satisfied. Several specific examples can be drawn from Hydrodynamic Instability Theory. The development of Tollmien–Schlichting waves in plane Poiseuille flow is described by eq. (5) in the case where $a_i > 0$, $c_d > 0$ and $c_n < 0$ [7]. Since the Landau constant a_i is positive, there exists a subcritical threshold instability mechanism which may lead to bursting solutions [8]. The nonlinear growth of convection rolls in the Rayleigh–Bénard problem [9] and the appearance of Taylor vortices in the flow between counter-rotating circular cylinders [10] is also governed by the Ginzburg–Landau equation with $a_i < 0$ and $c_d = c_n = 0$. In these situations, waves are not travelling and $\omega_c = \partial \omega/\partial k \big|_c = 0$. The amplitude A becomes a measure of the strength of the vortices or convection rolls. Monochromatic disturbances of a given wavenumber k_c then reach a supercritical equilibrium ($a_i < 0$). Finally, in the limit of infinite c_d and c_n, the evolution equation reduces to the cubic nonlinear Schrödinger equation (NLS) governing modulated deep water gravity waves [11],

$$i\partial A/\partial t + \partial^2 A/\partial x^2 + |A|^2 A = 0. \tag{6}$$

The NLS is integrable by the Inverse Scattering Transform [12] on the infinite interval $-\infty < \hat{x} < +\infty$ and admits envelope soliton solutions. When periodic boundary conditions are imposed on a finite interval of length L, the solutions of the NLS exhibit an FPU recurrence phenomenon [13] in which the evolution in wavenumber space leads to an almost periodic broadening and collapse on a long-time scale. Theoretical studies [14, 15] also seem to indicate that the motion is then quasi-periodic, the frequency spectrum displaying several incommensurate frequencies and their combination harmonics [see discussion in section 5].

The Ginzburg–Landau equation also arises naturally in the study of chemical systems. Reaction–diffusion equations of the form

$$\partial c/\partial t = F(c) + K\nabla^2 c \qquad (7)$$

in which c represents a vector of chemical concentrations, K a positive definite matrix of diffusivities, and $F(c)$ a nonlinear reaction mechanism, have been extensively studied in the last few years to elucidate the spatio-temporal structure of the Belousov–Zhabotinskii reaction [16, 17]. It has been shown by Kuramoto and Tsuzuki [18, 19] that the perturbation concentration $A = c - c^*$ away from the steady state solution c^* also satisfies the Ginzburg–Landau equation (5). Indeed, an earlier numerical investigation [20] conducted in this context has indicated that eq. (5) exhibits chaotic solutions for certain values of the parameters.

The pervasiveness of the Ginzburg–Landau equation in various fields of physics, as demonstrated in the present discussion, makes it ideally suited for an in-depth numerical investigation. Ultimately one would like to determine if the hydrodynamic equations of motion can support the complex dynamical behavior called turbulence. At an intermediate step, we examine, without relying on a truncation assumption, the case of a partial differential equation which is derivable from the Navier–Stokes equations: the Ginzburg–Landau equation.

2. General formulation

We shall solely consider the case of supercritical equilibrium, whence $\Delta r > 0$ and $a_i > 0$. In this instance, eq. (5) admits the particular finite amplitude Stokes solution $A_e(t) = \exp(-ic_n t)$. The sideband instability of $A_e(t)$ can be studied by writing $A(x, t)$ as

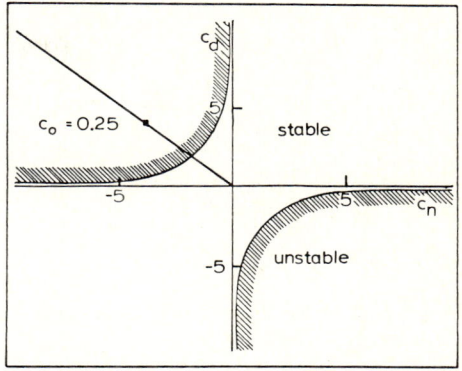

Fig. 2. Sideband instability domain of the Stokes solution in the c_n, c_d plane. The line $c_d = -c_n$, parametrized by c_0, is studied in sections 2–5. Regions of possible phase turbulence are hatched.

$$A(x, t) = A_e(t) + A_1 e^{iqx + \sigma t} + A_2 e^{-iqx + \sigma^* t}, \qquad (8)$$

where q and σ are the sideband wavenumber and sideband growth rate respectively. A star superscript denotes the complex conjugate. Substitution of (8) into (5) and subsequent linearisation results in the following Newell criterion [5] for sideband instability:

$$1 + c_d c_n < 0 \qquad (9)$$

as sketched in fig. 2. It is to be noted that the Ginzburg–Landau equation also admits a family of finite-amplitude travelling wave solutions which have, in the c_n–c_d plane, a wider domain of sideband instability than the Stokes solution $A_e(t)$. The reader is referred to [21] and [22] for a more detailed discussion of these cases. In the present discussion, we shall only consider perturbations from $A_e(t) = \exp(-ic_n t)$.

A comprehensive description of all possible asymptotic states of (5) in the c_n–c_d plane of fig. 2 is clearly a formidable task. For simplicity, we restrict our attention to the straight line $c_d = -c_n = 1/c_0$, the parameter c_0 defining the location of each point on that line. We shall return to the general case where c_d and c_n are distinct in

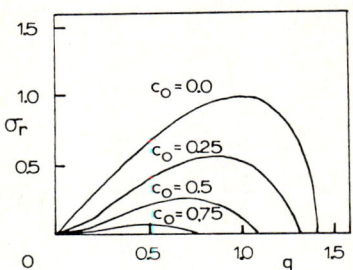

Fig. 3. Linear growth rate σ_r versus sideband wavenumber q for various c_0 values.

the discussion of section 5. Upon introducing a rescaled time $c_0^{-1}t$, the Ginzburg–Landau equation becomes

$$i\partial A/\partial t + (1 - ic_0)\partial^2 A/\partial x^2 = ic_0 A - (1 + ic_0)|A|^2 A, \quad (10)$$

where it is understood that t designates the new time variable. Note that (10) now depends on one control parameter only. If $c_0 = 0$, the conservative NLS (6) is recovered. For $c_0 \neq 0$, the equation becomes dissipative. In the limit $c_0 = \infty$, it reduces to the Newell–Whitehead equation [9] of Bénard convection. The variations of the linear growth rate σ_r of sideband fluctuations sketched in fig. 3 indicate that only a finite range of wavenumbers q is unstable. The goal of the present work is to investigate numerically the corresponding long-time dynamics of the solutions of (10) as a function of the two parameters q and c_0. Initial conditions of the form

$$A(x, 0) = 1 + 0.2 \cos qx \quad (11)$$

are chosen, which represent perturbations of sideband wavenumber q applied to the Stokes wave of initial amplitude unity. Furthermore, periodic boundary conditions are imposed on the interval $[-\pi/q, \pi/q]$. The character of the solutions is studied by varying q within the linearly unstable range of fig. 3 for a given c_0, $0 < c_0 < 1$.

A pseudo-spectral method [23] based on the Fast Fourier Transform algorithm is used to perform the calculations. Eq. (10) and the initial conditions (11) are invariant under $x \to -x$ reflection. Hence it is legitimate to represent the complex amplitude $A(x, t)$ by the spatial cosine Fourier series

$$A(x, t) = \sum_{n=0}^{N-1} [a_n(t) + ib_n(t)] \cos(nqx), \quad (12)$$

where a_n and b_n are real and N is the truncation parameter. Thus, in the numerical scheme, the infinite-dimensional system is actually replaced by a finite-dimensional system of $2N$ degrees of freedom $\{a_n(t), b_n(t)\}$ or N spatial modes. In all the cases to be considered in the next sections, at most six Fourier modes will be excited, which is much lower than the value $N = 64$ chosen for the truncation number. It was also checked that the results become independent of N for sufficiently large N. The time step was taken to be $\Delta t = 0.001$ and the equation was integrated for 134 basic Stokes periods. Also of interest is the evolution of $|A|^2(x, t)$, which admits a spatial Fourier decomposition of the form

$$|A|^2(x, t) = \sum_{n=0}^{N-1} \alpha_n(t) \cos(nqx). \quad (13)$$

Note that, if the complex amplitude $A(x_0, t)$, at a given location x_0, undergoes a quasi-periodic motion with M independent frequencies, $0 < \omega_0 < \omega_1 < \cdots < \omega_{M-1}$, then the motion of $|A|^2(x_0, t)$ is also quasi-periodic, but only contains $M - 1$ incommensurate frequencies $\omega_1 - \omega_0$, $\omega_2 - \omega_0, \ldots, \omega_{M-1} - \omega_0$. As shown in section 3, this property will be extremely useful to ascertain the existence of quasi-periodic motion with three independent frequencies.

A variety of techniques can be applied to characterize the motion. Qualitative changes in the long-time behavior are monitored in terms of the usual time series and frequency spectra of $A(x_0, t)$ and $|A|^2(x_0, t)$ taken at the midpoint $x_0 = 0$ of the interval. Poincaré sections of the trajectories of $A(x_0, t)$ in the reduced three-dimensional phase space (a_0, a_2, b_2) are constructed by making a cut in the plane $a_2 = 0$. In effect, b_2 is plotted versus a_0 whenever $a_2 = 0$ and $da_2/dt > 0$. Similarly, Poincaré sections of the motion of $|A|^2(x_0, t)$ are made by plotting the spatial mode amplitudes α_2 versus α_1, each time $\alpha_0 = 0.01$ and $d\alpha_0/dt > 0$. A Poincaré surface of section can further be reduced to a one-dimensional map by using a technique of Lorenz [1]. Each successive point in the section is assigned an angle θ_n relative to some particular interior reference point so as to generate a sequence of angles $\theta_1, \theta_2, \ldots$, etc. Members of this sequence are related via a difference equation of the form

$$\theta_{n+1} = F(\theta_n). \tag{14}$$

Properties of these return maps will help in relating the observed transitions to some of the currently studied "universal paths" to turbulence. Finally, information regarding the spatial structure of the flow can be gained by following the energy content of individual modes as a function of q. The time-averaged relative energy of the nth mode is defined as

$$E(nq) = \frac{\dfrac{1}{T}\displaystyle\int_0^T |a_n(t) + ib_n(t)|^2 \, dt}{\displaystyle\sum_{n=0}^{N-1} \frac{1}{T}\int_0^T |a_n(t) + ib_n(t)|^2 \, dt}, \tag{15}$$

where T is the integration time.

In the next two sections, we follow the sequence of bifurcations taking place at $c_0 = 0.25$ as the sideband wavenumber q is gradually decreased from the neutrally stable value $q = 1.3$ (see fig. 3). A tentative description of the motion in the (q, c_0) plane is given in section 5.

3. Transition to chaos via a three-torus

Details of the phenomena arising in this first transition have been reported in [24]. When q is

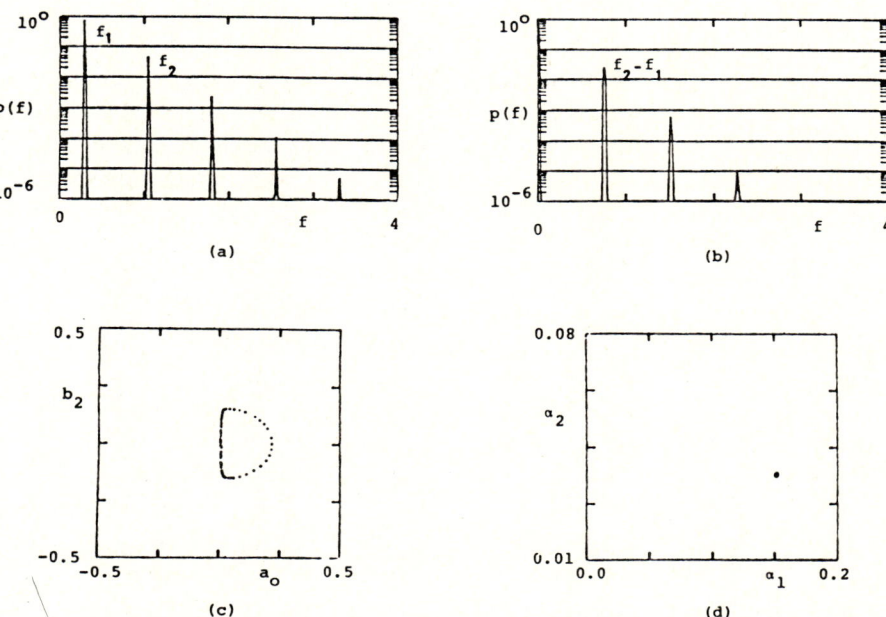

Fig. 4. Two-frequency régime, $q = 0.55$: (a) power spectrum of $A(0, t)$; (b) power spectrum of $|A|(0, t)$; (c) Poincaré section of $A(x, t)$ constructed by plotting b_2 versus a_0 whenever $a_2 = 0$ and $da_2/dt > 0$; (d) Poincaré section of $|A|^2(x, t)$ constructed by plotting α_2 versus α_1 whenever $\alpha_0 = 0.01$ and $d\alpha_0/dt > 0$.

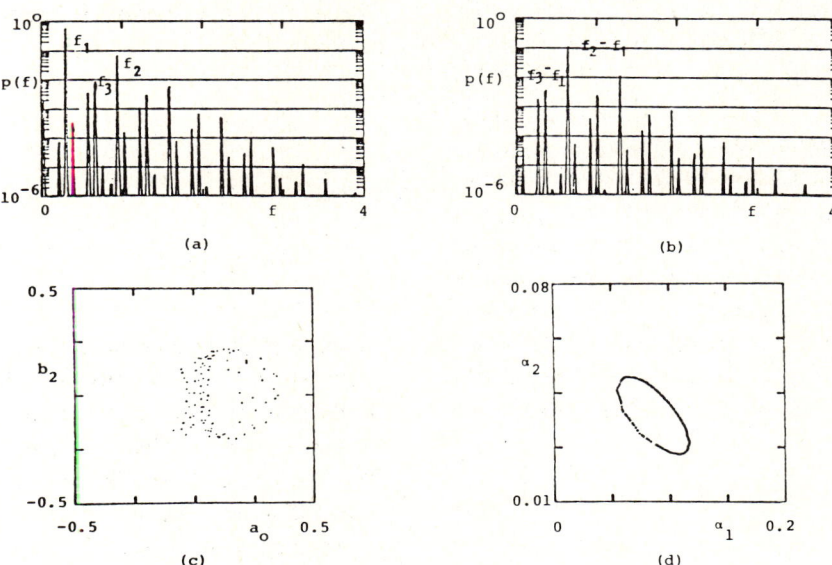

Fig. 5. Three-frequency régime, $q = 0.51$. Same as in fig. 4.

slightly above 0.60*, the power spectrum if A exhibits only one frequency f_1 which decreases continuously with q, and the motion is simply periodic. In the interval $0.52 < q < 0.60$, the flow becomes quasi-periodic with two independent frequencies f_1 and f_2 and their combination harmonics (fig. 4). The Poincaré section of A appears to be a closed curve, thereby suggesting that the motion is confined to a two-torus. In both régimes, only four spatial modes have any significant energy content. As q is decreased below 0.52, a third independent frequency f_3 (fig. 5) is introduced. The Poincaré section of fig. 5c does not allow us to confirm this conjecture. However, an examination of the Poincaré section of $|A|^2$ leads to the conclusion that its trajectory is confined to a two-torus and thus, that the motion of A lies on a three torus (see section 2). In this range, one frequency f_1 is associated with the phase of A, whereas the other two, f_2 and f_3, are associated with its amplitude $|A|$. It is important to notice that the three-torus persists throughout the finite interval $0.49 < q < 0.52$, as evidenced by the power spectra and Poincaré sections displayed in fig. 6. However, at $q = 0.49$, the motion of $|A|$ abruptly shrinks back to a limit cycle (fig. 7d). Correspondingly, the trajectories of A now lie on a two-torus (fig. 7c). The breakdown or locking of the three-torus into a two-torus is immediately followed by the onset of a chaotic régime (fig. 8), which is characterized by continuous power spectra for both A and $|A|$. In the three-frequency régime, frequency-locking stage and turbulent régime, five Fourier modes are excited, by comparison with only four modes in the simply periodic and two-torus states. The chaotic régime extends over the range $0.41 < q < 0.49$. The long-time behavior pertaining to lower values of the sideband wavenumber q will be discussed in section 4.

The occurrence of a three-torus prior to the onset of a turbulent state has been observed experimentally by Gollub and Benson [25] in a study of Rayleigh–Bénard convection for a layer of small aspect ratio. In both cases, the three-frequency quasi-periodic motion is present over a significant range of values of the control parameter. According to the general scenario proposed by Ruelle and

---

*Note added in proof: we have found another transition in the vicinity of $q = 1$, although the motion is single-periodic for q slightly larger than $q = 0.60$.

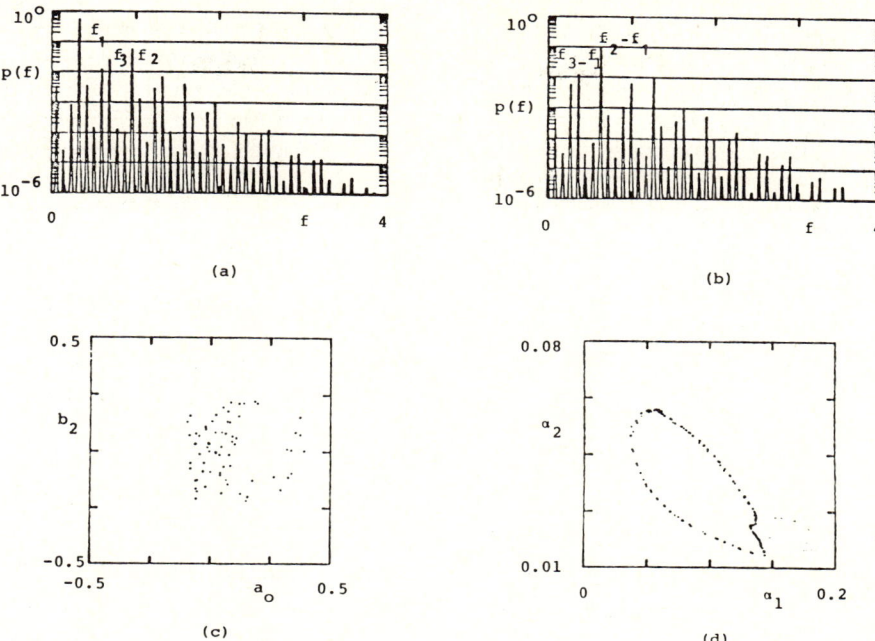

Fig. 6. Three-frequency régime, $q = 0.50$. Same as in fig. 4.

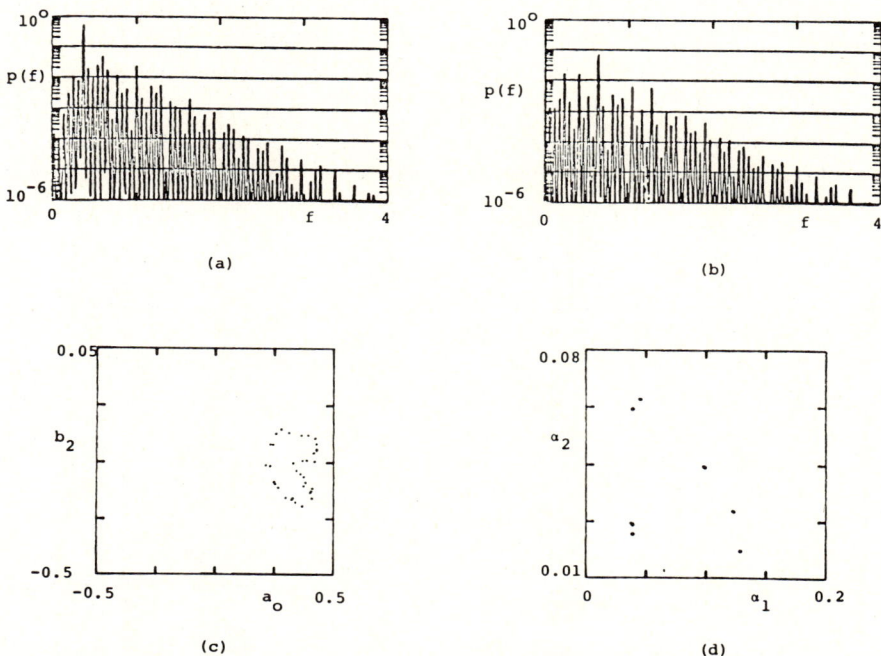

Fig. 7. Frequency locking to a two torus, $q = 0.49$. Same as in fig. 4.

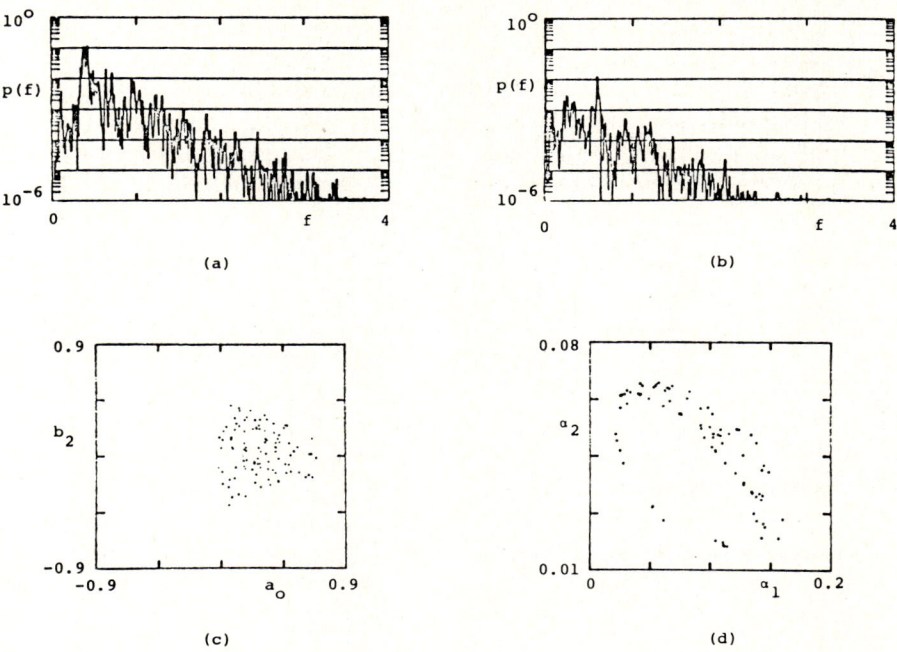

Fig. 8. Chaotic régime, $q = 0.48$. Same as in fig. 4.

Takens [26], one possible route to turbulence involves successive bifurcations from a limit cycle to a two-torus to chaos. The work of Newhouse, Ruelle and Takens [27] further indicates that an additional bifurcation to a triply-periodic régime before onset is unlikely since a small perturbation would destroy the three-torus and lead to chaos. The present result seems however to suggest that partial differential evolution equations may indeed sustain a stable three-torus motion.

Considerable progress has recently been made regarding the universal features of the transition from a two-torus to chaos [28, 29]. A convenient method of analysis [28] consists in studying the properties of maps of the circle which become noninvertible as a certain limiting value of the

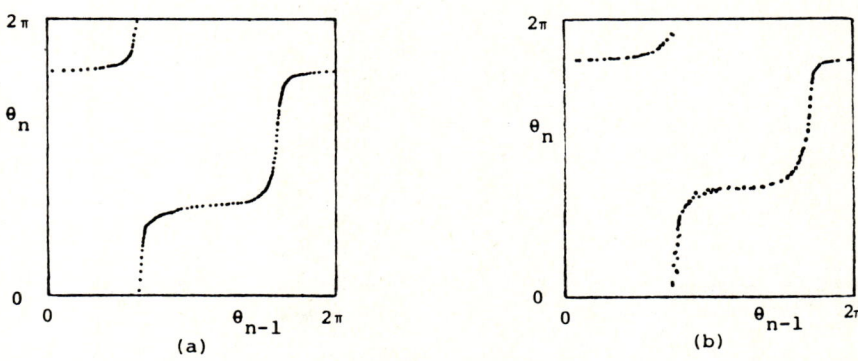

Fig. 9. Return maps of $|A|^2(x, t)$: (a) $q = 0.51$; (b) $q = 0.50$.

control parameter is reached Such a return map can be constructed for $|A|^2$ in the three-torus régime (figs. 9a and 9b) by following the technique described in the previous section. The appearance of a cubic singularity as q decreases is qualitatively consistent with the study of Shenker [29]. We may also note that the flattening of the inflexion point in the return map results from the development of an incipient fold in the Poincaré sections of fig. 5d and fig. 6d. Presumably, the noninvertibility of the map for values of q lower than 0.49 leads to the establishment of a turbulent régime.

4. Transition to chaos via a two-torus

A second transition to turbulence can be identified as the sideband wavenumber q is further decreased. In the range of $0.32 < q < 0.41$, the motion returns to a quasi-periodic state (fig. 10) with two incommensurate frequencies f_1 and f_2. The cross-section of the torus in fig. 10c is quite different from the two-torus which arose in the range $0.52 < q < 0.60$ (fig. 4c). The relative energy distribution of spatial Fourier modes (fig. 14a) indicates that only three modes participate in the motion. It is significant that the sideband wavenumber q which was initially set at a finite-amplitude level [eq. (11)], has lost all its energy to the $2q$-mode and the $4q$-mode in the limit $t \to \infty$. The Stokes mode $k = 0$ almost stays at its initial level. A reasonable explanation for this energy transfer can be given by comparing the linear growth rates σ_r of the q-mode and the $2q$-mode in fig. 3. At $q = 0.41$, the wavenumber $2q$ precisely becomes the linearly most amplified disturbance when $c_0 = 0.25$, whereas the growth rate of the q-component is only about half the maximum value of σ_r. The dominance of the wavenumber $2q$ in the interval $0.35 < q < 0.41$, as given by linear theory, is apparently preserved in the nonlinear régime and gives rise to the distribution of fig. 14a. It is conjectured that this mechanism is responsible for the disappearance of the chaotic

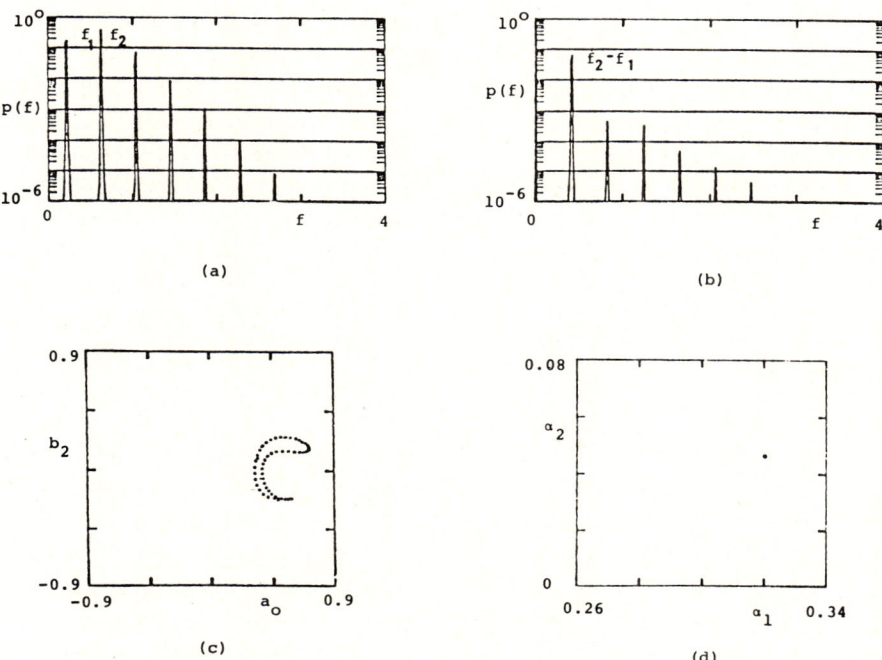

Fig. 10. Two-frequency régime, $q = 0.40$. Same as in fig. 4.

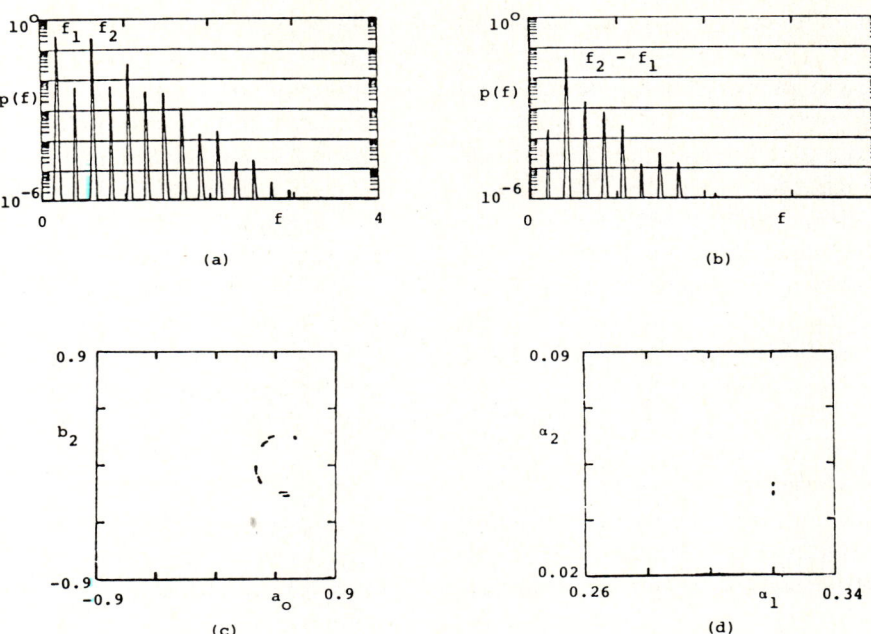

Fig. 11. Subharmonic bifurcation, $q = 0.378$. Same as in fig. 4.

régime at $q = 0.41$ and the emergence of a new two-torus.

As q decreases below the value 0.38, the doubly-periodic state with frequencies f_1 and f_2 loses its stability to a new two-torus with frequencies f_1 and $(f_1 + f_2)/2$ as shown in fig. 11. The single fixed point in the Poincaré section of fig. 10d at $q = 0.38$ has been replaced by two fixed points, as one could expect in a period-doubling bifurcation. The subharmonic frequency peak remains sharp in the short interval $0.377 < q < 0.380$ and rapidly gains energy to reach a maximum at $q = 0.380$. The presence of an additional peak at $(f_1 + f_2)/2$ is accompanied by the reappearance of the spatial Fourier components $k = q$ and $k = 3q$, as indicated in figs. 14a and 14c. In other words, as q crosses 0.380, subharmonic bifurcations in frequency space and wavenumber space occur simultaneously, leading to a doubling of the period and wavelength of $A(x, t)$. This phenomenon suggests a deep connection between the spatial and temporal structure which has to be further explored. We first note that the x-coordinate is in fact an angle, due to the choice of periodic boundary conditions. The system is axisymmetric and falls within the same class as Taylor–Couette flow berween counter-rotating concentric cylinders. Thus, the theory of Rand [30] concerning modulated waves in axisymmetric systems, which has been very successful in predicting the morphology of certain states in the Taylor–Couette system [31], can also be applied to the periodic Ginzburg–Landau equation. It is then hypothesized that, in the two-torus régime, the motion in (x, t) space is composed of finite-amplitude travelling waves moving at a constant velocity along the x-axis. A doubling in wavelength then necessarily results in a period-doubling as observed numerically.

The route to turbulence traced by Feigenbaum [32, 33] involves an infinite sequence of period-doubling bifurcations followed by a chaotic state. In the present case, however, only one single period-doubling event is detected before the onset of turbulence (fig. 12) at $q = 0.377$. This peculiarity

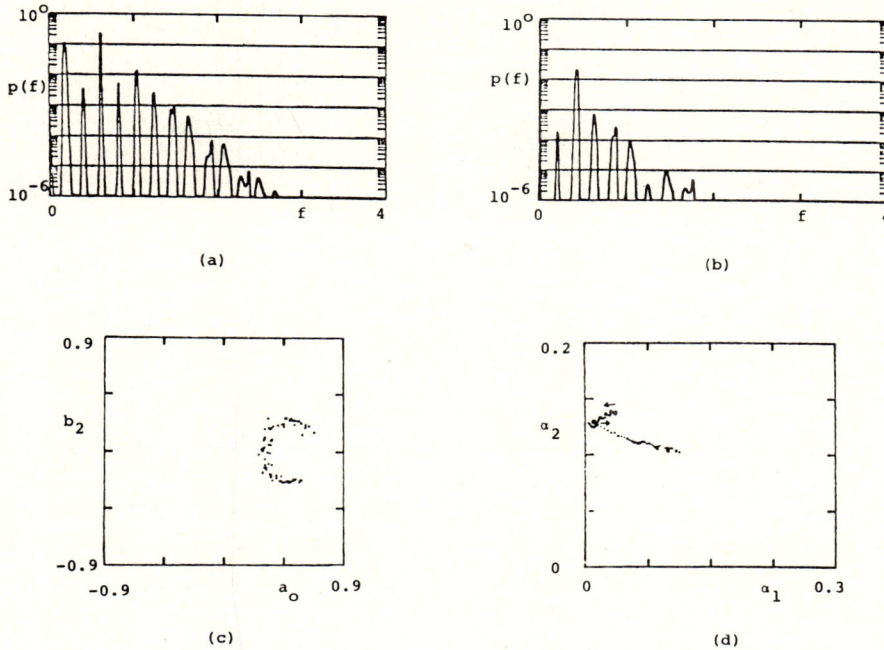

Fig. 12. Onset of chaotic régime, $q = 0.377$. Same as in fig. 4.

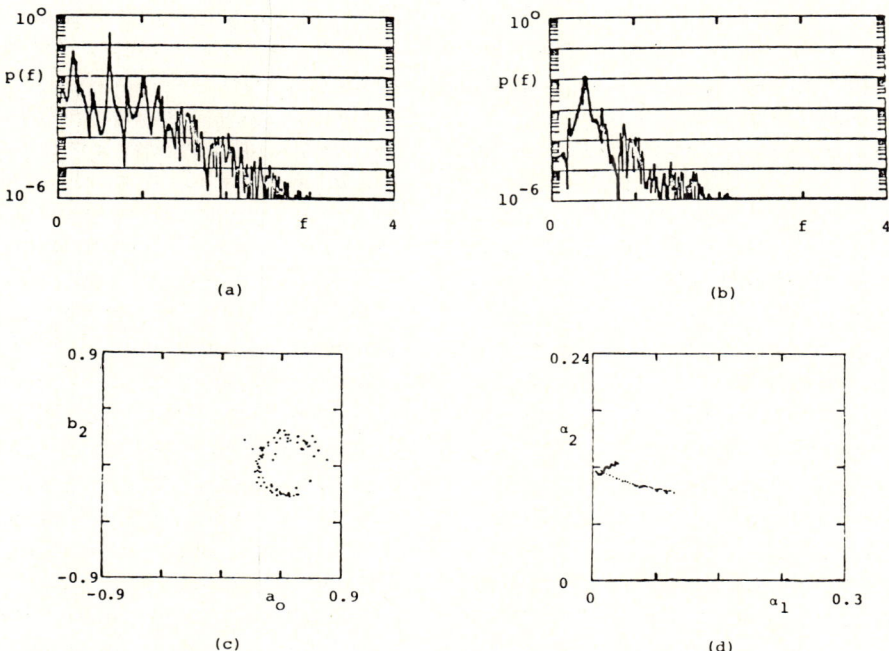

Fig. 13. Chaotic régime, $q = 0.376$. Same as in fig. 4.

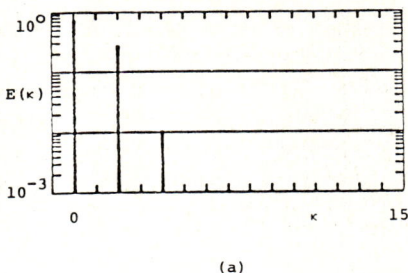 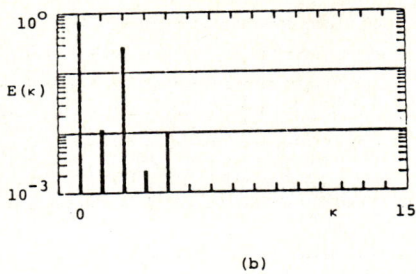

Fig. 14. Relative energy distribution of spatial Fourier modes [eq. (15)]: (a) two-frequency régime, $q = 0.40$; (b) subharmonic bifurcation, $q = 0.378$.

has also been noticed by Franceschini [34] in a study of a seven-mode truncation of the Navier–Stokes equations. In the two-torus range immediately preceding the strange attractor régime, his system undergoes only two successive period-doubling bifurcations from the incommensurate frequencies f_1 and f_2 to $f_1, f_2/2$, to $f_1, f_2/4$. The geometrical constraints which result from the application of periodic boundary conditions [30] necessarily restrict the number of period-doubling bifurcations to be finite. In other words, the quasi-periodic state with frequencies f_1 and f_2 and wavelength π/q (see fig. 14a) can sustain only one subharmonic bifurcation to $f_1, (f_1 + f_2)/2$, because the new wavelength of the spatial pattern already reaches its maximum allowable value $2\pi/q$ [see fig. 14b].

When the sideband wavenumber crosses $q = 0.377$, the peaks in the power spectra of A and $|A|^2$ start to broaden while the Poincaré section of A becomes "fuzzy" (fig. 12). These qualitative changes mark the beginning of the second chaotic state. The Poincaré section of fig. 12d reveals that the motion of $|A|^2$ slowly drifts away from its previous periodic state represented by the two larger dots on the figure. Comparisons between the autocorrelations of $|A|$ and Re A (see [22] for plots of these quantities) also imply that the phase exhibits a more rapid transition to turbulence than the amplitude $|A|$. Five spatial Fourier modes are found to participate in the dynamics.

For a slightly lower value $q = 0.376$, the spectrum becomes continuous (fig. 13) with many peaks exceeding the broadband component by as much as two decades. Additional computations show that this turbulent state persists for values of q as low as 0.20.

5. A tentative description in the c_d–c_n plane

As a summary of the previous two sections, which pertain to the case $c_0 = 0.25$, we conclude that, as q decreases from 0.60 i.e., as the length of the interval in x increases, the motion in phase space undergoes two distinct sequences of bifurcations to chaos. The first route involves a transition from a limit cycle to a two-frequency motion, to a three-frequency motion, and then to a chaotic state, the last step being preceded by frequency locking. The second route leads from a two-frequency motion to chaos via a single subharmonic bifurcation. Moreover, whenever the motion of A is quasi-periodic with M independent frequencies, the motion of $|A|$ is also quasi-periodic with only $(M-1)$ independent frequencies. A further check on the results is obtained by a plot of the ratios f_2/f_1 and f_3/f_1 versus q in fig. 15. The continuous variations of these ratios with q, in the régimes where they exist, provides additional evidence that the motion is indeed quasi-periodic.

A diagram of the various possible dynamical states in the q–c_0 plane is shown in fig. 16. The region of interest lies within the linearly unstable

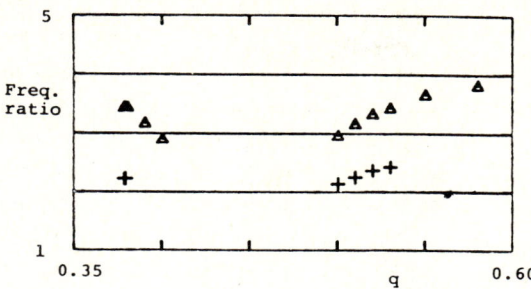

Fig. 15. Ratio of independent frequencies as a function of sideband wavenumber q when $c_0 = 0.25$. $\triangle$, f_2/f_1; $+$, f_3/f_1.

domain bounded by the elliptically-shaped curve on fig. 16. For each value of c_0 in the range $0.1 < c_0 < 0.50$, the sequence of bifurcations is the same as at $c_0 = 0.25$. But, as c_0 increases, the successive transitions occur in a very short interval of values of q. For example, at $c_0 = 0.50$, all the bifurcations take place in the range $0.35 < q < 0.40$. The dotted lines in fig. 16 indicate that the present conclusions may actually hold up to large c_0 although for very much lower values of q.

The numerical integration of the Ginzburg–Landau equation at very low values of q requires special care: many spatial Fourier modes $q, 2q, 3q, \text{etc.}\ldots$, which are linearly unstable (see fig. 3), are likely to participate in the long-time dynamics. Thus, the truncation number N needs to be considerably increased in order to faithfully reproduce the behavior of the partial differential equation. Alternatively, one may follow Kuramoto [20, 36] and obtain from the Ginzburg–Landau equation, other evolution equations which are valid at low wavenumbers. The results are conveniently summarized by applying the formalism of the method of multiple scales [35] to the more general equation (5) written in terms of the two independent parameters c_d and c_n, with $\Delta r > 0$ and $a_i < 0$. In the limit $q \to 0$, the sideband growth rate σ_r is found to admit the Taylor series expansion

$$\sigma_r \approx -(1 + c_d c_n)q^2 - \tfrac{1}{2}c_d^2(1 + c_n^2)q^4. \tag{16}$$

Let the parameter $\delta \ll 1$ denote the order of magnitude of the sideband wavenumber q. Two specific situations will be discussed.

First we assume that $1 + c_d c_n = 0(1)$ so that the first term in (16) dominates. In other words, we consider points of the (c_n, c_d) plane (fig. 2) which lie well within the interior of the linear stability boundary. Equivalently, we restrict our attention to a vertical strip of fig. 16 defined as $c_0 = 0(1)$, $q = 0(\delta)$. According to eq. (16), the growth rate is then $\mathcal{O}(\delta^2)$ and we should introduce the slow scales $\tau = \delta^2 t$ and $\chi = \delta x$. The wave amplitude $A(x, t)$ is taken to be of the form

$$A(x, t) \approx e^{ic_n t + i\theta(\chi, \tau)}. \tag{17}$$

Thus, to leading order, the Stokes wavetrain $\exp(-ic_n t)$ is phase-modulated by an unknown real phase function $\theta(\chi, \tau)$. Application of a suitable orthogonality condition at $\mathcal{O}(\delta)$ leads to the evolution equation

$$\frac{\partial \theta}{\partial \tau} = (1 + c_d c_n)\frac{\partial^2 \theta}{\partial \chi^2} - (c_d - c_n)\left(\frac{\partial \theta}{\partial \chi}\right)^2. \tag{18}$$

When written in terms of the new variable $u \equiv \partial\theta/\partial\chi$, eq. (18) reduces to Burgers' equation

$$\frac{\partial u}{\partial \tau} + 2(c_d - c_n)u\frac{\partial u}{\partial \chi} = (1 + c_d c_n)\frac{\partial^2 u}{\partial \chi^2}. \tag{19}$$

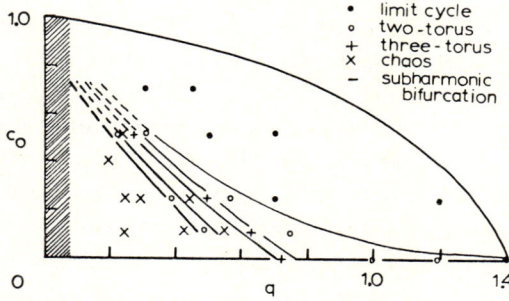

Fig. 16. Distribution of dynamical states in the q, c_0 plane. $\bullet$, limit cycle; 0, two-torus; $+$, three-torus; X, chaos; — subharmonic bifurcation; hatched region indicates possible extent of phase turbulence behavior.

Note that, in the linearly unstable domain, $1 + c_d c_n < 0$, which leads to negative diffusion. In the present context, Burgers' equation is therefore of limited utility since its linearized form does not yield a finite range of unstable wavenumbers. The difficulty arises because, in rescaling the variables, the higher-order term in (16) has been assumed negligible.

In the second case, the quantity $1 + c_d c_n$ is assumed to be $\mathcal{O}(\delta^2)$ in order to balance both terms in (16). Strictly speaking, the analysis is restricted to points of the (c_n, c_d) plane located within a band of $\mathcal{O}(\delta^2)$ of the stability boundary. In the more restricted case, $c_d = -c_n = 1/c_0$, studied in sections 2 and 3, the region of interest in the (q, c_0) plane (see fig. 16) lies in the neighborhood of the point $q = 0$, $c_0 = 1$. The spatial scale is left unchanged but the time scale is redefined as $\tau = \delta^4 t$, in accordance with an order of magnitude analysis of (16). Finally, the wave amplitude is rewritten as

$$A(x, t) \approx e^{-ic_n t + i\delta^2 \theta(\chi, \tau)}. \tag{20}$$

The same type of analysis then leads to the new amplitude equation

$$\frac{\partial \theta}{\partial \tau} = \frac{(1 + c_d c_n)}{\delta^2} \frac{\partial^2 \theta}{\partial \chi^2} - \frac{1}{2}(1 + c_d^2) \frac{\partial^4 \theta}{\partial \chi^4}$$
$$- (c_d - c_n) \left(\frac{\partial \theta}{\partial \chi}\right)^2, \tag{21}$$

which, in terms of u, becomes the modified Burgers equation

$$\frac{\partial u}{\partial \tau} + 2(c_d - c_n) u \frac{\partial u}{\partial \chi} = \frac{(1 + c_d c_n)}{\delta^2} \frac{\partial^2 u}{\partial \chi^2}$$
$$- \frac{1}{2}(1 + c_d^2) \frac{\partial^4 u}{\partial \chi^4}. \tag{22}$$

By comparison with eq. (18), this evolution model, when linearized, is well-behaved, a finite range of wavenumbers being unstable as relation (16) also indicates. Kuramoto's equation (21) arises in other contexts: it governs the hydrodynamic instability of viscous films flowing down a vertical wall at large values of surface tension [37]. The same equation also describes the spanwise deformations of initially plane wavefronts propagating in a two-dimensional medium, in the presence of reaction and diffusion mechanisms [38]. Computer simulations [20, 39, 40] of this equation with periodic boundary conditions on the interval [0, $2\pi/Q$] imply that, as Q decreases, $\theta(\chi, \tau)$ undergoes successive bifurcations from a limit cycle to a two-torus and finally to a turbulent régime. Correspondingly, we note that the motion of the magnitude of A, as given from (20), is, to leading order, simply periodic and uniform in space, while its phase $\delta^2 \theta(\chi, \tau)$ experiences fluctuations which may be chaotic. The peculiar nature of the long-time dynamics described by (20) and (21) at low values of q is referred by Kuramoto, as being *phase turbulence*. From the point of view of the spatio-temporal evolution, the system consists of waves moving along x without significant changes in amplitude. Each revolution over the interval [0, $2\pi/q$], however, requires a slightly different amount of time, due to the chaotic fluctuations in the phase. The word "*amplitude turbulence*" is then reserved for the more general phenomenology in which both $|A|$ and θ undergo turbulent fluctuations. The régimes described in sections 2 and 3 do not strictly belong to the phase turbulence region of fig. 16, and the detailed study of the spatio-temporal structure remains to be undertaken.

As mentioned in the introduction, the Ginzburg-Landau equation reduces to the NLS (6) in the limit of zero c_0 or infinite c_d and c_n. The integrability of the NLS on the infinite interval has been clearly established theoretically [12]. More specifically, arbitrary initial conditions $A(x, 0)$ evolve into a finite number of envelope solitons. The same general conclusion cannot be drawn for the case of periodic boundary conditions. However, for a restricted class of initial conditions, namely those which only give rise to an M-band potential in the associated scattering problem, the motion is quasi-periodic with M independent frequencies [15]. The application of singular perturbation techniques has also shown that the recurrent motion is indeed the result of the quasi-periodicity of the solutions [14]. Our numerical results clearly indicate that, for the class of initial conditions (11), the motion is doubly periodic (see fig. 16) in the range $0.75 < q < 1.4$. As q is lowered below 0.75, the time evolution becomes triply periodic. However, we have been unable to reliably determine the nature of the next dynamical state as the sideband wavenumber decreases below 0.70. A measure of the numerical errors is obtained by following the deviation from the conserved quantity $I = \int_0^{2\pi/q} |A|^2 \, dx$ during the course of the calculations. The results from the pseudo-spectral code were found to be in error by 6% at $q = 1.2$ and by 10% at $q = 0.70$ after about only 10^5 time steps. At lower values of q, the error grew even more rapidly. There remains to investigate the possible existence of higher dimensional tori for the periodic NLS ($c_0 = 0$) as q is lowered below 0.70. Also of interest would be the manner in which the nature of the solutions changes when a small but finite amount of dissipation, as measured by c_0, is introduced.

6. Concluding remarks

The Ginzburg–Landau partial differential evolution equation (5) with periodic boundary conditions, a rational model of many unstable dissipative systems close to critical, has been investigated numerically along the line $c_d = -c_n = 1/c_0$. It has been shown that, for a particular class of initial conditions, the long-time dynamical state may exhibit transitions to chaos in the manner depicted by Ruelle, Takens and Newhouse. Particular features may include a three-frequency stage or a subharmonic bifurcation. The survey of available results conducted in section 5 suggests a wealth of possible asymptotic states, only a few of which have been fully explored.

The study of partial differential equations, which formally possess an infinite number of degrees of freedom, allows the full determination of the spatial structure underlying the temporal evolution, without any a priori truncation assumption. In this regard, a recent investigation of the damped sine-gordon equation with external sinusoidal [41] forcing gives an illustration of the variety of spatial patterns which may be generated. The number of spatial modes participating in the long-time dynamics was, in the present study, relatively small (five modes, i.e. ten degrees of freedom). At low values of the sideband wavenumber q, many more components are likely to become excited according to the following heuristic mechanism. At a given q, nonlinear interactions transfer energy to the higher harmonics $q, 2q, 3q, \ldots$, etc. Among those, only a finite number lie within the linearly unstable range of fig. 3. In other words, the process of energy transfer does not continue indefinitely since dissipation becomes arbitrarily large at very high wavenumbers. Nonetheless, many components may lie within the unstable band of wavenumbers at very low q, and it can be expected that the dimension of the attractor will increase accordingly. Indeed, the investigation of a differential-delay equation [42] a system with infinitely many degrees of freedom, has revealed that the dimension of the attractor increases linearly with the magnitude of the time delay.

The particular values of the coefficients c_d and c_n chosen in this study do not pertain to a specific physical situation and comparison with available experiments should await a more detailed investigation of the possible asymptotic states in the c_d–c_n plane. We note, however, that many of the features reported here are qualitatively similar to the phenomena observed in Bénard convection experiments as the aspect ratio is varied [25]. Finally, a recent dynamical analysis of a baroclinic instability experiment, conducted in an axisymmetric container [43], shares many common characteristics with the transition sequence discussed in section 3.

Acknowledgements

The authors wish to thank Doyne Farmer and Norman Packard for many helpful discussions. This work was supported by the National Science Foundation under Grants No. CME-7916162 and No. MEA-8114780.

References

[1] E.N. Lorenz, J. Atmos. Sci. 20 (1963) 130.
[2] J.H. Curry, Commun. Math. Phys. 60 (1978) 193.
[3] J.H. Curry, Phys. Rev. Lett. 43 (1979) 1013.
[4] J. McLaughlin and P.C. Martin, Phys. Rev. A12 (1975) 186.
[5] A.C. Newell, Lect. Appl. Math. 15 (1974) 157.
[6] J.D. Gibbon and M.J. McGuinness, Proc. R. Soc. Lond. A377 (1981) 185.
[7] K. Stewartson and J.T. Stuart, J. Fluid Mech. 48 (1971) 529.
[8] L.M. Hocking, K. Stewartson and J.T. Stuart, J. Fluid Mech. 51 (1972) 705.
[9] A.C. Newell and J.A. Whitehead, J. Fluid Mech. 38 (1969) 279.
[10] S. Kogelman and R.C. DiPrima, Phys. Fluids 13 (1970) 1.
[11] H. Hasimoto and H. Ono, J. Phys. Soc. Japan 33 (1972) 805.
[12] V.E. Zakharov and A.B. Shabat, Sov. Phys. JETP 34 (1972) 62.
[13] H.C. Yuen and B.M. Lake, Ann. Rev. Fluid Mech. 12 (1980) 303.
[14] B. Hafizi, Phys. Fluids 24 (1981) 1791.
[15] Y.C. Ma and M.J. Ablowitz, Stud. Appl. Math. 65 (1981) 113.
[16] L.N. Howard, in Dynamics and Modelling of Reactive Systems (Academic Press, New York, 1980).
[17] N. Kopell and L.N. Howard, Stud. Appl. Math. 64 (1981) 1.
[18] Y. Kuramoto and T. Tsuzuki, Prog. Theor. Phys. 52 (1974) 1399.
[19] Y. Kuramoto and T. Tsuzuki, Prog. Theor. Phys. 54 (1975) 687.
[20] Y. Kuramoto, Suppl. Prog. Theor. Phys. 64 (1978) 346.
[21] J.T. Stuart and R.C. DiPrima, Proc. R. Soc. Lond. A362 (1978) 27.
[22] H.T. Moon, Transition to Chaos in the Ginzburg–Landau equation, Ph.D. Dissertation, University of Southern California, Los Angeles (1982).
[23] B. Fornberg and G.B. Whitham, Phil. Trans. R. Soc. Lond. 289 (1978) 373.
[24] H.T. Moon, P. Huerre and L.G. Redekopp, Phys. Rev. Lett. 49 (1982) 458.
[25] J.P. Gollub and S.V. Benson, J. Fluid Mech. 100 (1980) 449.
[26] D. Ruelle and F. Takens, Commun. Math. Phys. 20 (1971) 167.
[27] S. Newhouse, D. Ruelle and F. Takens, Commun. Math. Phys 64 (1978) 35.
[28] S.J. Shenker, "Scaling Behavior in a Map of the Circle onto itself: Empirical results", University of Chicago, preprint.
[29] D.A. Rand, S. Ostlund, J. Sethna and E.D. Siggia, "A universal transition from quasi-periodicity to chaos in dissipative systems", preprint.
[30] D.A. Rand, "Dynamics and symmetry. Predictions for modulated waves in rotating fluids", to appear in Arch. Rational Mech. Anal.
[31] M. Gorman, H.L. Swinney and D.A. Rand, Phys. Rev. Lett. 46 (1981) 992.
[32] M.J. Feigenbaum, J. Stat. Phys. 19 (1978) 25.
[33] M.J. Feigenbaum, Commun. Math. Phys. 77 (1980) 25.
[34] V. Franceschini, "Bifurcations of tori and phase locking in a dissipative system of differential equations", submitted to Physica D.
[35] A.H. Nayfeh, Perturbation Methods (Wiley, New York, 1973).
[36] Y. Kuramoto and T. Tsuzuki, Prog. Theor. Phys. 55 (1976) 356.
[37] D.J. Benney, J. Math. and Phys. 45 (1966) 150.
[38] Y. Kuramoto, Prog. Theor. Phys. 63 (1980) 1885.
[39] G.I. Sivashinsky and D.M. Michelson, Prog. Theor. Phys. 63 (1980) 2112.
[40] J.P. Crutchfield and N.H. Packard, private communication.
[41] D. Bennett, A.R. Bishop and S.E. Trullinger, "Coherence and chaos in the driven, damped sine-Gordon chain", to appear in Zeitschrift für Physik.
[42] J.D. Farmer, Physica 4D (1982) 366.
[43] J.D. Farmer, P.D. Weidman and J. Hart, "A phase space analysis of Baroclinic Flow", Phys. Lett. 91A (1982) 22.

CHAPTER 4

DIMENSION, FRACTAL STRUCTURES, AND COHERENCE VERSUS CHAOS

THE DIMENSION OF CHAOTIC ATTRACTORS

J. Doyne FARMER
Center for Nonlinear Studies and Theoretical Division, MS B258, Los Alamos National Laboratory, Los Alamos, New Mexico 87545, USA

Edward OTT
Laboratory of Plasma and Fusion Energy Studies, University of Maryland, College Park, Maryland, USA

and

James A. YORKE
Institute for Physical Science and Technology and Department of Mathematics, University of Maryland, College Park, Maryland, USA

Dimension is perhaps the most basic property of an attractor. In this paper we discuss a variety of different definitions of dimension, compute their values for a typical example, and review previous work on the dimension of chaotic attractors. The relevant definitions of dimension are of two general types, those that depend only on metric properties, and those that depend on the frequency with which a typical trajectory visits different regions of the attractor. Both our example and the previous work that we review support the conclusion that all of the frequency dependent dimensions take on the same value, which we call the "dimension of the natural measure", and all of the metric dimensions take on a common value, which we call the "fractal dimension". Furthermore, the dimension of the natural measure is typically equal to the Lyapunov dimension, which is defined in terms of Lyapunov numbers, and thus is usually far easier to calculate than any other definition. Because it is computable and more physically relevant, we feel that the dimension of the natural measure is more important than the fractal dimension.

Table of contents
1. Introduction 153
2. Definitions of dimension 157
3. Lyapunov numbers and dimension 161
4. Generalized baker's transformation: scaling properties 164
5. Distribution of probability 168
6. Computation of probabilistic dimensions 170
7. The core of attractors 174
8. An attractor that is a nowhere differentiable torus 175
9. Review of numerical experiments 176
10. Conclusions 178

1. Introduction

It is the purpose of this paper to discuss and review questions relating to the dimension of chaotic attractors. Before doing so, however, we should first say what we mean by the work "attractor".

1.1. Attractors

In this paper we consider dynamical systems such as maps (discrete time, n)

$$x_{n+1} = F(x_n),$$

or ordinary differential equations (continuous

time, t)

$$\frac{\mathrm{d}x(t)}{\mathrm{d}t} = G(x(t)),$$

where in both cases x is a vector. Thus, given an initial value of x (at $n = 0$ for the map or $t = 0$ for the differential equations) an orbit is generated $((x_1, x_2, \ldots, x_n, \ldots)$ for the map and $x(t)$ for the differential equations). We shall be interested in attractors for such systems. Loosely speaking, an attractor is something that "attracts" initial conditions from a region around it once transients have died out. More precisely, an *attractor* is a compact set, A, with the property that there is a neighborhood of A such that for almost every* initial condition the limit set of the orbit as n or $t \to +\infty$ is A. Thus, almost every trajectory in this neighborhood of A passes arbitrarily close to every point of A. The *basin of attraction* of A is the closure of the set of initial conditions that approach A.

We are primarily interested in *chaotic* attractors. We give a definition of chaos in section 3, but the reader may also wish to see the reviews given in references 1–4.

1.2. Why study dimension?

The dimension of an attractor is clearly the first level of knowledge necessary to characterize its properties. Generally speaking, we may think of the dimension as giving, in some way, the amount of information necessary to specify the position of a point on the attractor to within a given accuracy (cf. section 2). The dimension is also a lower bound on the number of essential variables needed to model the dynamics. For an extensive discussion of dimension in many contexts, see Mandelbrot [5, 6, 46].

---

* The phrase "almost every" here signifies that the set of initial conditions in this neighborhood for which the corresponding limit set is not A can be covered by a set of cubes of arbitrarily small volume (i.e. has Lebesgue measure zero).

† Mod 1 means that the values of x and y are truncated to be less than or equal to one and their integer part are discarded, so that the map is defined on the unit square.

For simple attractors, defining and determining the dimension is easy. For example, using any reasonable definition of dimension, a stationary time independent equilibrium (fixed point) has dimension zero, a stable periodic oscillation (limit cycle) has dimension one, and a doubly periodic attractor (2-torus) has dimension two. It is because their structure is very regular that the dimension these simple attractors takes on integer values.

Chaotic (strange) attractors, however, often have a structure that is not simple; they are often not manifolds, and frequently have a highly fractured character. For chaotic attractors, intuition based on properties of regular, smooth examples does not apply. The most useful notions of dimension take on values that are typically not integers.

To fully understand the properties of a chaotic attractor, one must take into account not only the attractor itself, but also the "distribution" or "density" of points on the attractor. This is more precisely discussed in terms of what we shall call the *natural measure* associated with a given attractor. The natural measure provides a notion of the relative frequency with which an orbit visits different regions of the attractor. Just as chaotic attractors can have very complicated properties, the natural measures of chaotic attractors often have complicated properties that make the relevant assignment of a dimension a nontrivial problem.

Precise definitions of such terms as "natural measure" follow, but we would first like to give an example in order to motivate the central questions we are addressing in this paper.

Consider the following two dimensional map†:

$$\begin{aligned} x_{n+1} &= x_n + y_n + \delta \cos 2\pi y_n \quad \text{mod } 1, \\ y_{n+1} &= x_n + 2y_n \quad \text{mod } 1. \end{aligned} \quad (1)$$

For small values of δ, Sinai [7] has shown that the attractor of this map is the entire square, and is thus of dimension 2. Therefore almost every initial condition generates a trajectory that eventually comes arbitrarily close to every point on the square. However, consider the typical trajectory

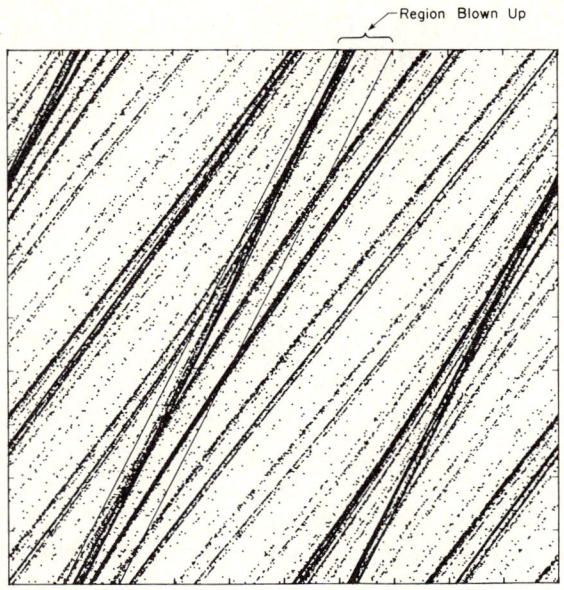

Fig. 1. Successive iterates of the initial point $x_0 = 0.5$, $y_0 = 0.5$ using eq. (1) with $\delta = 0.1$. 80,000 points are shown. Almost any initial condition gives a qualitatively similar plot; the location of the individual points of course changes, but the location of the dark bands does not. The density of these points is described by the *natural measure* of this attractor. (For example, the outlined parallelogram (which is blown up in fig. 2) contains approximately 27% of the points of a typical trajectory, and thus can be said to have a natural measure of approximately 0.27.)

Fig. 2. A blow-up of the strip marked in fig. 1. This strip was chosen in order to follow one of the dark bands; the blow-up was made by expanding the strip in a direction perpendicular to its long sides (roughly horizontally), and the top and bottom were trimmed to make the result square. What appears to be a single band in fig. 1 is now seen as a collection of bands.

shown in fig. 1. Certain regions are visited far more often than others. The natural measure of a given region is proportional to the frequency with which it is visited (see section 2.2.2), in this case the natural measure is highly concentrated in diagonal bands whose density of points is much greater than the average*. Furthermore, as shown in fig. 2, if a small piece of the attractor is magnified, the same sort of structure is still seen.

For this map we do not know if the value of δ chosen to construct fig. 1 is small enough to insure that the dimension of the attractor is two. For practical purposes, though, this may be irrelevant.

---

* In fact, for small values of δ, Sinai [7] has shown that for any $\epsilon > 0$, there exists a collection of tiny squares whose total area is less than ϵ, and such that almost every trajectory spends $1 - \epsilon$ of the time inside this collection of squares. These squares cover what is called the *core* of the attractor. (See section 7).

Even if a trajectory eventually comes arbitrarily close to any given point, the amount of time required for this to happen may be enormous. In order to assign a relevant dimension that will characterize the trajectories on the attractor, the natural measure must be taken into account. For this example the dimension that characterizes properties of the natural measure is between one and two.

These considerations are not as esoteric as they might seem. One may not be as interested in whether the dimension of a given attractor is 3.1 or 3.2 as in whether it is on the order of three or on the order of thirty. As we shall see, a proper understanding of *probabilistic* notions of dimension leads to an efficient method of computing the dimension of chaotic attractors, that provides the best known method of answering such questions.

The main points of this paper can be summarized as follows:

1) Although there are a variety of different definitions of dimension, the relevant definitions

Table I
Current evidence indicates that typically the first two dimensions take on the same value, called the fractal dimension, while the next five dimensions take on another typically smaller value, called the dimension of the measure.

Name of dimension	Symbol	Generic name	Symbol
Capacity	d_C	fractal	d_F
Hausdorff dimension	d_H	dimension	
Information dimension	d_I		
ϑ-capacity	$d_C(\vartheta)$		
ϑ-Hausdorff dimension	$d_H(\vartheta)$	dimension of the	d_μ
Pointwise dimension	d_p	natural measure	
Hausforff dimension of the core	$d_H(\text{core})$		
Lyapunov dimension	d_L		

are of two types, those which only depend on metric properties, and those which depend on metric and probabilistic properties (i.e., they involve the natural measure of the attractor).

2) Current evidence supports the conclusion that all of the metric dimensions typically take on the same value, and all of the frequency dependent dimensions take on another, typically smaller, common value.

3) Current evidence supports a conjectured relationship whereby the dimension of the natural measure can be found from a knowledge of the stability properties of an orbit on the attractor (i.e., knowledge of the Lyapunov numbers).

4) For typical chaotic attractors we conjecture that the distribution of frequencies with which an orbit visits different regions of the attractor is, in a certain sense, log-normal (section 5).

Points 1–3 are summarized in table I. The first two entries in the table are metric dimensions, while the next five are frequency dependent dimensions. Under the hypothesis that all the metric dimensions yield the same value (point 2), we call this value the *fractal dimension* and denote it d_F. Similarly, if all the probabilistic dimensions yield the same value, we call this value the *dimension of the natural measure*, and denote it d_μ. Although in special cases d_F equals d_μ, typically $d_F > d_\mu$. Finally, the last entry in table I, the Lyapunov dimension,

is by definition the predicted value of d_μ obtained from the Lyapunov numbers (cf. Point 3). The Lyapunov dimension is in a different category than the other dimensions listed, since it is defined in terms of dynamical properties of an attractor, rather than metric and natural measure properties.

1.3. Outline

This paper is organized as follows: In section 2 we give several definitions of dimension. Section 3 reviews conjectures relating Lyapunov numbers to dimension. These conjectures are particularly important because the Lyapunov numbers provide the only known efficient method to compute dimension. In sections 4, 5, 6, and 7, we compute all the dimensions discussed here for an explicitly soluble example, the generalized baker's transformation. In addition, based on this example, in section 5 we propose a new conjecture concerning the frequency with which different values of the probability occur. Section 7 gives a discussion of the "core" of attractors, and section 8 gives another example supporting the connection between Lyapunov numbers and dimension (an attractor which is topologically a torus but is nowhere differentiable). Section 9 reviews relevant results from numerical computations of the dimension of chaotic attractors. Concluding remarks are given in section 10.

In general terms, this paper has two functions. One is to present a review of the current status of research on the dimension of chaotic attractors. The other purpose is to present new results (sections 4–6).

2. Definitions of dimension

In this section we define and discuss six different concepts of dimension. The first two of these, the capacity and the Hausdorff dimension, require only a metric (i.e., a concept of distance) for their definition, and consequently we refer to them as "metric dimensions". The other dimensions we will discuss in this section are the information dimension, the ϑ-capacity, the ϑ-Hausdorff dimension, and the pointwise dimension. These dimensions require both a metric and a probability measure for their definition, and hence we will refer to them as "probabilistic dimensions".

In this paper we compute the values of these dimensions for an example that we believe is general enough to be "typical" of chaotic attractors, at least regarding the question of dimension. We find that the metric dimensions take on a common value. Whenever this is the case, we will refer to this common value d_F as the *fractal dimension**. For our example we also find that the probabilistic dimensions take on a common value d_μ, which we will refer to as the *dimension of the natural measure*. As we summarize in conjecture 1, we feel that this equality is a general property, true for typical cases.

Conjecture 1. For a typical chaotic attractor the capacity and Hausdorff dimensions have a common value d_F, and the information dimension, ϑ-capacity, ϑ-Hausdorff dimension, and pointwise dimensions have a common value d_μ, i.e., in the notation of table I,

$$d_C = d_H \equiv d_F$$

and

$$d_I = d_C(\vartheta) = d_H(\vartheta) = d_P \equiv d_\mu.$$

Note: For the case of diffeomorphisms† in two dimensions, L.S. Young has rigorously proven that information dimension, pointwise dimension, and the Hausdorff dimension of the core (see section 7) all take on the same value [12].

In addition to the dimensions defined in this section, we will also discuss three others‡. The Lyapunov dimension, the capacity of the core, and the Hausdorff dimension of the core. Lyapunov dimension is discussed in section 3, and the latter two dimensions are discussed in section 7. For our example the Lyapunov dimension and Hausdorff dimension of the core are equal to d_μ, while the capacity of the core is equal to d_F.

2.1. Metric dimensions

We begin by discussing two concepts of dimension which apply to sets in spaces on which a concept of distance, i.e., a metric is defined. In particular we begin by discussing the capacity and the Hausforff dimension.

2.1.1. Capacity

The capacity of a set was originally defined by Kolmogorov [14]. It is given by

$$d_C = \lim_{\epsilon \to 0} \frac{\log N(\epsilon)}{\log(1/\epsilon)}, \qquad (2)$$

---

* The term *fractal* was originally coined by Mandelbrot [5]. However, he uses "fractal dimension" as a synonym for Hausdorff dimension. We should also mention that in some of our previous papers on this subject [8–11], we used the term "fractal dimension" as a synonym for capacity, rather than our current usage as described in the text.

† A diffeomorphism is a differentiable invertible mapping whose Jacobian has non-zero determinant everywhere.

‡ Note that in this paper we will not discuss the concept of *topological dimension*, since its application to chaotic dynamics is not clear. Its value is an integer and it is generally equal to neither d_F nor d_μ. For discussions of topological dimension, we refer the reader to Hurwicz and Wallman [13].

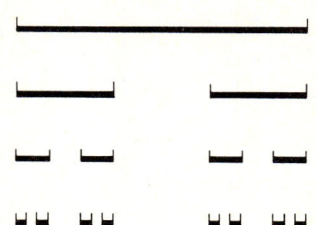

Fig. 3. The first few steps in the construction of the classic example of a Cantor set.

where, if the set in question is a bounded subset of a p-dimensional Euclidean space $\mathbb{R}^p$, then $N(\epsilon)$ is the minimum number of p-dimensional cubes of side ϵ needed to cover the set. For a point, a line, and an area, $N(\epsilon) = 1$, $N(\epsilon) \sim \epsilon^{-1}$, and $N(\epsilon) \sim \epsilon^{-2}$, and eq. (2) yields $d_C = 0, 1,$ and 2, as expected. However, for more general sets (dubbed *fractals* by Mandelbrot), d_C can be noninteger*. For example, consider the Cantor set obtained by the limiting process of deleting middle thirds, as, illustrated in fig. 3. If we choose $\epsilon = (1/3)^m$, then $N = 2^m$, and eq. (2) yields

$$d_C = \frac{\log 2}{\log 3} = 0.630 \ldots .$$

If one is content to know where the set lies to within an accuracy ϵ, then to specify the location of the set, we need only specify the position of the $N(\epsilon)$ cubes covering the set. Eq. (2) implies that for small ϵ, $\log N(\epsilon) \approx d_C \log(1/\epsilon)$. Hence, the dimension tells us how much information is necessary to specify the location of the set to within a given accuracy. If the set has a very fine-scaled structure (typical of chaotic attractors), then it may be advantageous to introduce some coarse-graining into the description of the set. In this case, ϵ may be thought of as specifying the degree of coarse-graining.

2.1.2. *Hausdorff dimension*

The capacity may be viewed as a simplified version of the Hausdorff dimension, originally introduced by Hausdorff in 1919 [15]. (We have reversed historical order and defined capacity before Hausdorff dimension because the definition of Hausdorff dimension is more involved.) We believe that for attractors these two dimensions are generally equal. While it is possible to construct simple examples of sets where the Hausdorff dimension and the capacity are unequal†, these do not seem to apply to attractors. (Although they may apply to the core of attractors. See section 7.)

To define the Hausdorff dimension of a set lying in a p-dimensional Euclidean space, consider a covering of it with p-dimensional cubes of variable edge length ϵ_i. Define the quantity $l_d(\epsilon)$ by

$$l_d(\epsilon) = \inf \sum_i \epsilon_i^d,$$

where the infimum (i.e. minimum) extends over all possible coverings subject to the constraint that $\epsilon_i \leq \epsilon$. Now let

$$l_d = \lim_{\epsilon \to 0} l_d(\epsilon).$$

Hausdorff showed that there exists a critical value of d above which $l_d = 0$ and below which $l_d = \infty$. This critical value, $d = d_H$, is the Hausdorff dimension. (Precisely at $d = d_H$, l_d may be either 0, ∞, or a positive finite number.) This concept of dimension will be used in sections 4, 6, and 7. It is easy to see that $d_C \geq d_H$‡.

---

* Sets can be constructed for which the limit of eq. (2) does not exist. We would then say that the capacity is not defined.

† For example, for the set of numbers 1, 1/2, 1/3, 1/4,....., the Hausdorff dimension is zero while (2) yields $d_C = \frac{1}{2}$.

‡ To show that $d_C \geq d_H$, consider a covering consisting of cubes of equal side $\epsilon_i = \epsilon$. Then due to the infimum in the definition of $l_d(\epsilon)$, we see that $\bar{l}_d(\epsilon) \equiv \Sigma_i \epsilon^d = N(\epsilon) \epsilon^d$ satisfies $\bar{l}_d(\epsilon) \geq l_d(\epsilon)$. Thus taking the limit $\epsilon \to 0$ and making use of eq. (2) we see that $d_H \leq d_C$.

2.2. Dimensions for the natural measure

2.2.1. The natural measure on an attractor

Note that, in computing d_C from eq. (2), all cubes used in covering the attractor are equally important even though the frequencies with which an orbit on the attractor visits these cubes may be very different. In order to take the frequency with which each cube is visited into account, we need to consider not only the attractor itself, but the relative frequency with which a typical orbit visits different regions of the attractor as well. We can say that some regions of the attractor are more probable than others, or alternatively we may speak of a measure on the attractor*. We define the natural measure of an attractor as follows: For each cube C and initial condition x in the basin of attraction, define $\mu(x, C)$ as the fraction of time that the trajectory originating from x spends in C†. If almost every such x gives the same value of $\mu(x, C)$, we denote this value $\mu(C)$ and call μ the *natural measure* of the attractor [16]. The natural measure gives the relative probability of different regions of the attractor as obtained from time averages, and therefore is the "natural" measure to consider. We will assume throughout that any attractor we consider has a natural measure, at least whenever C is one of the cubes we are using to cover the attractor.

The four definitions discussed in the remainder of this section are defined for attractors with a metric and a natural measure defined on them.

2.2.2. Information dimension

The *information dimension*, d_I, is a generalization of the capacity that takes into account the relative probability of the cubes used to cover the set. This dimension was originally introduced by Balatoni and Renyi [17].

The information dimension is given by

---

* Although there are many measures possible for a given attractor, we are only interested in one of them, the natural measure.

† $\mu(x, C) = \lim_{\tau \to \infty} \mu_\tau(x, C)$, where $\mu_\tau(X, C)$ is the fraction of time spent in C up to some finite time τ.

$$d_I = \lim_{\epsilon \to 0} \frac{I(\epsilon)}{\log(1/\epsilon)}, \qquad (3)$$

where

$$I(\epsilon) = \sum_{i=1}^{N(\epsilon)} P_i \log \frac{1}{P_i}$$

and P_i is the probability contained within the ith cube. Letting the ith cube of side ϵ be C_i, $P_i = \mu(C_i)$. Note that if all cubes have equal probability then $I(\epsilon) = \log N(\epsilon)$, and hence $d_C = d_I$. However, for unequal probabilities $I(\epsilon) < \log N(\epsilon)$. Thus, in general, $d_C \geq d_I$.

In information theory the quantity $I(\epsilon)$ defined in eq. (3) has a specific meaning [18]. Namely, it is the amount of information necessary to specify the state of the system to within an accuracy ϵ, or equivalently, it is the information obtained in making a measurement that is uncertain by an amount ϵ. Since for small ϵ, $I(\epsilon) \approx d_I \log(1/\epsilon)$, we may view d_I as telling how fast the information necessary to specify a point on the attractor increases as ϵ decreases. (For a more extensive discussion of the physical meaning of the information dimension, see refs. 9 and 10.)

2.2.3. ϑ-Capacity

Another definition of dimension which we shall be interested in is what we will call the ϑ-capacity, $d_C(\vartheta)$. Essentially, this quantity is the capacity of that part of the attractor of highest probability,

$$d_C(\vartheta) = \lim_{\epsilon \to 0} \frac{\log N(\epsilon; \vartheta)}{\log(1/\epsilon)}, \qquad (4)$$

where $N(\epsilon; \vartheta)$ is the minimum number of cubes of side ϵ needed to cover at least a fraction ϑ of the natural measure of the attractor. In other words, the cubes must be chosen so that their combined natural measure is at least ϑ. Thus $d_C(1) = d_C$. For the examples we study here, we find that for any value of $\vartheta < 1$, the ϑ-capacity is independent of ϑ, but that $d_C(\vartheta)$ for $\vartheta < 1$ may differ from its value at $\vartheta = 1$. In particular $d_C(\vartheta) = d_\mu$ for $\vartheta < 1$ and

$d_C(\vartheta) = d_C$ for $\vartheta = 1$. ϑ-capacity was originally defined by Frederickson et al. [8]. Similar quantities have also been defined by Ledrappier [18], and Mandelbrot [6, 45].

2.2.4. ϑ-Hausdorff dimension

In analogy with the relationship between capacity (a metric dimension) and ϑ-capacity (a probability dimension), we introduce here a probability dimension based on the Hausdorff dimension. We call this new dimension the ϑ-Hausdorff dimension and denote it $d_H(\vartheta)$. To define the ϑ-Hausdorff dimension, modify the definition of Hausdorff dimension as follows: Define $l_d(\epsilon, \vartheta)$ by

$$l_d(\epsilon, \vartheta) = \inf \sum_i \epsilon_i^d,$$

where now the infimum extends over all possible $\epsilon_i < \epsilon$ which cover a fraction ϑ of the total probability of the set. We define $d_H(\vartheta)$ as that value of d below which $l_d(\vartheta) = \infty$ and above which $l_d(\vartheta) = 0$, where $l_d(\vartheta) = \lim_{\epsilon \to 0} l_d(\epsilon, \vartheta)$. This concept of dimension will be used in section 6.

2.2.5. Pointwise dimension

Roughly speaking, the pointwise dimension d_p is the exponent with which the total probability contained in a ball decreases as the radius of the ball decreases. To make this notion more precise, let μ denote the natural probability measure on the attractor, and let $B_\epsilon(x)$ denote a ball of radius ϵ centered about a point x on the attractor. Roughly speaking, $\mu(B_\epsilon(x)) \sim \epsilon^{d_p}$. More precisely, define this dimension as

$$d_p(x) = \lim_{\epsilon \to 0} \frac{\log \mu(B_\epsilon(x))}{\log \epsilon}. \tag{5}$$

If $d_p(x)$ is independent of x for almost all x with respect to the measure μ^*, we call $d_p(x) = d_p$ the

---

* By "almost all x with respect to the measure μ" we mean that the set of x which does not satisfy this is a set of μ measure zero.

pointwise dimension. Similar definitions of dimension have also been given by Takens [20], Billingsley [31], Young [11], and Janssen and Tjon [21].

2.3. Using a grid of cubes to compute dimension

Some of the definitions we have used, such as the capacity, allow any location or orientation of the cubes used to cover the attractor. In a numerical experiment, however, it is much more convenient to select the cubes used to cover the attractor out of a fixed grid, as shown in fig. 4. For these dimensions (d_C, d_I, and $d_C(\vartheta)$) it can be shown that selecting from a fixed grid of cubes gives the same value of the dimension as an optimal collection of cubes. For example, for the case of an attractor in a two-dimensional space, using a fixed grid to compute $N(\epsilon)$ in eq. (2) results in an increase of at most a factor of four in $N(\epsilon)$, which has no effect on the value of the dimension. Note that this is *not* true for the Hausdorff dimension, which requires a more general cover.

In principle, the definitions of dimension given in this section and the use of a fixed grid provide specific prescriptions for obtaining capacity, information dimension, and ϑ-capacity. To find approximate values for these dimensions, one can generate an orbit on the attractor using a computer, and then divide the space containing the orbit into cubes of side ϵ in order to estimate the numbers $N(\epsilon)$, $I(\epsilon)$, or $N(\epsilon; \vartheta)$. By examining how

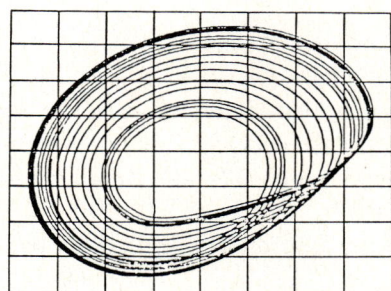

Fig. 4. The region of phase space containing an attractor can be divided with a fixed grid of cubes (in this case squares), which can be used to compute capacity, information dimension, or ϑ-capacity.

$N(\epsilon)$, $I(\epsilon)$, and $N(\epsilon; \vartheta)$ vary as ϵ is decreased the value of these dimensions can be estimated.

As discussed in section 9, however, in practice the agenda described above for computing dimension may be difficult, costly, or impossible. Thus it is of interest to consider other means of obtaining the dimension of chaotic attractors. The next section deals with this question. In particular, we discuss a conjecture that the dimension of chaotic attractors can be determined directly from the dynamics in terms of Lyapunov numbers.

3. Lyapunov numbers and Lyapunov dimension

The Lyapunov numbers quantify the average stability properties of an orbit on an attractor. For a fixed point attractor of a mapping, the Lyapunov numbers are simply the absolute values of the eigenvalues of the Jacobian matrix evaluated at the fixed point. The Lyapunov numbers generalize this notion for more complicated attractors. As we shall see, for a typical attractor there is a connection between average stability properties and dimension. The possibility of such a connection was first pointed out by Kaplan and Yorke [22] and later by Mori [23].

3.1. Definition of Lyapunov numbers

For expository purposes, for most of this paper we shall consider p-dimensional maps,

$$x_{n+1} = F(x_n),$$

where x is a p-dimensional vector. We emphasize, however, that similar considerations to those below apply to flows (e.g., systems of differential equations), including infinite-dimensional systems such as partial differential equations. To define the Lyapunov numbers, let $J_n = [J(x_n)J(x_{n-1})\ldots J(x_1)]$ where $J(x)$ is the Jacobian matrix of the map, $J(x) = (\partial F/\partial x)$, and let $j_1(n) \geq j_2(n) \geq \cdots \geq j_p(n)$ be the magnitudes of the eigenvalues of J_n. The Lyapunov numbers are

$$\lambda_i = \lim_{n \to \infty} [j_i(n)]^{1/n}, \quad i = 1, 2, \ldots, p, \tag{6}$$

where the positive real nth root is taken. The Lyapunov numbers generally depend on the choice of the initial condition x_1. The Lyapunov numbers were originally defined by Oseledec [24]. We have the convention

$$\lambda_1 \geq \lambda_2 \geq \cdots \geq \lambda_p.$$

For a two-dimensional map, for example, λ_1 and λ_2 are the average principal stretching factors of an infinitesimal circular arc (cf. fig. 5). For a chaotic attractor on the average nearby points initially diverge at an exponential rate, and hence at least one of the Lyapunov numbers is greater than one. This makes quantitative the notion of "sensitive dependence on ititial conditions". We will take $\lambda_1 > 1$ as our definition of *chaos*. (Note that many authors refer to *Lyapunov exponents* rather than Lyapunov numbers. The Lyapunov exponents are simply the logarithms of the Lyapunov numbers.)

In this paper we assume that *almost every* initial condition in the basin of any attractor that we consider has the same Lyapunov numbers. Thus, the spectrum of Lyapunov numbers may be considered to be a property of an attractor. This assumption is supported by numerical experiments [25]. Exceptional trajectories, such as unstable fixed points on the attractor, typically do not sample the whole attractor and thus typically have Lyapunov numbers that are different from those of the attractor. Those points in the basin of attraction that have different Lyapunov numbers or for

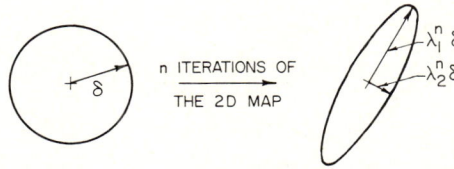

Fig. 5. n iterations of a two-dimensional map transform a sufficiently small circle of radius δ approximately into an ellipse with major and minor radii $(\lambda_1)^n \delta$ and $(\lambda_2^n)\delta$, where λ_1 and λ_2 are the Lyapunov numbers.

which Lyapunov numbers do not exist are here assumed to be of measure zero. (In other words, they may be covered by a collection of cubes of varying size having arbitrarily small total volume).

3.2. Definition of Lyapunov dimension

The following discussion contains a heuristic argument that motivates a connection between Lyapunov numbers and dimension. Consider a two-dimensional map. Suppose we wish to compute the capacity of a chaotic attractor, for which $\lambda_1 > 1 > \lambda_2$. Cover the attractor with $N(\epsilon)$ squares of side ϵ. Now, iterate the map q times. For q fixed and ϵ small enough, the action of the mapping is roughly linear over the square, and each square will be stretched into a long thin parallelogram. From the definition of the Lyapunov numbers, the average length of these parallelograms is $(\lambda_1)^q \epsilon$, and the average width is $(\lambda_2)^q \epsilon$. Now, suppose we had used a finer cover of squares of side $(\lambda_2)^q \epsilon$. (See fig. 6.) To cover each parallelogram takes about $(\lambda_1/\lambda_2)^q$ smaller squares. Thus, *if it is supposed* that all squares on the attractor behave in this typical way, then one is lead to the estimate

$$N(\lambda_2^q \epsilon) \approx \left(\frac{\lambda_1}{\lambda_2}\right)^q N(\epsilon). \tag{7}$$

Motivated by eq. (2), assume $N(\epsilon) \approx k(1/\epsilon)^{d_C}$, and

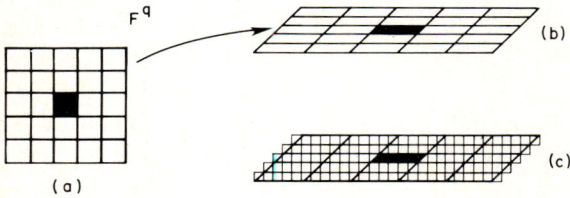

Fig. 6. A schematic illustration of the heuristic argument for the Lyapunov dimension. The image of each small square in (a) is approximately a parallelogram which has been stretched horizontally be a factor of λ_1^q and contracted vertically by a factor λ_2^q. The images in (b) thus have a smaller cover of squares as shown in (c).

substitute into both sides of eq. (7). This gives

$$k\left(\frac{1}{\lambda_2^q \epsilon}\right)^{d_C} \approx k\left(\frac{\lambda_1}{\lambda_2}\right)^q \left(\frac{1}{\epsilon}\right)^{d_C}.$$

Collecting terms, taking logarithms, and solving for d_C gives

$$d_C = 1 + \frac{\log \lambda_1}{\log(1/\lambda_2)}.$$

We will see that this expression is often meaningful even when this heuristic derivation is invalid, so we will call it the *Lyapunov dimension* d_L.

$$d_L = 1 + \frac{\log \lambda_1}{\log(1/\lambda_2)}. \tag{8}$$

Generalization of the above heuristic argument to p-dimensional maps gives (cf. ref 7)

$$d_L = k + \frac{\log(\lambda_1 \lambda_2 \ldots \lambda_k)}{\log(1/\lambda_{k+1})}, \tag{9}$$

where k is the largest value for which $\lambda_1 \lambda_2 \ldots \lambda_k \geq 1$. If $\lambda_1 < 1$, define $d_L = 0$; if $\lambda_1 \lambda_2 \ldots \lambda_p \geq 1$, define $d_L = p$. We shall refer to d_L as the *Lyapunov dimension*. This quantity was originally defined by Kaplan and Yorke [22], who originally gave it as a lower bound on the fractal dimension.

From the above argument one might be tempted to guess that $d_C = d_L$. The Lyapunov numbers are *average* quantities, however, and to compute an average, each cube must be weighted according to its probability. The capacity does not distinguish between probable and improbable cubes. To understand how some cubes might have vastly different probabilities than others, consider an atypical square of a two-dimensional map. If the area of the images of this square decreases half as fast as the average for k iterations, then its kth image will be 2^k times larger than the image of a typical square, and the number of squares needed to cover it will be 2^k times greater than the typical value. In fact, as will be evident from considerations of explicit examples (cf. section 5), it is commonly the case that the vast majority of cubes needed to cover the attractor are atypical, and do not represent the properties of time averages. By this we mean that all the atypical cubes taken together contain an extremely small fraction of the total probability on the attractor yet account for almost all of $N(\epsilon)$. Furthermore, this tendency increases as ϵ decreases. The behavior of the atypical cubes under iteration is in general not described by the Lyapunov numbers. It is clear, then, that in order for this estimate to be valid, we must consider only the more probable cubes, i.e., the estimate should be in terms of the dimension of the natural measure rather than the capacity. Assuming the equality of probabilistic dimensions (conjecture 1), we are led to the following conjecture:

Conjecture 2. For a typical* attractor $d_\mu = d_L$.

In the following six sections we present evidence supporting this conjecture. Also, L.S. Young has proved some rigorous results along these lines, which are reviewed in the next subsection.

In the special case that *every* initial condition on the attractor generates the same Lyapunov numbers, we will say that the attractor has *absolute* Lyapunov numbers. In this case it is not necessary to distinguish probable from improbable cubes, and the above conjecture can be made in terms of

---

* The reason for the word "*typical*" is that there exist examples of maps that do not satisfy $d_\mu = d_L$. These maps are exceptional, however, in that arbitrarily small perturbations of them restore the conjectured equality of d_μ and d_L. An example of such an atypical case is where a point x_0 is attracting and yet has $\lambda_1 = 1$ (i.e., the Jacobian matrix $\partial F/\partial x$ has an eigenvalue $+1$ at x_0). The attraction here is due to higher order terms. The attractor is a point and so has dimension zero, yet $d_L \geq 1$. Small perturbations, however, will destroy this delicate balance. For example, the one-dimensional map $x_{i+1} = F(x_i) \equiv \alpha x_i - x_i^3$ has a fixed point at $x = 0$ with $\lambda_1 = 1$ for $\alpha = 1$ yet $x = 0$ is attracting. This situation is changed, however, as soon as $\alpha \neq 1$. When $|\alpha| < 1$, $d_L = 0$, and when $|\alpha| > 1$, $x = 0$ is no longer attracting.

the fractal dimension rather than the dimension of the natural measure. We call this conjecture 3,

Conjecture 3. If *every* (not just almost every) initial condition generates the same set of p Lyapunov numbers $\lambda_1, \lambda_2, \ldots \lambda_p$, and if $\lambda_1 > 1$, then for a typical attractor of this type $d_F = d_L = d_\mu$.

The requirement of conjecture 3 that every initial condition on the attractor generate the same Lyapunov numbers is very restrictive and only holds for special cases. For example, it holds if the Jacobian matrix of the map is independent of x. In more general cases, the requirement of conjecture 3 would be expected to fail because of the existence of unstable fixed and periodic points on the attractor. For example, if x_1 is chosen to be precisely on an unstable fixed point, the Lyapunov numbers generated will simply be the eigenvalues of $J(x_1)$. These will typically be different from those generated by a chaotic orbit on the attractor. Examples for which conjecture 3 is valid will be special cases of the more general example presented in the following section. In addition, an example for which conjecture 3 can be proven to hold is given in section 8.

3.3. *Review of rigorous results concerning Lyapunov dimension*

In addition to the analytic and numerical evidence we will give for conjectures 1–3 in the remainder of this paper, there are several rigorous results supporting these statements which are reviewed in this section. For example, Ledrappier [19] has proven an inequality that is somewhat similar to conjecture 2. In particular, he defines a dimension that we will call d_{Led}, which is the ϑ-capacity in the limit as ϑ goes to one, i.e.

$$d_{\text{Led}} = \lim_{\vartheta \to 1} d_C(\vartheta).$$

For C^2 diffeomorphisms he has shown that

$$d_L \geq d_{\text{Led}}.$$

The proof is a rigorous version of the heuristic argument that we have given (fig. 6). Also, Douady and Oesterle [26] have proven that an upper bound for the fractal dimension can be obtained yielding an expression like eq. (8), where the numbers they use are basically upper bounds for the Lyapunov numbers.

L.S. Young [12] has proven several results that strongly support conjectures 1 and 2. Particularly relevant are the following two theorems*.

1. If d_p exists then

$$d_p = d_I = d_H(\text{core}) = d_{\text{Led}}. \tag{10}$$

2. For two-dimensional C^2 diffeomorphisms with $\lambda_1 > 1 > \lambda_2$, d_p exists, and

$$d_p = \frac{h_\mu}{\log \lambda_1}\left(1 + \frac{\log \lambda_1}{\log(1/\lambda_2)}\right). \tag{11}$$

(See section 7 for a definition of $d_H(\text{core})$.) h_μ denotes the Kolmogorov entropy† of the attractor taken with respect to the measure μ, and λ_1 and λ_2 are the Lyapunov numbers with respect to μ. (More precisely, almost every initial condition x with respect to μ give λ_1 and λ_2 as the Lyapunov numbers.)

For Axiom-A attractors Bowen and Ruelle [16] have shown that there is a natural measure such that h_μ with respect to this measure is the sum of the logarithms of the Lyapunov numbers that are greater than one. For attractors with only one Lyapunov number greater than one, this implies that $h_\mu = \log \lambda_1$. Thus, for axiom-A attractors of two-dimensional maps, eqs. (9)–(11) yield $d_\mu = d_L$. Therefore Young has shown that conjecture 2 holds for this case. (It has been conjectured that the relationship between h_μ and the positive λ_i holds for non-axiom-A attractors that have a natural measure.) This result for the case of axiom-A attractors of two-dimensional maps has also been obtained independently by Pelikan [30].

4. Generalized baker's transformation: scaling

4.1. Definition of generalized baker's transformation

In this section we define the example which we will study in detail in this and the following four sections. Although we feel that this example is general enough to be typical of low-dimensional chaotic attractors (at least concerning its dimensional properties), it is also simple enough that all of the dimensions discussed in this paper can be analytically calculated‡. Thus, for this example, we shall be able to verify conjectures 1–3 in a case where generally $d_F \neq d_\mu$. As we shall show in section 5, another nice property of this map is that it allows us to investigate certain properties of the natural probability distribution in detail.

The map to be considered is

$$x_{n+1} = \begin{cases} \lambda_a x_n, & \text{if } y_n < \alpha, \\ \frac{1}{2} + \lambda_b x_n, & \text{if } y_n > \alpha; \end{cases} \tag{12a}$$

$$y_{n+1} = \begin{cases} \dfrac{1}{\alpha} y_n, & \text{if } y_n < \alpha, \\ \dfrac{1}{1-\alpha}(y_n - \alpha), & \text{if } y_n > \alpha; \end{cases} \tag{12b}$$

where we shall assume $0 \leq x_n \leq 1$ and $0 \leq y_n \leq 1$. If this condition is satisfied initially it is also satisfied

---

* For these results Young does not require the existence of a natural measure, but rather assumes simply the existence of some invariant measure μ. In this case the Lyapunov numbers are those obtained when starting at almost every initial point with respect to μ.

† The Kolmogorov entropy, originally defined by Shannon [18] and applied to dynamical systems by Kolmogorov [27] and Sinai [28], puts a quantitative value on the average amount of new information obtained from a sequence of measurements. See [10] or [29] for physically motivated reviews. Note that this is also called metric entropy. The name *metric* entropy derives from the invariance properties of this quantity; in fact, the definition of metric entropy does not require a metric (but does require a measure).

‡ Except for the ϑ-Hausdorff dimension, for which we only obtain an upper bound.

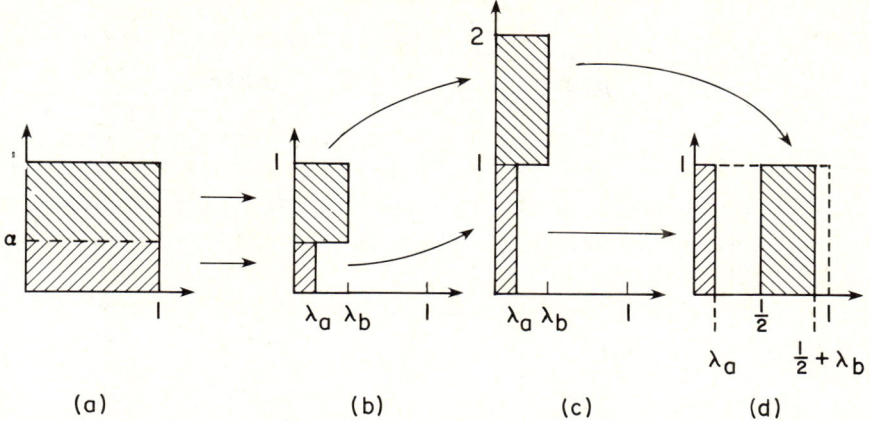

Fig. 7. The generalized baker's transformation. One iteration of the map takes us from (a) to (d). Steps (b) and (c) are conceptual intermediate stages.

at all subsequent iterates. Fig. 7 illustrates the action of this map on the unit square. As shown in fig. 7, we take α, λ_a, $\lambda_b \leq \frac{1}{2}$, and $\lambda_b \geq \lambda_a$. Fig. 8 shows the result of applying the map two times to the unit square. From fig. 8 it is seen that, if the x interval $[0, \lambda_a]$ is magnified by a factor $1/\lambda_a$, it becomes a precise replica of fig. 7d. Similarly, if the x interval $[\frac{1}{2}, \frac{1}{2} + \lambda_b]$ is magnified by $1/\lambda_b$, a replica of fig. 7d again results. This self similarity property of eq. (12) will subsequently be used to obtain d_C, d_I, d_H, and d_p.

4.2. Lyapunov numbers of generalized baker's transformation

Now we consider the Lyapunov numbers. Eq. (12b) involves y alone and consists of a linear stretching on each of the y intervals $[0, \alpha]$ and $[\alpha, 1]$. Thus almost every y initial condition in $[0, 1]$ will generate an ergodic orbit in y with uniform density in $[0, 1]$. The Jacobian of eq. (12) is diagonal and depends only on y.

$$J = \begin{pmatrix} L_2(y) & 0 \\ 0 & L_1(y) \end{pmatrix},$$

where

$$L_2(y) = \begin{cases} \lambda_a, & \text{if } y < \alpha, \\ \lambda_b, & \text{if } y > \alpha, \end{cases}$$

and

$$L_1(y) = \begin{cases} \dfrac{1}{\alpha}, & \text{if } y < \alpha, \\ \dfrac{1}{1-\alpha}, & \text{if } y > \alpha. \end{cases}$$

Thus applying eq. (6) we have

$$\lambda_1 = \lim_{n \to \infty} [L_1(y_n) \ldots L_1(y_1)]^{1/n},$$

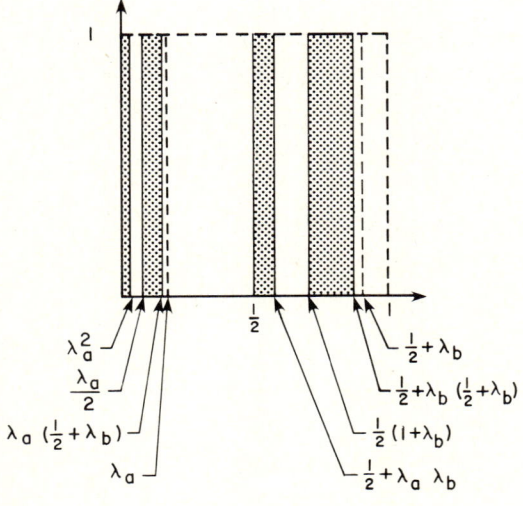

Fig. 8.

or

$$\log \lambda_1 = \lim_{n \to \infty} \left\{ \frac{n_\alpha}{n} \log \frac{1}{\alpha} + \frac{n_\beta}{n} \log \frac{1}{\beta} \right\},$$

where $\beta = 1 - \alpha$. n_α is the number of times the orbit has been in the set $y < \alpha$, and n_β is the number of times the orbit has been in the set $y > \alpha$. Since for *almost* any y_1, the orbit in y is ergodic with uniform density in $[0, 1]$, $\lim_{n \to \infty} n_\alpha/n = \alpha$, and similarly $\lim_{n \to \infty} n_\alpha/n = \beta$. Thus

$$\log \lambda_1 = \alpha \log \frac{1}{\alpha} + \beta \log \frac{1}{\beta}. \tag{13}$$

Similarly, we obtain for λ_2

$$\log \lambda_2 = \alpha \log \lambda_a + \beta \log \lambda_b. \tag{14}$$

To simplify notation in this and subsequent expressions, let

$$H(\alpha) = \alpha \log \frac{1}{\alpha} + (1 - \alpha) \log \frac{1}{1-\alpha}. \tag{15}$$

$H(\alpha)$ is called the *binary entropy function* and is the amount of information contained in a coin-toss where heads has a probability α.

The Lyapunov dimension of the attractor of the generalized Baker's transformation (eq. (12)) is

$$d_L = 1 + \frac{H(\alpha)}{\alpha \log(1/\lambda_a) + \beta \log(1 + \lambda_b)}. \tag{16}$$

In the following sections we compute the values of the dimensions defined in this paper, and show that

---

* To see that for almost all parameter values the Lyapunov numbers of the generalized baker's transformation are not absolute, consider the special initial condition on the attractor with y-value $y_1 = y_a$ where $y_a = \alpha^2(1 - \alpha + \alpha^2)^{-1}$. This initial condition corresponds to one of the points on the unstable period 2 orbit, $(y_a, y_b, y_a, y_b, \ldots)$, where $y_b = \alpha^{-1} y_a$. Since $0 < y_a < \alpha < y_b < 1$, we have $n_\alpha = n_\beta = \frac{1}{2}$, and the Lyapunov numbers generated by this initial condition are $\lambda_1 = (\alpha \beta)^{-1/2}$ and $\lambda_2 = (\lambda_a \lambda_b)^{1/2}$, rather than those given by eq. (13) and (14).

all the probabilistic dimensions take on the value given in eq. (16).

For all but special values of λ_a, λ_b, and α, there exist unstable periodic orbits whose Lyapunov numbers are different from those given in eqs. (13) and (14)*. Thus, in general we expect that conjecture 2 rather than conjecture 3 applies and $d_F \neq d_\mu$.

4.3. *Capacity of generalized Baker's transformation*

To calculate d_C we first note that the attractor is a product of a Cantor set along x and the interval $[0, 1]$ along y. Thus the capacity, or any of the other dimensions, are in the form $d_C \equiv 1 + \bar{d}_C$, where $\bar{d}_C$ is the dimension of the attractor in the x-direction. We will generally use a bar over a dimension to refer to the dimension along the x-direction.

We now obtain $\bar{d}_C$ by making use of the scaling property of the generalized Baker's transformation, discussed at the end of section 4.1. We write $N(\epsilon)$ as

$$N(\epsilon) = N_a(\epsilon) + N_b(\epsilon),$$

where $N_a(\epsilon)$ is the number of x-intervals of length ϵ needed to cover that part of the attractor which lies in the x-interval $[0, \lambda_a]$, and $N_b(\epsilon)$ is the analogous quantity for the x-interval $[\frac{1}{2}, \frac{1}{2} + \lambda_b]$. From the scaling property, $N_a(\epsilon) = N(\epsilon/\lambda_a)$, and similarly $N_b(\epsilon) = N(\epsilon/\lambda_b)$. Thus

$$N(\epsilon) = N(\epsilon/\lambda_a) + N(\epsilon/\lambda_b). \tag{17}$$

Assuming heuristically that $N(\epsilon) \approx k\epsilon^{-\bar{d}_C}$ for small ϵ, substituting into eq. (17) gives

$$k\left(\frac{1}{\epsilon}\right)^{\bar{d}_C} = k\left(\frac{\lambda_a}{\epsilon}\right)^{\bar{d}_C} + k\left(\frac{\lambda_b}{\epsilon}\right)^{\bar{d}_C},$$

implying that

$$1 = \lambda_a^{\bar{d}_C} + \lambda_b^{\bar{d}_C}, \tag{18}$$

which is a transcendental equation for $\bar{d}_C$. As

expected, eqs. (16) and (18) show that, in general, $1 + \bar{d}_C = d_C \neq d_L$. However, for the special choice $\lambda_a = \lambda_b$, $\alpha = \frac{1}{2}$, corresponding to eq. (12) with $\lambda_a = \lambda_1 = 2$, the two agree. Note that for this case the Jacobian matrix is constant, the Lyapunov numbers are therefore absolute, and conjecture 3 applies.

In obtaining eq. (18), in order to keep the argument simple, we have made the strong assumption that $N(\epsilon) \approx k\epsilon^{-\bar{d}_C}$ for small ϵ, which implies the existence of the limit given in the definition of capacity, eq. (2). We can, however, show that the limit given in eq. (2) exists and $\bar{d}_C$ must satisfy eq. (18) in a rigorous manner, as follows:

Define $E_C(\epsilon)$ by

$$N(\epsilon) = E_C(\epsilon)\epsilon^{-\bar{d}},$$

where $\bar{d}$ is defined by $1 = \lambda_a^{\bar{d}} + \lambda_b^{\bar{d}}$. Substituting this into eq. (17) then yields

$$E_C(\epsilon) = \bar{\alpha} E_C\left(\frac{\epsilon}{\lambda_a}\right) + \bar{\beta} E_C\left(\frac{\epsilon}{\lambda_b}\right), \tag{19}$$

where $\bar{\alpha} = \lambda_a^{\bar{d}}$ and $\bar{\beta} = \lambda_b^{\bar{d}}$, and are independent of ϵ. Notice that by definition $\bar{\alpha} + \bar{\beta} = 1$, so the above expression says that $E_C(\epsilon)$ is a weighted average of its values at ϵ/λ_a and ϵ/λ_b. Choose ϵ_1 and ϵ_2 so that $\epsilon_1 > \epsilon_2 > 0$. Since $N(\epsilon)$ and hence $E_C(\epsilon)$ are finite and positive for any finite ϵ, there exist finite non-zero numbers $B_1 > B_2 > 0$ such that $B_2 < E_C(\epsilon) < B_1$ for $\epsilon_1 > \epsilon > \epsilon_2$. We can assume that ϵ_1 and ϵ_2 are chosen so that ϵ_1/ϵ_2 is large. Since $\bar{\alpha} + \bar{\beta} = 1$, eq. (19) implies that $B_2 < E_C(\epsilon) < B_1$ also applies to the wider interval $\epsilon_1 > \epsilon > \lambda_b \epsilon_2$. Repeating this argument increases the domain of validity of the bound to $\epsilon_1 > \epsilon > \lambda_b^2 \epsilon_2$, and so on. Hence $E_C(\epsilon)$ is bounded uniformly from above and below for arbitrarily small ϵ. Thus the limit of eq. (2) exists and $\bar{d}_C = \bar{d}$. (In fact it can be shown that eq. (19) implies that $\lim_{\epsilon \to 0} E_C(\epsilon)$ is a constant if $\log \lambda_a / \log \lambda_b$ is an irrational number.) Note that in eq. (18), since both terms on the right-hand side are monotonically decreasing, d_C obtained from solving this equation is unique.

4.4. Computation of Hausdorff dimension

The Hausdorff dimension d_H can be calculated by an argument that is very similar to the one used above in computing the capacity. Let $\bar{d}_H \equiv d_H - 1$, the Hausdorff dimension along x. Applying the scaling property of the map to the quantity $l_d(\epsilon)$ (defined in section 2), we obtain

$$l_d(\epsilon) = (\lambda_a)^d l_d\left(\frac{\epsilon}{\lambda_a}\right) + (\lambda_b)^d l_d\left(\frac{\epsilon}{\lambda_b}\right).$$

Substituting $l_d(\epsilon) = E_H(\epsilon) \epsilon^{-(\bar{d}-d)}$ into the above equation, we again find that $E_H(\epsilon)$ satisfies eq. (19). Thus the limit $\epsilon \to 0$ yields $l_d = \infty$ or $l_d = 0$ for $d < \bar{d}_C$ or $d > \bar{d}_C$, respectively. Hence, as predicted in section 2, the Hausdorff dimension and capacity are equal, $d_H = d_C$.

4.5. Calculation of information dimension

The information dimension d_I can also be calculated by a scaling argument similar to that used above in computing the capacity. Once again, let $d_I = 1 + \bar{d}_I$ and express the summation for $I(\epsilon)$ in eq. (3) as the sum of contributions from the two strips in fig. 7d,

$$I(\epsilon) = I_a(\epsilon) + I_b(\epsilon). \tag{20}$$

The total probability contained in strip $[0, \lambda_a]$ is α, and that in strip $[\frac{1}{2}, \lambda_b + \frac{1}{2}]$ is β. Assuming that it takes $N(\epsilon)$ strips of width ϵ to cover the whole attractor, then from the scaling property of eq. (12), covering the strip $[0, \lambda_a]$ at resolution $\epsilon \lambda_a$ also requires $N(\epsilon)$ strips. Thus

$$I_a(\epsilon \lambda_a) = \sum_{i=1}^{N(\epsilon)} \alpha P_i \log \frac{1}{\alpha P_i}$$

$$= \alpha \left[\log \frac{1}{\alpha} + I(\epsilon)\right].$$

Hence, replacing $\epsilon\lambda_a$ by ϵ in the above,

$$I_a(\epsilon) = \alpha \log \frac{1}{\alpha} + \alpha I\left(\frac{\epsilon}{\lambda_a}\right),$$

$$I_b(\epsilon) = \beta \log \frac{1}{\beta} + \beta I\left(\frac{\epsilon}{\lambda_b}\right).$$

Thus

$$I(\epsilon) = \alpha I\left(\frac{\epsilon}{\lambda_a}\right) + \beta I\left(\frac{\epsilon}{\lambda_b}\right) + H(\alpha), \quad (21)$$

where $H(\alpha)$ is given by eq. (15). Motivated by eq. (3), if we assume that $I(\epsilon) = \bar{d}_I \log(1/\epsilon)$ for small ϵ, and substitute for $I(\epsilon)$, $I(\epsilon/\lambda_a)$, and $I(\epsilon/\lambda_b)$ in the above equation we obtain

$$\bar{d}_I = \frac{H(\alpha)}{\alpha \log(1/\lambda_a) + \beta \log(1/\lambda_b)},$$

which is in turn equal to $\bar{d}_L$. The assumption that $I(\epsilon) = \bar{d}_I \log(1/\epsilon)$ can be made rigorous in the limit as $\epsilon \to 0$ using an argument that is completely analogous to that used in deriving the capacity in the last part of subsection 4.3.

We should mention that Alexander and Yorke [11] have computed the Lyapunov and information dimensions of the generalized baker's transformation for the special case $\alpha = \frac{1}{2}$, $\lambda = \lambda_a = \lambda_b$, where $\lambda > \frac{1}{2}$. In this case $d_L = 2$. For uncountably many values of λ they find that also $d_I = 2$, although there are certain special values of λ for which $d_I < 2$.

In order to calculate the other probability dimensions listed in table I more information concerning the probability distribution is required. This is dealt with in section 5, and we therefore defer calculation of the remaining dimensions to the sections following section 5.

5. Distribution of probability

In this section we derive the form of the probability distribution $\{P_i(\epsilon)\}$ associated with the natural measure μ of the generalized baker's transformation. Here P_i denotes the probability of the ith cube C_i of edge ϵ, i.e., $P_i = \mu(C_i)$. The collection of numbers $\{P_i(\epsilon)\}$ may be also be thought of as the result of coarse graining the natural measure. This probability distribution is interesting both for its own sake, and because it is needed to compute some of the dimensions that we are interested in. In what follows we restrict ourselves to the case in which $\lambda_a = \lambda_b \equiv \lambda_2$, which keeps the width of all the strips the same. Thus a particularly convenient partition for computing $\{P_i\}$ is the set of 2^n nonempty strips obtained by iterating the unit square n times.

Starting with a uniform probability distribution, on one application of the map two strips are produced, one with total probability α and the other with total probability β. (See fig. 7d.) If the map is applied again (fig. 8), there results one strip of probability α^2, one of probability β^2, and two of probability $\alpha\beta$. In general, after n applications of the map, there result 2^n strips of width $(\lambda_2)^n$ and probabilities $\alpha^m \beta^{(n-m)}$, $m = 0, 1, 2, \ldots, n$. The number of strips with probability $\alpha^m \beta^{(n-m)}$ is

$$Z(n, m) = \frac{n!}{(n-m)!m!}, \quad (22)$$

i.e., the binomial coefficient. Since we take $\alpha < \frac{1}{2} < \beta$, lower m corresponds to more probable strips, i.e. strips of greater natural measure. The total probability contained in these $Z(n, m)$ strips is

$$W(n, m) \equiv \alpha^m \beta^{(n-m)} Z(n, m). \quad (23)$$

Note the similarity to a sequence of coin tosses. Using a coin with probability α of heads and β of tails, for a sequence of n flips the total number of sequences with m occurrences of heads is given by eq. (22), and the likelihood of all such sequences is given by eq. (23).

For large n (small ϵ) it is convenient to have smooth estimates for $Z(n, m)$ and $W(n, m)$. Using

Sterling's approximation, i.e.

$$\log n! = (n + \tfrac{1}{2})\log(n+1) - (n+1) + \log(2\pi)^{1/2}$$
$$+ \mathcal{O}(n^{-1}),$$

we obtain from eq. (22)

$$\log Z \approx (n + \tfrac{1}{2})\log(m+1) - \log(2\pi)^{1/2} + 1.$$

Expanding this expression in a Taylor series about its maximum value, $m = n/2$, yields

$$Z(n, m) \approx \frac{2^n}{\sqrt{2\pi}}\sqrt{\frac{4}{n}} \exp\left\{-\frac{1}{2}\left[\frac{4}{n}\left(m - \frac{n}{2}\right)^2\right]\right\}. \quad (24)$$

Similarly, from eq. (23), $W(n, m)$ is

$$W(n, m) \approx \frac{1}{\sqrt{2\pi n \alpha \beta}} \exp\left\{-\frac{(m - n\alpha)^2}{2n\alpha\beta}\right\}. \quad (25)$$

Note that, since these expressions were obtained by Taylor series expansion, eq. (24) is only valid for $|m/n - \tfrac{1}{2}| \ll 1$, and eq. (25) is only valid for $|m/n - \alpha| \ll 1$. However, since the width of these Gaussians is $\mathcal{O}(1/n^{1/2})$, eq. (24) is valid for most of the strips, and eq. (25) is valid for most of the probability.

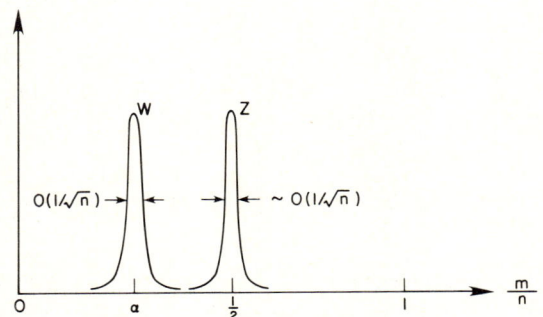

Fig. 9. A schematic representation of the distribution of probabilities on the attractor. $Z(n, m)$ is the number of cubes with probability $p = \alpha^m \beta^{(n-m)}$, and $W(n, m)$ is the sum of the probability contained in cubes of probability p. For large n and m/n close to its mean value, these are both approximately Gaussian distributions in m/n whose width is proportional to n. In the limit as $n \to \infty$, W and Z become delta functions, and no longer overlap.

Fig. 9 shows a schematic plot of Z and W. It is clear from this figure that, for large n, almost all of the probability is contained in a very small fraction of the total number of strips. Furthermore, the situation is accentuated as ϵ gets smaller (n gets larger), since the width of the Gaussians given in eqs. (24) and (25) decreases according to $n^{1/2}$. In the limit as $\epsilon \to 0$ these Gaussians approach delta functions, and they do not overlap. We feel that the above properties are typical features of chaotic attractors.

5.1. *Log-normal distribution of probabilities*

It is instructive to rewrite eq. (25) in another form. Let $p = \alpha^m \beta^{(n-m)}$ denote the probability of a strip, and reexpress eq. (25) in terms of $u = \log(1/p)$ rather than m. Noting that m is proportional to u, and letting $\epsilon = \lambda_2^n$ $W(n, m)$ becomes

$$F(u) = \frac{1}{\sqrt{2\pi}\sigma} e^{-(u-u_0)^2/2\sigma^2}, \quad (26)$$

where

$$\sigma^2 = \frac{[\alpha\beta(\log(\beta/\alpha))^2 \log(1/\epsilon)]}{\log(1/\lambda_2)}$$

and

$$u_0 = \bar{d}_L \log\frac{1}{\epsilon}, \quad (27)$$

with d_L given by eq. (16). Eq. (26) is only valid if

$$\frac{(u - u_0)^2}{\sigma^2} \ll \log\frac{1}{\epsilon}, \quad (28)$$

corresponding to $|m/n - \alpha| \ll 1$. $F(u)du$ is the total probability contained in strips whose values of $u = \log(1/p)$ fall between u and $u + du$. Thus we see that the values the logarithm of p asymptotically have a Gaussian distribution, or in other words, the values of p asymptotically have a log-normal distribution. We believe that this is typically true of chaotic attractors. In particular, we offer the following conjecture*:

Conjecture 4. Let A be a chaotic attractor of a p-dimensional *invertible* dynamical system, and assume that this attractor has a natural measure μ. Cover A with a fixed grid of p-dimensional cubes of side length ϵ. Assign each nonempty cube C_i probability $P_i = \mu(C_i)$, and let $U_i = \log(1/P_i)$. Let u_0 be the mean of the numbers U_i, and let σ^2 be the variance. For typical chaotic attractors, in the limit as $\epsilon \to 0$, values of U_i sufficiently close to the mean (in the sense of eq. (28)) approach a Gaussian distribution. In other words, the corresponding values of P_i approach a log-normal distribution.

Note that U_i is the information obtained in a measurement that finds the orbit inside of the ith cube [1, 9, 10]. Thus, conjecture 4 states that for chaotic attractors the information is approximately normally distributed for small ϵ.

The function $Z(n, m)$ given in eq. (24), can also be reexpressed in terms of p rather than m. When this is done, with similar restrictions to those of eq. (28), the result is also a Gaussian in terms of $\mu = \log(1/p)$. When recast in the more general setting of conjecture 4, this says that the number of cubes C_i whose values U_i lie between u and $u + du$ are given by a Gaussian distribution. (Similar restrictions to those given in conjecture 4 apply.)

6. Computation for the natural measure dimensions

In this section we verify conjectures 1 and 2 for the generalized baker's transformation by explicitly computing all of the probability dimensions defined in section 2. In order to simplify the computations, for all but the ϑ-Hausdorff dimension we restrict ourselves to the case in which $\lambda_a = \lambda_b \equiv \lambda_2$. For the ϑ-Hausdorff dimension we treat the most general case in which $\lambda_a \neq \lambda_b$, but are

---

* The form of this conjecture was developed in collaboration with Erica Jen.

only able to obtain an upper bound for the dimension.

6.1. Alternate derivation of information dimension

Now that we know the probability distribution for the generalized baker's transformation for $\lambda_a = \lambda_b \equiv \lambda_2$, we can obtain the information dimension directly from its definition. From eq. (3) and eq. (26), $I(\epsilon)$ is the average value of $\log(1/P_i)$ or

$$I(\epsilon) = \int u F(u) \, du = u_0.$$

Since from eq. (27) $u_0 = d_L \log(1/\epsilon)$, eq. (3) yields $d_I = d_L$ (previously shown in section 4 for the more general case $\lambda_a \neq \lambda_b$). Thus the mean value of the log-normal distribution is simply the information contained in the probability distribution, and its scaling rate is the dimension of the nature measure, i.e., $I(\epsilon) \approx d_\mu \log(1/\epsilon)$.

6.2. Determination of ϑ-capacity

Here we calculate $d_C(\vartheta)$ for $\lambda_a = \lambda_b = \lambda_2$. We choose ϵ equal to the width of a strip, $\epsilon = \lambda_2^n$. As usual, for convenience we compute the ϑ-capacity of the attractor projected onto the x-axis, i.e. $\bar{d}_C(\vartheta) = d_C(\vartheta) - 1$. The ϑ-capacity $\bar{d}_C(\vartheta)$ is defined in terms of the minimum number of intervals $N(\epsilon; \vartheta)$ of width ϵ that have total natural measure at least ϑ,

$$N(\epsilon; \vartheta) = \sum_{m=0}^{m_\vartheta} Z(n, m), \tag{29}$$

where m_ϑ is the largest integer such that

$$\sum_{m=0}^{m_\vartheta - 1} W(n, m) \leq \vartheta, \tag{30}$$

To find m_ϑ we use eq. (25) and approximate the sum in eq. (30) by an integral,

$$\vartheta \approx \frac{1}{\sqrt{2\pi\alpha\beta n}} \int_{-\infty}^{m_\vartheta} e^{-(m - n\alpha)^2 / 2n\alpha\beta} \, dm.$$

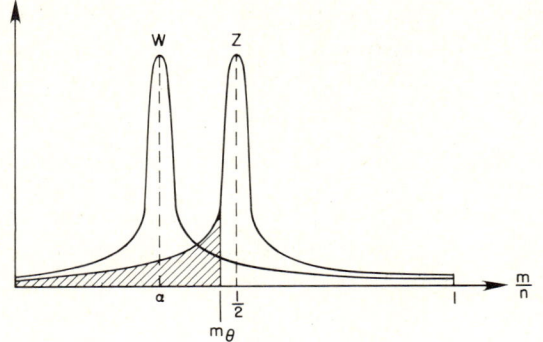

Fig. 10. The principal contribution to the sum needed to compute $N(\epsilon; \vartheta)$ (eq. (29)) comes from values of m near m_ϑ.

Thus for fixed ϑ we obtain

$$\frac{m_\vartheta}{n} \approx \alpha + \text{erfc}^{-1}(\vartheta)\sqrt{\frac{\alpha\beta}{n}}, \qquad (31)$$

where $\text{erfc}(x) = (1/\sqrt{2\pi})\int_{-\infty}^{x} e^{-x^2/2}\,dx$. Now, consider eq. (29). The principal contribution to the sum will come from m values very close to m_ϑ, as depicted in fig. 10. Thus we use eqs. (23) and (25) to approximate $Z(n, m)$ as

$$Z(n, m) \approx \frac{\beta^{-n}(\beta/\alpha)^m}{\sqrt{2\pi n\alpha\beta}} e^{-(m-n\alpha)^2/2n\alpha\beta}. \qquad (32)$$

The term $(\beta/\alpha)^m$ decreases as m decreases away from m_ϑ, and this decrease is very rapid compared to the variation of $e^{-(m-n\alpha)^2/2n\alpha\beta}$. Thus in performing the sum in eq. (29), we may approximate $e^{-(m-n\alpha)^2/2n\alpha\beta}$ as being constant and equal to its value at $m = m_\vartheta$. Hence the only m dependent term in the sum is $(\beta/\alpha)^m$. Since

$$\sum_{m=0}^{m_\vartheta}\left(\frac{\beta}{\alpha}\right)^m \approx \left(\frac{\beta}{\alpha}\right)^{m_\vartheta}\frac{\beta}{(\beta-\alpha)},$$

we find that

$$N(\epsilon; \vartheta) \sim \beta^{-(n-m_\vartheta)}\alpha^{-m_\vartheta}n^{-1/2}.$$

From eq. (4) and $n = \log(1/\epsilon)/\log(1/\lambda_2)$, the above estimate of $N(\epsilon; \vartheta)$ yields $d_C(\vartheta) = d_L$, in agreement with conjecture 2.

6.3. Computation of ϑ-Hausdorff dimension

In this section we obtain an upper bound on the ϑ-Hausdorff dimension of the generalized baker's transformation with $\lambda_a \neq \lambda_b$. (Recall that for our work in the previous section we took $\lambda_a = \lambda_b$.) We obtain an inequality for the ϑ-Hausdorff dimension by using a specific covering along x to compute the sum

$$l_d^*(\epsilon, \vartheta) = \sum_i \epsilon_i^d,$$

where the $\epsilon_i < \epsilon$ cover a fraction ϑ of the natural measure of the attractor. Our choice for the ϵ_i is specified below. Taking the limit as $\epsilon \to 0$, we find that there is a value of d at which $l_d^*(\epsilon, \vartheta)$ crosses over from ∞ to 0. For the partition we have chosen, we find that crossover occurs at $d = \bar{d}_L$. We believe that the value we obtain is in fact the true ϑ-Hausdorff dimension. However, we cannot be sure that the particular covering we have chosen gives the lowest possible value of d, and thus we can only say that we have obtained an upper limit.

After n iterations of the map, an initially uniform probability distribution in the unit square is transformed to 2^n strips with widths $\lambda_a^m\lambda_b^{(n-m)}$ and probabilities $\alpha^m\beta^{(n-m)}$, $m = 0, 1, 2, \ldots, n$. As shown in eq. (22), the number of such strips is $Z(n, m)$. We shall choose the ϵ_i to cover the most probable strips so that ϑ of the total probability is covered. ϵ for our covering is equal to the width of the widest strip, which is either $(\lambda_a)^n$ or $(\lambda_b)^n$, whichever is larger. Letting $U_d(n, m)$ be

$$U_d(n, m) = (\lambda_b^{m-n}\lambda_a^m)^d Z(n, m),$$

we have that

$$l_d^*(\epsilon, \vartheta) = \sum_i \epsilon_i^d = \sum_m U_d(n, m), \qquad (33)$$

We still have yet to specify which m values are to be included in the sum. To do this, we expand $U_d(n, m)$ about its maximum value (as done for Z and W in section 5), and obtain

$$U_d(n,m) \approx \frac{[\lambda_a^d + \lambda_b^d]^n}{\sqrt{2\pi n \frac{\lambda_a^d \lambda_b^d}{(\lambda_a^d + \lambda_b^d)(\lambda_a^d + \lambda_b^d)}}}$$

$$\times \exp -\frac{1}{2} \left\{ \frac{\left[\frac{m}{n} - \frac{\lambda_a^d}{\lambda_a^d + \lambda_b^d}\right]^2}{\frac{\lambda_a^d \lambda_b^d}{n(\lambda_a^d + \lambda_b^d)(\lambda_a^d + \lambda_b^d)}} \right\}. \quad (34)$$

In order to compute $l_d^*(\epsilon, \vartheta)$, we must consider the natural measure as well as $U_d(n, m)$. Note that for the general case we are considering now with $\lambda_a \neq \lambda_b$, $W(n, m)$ obtained in eq. (25) continues to be the correct expression for the distribution of probabilities in each strip. Depending on the values of α, d, λ_a, and λ_b, W may peak at a value of m that is smaller, larger, or equal to the value of m at the peak of U_d. Comparing the location of the peaks of the Gaussians in eq. (34) (for U) and in eq. (25) (for W), we see that there are three cases:

Case 1: $\alpha < \dfrac{\lambda_a^d}{(\lambda_a^d + \lambda_b^d)}$,

Case 2: $\alpha > \dfrac{\lambda_a^d}{(\lambda_a^d + \lambda_b^d)}$,

Case 3: $\alpha = \dfrac{\lambda_a^d}{(\lambda_a^d + \lambda_b^d)}$.

Cases 1 and 2 may be shown to be equivalent as follows. From the case 2 inequality and the fact that $\alpha + \beta = 1$, we obtain $\beta < \lambda_b^d/(\lambda_a^d + \lambda_b^d)$. But, if we define m' by $m = n - m'$, and change the sums over m to sums over m', then the roles of (α, λ_a) and (β, λ_b) are interchanged, and case 2 is converted to case 1. We shall not consider case 3 here; suffice it to say that it does not alter the results obtained from consideration of cases 1 and 2. Therefore it is sufficient to compute the ϑ-Hausdorff dimension for case 1.

For case 1, selecting the best covering of intervals that contain ϑ of the total probability is easy. Since W remains valid, we get a covering that includes ϑ of the total natural measure by including intervals whose value of m is less than m_ϑ, just as we did for the computation of ϑ-capacity. Furthermore, since U_d peaks at a larger value of m/n than W does, this selection gives the smallest value of l_d^*. The situation is analogous to the computation of ϑ-capacity, except that here the role of Z is played by U_d (cf. fig. 10). To evaluate

$$l_d^* = \sum_{m=0}^{m_\vartheta} U_d(n, m), \quad (35)$$

we note that, as for the analogous evaluation for ϑ-capacity in the previous subsetion, the principal contribution to the sum comes from m-values close to m_ϑ. Thus we approximate $U_d(n, m)$ as

$$U_d(n,m) \approx \frac{\lambda_a^m \lambda_b^{n-m}}{\alpha^m \beta^{(n-m)}} W(n,m),$$

with W approximated by eq. (25). Proceeding as in section 6.2 we obtian an estimate for $l_d^*(\epsilon, \vartheta)$,

$$l_d^*(\epsilon, \vartheta) \sim n^{-1/2} \left(\frac{\beta}{\lambda_b^d}\right)^{m_\vartheta - n} \left(\frac{\alpha}{\lambda_a^d}\right)^{-m_\vartheta},$$

or

$$\log[l_d^*(\epsilon, \vartheta)] \approx -n[d - (\bar{d}_L)] \log\left(\frac{1}{\lambda_2}\right).$$

For $\epsilon \to 0$ (i.e., $n \to \infty$) we obtain $l_d^*(\vartheta) = 0$ for $d > \bar{d}_L$ and $l_d^*(\vartheta) = \infty$ for $d < \bar{d}_L$. Thus remembering that $d_H(\vartheta) = \bar{d}_H(\vartheta) + 1$,

$$d_H(\vartheta) \leq d_L. \quad (36)$$

As already mentioned, we expect that the above inequality is really an equality. This expectation is reinforced by the fact that when $\vartheta = 1$ we recover the exact expression for the Hausdorff dimension computed in eq. (18). To see that this is true, replace m_9 in eq. (35) by n. From the form of U_d, this sum is simply the binomial expansion of $(\lambda_a^d + \lambda_b^d)^n$. As $n \to \infty$, this quantity is 0 or ∞ for $d > \bar{d}_H$ or $d < \bar{d}_H$, where $\bar{d}_H$ satisfies $\lambda_a^{\bar{d}_H} + \lambda_b^{\bar{d}_H} = 1$, which is the same as eq. (18). That is, for the specific choice of ϵ_i that we have used, we obtain the correct value of d_H. Since the same choice of the ϵ_i was used in obtaining $d_H(\vartheta)$, it seems plausible that the equality might apply in eq. (36).

6.4. Computation of the pointwise dimension

We now consider the pointwise dimension for the generalized baker's transformation with $\lambda_a = \lambda_b < \frac{1}{2}$, and we show that d_p exists and is equal to d_L.

As previously noted in section 5, application of the map n times to the unit square produces 2^n strips of widths $(\lambda_a)^n$. (Recall that we are assuming $\lambda_a = \lambda_b$.) In order to compute the pointwise dimension, we choose a point x at random with respect to the natural measure μ, compute the natural measure contained in an ϵ ball centered about x, (i.e. $\mu(B_\epsilon(x))$), and compute the ratio of $\log \mu(B_\epsilon(x))$ to $\log \epsilon$ in the limit as ϵ goes to zero (cf. eq. (5)). The simplest case for this computation occurs when $\lambda_a < \frac{1}{4}$, so that the gaps between strips are bigger than the strips themselves, as pictured in fig. 11a. Choosing a point x at random with respect to the natural measure μ, let S_n denote the nth order strip of width $(\lambda_a)^n$ that the point x lies in. Letting $\epsilon = (\lambda_a)^n$, the natural measure contained in a ball of radius ϵ around x (i.e., the x-interval $[x - (\lambda_a)^n, x + (\lambda_a)^n]$) will be equal to the natural measure of the strip S_n, regardless of where in the strip x lies. (See fig. 11a.) The natural measure contained in a given strip is $\alpha^m \beta^{(n-m)}$, where $n \geq m \geq 0$, where m depends on the particular strip that x happens to lie in. (See section 5.) Thus, we have

$$\lim_{n \to \infty} \frac{\mu(B_\epsilon(x))}{\log \epsilon} = \lim_{n \to \infty} \frac{\log \mu(S_n)}{n \log \lambda_a}$$

$$= \lim_{n \to \infty} \frac{m \log \alpha + (n - m) \log \beta}{n \log \lambda_a}. \quad (37)$$

In the limit as n grows large, as shown in section 5 (see fig. 9), the total probability $W(n, m)$ contained in strips of a given m value is distributed as a Gaussian centered about $m/n = \alpha$. Thus, in the limit as $n \to \infty$ it becomes overwhelmingly likely that $m/n = \alpha$. Thus for almost every x with respect to the natural measure μ, $\lim_{n \to \infty} m/n = \alpha$. (This is just a statement of the law of large numbers.) Putting this into eq. (37) gives

$$d_p = \lim_{n \to \infty} \frac{\mu(B_\epsilon(x))}{\log \epsilon} = \frac{\alpha \log \alpha + \beta \log \beta}{\log \lambda_a}$$

$$= \frac{H(\alpha)}{\log(1/\lambda_d)} = \bar{d}_L. \quad (38)$$

(See eqs. (15) and (16).)

To extend this computation of the pointwise dimension to the case that $\frac{1}{2} > \lambda_a > \frac{1}{4}$, for any $\lambda_a < \frac{1}{2}$ choose a k such that $\lambda_a^{k+1} \leq \frac{1}{2} - \lambda_a$ (e.g., for $\lambda_a \leq \frac{1}{4}$ this relation is satisfied for any $k \geq 0$; for $\lambda_a \leq 0.365\ldots$, for any $k \geq 1$; etc.). Then we can show $\mu(B_\epsilon(x)) \leq \alpha^{-k} \mu(S_n)$, where without loss of generality, we have assumed $\alpha \leq \beta$. Since $B_\epsilon(x) \supset S_n$, we have also $\mu(B_\epsilon(x)) \geq \mu(S_n)$. Thus $\mu(S_n) \leq \mu(B_\epsilon(x)) \leq \alpha^{-k} \mu(S_n)$ which with eq. (37) yields $d_p = \bar{d}_L$. (Our evaluation of eq. (37) holds not for $\epsilon \to 0$ but rather holds for the restricted set of $\epsilon = \lambda_a^n$, $n = 1, 2, \ldots$, however, it is not hard to show that that in fact implies eq. (38) for every sequence of ϵ going to 0.)

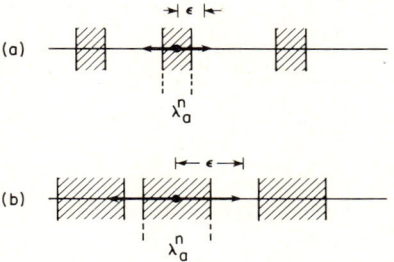

Fig. 11. (a) Computing the pointwise dimension for the case that $\lambda_a < \frac{1}{4}$. (b) The case $\lambda_a > \frac{1}{4}$, in which the computation is a little more complicated.

Thus we have shown that for the generalized baker's transformation the pointwise dimension is equal to the dimension of the natural measure. (Although we have only shown this for $\lambda_a = \lambda_b$, it is not hard to extend this result to $\lambda_a \neq \lambda_b$.)

7. The core of attractors

As shown in section 5, for the generalized baker's transformation, typically almost all of the probability is contained in a very small fraction of the total number of cubes needed to cover the attractor. In the limit as ϵ goes to zero, this fraction goes to zero. Thus, the natural measure of the attractor is *concentrated* on a subset of the attractor. We will call this subset the *core* of the attractor.

To get a better feel for why this comes about, and to see how the properties of the core are related to those of the attractor and its natural measure, consider the special case of the generalized baker's transformation where $\lambda_a = \lambda_b = \frac{1}{2}$. As we have already seen, at the nth level of approximation the natural measure consists of 2^n vertical strips of probability $\alpha^m \beta^{n-m}$. For large n and $\beta > \alpha$, a small fraction of the strips whose m values are close to αn contain much more of the natural measure than all other strips. Fig. 12 shows a plot

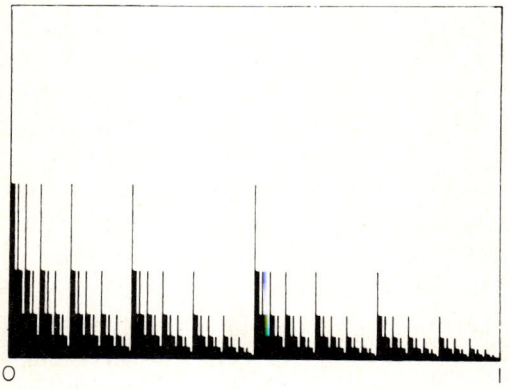

Fig. 12. The natural probability distribution of the generalized baker's transformation projected onto the x-axis, and coarse grained using intervals of width $\epsilon = 2^{-10}$. In this case $\alpha < \frac{1}{2}$, and $\lambda_a = \lambda_b = \frac{1}{2}$.

of the nth level approximation to the probability distribution as a function of x with $\lambda_a = \lambda_b = \frac{1}{2}$, and $\alpha < \frac{1}{2}$ and $n = 10$. The probability distribution looks as though it were made up of spikes, showing that already at $n = 10$ the natural measure has become quite concentrated in certain cubes (in this case intervals).

To understand the form of this probability distribution, it is instructive to represent the probability distribution of these strips in terms of x rather than m. To do this, approximate x using its first n binary digits, i.e. as a binary decimal truncated after n digits. Let m be the number of ones contained in the first n digits of the binary expansion of x. The natural measure of the strip $S_n(x)$ containing x is then $\mu(S_n(x)) = \alpha^m \beta^{(n-m)}$. (See the discussion at the beginning of section 5.) As we have already shown (see fig. 9), when written in terms of m, for large n the natural measure is approximately a Gaussian centered about αn, and in the limit where n is large almost all the measure is contained in strips with $m \approx \alpha n$. In other words, the natural measure of the generalized baker's transformation for $\lambda_a = \lambda_b = \frac{1}{2}$ is concentrated on those values of x that have 1's in their binary expansions in the fraction α, or equivalently, 0's in the fraction β. In the limit $n \to \infty$, *all* the natural measure is contained in this set, which we will call the core of this attractor.

For this case ($\lambda_a = \lambda_b = \frac{1}{2}$) the attractor is the entire unit square. The core of this attractor is dense on the attractor. In other words, any point of the attractor has points of the core arbitrarily close to it. Hence any covering of the core must also be a covering of the attractor, and vice versa. Thus the capacity of the core is the same as that of the attractor. The Hausdorff dimension, in contrast, is more subtle, and in fact, computing the Hausdorff dimension of the set of numbers whose binary expansions have a given fraction of ones is a classic problem in the study of Hausdorff dimension [31]. The Hausdorff dimension of this set is

$$d_H = \frac{H(\alpha)}{\log 2}.$$

(See eq. (15).) This result was conjectured by Good in 1941 [32] and proved by Eggleston in 1949 [33]. Also, the Hausdorff dimension of a very similar example (involving ternary rather than binary expansions) was proven by Besicovitch in 1931 [34].

Thus, for this example we see that the Hausdorff dimension of the core is equal to the dimension of the natural measure, and the capacity of the core is equal to the fractal dimension of the attractor (cf. eq. (16)). For the case of diffeomorphisms of the plane, the former result has been proven by Young [12]. We suspect that this is a property of typical attractors.

8. An attractor that is a nowhere differentiable torus

This section contains a review of the work of Kaplan, Mallet-Paret, and Yorke [35] on the dimension of a chaotic attractor in a setting that is quite different from that of the generalized baker's transformation. The attractor described below has the same topological form as a torus, and yet is nowhere differentiable, thus providing an interesting example of the nonanalytic forms that can be produced by chaotic dynamics.

Consider the following map:

$$x_{n+1} = 2x_n + y_n \mod 1,$$
$$y_{n+1} = x_n + y_n \mod 1, \qquad (39)$$
$$z_{n+1} = \lambda z_n + p(x_n, y_n).$$

where x and y are taken mod 1, z can be any real number, and p is periodic in x and y with period 1 and is at least five times differentiable. (For example, $p(x, y) = \cos 2\pi x$.) In order to keep z bounded, λ is chosen between 0 and 1. Note that the eigenvalues and eigenvectors of the Jacobian matrix of eq. (39) are independent of x, y, and z. Thus *every* initial condition has the same Lyapunov numbers, i.e., the Lyapunov numbers are absolute, so that in this case conjecture 3 is relevant, and we expect that the fractal dimension and the dimension of the natural measure should be equal.

The equations for x and y are independent of z, and in fact are the classic Anasov or "cat" map [36],

$$\begin{pmatrix} x_{n+1} \\ y_{n+1} \end{pmatrix} = A \begin{pmatrix} x_n \\ y_n \end{pmatrix} \mod 1,$$

where

$$A = \begin{pmatrix} 2 & 1 \\ 1 & 1 \end{pmatrix}.$$

Thus, the x–y dynamics are chaotic, and are unaffected by the value of z.

To understand the shape of the attractor in the z-direction, put a sample initial condition into eq. (39). For example, consider $(x_0, y_0, 0)$. z_n takes on the form

$$z_n = \sum_{k=1}^{n} \lambda^{k-1} p(x_{n-k}, y_{n-k}).$$

Making use of the fact that $\begin{pmatrix} x_{n-k} \\ y_{n-k} \end{pmatrix} = A^{-k} \begin{pmatrix} x_n \\ y_n \end{pmatrix}$, and letting n go to infinity, it can be shown that the surface given by

$$z(x, y) = \sum_{k=1}^{\infty} \lambda^{k-1} p\left(A^{-k} \begin{pmatrix} x \\ y \end{pmatrix}\right),$$

is invariant and is the unique attractor of this dynamical system.

$z(x, y)$ has some very interesting properties. For $\lambda < 1/R$, where $R = (3 + \sqrt{5})/2$, $z(x, y)$ is smooth and has dimension 2. If $\lambda > 1/R$, however, for most choices of p, $z(x, y)$ is nowhere differentiable. A typical cross section of $z(x, y)$ is shown in fig. 13. To understand intuitively how the nondifferentiability of $z(x, y)$ comes about, notice that $z(x, y)$ is the sum of an infinite number of periodic functions whose arguments are the successive iterates of the cat map. Unless λ is small enough to diminish the effect of higher order iterates, the value of the sum can swing wildly as x or y change.

The Lyapunov numbers of the map given in eq.

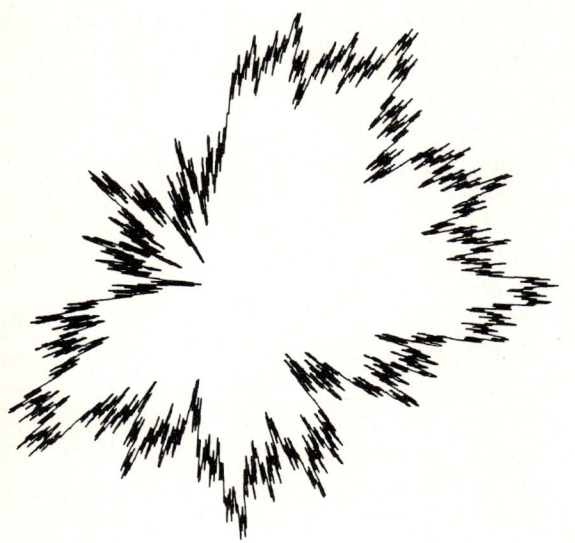

Fig. 13. A cross section of a nowhere differentiable torus, made using eq. (39) with $p(x, y) = \cos 2\pi x$ and λ chosen so that $d_C = 2\frac{1}{2}$.

(39) are $\lambda_1 = R$, $\lambda_2 = \lambda$, and $\lambda_3 = 1/R$, where $R = (3 + \sqrt{5})/2$, as given above. Kaplan, Mallet-Paret, and Yorke [35] have shown that there are two possibilities for the dimension of $z(x, y)$: Either

(i) $z(x, y)$ is nowhere differentiable and

$$d_C = d_L$$

or

(ii) $z(x, y)$ is differentiable and $d_C = 2$.

For given p, the nowhere differentiable case occurs for nearly every choice of λ. Thus we see that conjecture 3 is satisfied for this example.

9. Numerical computations

In this section we discuss some aspects of the numerical computation of dimension. First we will discuss the basic ideas behind numerical computations of dimension, secondly we will discuss some of the problems encountered in such computations, and finally we will review some previous numerical work.

The methods to compute dimension vary considerably depending on the dimension that one wishes to compute. Thus far, we are aware of numerical computations only of capacity [37–41], Lyapunov dimension [37–40], and Hausdorff dimension [42]. Of these, only the studies involving the capacity and the Lyapunov dimension were applied to attractors of dynamical systems. In each case, the computations follow from the definitions. As we shall see, the capacity is (in principle) straightforward to compute, but is in practice unfeasible to compute for all but very low dimensional attractors. The Lyapunov dimension, in contrast, is much more feasible to compute. We will begin the discussion with a description of the computation of Lyapunov dimension, and then go on to discuss the computation of capacity.

9.1. Numerical computation of Lyapunov dimension

The Lyapunov dimension is defined in terms of the Lyapunov numbers. (See section 3.) Thus, the work involved in computing Lyapunov dimension is in computing the Lyapunov numbers. Numerical methods for doing this have been discussed by Bennetin et al. [43], Shimada and Nagashima [44], and in infinite dimensions by Farmer [38]. With appropriate numerical caution, the largest k Lyapunov numbers can be computed by following the evolution of k nearby trajectories simultaneously and measuring their rate of separation. There are various numerical problems with this method, however, and a better method is to follow only one trajectory, but also follow k trajectories of the associated equations for the evolution of vectors in the tangent space. These methods have been successfully used in a variety of numerical studies.

For low-dimensional cases, such as two-dimensional maps or systems of three autonomous ordinary differential equations, with a modern computer and plenty of computer time, numerical computation of the dimensions we discuss here directly from their definitions is feasible, as discussed in the next subsection. Even in such low-dimensional cases, however, the computation of Lyapunov dimension is by far less costly in terms of computer time and memory than the computation of other dimensions. For higher dimensional attractors it appears that only the Lyapunov dimension is computationally feasible. The key reason that the Lyapunov dimension is feasible to compute numerically even for attractors of rather high dimension (e.g. $d_L \approx 10$) is that the difficulty of the computation scales linearly with the dimension of the attractor times the dimension of the space it lies in, rather than exponentially as it does for a computation of the fractal dimension, or any of the other dimensions discussed in this paper. The memory needed to compute the largest j Lyapunov numbers is equal to the memory needed to numerically integrate the equations under study, multiplied by $j + 1$. (Memory requirements are usually a problem only in computations involving partial differential equations.) The computer time needed is the time needed to compute a time average to the desired accuracy (which depends, among other things, on the irregularity of the natural measure of the attractor), multiplied by $j + 1$. Fortunately it is only necessary to compute the largest Lyapunov numbers, and the number of these needed depends on the dimension of the attractor rather than the dimension of the phase space. (See eq. (9).) This linear dependence on the dimension of the attractor has allowed computation of the Lyapunov dimension for attractors of dimension as large as twenty [38].

We should mention one disadvantage concerning Lyapunov dimension. Namely, it is not presently known how the Lyapunov dimension can be determined directly from a physical experiment. The difficulty comes about because, in some sense, in order to determine Lyapunov numbers it is necessary to be able to follow adjacent trajectories. To determine all the necessary Lyapunov numbers, it is necessary to follow some trajectories (at least one) which are not on the attractor. Thus it is not possible to compute the Lyapunov dimension by simply observing behavior on the attractor; one must perturb the system from the attractor, and do so in a very well defined way. This poses a very severe problem in the computation of dimension from experimental data, one that is not present in the computation of other dimensions.

9.2. *Computation of fractal dimension*

In principle, it is quite straightforward to use the definition of capacity, eq. (2), to compute the fractal dimension. The region of phase space surrounding the attractor is divided up into a grid of cubes of size ϵ, the equations are iterated, and the number of cubes $N(\epsilon)$ that contain part of the attractor are counted. ϵ is decreased and the process is repeated. If $\log N(\epsilon)$ is plotted against $\log \epsilon$, in the limit as ϵ goes to zero the slope is the fractal dimension.

The difficulty with this method is that one must use values of ϵ small enough to insure that the asymptotic scaling has been reached. (See Froehling et al. [40] and Greenside et al. [39].) The total number of cubes containing part of the attractor scales roughly as

$$N(\epsilon) \sim \epsilon^{-d_C}. \tag{40}$$

Thus, the number of cubes increases *exponentially* with the fractal dimension of the attractor. To get a feel for the seriousness of this problem, plug in some typical numbers: If $\epsilon = 0.01$ and $d_C = 3$, then $N \approx 10^6$, exceeding the core memory of all but the biggest current computers. Thus, computations of fractal dimension are currently not feasible for attractors of dimension significantly greater than three.

In addition, there is another potential problem involved in computing capacity. In counting cubes, how can one be sure that all the nonempty cubes have been counted? This problem is compounded by the highly nonuniform distribution of probability on an attractor. In particular, if our hypothesis that the probability is distributed log-normally is correct, in order to count the highly improbable cubes present in the wings of the distribution requires that a large number of points on the attractor must be generated. Furthermore, this number increases rapidly as ϵ decreases.

The conclusion is that a great deal of care must be taken in the computation of fractal dimension, and in particular, a sufficiently large number of points on the attractor must be generated to insure that low probability cubes are not left out in the determination of $N(\epsilon)$.

Although there are as yet no extensive results on direct computations of the dimension of the natural measure, it may be easier to reliably compute than the fractal dimension. The reason for this is that very improbable cubes are irrelevant for a computation of the dimension of the natural measure. Numerical experiments on this topic are currently in progress.

9.3. Summary of past numerical experiments

In this section we summarize previous numerical experiments on dimension computation. The two studies most relevant to the topic under discussion are those of Russel et al. [37] and Farmer [38]. Both of these were made in an attempt to test the Kaplan–Yorke conjecture [8, 22]. (See section 3.) In both of these studies, the capacity of chaotic attractors was computed directly from the definition. The Lyapunov dimension was also computed, and compared to the capacity.

In the study of Russel et al., five examples were examined. In each case, the computed capacity agree with the computed Lyapunov dimension to within experimental accuracy. These computations were done on the Cray1, a state of the art mainframe computer; at the smallest value of $\epsilon = 2^{-14}$, more than 10^5 cubes were counted.

The numerical experiments of Farmer were done using high-dimensional approximations to an infinite dimensional dynamical system. Because the equations under study were more time consuming to integrate, and because the capacity computations were done on a minicomputer, it was only possible to achieve about two significant figures of accuracy. The computed capacity and Lyapunov dimension agreed to this accuracy at the two parameter values tested.

In 1980, Mori [23] conjectured an alternate formula relating the fractal dimension to the spectrum of Lyapunov numbers. For attractors in a low-dimensional phase space, such as those studied by Russel et al. [37], Mori's formula and the Kaplan–Yorke formula (eq. (9)) predict the same value. For higher dimensional phase spaces, however, the two formulas no longer agree. Farmer's results support the Kaplan–Yorke formula.

One puzzling aspect of both of these numerical experiments is the striking agreement between the computed value of capacity and the Lyapunov dimension. The Kaplan–Yorke conjecture equates the Lyapunov dimension to the dimension of the natural measure, and therefore only gives a lower bound on the fractal dimension. Why, then, was such good agreement obtained between the computed capacity and the computed Lyapunov dimension? We do not yet understand the answer to this question, though further numerical experiments may resolve the question.

10. Conclusions

We have given several different definitions of dimension. These divide into two types, those that require a probability measure for their definition, and those that do not. (Refer back to table I.) For an example that we believe is typical of chaotic attractors, i.e., the generalized baker's transformation, our computations of dimension show that all of the probabilistic definitions take on one value, which we call the dimension of the natural measure, while the definitions that do not require a probability measure take on another value, which we call the fractal dimension of the attractor. We believe that this is true for typical attractors.

If the probability distribution on the attractor is "coarse grained" by covering the attractor with cubes, for the generalized baker's transformation we find that the probability contained in these cubes is distributed nearly log-normally when the cubes are sufficiently small. In other words, the total probability contained in cubes whose natural measure is between $u = \log p_i$ and $u + du$ has a distribution that is nearly Gaussian, and as the size of the cubes is decreased, it becomes more nearly Gaussian. Furthermore, the number of cubes in a given interval of u also has a Gaussian distribution, but with a different mean and variance. (See fig. 9.) As ϵ decreases, both of these distributions become narrower in a relative sense, in that the ratio of their variance to their mean decreases. In the limit as ϵ goes to zero, both distributions approach delta functions; since their means are different, in this limit the two distributions typically do not overlap. Thus, almost all of the natural measure is contained in almost none of the cubes, and the natural measure is concentrated on a core set. The capacity of the core is the fractal dimension of the attractor, while the Hausdorff dimension of the core is the dimension of the natural measure. Once again, although we have demonstrated the results mentioned in this paragraph only for the generalized baker's transformation, we feel that they are true for typical chaotic attractors.

Most of the dimensions that we have defined are difficult to compute numerically. The Lyapunov dimension, however, is much easier to compute numerically than any of the other dimensions. We compute the Lyapunov dimension for the generalized baker's transformation, and show that it is equal to the dimension of the natural measure obtained from any of the other probabilistic dimensions that we have investigated. This supports the conjecture of Kaplan and Yorke [22].

References

[1] R. Shaw, "Strange Attractors, Chaotic Behavior, and Information Flow", Z. Naturforsch. 36a (1981) 80.
[2] E. Ott, "Strange Attractors and Chaotic Motions of Dynamical Systems", Rev. Mod. Phys. 53 (1981) 655.
[3] R. Helleman, "Self-Generated Chaotic Behavior in Nonlinear Mechanics", Fundamental Problems in Stat. Mech. 5, E.G.D. Cohen, ed. (North-Holland, Amsterdam and New York, 1980) 165–233.
[4] J.A. Yorke and E.D. Yorke, "Chaotic Bahavior and Fluid Dynamics", Hydrodynamic Instabilities and the Transition to Turbulence, H.L. Swinney and J.P. Gollub eds., Topics in Applied Physics 45 (Springer, Berlin, 1981) pp. 77–95.
[5] B. Mandelbrot, Fractals: Form, Chance, and Dimension (Freeman, San Francisco, 1977).
[6] B. Mandelbrot, The Fractal Geometry of Nature (Freeman, San Francisco, 1982.)
[7] Ja. Sinai, "Gibbs Measure in Ergodic Theory", Russ. Math. Surveys 4 (1972) 21–64.
[8] P. Frederickson, J. Kaplan, E. Yorke, and J. Yorke, "The Lyapunov Dimension of Strange Attractors", J. Diff. Eqns. in press.
[9] J.D. Farmer, "Dimension, Fractal Measures, and Chaotic Dynamics", Evolution of Order and Chaos, H. Haken, ed., (Springer, Berlin, 1982), p. 228.
[10] J.D. Farmer, "Information Dimension and the Probabilistic Structure of Chaos", first chapter of UCSC doctoral dissertation (1981) and Z. Naturforsch. 37a (1982) 1304–1325.
[11] J. Alexander and J. Yorke, "The Fat Baker's Transformation", U. of Maryland preprint (1982).
[12] L.S. Young, "Dimension, Entropy, and Lyapunov Exponents", to appear in Ergodic Theory and Dynamical Systems.
[13] W. Hurwicz and H. Wallman, Dimension Theory (Princeton Univ. Press, Princeton, 1948).
[14] A.N. Kolmogroov, "A New Invariant for Transitive Dynamical Systems", Dokl. Akad. Nauk SSSR 119 (1958) 861–864.
[15] Hausdorff, "Dimension und Außeres Maß", Math. Annalen. 79 (1918) 157.
[16] R. Bowen and D. Ruelle, "The Ergodic Theory of Axiom-A Flows", Inv. Math. 29 (1975) 181–202.
[17] J. Balatoni and A. Renyi, Publ. Math. Inst. of the Hungarian Acad. of Sci. 1 (1956) 9 (Hungarian). English translation, Selected Papers of A. Renyi, vol. 1 (Academiai Budapest, Budapest, 1976), p. 558. See also A. Renyi, Acta Mathematica (Hungary) 10 (1959) 193.
[18] C. Shannon, "A Mathematical Theory of Communication", Bell Tech. Jour. 27 (1948) 379–423, 623–656.
[19] F. Ledrappier, "Some Relations Between Dimension and Lyapunov Exponents" Comm. Math. Phys. 81 (1981) 229–238.
[20] F. Takens, "Invariants Related to Dimension and Entropy", to appear in Atas do 13 Colognio Brasiliero de Mathematica.
[21] T. Janssen and J. Tjon, "Bifurcations of Lattice Structure", Univ. of Utrecht preprint (1982).
[22] J. Kaplan and J. Yorke, Functional Differential Equations and the Approximation of Fixed Points, Proceedings, Bonn, July 1978, Lecture Notes in Math. 730, H.O. Peitgen and H.O. Walther, eds., (Springer, Berlin, 1978), p. 228.
[23] H. Mori, "Prog. Theor. Phys. 63 (1980) 3.

[24] V.I. Oseledec, "A Multiplicative Ergodic Theorem. Lyapunov Characteristic Numbers for Dynamical Systems", Trans. Moscow Math. Soc. 19 (1968) 197.
[25] C. Grebogi, E. Ott, and J. Yorke, "Chaotic Attractors in Crisis", in this volume.
[26] A. Douady and J. Oesterle, "Dimension de Hausdorff des Attracteurs, Comptes Rendus des Seances de L'academie des Sciences 24 (1980) 1135–38.
[27] A.N. Kolmogorov, Dolk. Akad. Nauk SSSR 124 (1959) 754. English summary in MR 21, 2035.
[28] Ya. G. Sinai, Dolk. Akad. Nauk SSSR 124 (1959) 768. English summary in MR 21, 2036.
[29] J. Crutchfield and N. Packard, "Symbolic Dynamics of One-Dimensional Maps: Entropies, Finite Precision, and Noise", Int'l. J. Theo. Phys. 21 (1982) 433.
[30] S. Pelikan, private communication.
[31] P. Billingsley, Ergodic Theory and Information, New York, 1965).
[32] I.J. Good, "The Fractional Dimensional Theory of Continued Fractions", Proc. Camb. Phil. Soc. 37 (1941) 199–228.
[33] H.G. Eggleston, "The Fractional Dimension of a Set Defined by Decimal Properties", Quart. J. Math. Oxford Ser. 20 (1949) 31–36.
[34] A. Besicovitch, "On the Sum of Digits of Real Numbers Represented in the Dyadic System", Math. Annalen. 110 (1934) 321.
[35] J.L. Kaplan, J. Mallet-Paret, and J.A. Yorke, "The Lyapunov Dimension of a Nowhere Differentiable Attracting Torus", Univ. of Maryland Preprint (1982).

[36] V.I. Arnold and Avez, Ergodic Theory in Classical Mechanics (New York, 1968).
[37] D. Russel, J. Hansen, and E. Ott, "Dimensionality and Lyapunov Numbers of Strange Attractors", Phys. Rev. Lett. 45 (1980) 1175.
[38] J.D. Farmer, "Chaotic Attractors of an Infinite Dimensional Dynamical System", Physica 4D (1982) 366–393.
[39] H. Greenside, A. Wolf, J. Swift, and T. Pignataro, "The Impracticality of a Box Counting Algorithm for Calculating the Dimensionality of Strange Attractors", Phys. Rev. A25 (1982) 3453.
[40] H. Froehling, J. Crutchfield, J.D. Farmer, N. Packard, and R. Shaw, "On Determining the Dimension of Chaotic Flows", Physica 3D (1981) 605.
[41] R. Kautz, private communication.
[42] A.J. Chorin, "The Evolution of a Turbulent Vortex", Comm. Math. Phys. 83 (1982) 517–535.
[43] G. Bennètin, L. Galgani and J. Strelcyn, Phys. Rev. A 14 (1976) 2338; also see G. Benettin, L. Galgani, A. Giorgilli and J. Strelcyn, Meccanica 15 (1980) 9.
[44] I. Shimada and T. Nagashima, Prog. Theor. Phys. 61 (1979) 228.
[45] B.B. Mandelbrot, "Intermittent turbulence in self-similar cascades: Divergence of high moments and dimension of the carrier". J. Fluid Mechanics 62 (1974) 331–358. See 351, 354 for a discussion of a frequency dependent dimension.
[46] B.B. Mandelbrot, "Fractals and turbulence: attractors and dispersion". Turbulence Seminar, Berkeley 1976/7. Lecture Notes in Mathematics 615, 83–93 (Springer, New York).

CRISES, SUDDEN CHANGES IN CHAOTIC ATTRACTORS, AND TRANSIENT CHAOS

Celso GREBOGI and Edward OTT
Laboratory for Plasma and Fusion Energy Studies, University of Maryland, College Park, Md. 20742, USA

and

James A. YORKE
Institute for Physical Science and Technology and Department of Mathematics, University of Maryland, College Park, Md. 20742, USA

The occurrence of sudden qualitative changes of chaotic dynamics as a parameter is varied is discussed and illustrated. It is shown that such changes may result from the collision of an unstable periodic orbit and a coexisting chaotic attractor. We call such collisions *crises*. Phenomena associated with crises include sudden changes in the size of chaotic attractors, sudden appearances of chaotic attractors (a possible route to chaos), and sudden destructions of chaotic attractors and their basins. This paper presents examples illustrating that crisis events are prevalent in many circumstances and systems, and that, just past a crisis, certain characteristic statistical behavior (whose type depends on the type of crisis) occurs. In particular the phenomenon of chaotic transients is investigated. The examples discussed illustrate crises in progressively higher dimension and include the one-dimensional quadratic map, the (two-dimensional) Hénon map, systems of ordinary differential equations in three dimensions and a three-dimensional map. In the case of our study of the three-dimensional map a new route to chaos is proposed which is possible only in invertible maps or flows of dimension at least three or four, respectively. Based on the examples presented the following conjecture is proposed: almost all sudden changes in the size of chaotic attractors and almost all sudden destructions or creations of chaotic attractors and their basins are due to crises.

Contents
1. Introduction 181
2. The one-dimensional quadratic map 183
3. The Hénon map 187
3.1. An example of a boundary crisis 187
3.2. Protohorseshoes become horseshoes via crises in the Hénon map 191
3.3. The effect of small two dimensionality on the one dimensional quadratic map 193
4. Ordinary differential equations 196
5. Fractal torus crises 197
6. Conclusions 199

1. Introduction

The study of attractors and how they change as a parameter of the system is varied is of great interest and has received much attention. Topics in this area include bifurcation theory [1], period-doubling cascades resulting in chaos [2], etc. In this paper we review and discuss some aspects of sudden changes in chaotic attractors which occur as a system parameter is varied. In particular, we discuss certain events which we call *crises* [3]. We define a crisis to be *a collision between a chaotic attractor and a coexisting unstable fixed point or periodic orbit*. In the following four sections we offer examples and discussion illustrating our main points:

(1) crises appear to be the cause of most sudden changes in chaotic dynamics (with the exception of "subductions" to be described later);

(2) such events occur in many circumstances and systems; and

(3) following the occurrence of a crisis, certain

0167-2789/83/0000–0000/$03.00 © 1983 North-Holland

characteristic statistical behavior (which depends on the type of crisis) occurs.

In general, to illustrate the above, the material of this paper consists of a combination of review and interpretation of past work, as well as new research results.

In section 2 we treat the case of the one-dimensional quadratic map. This simple paradigmatic example is used to introduce and distinguish two types of crises, the *boundary crisis* and the *interior crisis*. The former leads to sudden destruction of the chaotic attractor and its basin of attraction, while the latter can cause sudden changes in the size of the chaotic attractor. Section 2 also discusses the associated characteristic statistical behavior occurring for parameter values just past that at which the crisis takes place. In the case of the boundary crisis, for parameter values just past the crisis point, the attractor no longer exists. Nevertheless, typical trajectories initialized in the region formerly occupied by the destroyed attractor appear to move about in this region chaotically, as before the crisis occurred, but only for a finite time after which the orbit rapidly leaves the region (a *chaotic transient*). In the case of an interior crisis, the attractor suddenly increases in size, and we study the dependence of the fraction of time a typical orbit on an attractor of increased size spends in the formerly empty region as a function of closeness of the parameter to its crisis value. In this case a new type of scaling phenomenon is found and discussed. [The example of an interior crisis used in section 2, the sudden widening of three chaotic bands to form one broader band, was first extensively discussed by Chang and Wright [4] (see also Grebogi et al. [3]).]

In section 3 we give two examples from the Hénon map. In the first, previously noted by Simo' [5], a range of the parameter occurs in which two strange attractors, each with its own basin of attraction, coexist. As the parameter is raised, one of the attractors experiences a boundary crisis and is destroyed. Past this point a chaotic transient occurs, and we investigate its characteristics, including the average lifetime of the transient. In addition, the connection between crises and the formation of horseshoes for this example is discussed and illustrated. As a byproduct of our study of this example, we have carried out an extensive numerical examination of the dependence of Lyapunov numbers on initial conditions. To our knowledge, this represents the most systematic and rigorous numerical test and confirmation yet done of the proposition that the same Lyapunov numbers result from almost all initial conditions in some basin of attraction. In our second example utilizing the Hénon map, we consider the effect of having the Jacobian nearly zero (so that the map is nearly one-dimensional) on the intermittent chaos associated with the tangent bifurcations which occur for many parameter values within the chaotic range of the one-dimensional quadratic map. We find that, for small positive Jacobian, the intermittency and tangent bifurcation phenomenon is replaced by a sequence of events involving the coexistence of two attractors and a boundary crisis.

In section 4 we consider examples involving differential equations in three dimensions. We show that, for several examples appearing in the literature, boundary crises are probably present. In particular, the Lorenz attractor provides a clear case [6], in which the onset of chaos takes the form of a sequence of events involving the coexistence of two attractors and a boundary crisis.

In section 5 we describe a special kind of boundary crisis and present an illustrative example. For the example considered the chaotic attractor takes the form of a "fractal" (nowhere differentiable) torus lying in three-dimensional space. There is another fractal torus which forms the boundary of the basin of attraction for the chaotic attractor. As a parameter of the system is raised a boundary crisis occurs in which unstable fixed points on the two tori collide. Above this value of the parameter almost all initial conditions asymptote to infinity.

We wish to emphasize that, although the discussion of boundary crises has so far concerned with the destruction of strange attractors, the inverse process is also necessarily implied. That is, if a parameter is varied from a to b and the strange attractor is destroyed, then when the parameter is varied in the opposite direction, from b to a, the attractor is born. Thus the *sudden* appearance of a strange attractor and its basin is a possible "route to chaos". This sudden appearance is to be contrasted with the routes discussed by Eckmann [7] and called by him "scenarios". In a scenario a non-chaotic attractor changes its character so that it becomes chaotic. In contrast, in the inverse crisis route to chaos, a chaotic transient is converted into a chaotic attractor. For example, the route to chaos, (discussed by Kaplan and Yorke [6]) for the Lorenz attractor is of this type. Also, our discussion in section 5 can be viewed as illustrating a new route to chaos in which chaos appears due to a type of bifurcation which can only occur in invertible maps (flows) of at least three (four) dimensions.

To conclude this section, we note that the Hénon map crisis in section 3 occurs at a parameter value where there is a tangency of the stable and unstable manifolds as illustrated in fig. 13. However, the crisis described in section 5 is different in that no tangency of stable and unstable manifolds occurs (as shown in fig. 29). Thus, the concept of crises (collisions of attractors with unstable fixed points or unstable periodic orbits) cannot be replaced by the older concept of tangencies of stable and unstable manifolds.

2. The one-dimensional quadratic map

The one-dimensional quadratic map,

$$x_{n+1} = C - x_n^2 \equiv F(x_n, C), \qquad (1)$$

exhibits the following well-known behavior. For $C < -\frac{1}{4}$ no fixed point of the map exists, and all orbits asymptote to $x = -\infty$. At $C = -\frac{1}{4}$ a tangent bifurcation occurs at which a stable and unstable fixed point are created. For $C > -\frac{1}{4}$ these points are $x = -x_* \equiv -\frac{1}{2} - [\frac{1}{4} + C]^{1/2}$ (unstable) and $x = -\frac{1}{2} + [\frac{1}{4} + C]^{1/2}$ (stable). As C increases

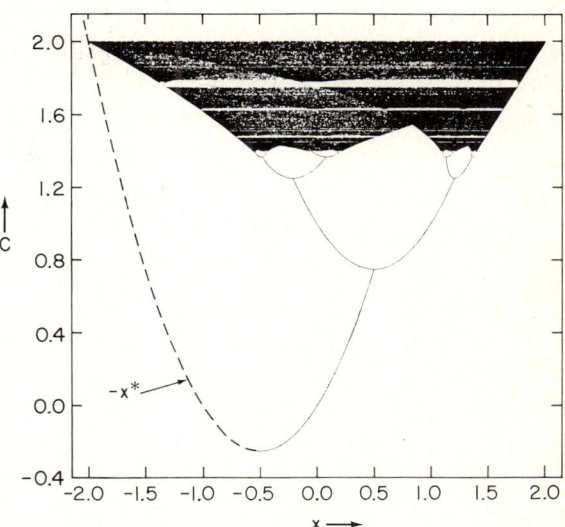

Fig. 1. Bifurcation diagram for the map, eq. (1). The dashed curve is the unstable fixed point. For a given value of C, an initial condition x is chosen and its orbit is plotted after the first several thousand iterates. The first several thousand iterates are discarded in order to ensure that the transient has died away. The figure is generated by doing this for many different values of C.

past $-\frac{1}{4}$, the stable fixed point undergoes period doubling followed by chaos [2]. The basin of attraction for these orbits for $-\frac{1}{4} \leq C \leq 2$ is $|x| \leq x_*$. Thus the unstable orbit, $x = -x_*$, is on the boundary of the basin of attraction. As C is incrased past $C = 2$, the chaotic attracting orbit is destroyed, and almost all initial conditions lead to orbits which approach $x = -\infty$. Fig. 1 gives a bifurcation diagram corresponding to this situation. In this figure the unstable fixed point is plotted as a dashed curve. Note that the destruction of the chaotic attractor and its basin as C increases through $C = 2$ coincides with the intersection of the chaotic band with the unstable fixed point. For C slightly larger than two, a point initialized in the formerly chaotic region will generate a chaotic looking orbit (a chaotic transient) until "by chance" x_n falls into a loss region [cf. discussion following eq. (3)]. After this happens, the orbit rapidly accelerates to large negative values of x. It has been shown [8, 9] that, for a smooth distribution of initial conditions in the formerly chaotic region of x, the length of a chaotic transient, τ, is exponentially distributed [10],

$$\phi(\tau) = \langle \tau \rangle^{-1} \exp -(\tau/\langle \tau \rangle). \qquad (2)$$

Letting $\langle \tau \rangle$ denote the expected value of the transient time, eq. (2) means that the probability of observing a transient time in an interval between τ and $\tau + d\tau$ is $\phi(\tau) d\tau$ (when an initial condition is chosen at random). Numerical evidence for (2) in higher-dimensional cases will be given in section 3. Furthermore, for $1 \gg C - 2 > 0$, the average length of a chaotic transient, $\langle \tau \rangle$, scales as [9]

$$\langle \tau \rangle \sim (C - 2)^{-1/2}. \qquad (3)$$

That this should be so can be se seen as follows. For $C - 2 > 0$ there is a region $|x| < (C - x_*)^{1/2} \approx [\frac{2}{3}(C - 2)]^{1/2}$, which, on one iterate, maps to $x > x_*$. On one more iterate x is mapped to $x < -x_*$, after which it accelerates to $-\infty$. Thus points which fall in this region are lost. Since the length of the loss region is approximately proportional to $(C - 2)^{1/2}$, we expect $\langle \tau \rangle \sim (C - 2)^{-1/2}$. It will be demonstrated numerically in section 3 that (3) holds for a two-dimensional case occurring in the Hénon map. The type of crisis illustrated in fig. 1 (namely, one in which the unstable orbit which collides with the chaotic attractor is on the boundary of the basin) we call a *boundary crisis*.

A second type of crisis in which the collision with an unstable orbit occurs within the basin of attraction, herein called an *interior crisis*, is illustrated by the bifurcation diagram shown in fig. 2. Fig. 2 is a magnification of the region in fig. 1 encompassing the tangent bifurcation at which a stable period three orbit is born within the chaotic regime. Also shown in fig. 2 are dashed curves denoting the *unstable* period-three orbit created at the tangent bifurcation. Note from fig. 2 that for a range of C less than a certain critical value, C_{*3}, the chaotic attractor lies within three distinct bands, but that, when C increases past $C_{*3} \approx 1.79$, the three chaotic bands suddenly widen to form a single band. Furthermore, this coincides precisely with the intersection of the unstable period-three

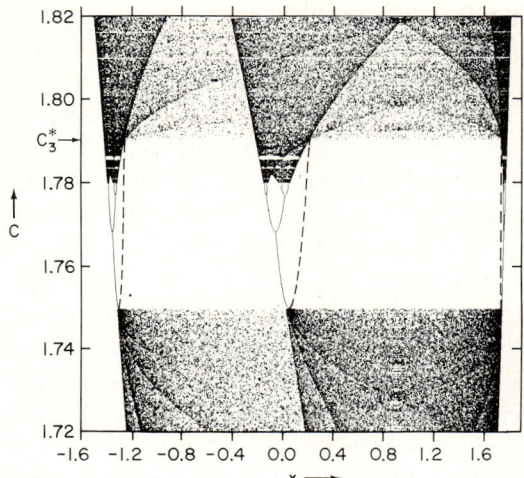

Fig. 2. Blow up of the bifurcation diagram of fig. 1 in the region of the period three tangent bifurcation. The dashed curves denote the unstable period-three orbit created at the tangent bifurcation.

orbit created at the original tangent bifurcation with the chaotic region. Similar crisis-induced widenings are associated with the other tangent bifurcations occurring in the chaotic range (i.e., $C_\infty < C \leq 2$, where C_∞ is the accumulation point for period doubling bifurcations of the original stable fixed point). As another example of this, fig. 3 shows a bifurcation diagram for the period five window.

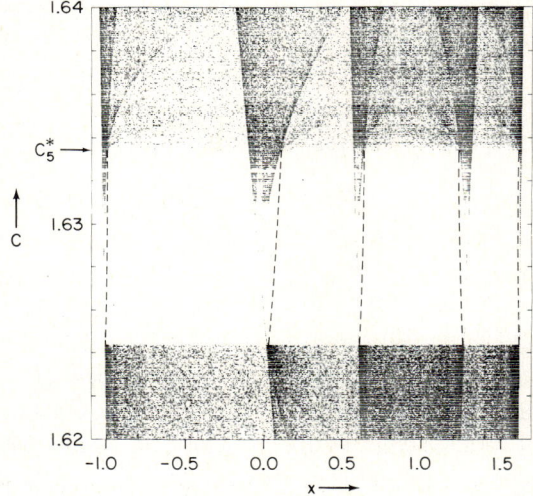

Fig. 3. Bifurcation diagram for the region of the period five tangent bifurcation.

In addition, we also note that figs. 2 and 3 provide examples illustrating that boundary crises are not the only means by which a chaotic attractor may suddenly terminate. Namely, a process that might be called *subduction* can occur. In a subduction a nonchaotic attractor appears within a chaotic attractor, and the chaotic attractor is replaced by the nonchaotic attractor. The effect of a subduction differs from that of a boundary crisis in that a subduction does not destroy the basin of the chaotic attractor, but rather the basin remains unchanged as the chaotic attractor converts to a nonchaotic attractor. Subductions are well-known phenomena in the case of the one-dimensional quadratic map, and correspond to the tangent bifurcations in which non-chaotic periodic orbits appear within the chaotic range. The points $C \approx 1.7499$ and $C \approx 1.6244$ at which the period-three and five orbits are born correspond to subductions (cf. figs. 2 and 3). Examples of subductions occurring in a two-dimensional map will be given in subsection 3.3.

Returning now to our discussion of the interior crisis at $C = C_{*3}$, consider fig. 4a which shows the

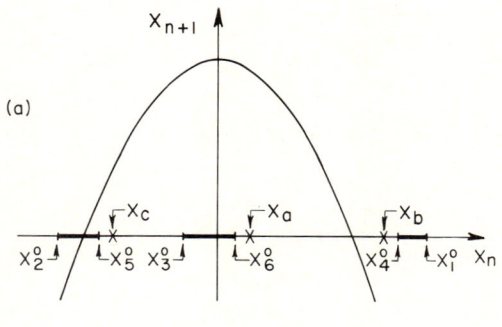

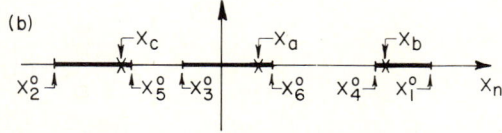

Fig. 4. (a) Schematic illustration of the quadratic map, eq. (1), for a value of C slightly less than C_{*3}. The three chaotic bands are indicated on the x_n-axis with boundary points $x_1^0, x_2^0, x_3^0, x_4^0, x_5^0, x_6^0$. Also shown as crosses are the components of the unstable period three orbit, x_a, x_b, x_c. (b) Schematic illustration of the x_n-axis for C slightly larger than C_{*3}.

map (1) for a value of C slightly less than C_{*3}. The three chaotic bands are indicated schematically on the x_n axis as the intervals $[x_3^0, x_6^0]$, $[x_4^0, , x_1^0]$, and $[x_2^0, x_5^0]$. Also, the unstable period three points, x_a, x_b, x_c, are indicated as crosses. The rightmost boundary of the chaotic region, x_1^0, is clearly the image of $x = 0$, since $F(x, C)$ is maximum at $x = 0$. Thus, $x_1^0 = F(0, C)$. We denote F composed with itself n times by $F^{(n)}(x, C)$; i.e., $F^{(n)}(x, C) = F[F^{(n-1)}(x, C), C]$, and $F^{(1)}(x, C) \equiv F(x, C)$. Examination of fig. 4 then shows that $x_n^0 = F^{(n)}(0, C)$. Now consider x_4^0. At $C = C_{*3}$, $x_4^0 = x_b$, and hence $x_4^0 = x_7^0$, or $F^{(4)}(0, C_{*3}) = F^{(7)}(0, C_{*3})$. This relation provides a means for the accurate numerical determination of C_{*3}. We obtain $C_{*3} = 1.790327492\ldots$.

For C slightly larger than C_{*3}, the unstable orbit x_a, x_b, x_c will lie within the bands, $[x_3^0, x_6^0]$, $[x_4^0, x_1^0]$, $[x_2^0, x_5^0]$, [cf. fig. 4b]; x_a will be slightly less than x_6^0, x_b will be slightly greater than x_4^0, and x_c will be slightly less than x_5^0. An orbit started within one of the regions, $[x_3^0, x_a]$, $[x_b, x_1^0]$, $[x_2^0, x_c]$, will typically *initially* move about in a chaotic way, cycling among the three regions, as in the case $C < C_{*3}$. However (for C slightly larger than C_{*3}), the orbit will eventually fall within one of the small regions $[x_a, x_6^0]$, $[x_4^0, x_b]$, $[x_c, x_5^0]$. It will then be repelled by the unstable period-three orbit and be pushed into the formerly empty region, namely $[x_6^0, x_4^0]$ and $[x_5^0, x_3^0]$. The discussion of fig. 4 just given parallels the treatments in refs. 3 and 4 (cf. also ref. 11).

Figs. 1 and 2 illustrate three examples of general types of sudden changes of chaotic orbits: the tangent bifurcation (in which a chaotic orbit is converted to a periodic one; this is not a crisis but a subduction), the boundary crisis (in which the chaotic attractor and its basin are destroyed), and the interior crisis (in which a sudden widening of the chaotic attractor occurs). What is the characteristic behavior of an orbit when the parameter is near a point at which a sudden change in chaotic dynamics occurs? For a tangent bifurcation the answer is intermittency [12]. For a boundary crisis the answer is a chaotic transient [8–10]. To answer the question for the case of an interior crisis, consider the fraction of time which an orbit spends in the formerly empty region, $[x_5^0, x_3^0]$ and $[x_6^0, x_4^0]$. We let $c \equiv C - C_{*3}$ and denote the fraction by $f(c)$. We now attempt to predict the form of $f(c)$ for $c \ll 1$. The trajectory generally spends a long time (denote the average τ_2) in the old regions (namely $[x_3^0, x_6^0]$, $[x_4^0, x_1^0]$ and $[x_2^0, x_5^0]$) before being pushed into the new region. There it spends an average time τ_1 before returning to the old region, and the process continues back and forth. Since we are concerned with an invariant distribution, we balance the transfer rates between the two regions, $f/\tau_1 = (1-f)/\tau_2$. Since $c \ll 1$, we have $f \ll 1$, and thus $f(c) \approx \tau_1/\tau_2$. To estimate τ_2, we follow the argument leading to eq. (3) and consider the interval $[x_3^0, x_a]$. For $c > 0$ an interval about $x = 0$ will be mapped by $F^{(3)}$ to $[x_a, x_6^0]$ and then be repelled into the formerly empty region. Since F and hence $F^{(3)}$ have quadratic extrema about $x = 0$, the length of this loss region centered at $x = 0$ is proportional to $c^{1/2}$. Hence we estimate that $\tau_2 \sim c^{-1/2}$. To estimate τ_1, consider the action of the map on the interval $[x_a, x_b]$. Since $x_a \to x_b$ and $x_b \to x_c$, the interval is inverted and stretched between x_c and x_b. Thus some range near the middle of the formerly empty region $[x_6^0, x_4^0]$ is mapped into $[x_3^0, x_a]$. We denote this range near the middle of $[x_6^0, x_4^0]$ by $[x_\alpha, x_\beta]$. The lifetime of an orbit in the formerly empty region is essentially determined by the time it takes to land in $[x_\alpha, x_\beta]$ after which it gets kicked back into the old region. Let λ denote the linear stability parameter of the unstable period-three orbit, $\lambda = F^{(3)\prime}(x_a) = F'(x_a)F'(x_b)F'(x_c)$, where $F'(x) \equiv dF(x)/dx$. At $c = 0$, $\lambda = 3.714\ldots$. Consider iterates of the point $x = 0$ under the application of $F^{(3)}$. For $1 \gg c > 0$, on two applications of $F^{(3)}$ the image of $x = 0$ falls just to the right of x_a [cf. fig. 4b]. Since, for small c, this image x_6^0, is close to x_a, $(x_6^0 - x_a \sim c)$ the linear approximation is valid. On n further applications of $F^{(3)}$, the distance from x_a increases as λ^n. Thus the time it takes the orbit to exceed x_α is approximately $(\ln c^{-1})/(\ln \lambda)$, that is $x_\alpha - x_a \sim c\lambda^n$. Say that, for a particular value of c, after $\bar{\tau}$ iterates, x_6^0 falls exactly

on x_α. As c is increased further the orbit will fall into $[x_\alpha, x_\beta]$ and be lost. On further increase of c the $\bar{\tau}$ iterate of x_6^0 will exceed x_β with the $\bar{\tau} - 1$ iterate of x_6^0 still less than x_α. When this happens τ_1 increases above $\bar{\tau}$. On still further increase of c, the $\bar{\tau} - 1$ iterate of x_6^0 will fall on x_α, and the whole process will repeat. Based on the above picture of the process whereby orbits get fed back into the region of high density, we estimate that $\tau_1 \approx \tau_\alpha + P(\ln c)$, where $\tau_\alpha = \eta^{-1} \ln c^{-1}$, $\eta = \ln \lambda$, and P is a periodic function of period η. Hence, from $f(c) \approx \tau_1/\tau_2$, we have

$$f(c) \approx K[c^{1/2}P(\ln c^{-1}) + \eta^{-1}c^{1/2} \ln c^{-1}]. \quad (4)$$

Note that the first term in eq. (4), $f_1(c) \equiv Kc^{1/2}P(\ln c^{-1})$, possesses scale invariance, $f_1(\lambda c) = \lambda^{1/2} f_1(c)$. That is, if we look at a graph of $f_1(c)$ and then properly magnify it about the origin, it will look the same. Fig. 5 shows a graph of $f(c)/c^{1/2}$ versus $\ln c$ which has been numerically generated by iterating the map. The result support eq. (4) including the predicted period $\eta \approx 1.312$. [It may be that a renormalization group analysis of this scaling behavior could succeed in obtaining the function P of eq. (4).]

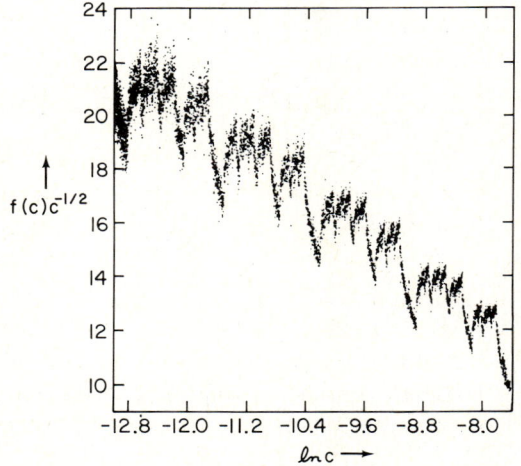

Fig. 5. $f(c)/c^{1/2}$ versus $\ln c$. $c = C - C_{*3}$.

3. The Hénon map

In this section we shall discuss two examples of crisis within the context of the well-known quadratic, two-dimensional, invertible map introduced by Hénon,

$$x_{n+1} = 1 - \alpha x_n^2 + y_n, \tag{5a}$$

$$y_{n+1} = -Jx_n. \tag{5b}$$

Note that the parameter J is the determinant of the Jacobian matrix of the map and that, for $J = 0$, eq. (5a) reduces to the one-dimensional quadratic map, $x_{n+1} = 1 - \alpha x_n^2$ (which can be transformed into (1) by the change of variables $\alpha x \to x$).

In subsection 3.1 we discuss a boundary crisis occuring for (5) for a case with large Jacobian determinant ($J = -0.3$). In subsection 3.2 we investigate the connection between crises and horseshoe formation. Then, following that, subsection 3.3 discuss how certain results from the one-dimensional map ($J = 0$) are modified by a small amount of two dimensionality (i.e., small J), and the role of crises in this phenomenon is pointed out.

3.1. An example of a boundary crisis

We consider the Hénon map for $J = -0.3$. It has been pointed out [5] that, for this J and a range of α, two attractors can exist simultaneously. Numerical results for $\alpha = 1.0807000$ are shown in figs. 6–10. For this value of α the two coexisting attractors are both chaotic. One attractor consists of four pieces on which an orbit on the attractor cycles sequentially. The other attractor consists of six pieces on which its orbit similarly cycles. Fig. 6 shows the basins of attraction for the six-piece chaotic attractor in black and that for the four piece chaotic attractor in white. Also shown in fig. 6 is the four-piece attractor. Fig. 7 shows both the four-piece attractor (A1, ..., A4) and the six-piece attractor (B1, ..., B6). Note that an unstable (saddle) fixed point of the map is located at $x = 0.69113..., y = 0.20734...$ (labeled FP in fig.

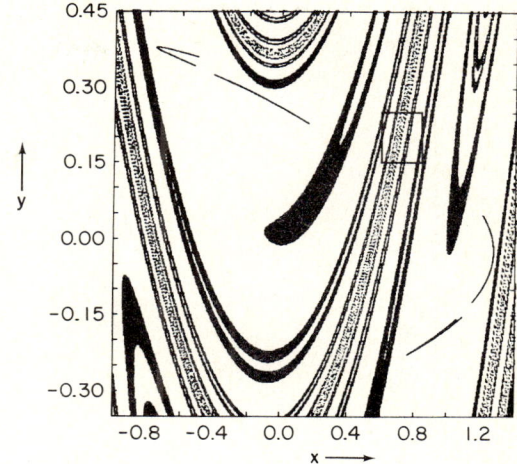

Fig. 6. Basins of attraction for the six-piece chaotic attractor (black) and the four-piece chaotic attractor (blank) for $\alpha = 1.0807000$ and $J = -0.3$. Also shown is the four-piece chaotic attractor.

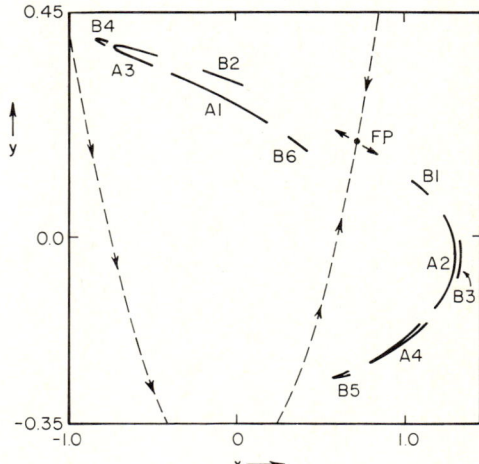

Fig. 7. Orbits on the four-piece attractor cycle from segment A1 to A2 to A3 to A4 and back to A1. Orbits on the six-piece attractor also cycle (B1→B2→···→B6→B1). The parameter value is the same as in fig. 6. The stable manifold of the unstable saddle fixed point FP is shown (as a dashed line) and the unstable directions are indicated.

7). The stable and unstable manifolds of this fixed point are shown in fig. 8. Comparison of figs. 7 and 8 suggests that the two attractors lie on the closure of the unstable manifold of the fixed point. From fig. 6 fine scale structure in the basins of attraction

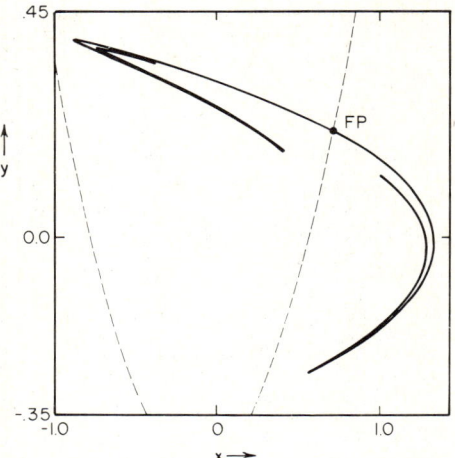

Fig. 8. The stable manifold (dashed line) and unstable manifold (continuous line) of the fixed point FP for the same parameter value as in figs. 6 and 7.

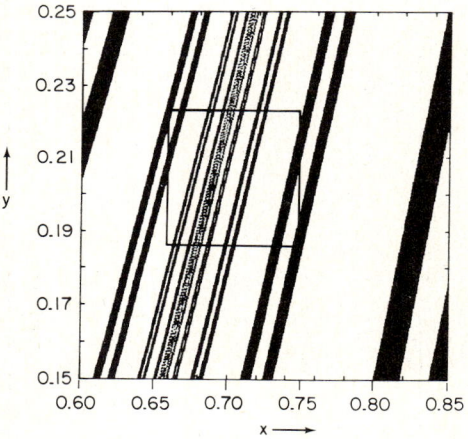

Fig. 9. Blow up of the boxed region in fig. 6.

is apparent. To see this more clearly, fig. 9 presents a blow up of the boxed region in fig. 6. This box was chosen so that the fixed point of the map at $x = 0.69113\ldots, y = 0.20734\ldots$ is at the box's center. The eigenvalues of the Jacobian matrix of the map at the saddle fixed point are $\bar{\lambda}_u = -1.6731\ldots$ and $\bar{\lambda}_s = 0.17931\ldots$. The eigenvector associated with $\bar{\lambda}_u$ is of course tangent to the unstable ($|\bar{\lambda}_u| > 1$) manifold of the fixed point, while the eigenvector associated sigh $\bar{\lambda}_s$ is tangent to the stable manifold of the fixed point. The stable manifold (also shown as a dashed line in fig. 7) is in the direction along the bands of fig. 9, while the unstable manifold cuts across them. Now consider what happens if we apply the inverse map to fig. 9. Bands crossing the unstable manifold will be drawn in towards the fixed point, shrunk in width (by approximately $|\bar{\lambda}_u|^{-1}$), flipped to the other side of the fixed point (because $\bar{\lambda}_u$ is negative), and stretched out along the stable manifold (by $1/\bar{\lambda}_s$). (Thus the fine scale structure seen in fig. 6 aligns along the stable manifold of the fixed point.) To further demonstrate the above, we magnify the region about the fixed point in fig. 9 by a factor $\bar{\lambda}_u^2$. The result, fig. 10, reproduces fig. 9, thus showing clearly the scaling of the basins of attraction about the fixed point.

We now discuss how the basin of attraction diagrams, figs. 6, 9 and 10, were constructed. Numerically it is found that the positive Lyapunov exponents for the map are $h_4 = 0.1391\ldots$ (corresponding to the four-piece attractor) and $h_6 = 0.1097\ldots$ (corresponding to the six-piece attractor). We choose an x–y grid of 1.09×10^4 initial conditions. We then iterate the map and its Jacobian and calculate the resulting Lyapunov exponent. For every initial condition tested we find that the computed Lyapunov number falls within either the interval $[h_4 - 5 \times 10^{-4}, h_4 + 5 \times 10^{-4}]$ or

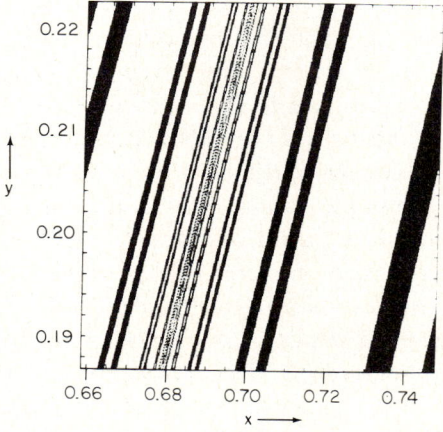

Fig. 10. Blow up of the boxed region in fig. 9.

the interval $[h_6 - 5 \times 10^{-4}, h_6 + 5 \times 10^{-4}]$. (Presumably the tolerances would be reduced if the number of iterates of the map was increased.) When a Lyapunov exponent falls within the latter interval, we plot its initial condition as a dot. Because of the high density of initial conditions, the six-piece attractor basin is completely filled in as the black regions of figs. 6, 9 and 10. In no case does an isolated black dot occur in a white region and vice versa.

The above described result indicates that, from a practical numerical point of view, the Lyapunov numbers are essentially independent of the choice of initial point in the basin (i.e., we saw only two ranges rather than a much larger class of Lyapunov numbers). While this appears to be a reasonable a priori assumption and is basic to a number of previous studies, its validity does not appear to have been previously tested as intensively as was done here.

Returning now to the issue of how the dynamics change as a parameter is varied, fig. 11 shows a bifurcation diagram (α versus x) for a range of x which encompasses one segment each of the four piece and the six-piece attractors. The six-piece attractor is created at $\alpha = \alpha_0 \equiv 1.062371838...$ by a saddle-node bifurcation in which an attracting period-six orbit and an unstable saddle type period-six orbit simultaneously appear. The position of the saddle is shown as a dashed line in the figure. The stable manifold emanating from the saddle forms the boundary separating the basins of attraction for the two attractors. As α is increased past α_0, the stable period-six orbit undergoes period doubling followed by chaos. Finally at $\alpha = \alpha_c = 1.080744879...$ a crisis occurs, and the six-piece attractor ceases to exist for $\alpha > \alpha_c$. The evolution of a segment of the six-piece attractor (B5 of fig. 7) and the basin boundary, as α approaches α_c from below, is illustrated in fig. 12. The location of the unstable period-six saddle (which was created at $\alpha = \alpha_0$) is indicated in fig. 12. The collision of the chaotic attractor and the saddle on the basin boundary is evident. From fig. 12c we see that at the crisis, $\alpha = \alpha_c$, a "finger"-shaped region of the four-piece attractor basin comes down and touches the six-piece attractor. This results from a backwards iterate of the region around the apparent tangency point of the attractor and the basin boundary at the left of the figure. Furthermore, there are, in fact, an infinite number of such fingers, accumulating at the unstable saddle. These fingers correspond to higher order backwards iterates of the tangency region at the left of the figure, but are so narrow that they are not visible in fig. 12c.

Generally, for the examples discussed in this paper, we find that, when a crisis approaches, it seems that the attractor lies in the closure of the unstable manifold of the periodic orbit that it is about to collide with. Thus the evolution of the attractor shown in fig. 12 may also be discussed in terms of stable and unstable manifolds of the saddle [5]. The six-piece attractor apparently lies on the unstable manifold of the saddle (compare figs. 7 and 8). Thus, as the crisis is approached, the stable manifold (basin boundary) and the unstable manifold approach each other, becomes tangent ($\alpha = \alpha_c$), and then cross ($\alpha > \alpha_c$). The situation for $\alpha = \alpha_c$ is schematically illustrated in fig. 13. Near the crisis ($\alpha = \alpha_c$), the results of Newhouse [13] imply that there can be an infinite number of attractors. Our numerical results, however, have only succeeded in finding the original six- and

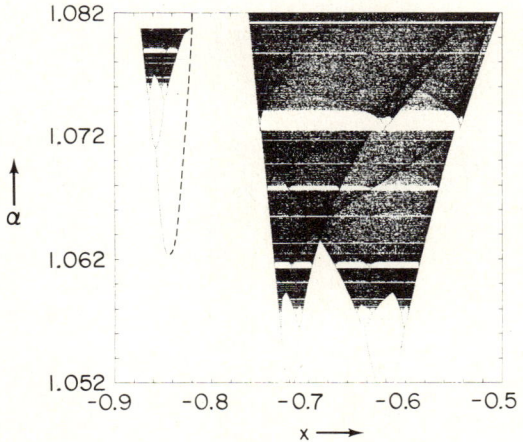

Fig. 11. Bifurcation diagram for eq. (5) with $J = -0.3$.

four-piece attractors. Thus any other attractors must be difficult to detect (e.g., their basins may be very small).

Fig. 14 shows successive iterates of an orbit initialized in the former basin of attraction of the six-piece attractor for α slightly larger than α_c ($\alpha = 1.0807455$). The numbers indicate the sequence in which points appear in the figure. Thus after the first iterate the orbit rapidly approaches the remnant of the destroyed attractor. The orbit then bounces around on this remnant roughly 2,000 times, until it finds its way out, and then rapidly goes to the four-piece attractor. Fig. 15

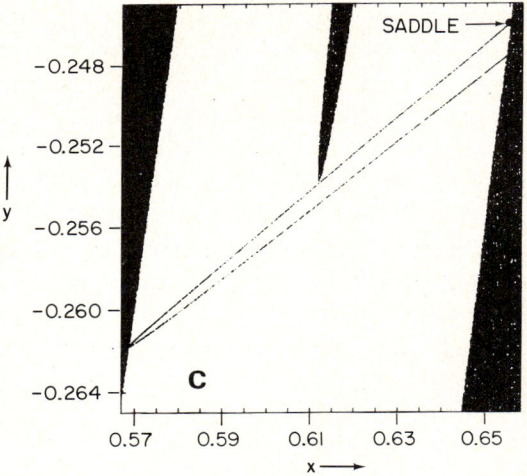

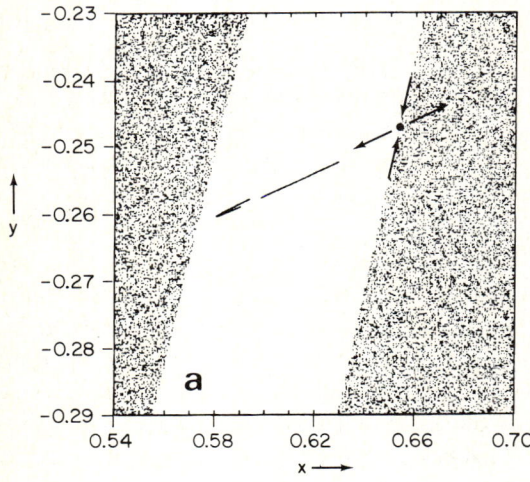

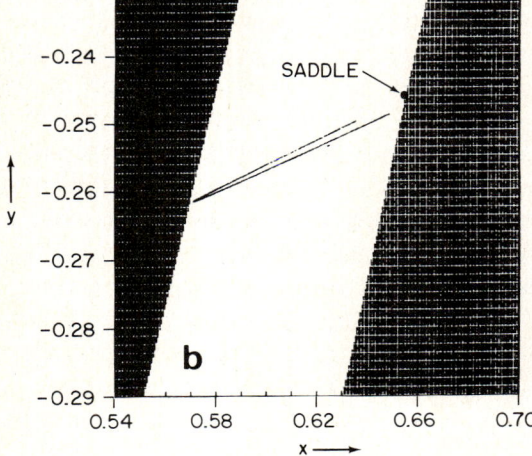

Fig. 12. (a) A segment of the six-piece attractor in its basin for $\alpha = 1.0770000$, $J = -0.3$. The location of one component of the period-six saddle and its stable and unstable directions are indicated. The region filled in with dots denotes the basin of the four-piece attractor. (For this value of α each segment of the "six-piece" attractor consists of two pieces so it is really a twelve-piece attractor). (b) $\alpha = 1.0800000$. (c) $\alpha = 1.0807430$.

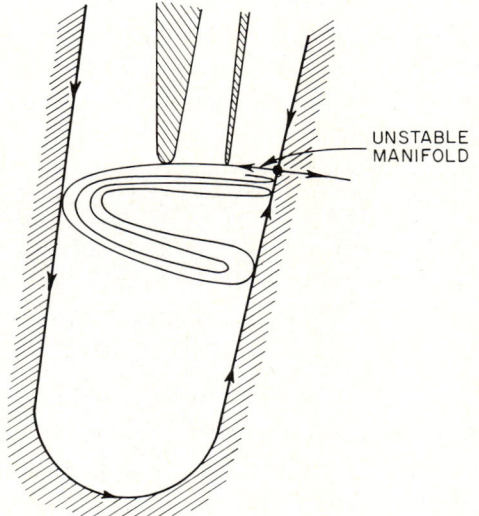

Fig. 13. Schematic diagram of the stable and unstable manifolds of the saddle for $\alpha = \alpha_c$. The stable and unstable manifolds of the saddle become tangent at an infinite number of points.

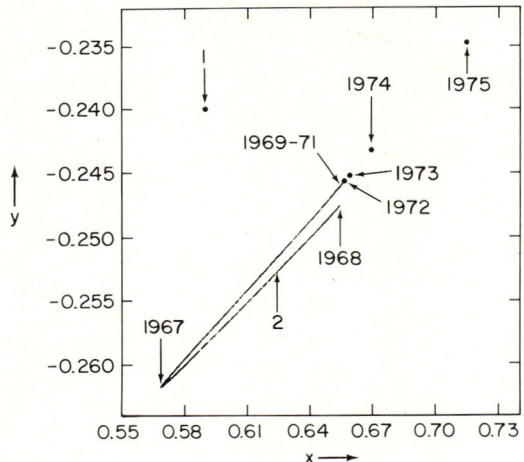

Fig. 14. Iterates of the map for $\alpha = 1.0807455 > \alpha_c$.

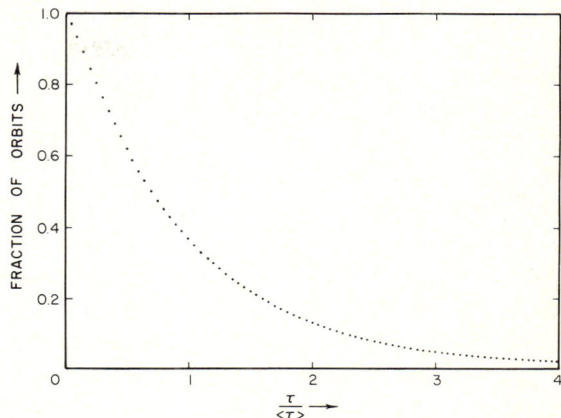

Fig. 16. Numerical results for fraction of orbits remaining on the remnant of the chaotic attractor as a function of time. These results were obtained for eq. (5) for $J = -0.3$ and $\alpha = 1.08007500$, using data from 10^5 initial conditions randomly chosen within the rectangular region $x \in (0.59000, 0.59125)$ and $y \in (-0.64000, -0.63875)$. $\langle \tau \rangle = 1753.6$.

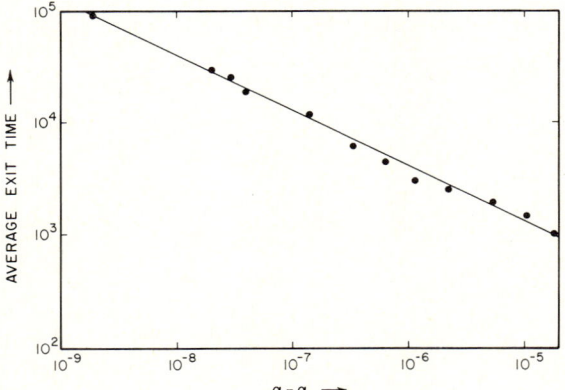

Fig. 15. Average lifetime of a chaotic transient obtained by numerical iteration of eq. (5) with $J = -0.3$ (dots) and theoretical predicted proportionality to $(\alpha - \alpha_c)^{-1/2}$ eq. (3) (straight line). Each point was obtained by averaging 10^2 initial conditions randomly chosen into the rectangular region $x \in (0.59000, 0.59125)$ and $y \in (-0.64000, -0.63875)$.

shows results from a numerical test (dots) of eq. (3) (straight line) (see captions for details). As can be seen from the figure the agreement is quite good. Fig. 16 shows numerical results (dots) testing eq. (2). On the scale of this figure there is no discernible difference between the numerical data and the exponential decay predicting of eq. (2).

3.2. Protohorseshoes become horseshoes via crises in the Hénon map

In this subsection we discuss the formation of a horseshoe associated with the six-piece attractor. The formation of horseshoes is sometimes thought to be associated with the appearance of chaotic attractors. In contrast, for our example we find that the horseshoe which we consider forms at some value of α greater than α_c. That is, it forms after the chaotic attractor has already been destroyed. In fact, as subsequent discussion will show, the destruction of the attractor appears to be a natural consequence of the formation of the horseshoe. On the other hand, we also find that before the horseshoe forms another configuration, which we call a *protohorseshoe*, exists, and that the attractor is contained within the protohorseshoe. Further, some of the observed aspects of the evolution of the six-piece attractor as α increases can be understood by considering the evolution of the protohorseshoe as it changes into a horseshoe.

A horseshoe [14] is a rectangle (or topological rectangle) whose image forms a horseshoe-like shape as shown in fig. 17(i). In this figure the rectangle has sides ABCD and its image sides A'B'C'D'. We have numerically taken the image of a rectangle in the region shown in fig. 12. The result is shown in fig. 17(iv), for $\alpha = 1.085 > \alpha_c$. It is seen that a horseshoe exists. Subject to certain regularity hypotheses it may be shown that there are infinitely many unstable periodic points in a horseshoe. Horseshoes do not contain attractors. As we vary parameters, the image of the rectangle will change, and we may no longer find a horseshoe. However, herein lies a practical problem: as the parameter is varied the sides of the rectangle can be varied in quite a few ways, and making a non-optimal choice of the sides will sometimes be the reason we no longer find the horseshoe. To aid in the procedure we will describe a way of choosing an optimal placement of sides A and C in fig. 17(i). In this figure, part of the rectangle at side A could be peeled off the rectangle and discarded. Then, the end of the image at A' would also be shortened. We could imagine whittling down side A, peeling off slices from the rectangle until the new A had an image A' which would lie on A as shown in fig. 17(ii). We may similarly peel from side C so that the new image C' lies on A as in fig. 17(iii). We will call this configuration a *reduced* horseshoe. We pay less attention to the placement of B and D. As the parameter of the map is varied, the sides A and C should be moved continuously so that we have a reduced horseshoe for each parameter value. Results proved for horseshoes, about the existence of unstable periodic orbits by Smale [14], are also valid for reduced horseshoes. Notice that in fig. 17(iii), side A is mapped into itself and so generally has a fixed point P. The contraction of A leads to the realization that A lies on the stable manifold of P, and, since C is mapped onto the stable manifold, C is also a segment of the same stable manifold of P (cf. fig. 13). Also, from the action of the map illustrated in fig. 17, generally there will be another fixed point, labeled Q in fig. 17(iii). For the example of subsection 3.1, the points P and Q correspond to those originally created by the saddle-node bifurcation at $\alpha = \alpha_0$.

In fig. 18(i) we show what we call a *proto-*

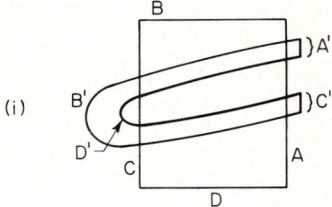

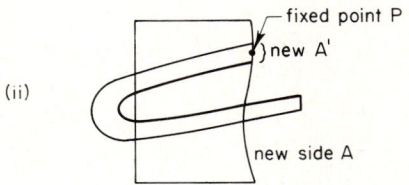

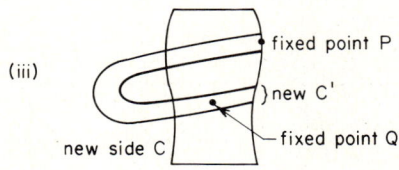

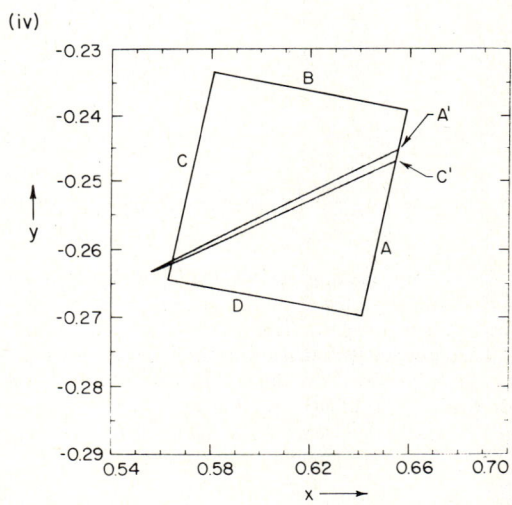

Fig. 17. A horseshoe is a topological rectangle ABCD whose image A'B'C'D' lies as shown in (i). The lettering of the sides is done for the case when the Jacobian is positive since otherwise B' and D' would be reversed. (ii) and (iii) show the steps in constructing a reduced horseshoe. (iv) Horseshoe generated numerically for the piece of attractor shown in fig. 12 ($\alpha = 1.085 > \alpha_c$). The horseshoe appears to be only a line because of the strong contraction along the stable direction.

(i)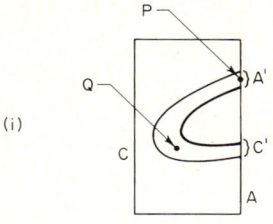

(ii)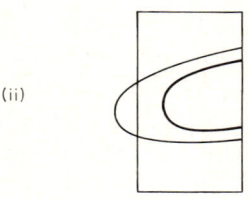

(iii)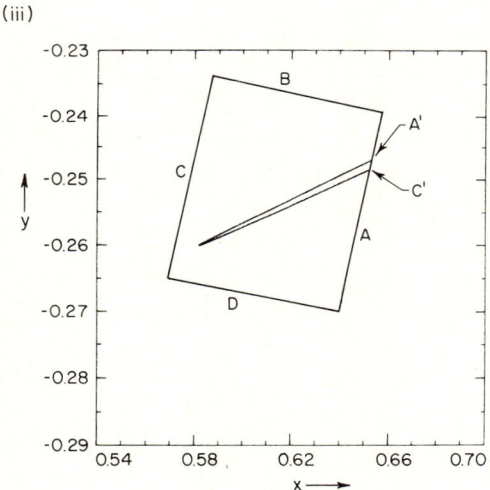

Fig. 18. (i) Protohorseshoe containing two fixed points P and Q with P on side A' and Q in the interior. (ii) Intermediate stage. (iii) Numerically generated protohorseshoe for the piece of the attractor and the same map parameter ($\alpha = 1.077 < \alpha_c$) as shown in fig. 12.

horseshoe. Since the original ABCD is mapped into itself the protohorseshoe must contain an attractor. Protohorseshoes for the sixth iterate of the Hénon map can be found for $\alpha_0 < \alpha < \alpha_c$ (cf. fig. 18(iii)). The protohorseshoe of the sixth iterate of the Hénon map contains a fixed point [Q in fig. 18(i)] that is attracting for $\alpha_0 < \alpha < 1.0709...$ Thereafter follows a period-doubling sequence and chaos, all of these periodic and chaotic attracting orbits lying in the protohorseshoe. Finally, past $\alpha = \alpha_c$, a boundary crisis occurs, and it is no longer possible to find sides B and D to construct a protohorseshoe. On further increase of α sufficiently past α_c the protohorseshoe of fig. 18(i) evolves into the reduced horseshoe shown in fig. 17(iii) (an intermediate stage in this process is shown in fig. 18(ii)). [However, it should also be noted that, as soon as $\alpha > \alpha_c$, it may be possible to construct a horseshoe (or reduced horseshoe) for some rectangle and some *sufficiently large* iterated image of it. In contrast, the horseshoe and protohorseshoe of figs. 17 and 18 pertain to the rectangle and its first image (or, in the case of the Hénon six-piece attractor, the sixth iterate of the map).]

3.3. *The effect of adding small two-dimensionality to the one-dimensional quadratic map*

We have investigated the effect of small J in eqs. (5). Since $J = 0$ corresponds to the one-dimensional quadratic map, small J may be viewed as a two-dimensional perturbation of the one-dimensional case. Numerically, the most striking effect occurring for small J is the modification of the tangent bifurcation phenomenon (subduction) which for $J = 0$ occurs throughout the chaotic range of α. Fig. 19 shows a bifurcation diagram for $J = 0.015$ and α in a range corresponding to the period-three window in the one-dimensional case (cf. fig. 2). Note that for $1.746... > \alpha > 1.713...$ two attractors are present, the original attractor and one which is born by a period-three saddle-node bifurcation at $\alpha \approx 1.713$. As α is increased to $\alpha \approx 1.746$ a boundary crisis ocurs in which the unstable period three saddle (dashed line in fig. 19) collides with the original chaotic attractor. Fig. 20 shows the basin of attraction for the original chaotic attractor (white region), the original chaotic attractor itself, the basin of attraction of the newly created attractor (black region), and the position on the basin boundary where the period

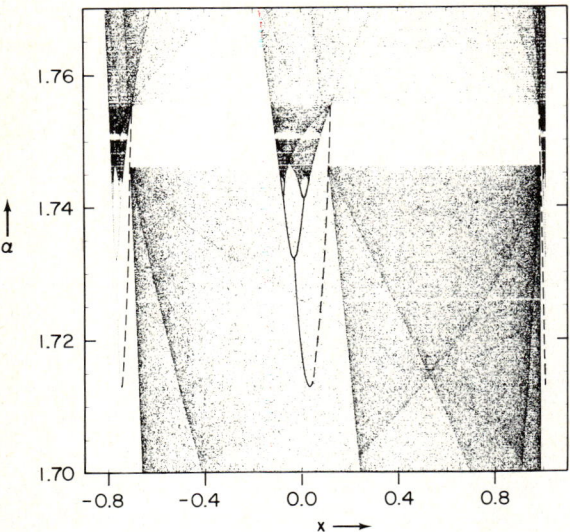

Fig. 19. Bifurcation diagram for $J = 0.015$.

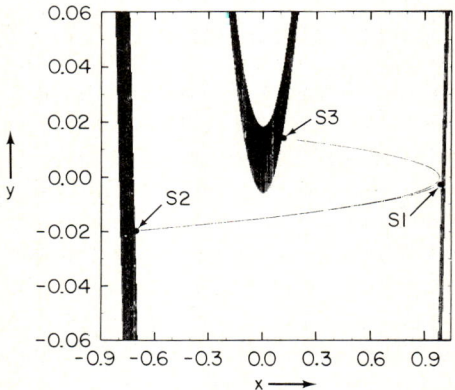

Fig. 20. Original chaotic attractor, period-three unstable saddle $\{S1, S2, S3\}$, and basin of the attractor created at the saddle-node bifurcation (dark region). $\alpha = 1.7300000$, $J = 0.020$.

illustrated in fig. 19 has been found experimentally [15].

Figs. 19 and 20 correspond to a case with small positive J, and it is evident that, for the case shown, the tangent bifurcation and intermittency phenomenon (or subduction) is replaced by a sequence involving first a saddle-node bifurcation then hysteresis (i.e., two attractors exist simultaneously), then a boundary crisis, and then an interior crisis. Actually, we can show with heuristic arguments that the sequence described is only expected if J exceeds some critical value, J_*, where J_* is small and positive (numerically we find that $J_* < 10^{-3}$ for the period-three window). For $J < J_*$ the original one-dimensional subduction sequence still occurs (see fig. 21 which corresponds to $J = -0.025$).

Also we note that as J increases and passes a second critical value $J'_* > J_*$, the sequence of events illustrated in fig. 19 again alters. Namely, the order in which the unstable saddle orbit hits the two chaotic attractors is interchanged. The result, illustrated in fig. 22 for $J = 0.025$, is first a boundary crisis destroying the now-chaotic three-piece attractor produced by the saddle-node bifurcation

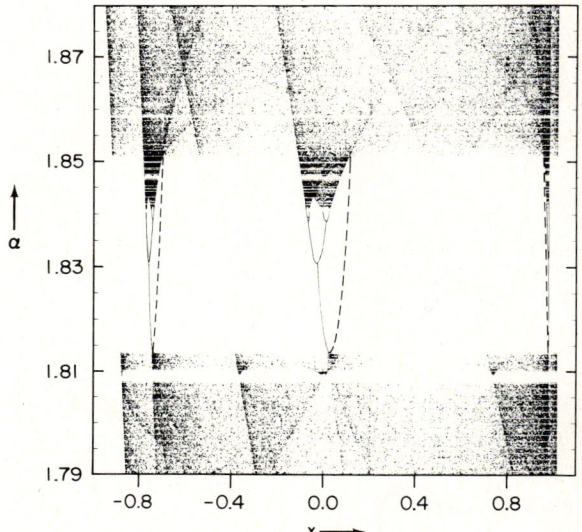

Fig. 21. Bifurcation diagram for $J = -0.025$.

three saddle $\{S1, S2, S3\}$ is located, for a case in which both attractors exist simultaneously. As α increases further, the period-three saddles collides with the now chaotic three-piece attractor originally created at $\alpha \approx 1.713$. This collision at $\alpha \approx 1.756$ is an interior crisis similar in effect to what happens in the one-dimensional case (i.e., a sudden widening of the three chaotic bands to form one broader chaotic band). The phenomenon

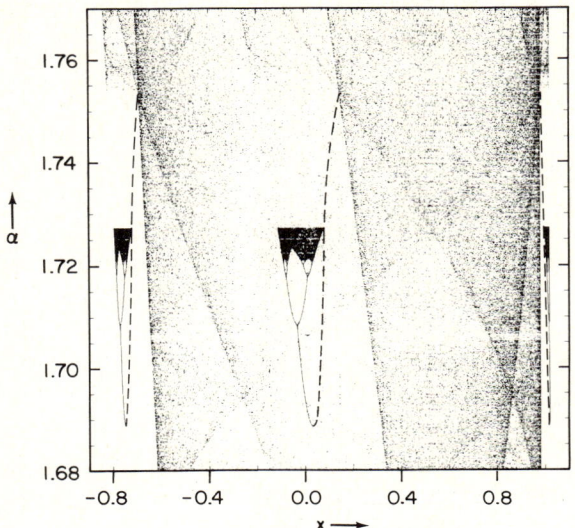

Fig. 22. Bifurcation diagram for $J = 0.025$.

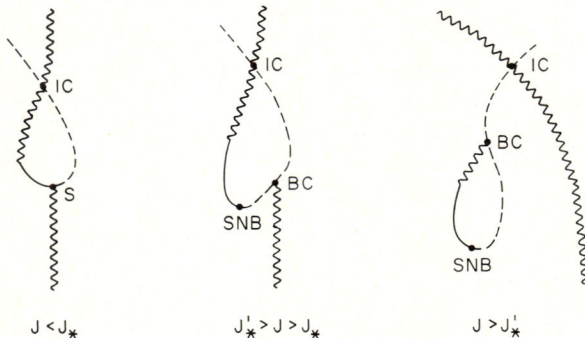

Fig. 23. Crisis diagrams for $J < J_*$, $J'_* > J > J_*$ and $J > J'_*$. IC, BC, SNB and S denote, interior crisis, boundary crisis, saddle node bifurcation and subduction, respectively.

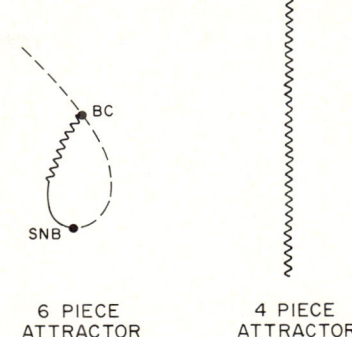

Fig. 24. Crisis diagram for phenomena described in subsection 3.1.

24 gives the crisis diagram for the evolution of the sequence involving the six-piece and four-piece attractors discussed in subsection 3.1 (cf. fig. 11).

In order to explain the results just shown, consider the schematic diagram, fig. 25, corresponding to $J > J_*$ and α slightly larger than its value at which the saddle-node bifurcation occurs (similar to fig. 20). In this figure a strange attractor and an attracting periodic orbit denoted {P1, P2, P3}, where P1→P2→P3→P1, exists simultaneously. Notice that the point P1 lies outside the original attractor's "elbow". This figure illustrates a common geometric inter-relationship that often occurs when two attractors coexist. For example, a situ-

and then an interior crisis giving a sudden widening of the original chaotic single band attractor.

Fig. 23 schematically summarizes the sequences of events occurring for the three cases, $J < J_*$, $J_* < J < J'_*$, and $J > J'_*$. We call the symbolic representations given in fig. 23 "crisis diagrams". In these crisis diagrams the parameter α is taken to increase vertically, a chaotic attracting orbit is represented by a wiggly line, an unstable periodic orbit by a dashed line, and a stable non-chaotic orbit is represented by a solid smooth line. Also fig.

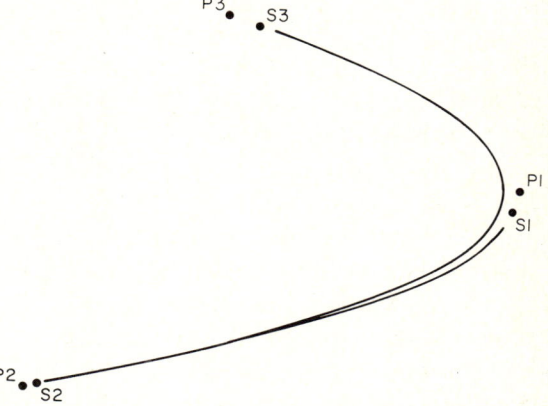

Fig. 25. Schematic representation of coexisting chaotic and periodic {P1, P2, P3} attractors (positive Jacobian). {S1, S2, S3} is the period-three unstable saddle.

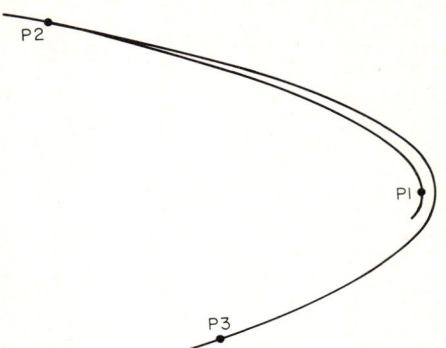

Fig. 26. Schematic diagram of the chaotic attractor at subduction (negative Jacobian).

ation similar to that in fig. 25 is seen in fig. 7, where the attractor piece B3 lies beyond the elbow of attractor A. Had fig. 7 been drawn at $\alpha = 1.065$ instead of 1.0807, the attractor piece B3 would have been a single point outside the elbow. For fig. 7, the periodicity is 6 (even) and the Jacobian of the map of the sixth iterate of the Hénon map is $(-0.3)^6$, a positive number. The simultaneous existence of the two attractors depends crucially on the orbit being beyond the elbow. Figs. 21 and 26 show what happens when the orbit appears inside the elbow. Here J is negative. Fig. 26 is a schematic representation of the chaotic attractor precisely at subduction ($\alpha \approx 1.813$ in fig. 21). The chaotic attractor becomes the periodic attractor {P1, P2, P3} due to a saddle-node bifurcation. In this case, the nature of the elbow of the chaotic attractor near P1 is that it bends tight on iteration. In fig. 7, the elbow piece A2 bends to become A3, and B3 to become B4, the piece B4 lying beyond the end of attractor A. Referring now to fig. 26, if it is assumed that the periodic orbit lies inside the elbow, but off the attractor, the bending elbow entangles the periodic points within the attractor as the map is iterated. Thus there is no separation between the orbit and the attractor, and the orbit must actually lie on the attractor as in fig. 26. The apparently contradictory idea of {P1, P2, P3} being an attractor and being in the chaotic attractor is resolved by noting that the chaotic attractor is actually destroyed by the appearance of the attracting periodic orbit. The distinction between the inside and outside cases lies in the Jacobian. When the Jacobian (or rather the Jacobian raised to the pth power, where p is the period) is negative, it can be argued that the periodic orbit lies inside and its appearance instantly destroys the chaotic attractor (so we have subduction). When the Jacobian is positive, however, it is possible for the saddle-node bifurcation to occur outside the attractor, in which case subduction no longer occurs. (Actually, as previously mentioned, subduction still takes place for J positive but small i.e., $0 < J < J_*$.)

4. Ordinary differential equations

The occurrence of crisis in differential equations is illustrated by at least three examples occurring in the literature: (a) the Lorenz attractor [6, 10, 16], (b) an example occurring in plasma physics [17], and (c) an example involving a study of Josephson junction diodes [18]. Examples (a) and (b) deal with autonomous systems of three first order ordinary differential equations, while example (c) concerns a nonautonomous second order ordinary differential equation. Thus all three cases may be viewed as describing flows in a three-dimensional phase space. For examples (b) and (c), the existence of crises has not been explicitly demonstrated, but we view the evidence as strongly suggestive of crises. For the more well-studied Lorenz attractor, however, the work of Kaplan and Yorke [6] directly implies the existence of a crisis. In all three cases a common pattern occurs. Namely, in some range of a parameter p, say $p_a > p > p_b$, a nonchaotic attractor exists; while in some other range $p_2 > p > p_1$ a chaotic attractor exists. Furthermore for all three cases, $p_a > p_1$. The situation is as illustrated schematically in fig. 27, which shows the time average of some dependent variable of the system versus p (cf. fig. 4 of ref. 17 and fig. 1 of Ref. 18. For $p_a > p > p_1$ both attractors occur simultaneously with different basins of attraction. In the

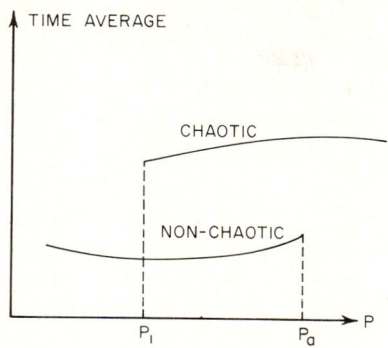

Fig. 27. Schematic illustration of the time average of a dependent variable versus a parameter p.

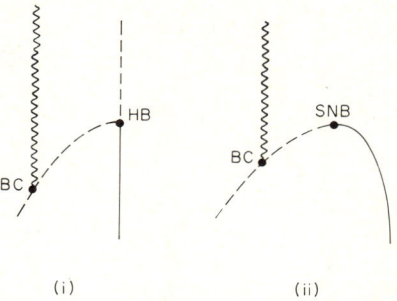

Fig. 28. (i) Crisis diagram for the Lorenz attractor case and (ii) for cases (b) and (c) of the text. BC, HB, and SNB denote boundary crisis, inverted Hopf bifurcation and saddle node bifurcation of periodic orbits.

case of the Lorenz attractor the non-chaotic attractor is a fixed point, while for cases (b) and (c) the non-chaotic attractors are limit cycles. As p is increased past p_a the non-chaotic attractors are destroyed by bifurcations. For the Lorenz case the bifurcation is of the inverse Hopf type. Namely, for $p < p_a$ an unstable limit cycle exists, which, as p approaches p_a, collapses to the stable fixed point, converting it to an unstable fixed point for $p > p_a$. For cases (b) and (c), we beliefe that the bifurcation is of the saddle-node type. That is, for $p < p_a$ an unstable (saddle) limit cycle again exists; as p approaches p_a the attracting stable limit cycle (node) and the unstable limit cycle coalesce and annihilate each other. Thus in all cases an unstable periodic orbit exists in the range $p_a > p > p_1$. The stable manifold of this unstable periodic orbit forms the boundary separating the basins of attraction of the chaotic and non-chaotic attractors. As p decreases toward p_1, the chaotic attractor moves closer to the unstable orbit on the boundary, until, at $p = p_1$, a crisis occurs. If one examined the surface of section, the situation would be similar to that illustrated in fig. 12 for the boundary crises for the Hénon map. Fig. 28 gives a crisis diagram summarizing the sequence of events for the three cases (p increases vertically) [19]. Furthermore, Yorke and Yorke [10] have numerically reduced the Lorenz system of equations to a one-dimensional map and have used this to examine the statistical characteristics of the resulting chaotic transients and their dependence on a system parameter. Their work also confirms the existence of an exponential distribution of lifetimes for these transients.

5. Fractal torus crises

Here we wish to consider a particular type of crisis which occurs only in invertible maps whose dimension is at least three. Our discussion will be brief, with further results and details to be given in a future publication [20]. To begin, we consider a bifurcation of a three-dimensional map in which two unstable fixed points, R and A, are created as a parameter δ decreases through one (cf. fig. 29). The fixed point R has one stable direction (s in fig. 29) and two unstable directions (u and ξ in fig. 29), while A has two stable directions (s and ξ) and one unstable direction (u). When R and A are close together (δ slightly less than one), we represent the map locally in the region near R and A as $u_{n+1} = \lambda_u u_n$, $s_{n+1} = \lambda_s s_n$, $\xi_{n+1} = \xi_n^2 + \delta/2$, where $\lambda_u > 1 > \lambda_s > 0$. The two fixed points of this map corresponding to R and A for $1 > \delta > 0$ are $u = s = 0$ and $\xi = 1 \pm (1 - \delta)^{1/2}$. As δ increases toward one, R and A move toward each other and coalesce at $\delta = 1$. For $\delta > 1$, the two fixed points no longer exist.

The local picture just given is elementary. However, the global consequences of this bifurcation

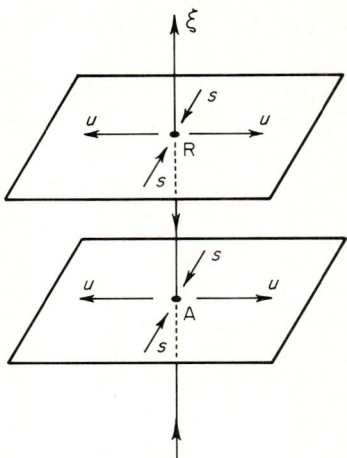

Fig. 29. Local illustration of an unstable pair of fixed points.

are not so clear. In particular, we suggest the possibility that the closure of the unstable manifold emanating from A forms a chaotic attractor. Similarly, since time reversal interchanges the roles of R and A, we also consider the possibility that the closure of the stable manifold emanating from R is a chaotic repellor (i.e. it is a chaotic attractor on time reversal). Furthermore, it seems reasonable from fig. 29 that this repellor might form the boundary of the basin of attraction for the strange attractor. If the set of hypotheses just given do in fact occur, and the point A is in the strange attractor, then when A coalesces with R we expect that the strange attractor will be destroyed. Since R is on the boundary of the basin of attraction, the destruction of the attractor is due to a boundary crisis.

To demonstrate that this picture can in fact occur, it suffices to give an example. The example which we choose is a modification of a map previously studied in ref. 21,

$$x_{n+1} = (x_n + y_n) \mod 1, \quad (6a)$$
$$y_{n+1} = (x_n + 2y_n) \mod 1, \quad (6b)$$
$$z_{n+1} = \lambda z_n + z_n^2 + \epsilon \cos(2\pi x_n), \quad (6c)$$

where we assume $\epsilon, \lambda > 0$. Below, we shall demonstrate numerically that this example does, in fact, yield an attractor with the required behavior. Even so, the example appears to be a very special one; e.g., the x and y equations do not depend on z and are area preserving (in fact, the x–y component of the map is the well-known "cat map" of Arnol'd). Thus, even if this map demonstrates the phenomena, one might still question whether these phenomena are to be expected in typical dynamical systems. To answer this question we have numerically studied the stability of our qualitative results under perturbations. That is, we have added to the x, y and z equations perturbing functions of x, y and z which are of a fairly arbitrary nature. We find that the qualitative results appear to survive these perturbations (cf. ref. 20), and thus we believe that our picture should be added to the list of events possible in typical dynamical systems.

For the map (6) we may think of x and y as angle coordinates (due to the mod 1) and z as a radial coordinate. Thus we are essentially dealing with a toroidal coordinate system. We note that the system (6) has a pair of fixed points at $x = y = 0$, $z = z_{R,A} = \frac{1}{2}\{(1-\lambda) \pm [(1-\lambda)^2 - 4\epsilon]^{1/2}\}$, corresponding to R and A of fig. 29. Since the term $\cos(2\pi x)$ in (6b) is maximum at the fixed point (i.e., at $x = 0$), we expect that the attractor will assume its maximum z-value at this fixed point. Similarly, the repellor will assume its minimum z-value at this fixed point. Thus, if the two touch, they will first do so at $x = y = 0$, and the attractor will cease to exist after this happens. From the expression for $z_{R,A}$ we see that the strange attractor is expected to exist for $\lambda \leq \lambda_c \equiv 1 - 2\epsilon^{1/2}$ with a crisis occurring at $\lambda = \lambda_c$. Fig. 30 shows a $y = 0$ cross-section of the attractor (inner curve) and the encircling repellor for eqs. (6) obtained by numerical iteration with $\epsilon = 0.04$ and $\lambda = 0.58 < \lambda_c = 0.60$. For the attractor, the cross-section is obtained by numerically iterating eqs. (6), and, after transients have died away, plotting points whose y-values lie between $y = \pm 10^{-3}$. The repellor curve is obtained similarly by iteration of the inverse map (to do this one must choose the plus sign when inverting the quadratic z equation). The z and x values in this

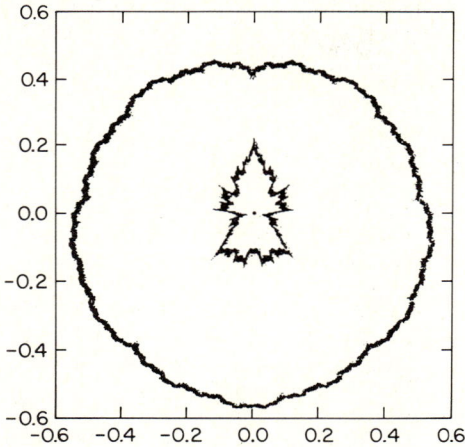

Fig. 30. $y = 0$ cross-section of the attractor and its basin boundary for eq. (6) with $\lambda = 0.58$ and $\epsilon = 0.04$.

figure have been plotted in polar coordinates to emphasize that what is shown represents cross-sections of toroidal surfaces. The origin of the polar coordinate system is signified by a dot on the figure; $2\pi x$ represents the polar angle from the vertical; and $z + 0.1$ is the radial distance from the origin. The cross-section of both the attractor and the repellor appear to be fractal curves which are continuous but nowhere differentiable. Orbits initialized at radial positions falling within the region bounded by the repellor are observed to be attracted to the attractor, while those initialized outside this region are observed to approach $z = \infty$ as

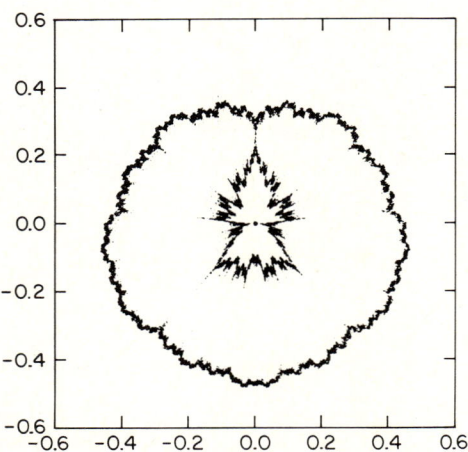

Fig. 31. Same as in fig. 30 but $\lambda = 0.68$.

$n \to \infty$. Thus the repellor surface is the boundary of the basin of attraction for the attractor. As λ is raised the z-coordinates at $x = y = 0$ of the repellor and attractor appear to move towards each other. After the crisis occurs, the attractor and repellor remnants are still visible on our plots because the chaotic transients can be quite long. Fig. 31 shows such a plot for $\epsilon = 0.04$ and $\lambda = 0.68 > \lambda_c$. Note the apparent joining of the attractor and repellor at $x = 0$.

6. Conclusions

In this paper we have illustrated crisis phenomena for chaotic attractors in a one-dimensional noninvertible map, a two-dimensional invertible map, three-dimensional flows, and three-dimensional maps. Our results show that crises are prevalent in many circumstances and systems and serve to illustrate their accompanying characteristic statistical behavior (e.g., chaotic transients). Based on these results, we are lead to propose the following two conjectures:

1) Almost all sudden changes in the size of chaotic attractors are due to crises.
2) Almost all sudden creations and destructions of chaotic attractors with their basins are due to crises.

For the purposes of these conjectures, and depending on the results of future research, it may be that the definition of a crisis (section 1) should be widened to include collisions of chaotic attractors with either unstable fixed points, unstable periodic orbits or unstable *multiply* periodic orbits (i.e. tori).

Note that conjecture 2 is not violated, for example, by the existence of a tangent bifurcation transition from a chaotic to a non-chaotic orbit (see, for example, fig. 2), because the attractor basin is neigher created nor destroyed in such a transition. Thus a single identifiable attractor remains throughout the transition but changes its character (subduction).

Also note in the statements of these conjectures the phrase "almost all". By this we mean to indicate that there may be special examples of dynamical systems which violate our conjectures, but which we believe are atypical in the sense that arbitrarily small perturbations of these systems restore the validity of our conjectures.

In conclusion, we wish to emphasize that the study of sudden changes in chaotic dynamics is just at its beginning, and that we believe that many new and intriguing phenomena await discovery. This will be particularly so as phenomena in higher dimensions begin to be explored (cf., for example, our section 5).

Acknowledgements

This work was supported by the Office of Basic Energy Sciences of the U.S. Department of Energy under contract No. DE-AS05-82ER-12026. One of us (J.A.Y.) was also partly supported by AFOSR, Air Force Systems Command under Grant No. AF0FR-81-0217A.

References

[1] J.E. Marsden and M. McCracken, The Hopf Bifurcation and Its Applications (Springer, Berlin, 1976). D. Ruelle and F. Takens, Comm. Math. Phys. 20 (1971) 167.
[2] M.J. Feigenbaum, J. Stat. Phys. 19 (1978) 25. P. Collet and J.-P. Eckmann, Iterated Maps on the Interval as Dynamical Systems (Birkhauser, Boston, 1980). R.M. May, Nature 261 (1976) 459. E. Ott, Rev. Mod. Phys. 53 (1981) 655. J.-P. Eckmann [7].
[3] C. Grebogi, E. Ott and J.A. Yorke, Phys. Rev. Lett. 48 (1982) 1507.
[4] S.-J. Chang and J. Wright, Phys. Rev. A 23 (1981) 1419.
[5] C. Simo', J. Stat. Phys. 21 (1979) 465.
[6] J.L. Kaplan and J.A. Yorke, Comm. Math. Phys. 67 (1979) 93.
[7] J.-P. Eckmann, Rev. Mod. Phys. 53 (1981) 643.
[8] G. Pianigiani and J.A. Yorke, Trans. Amer. Math. Soc. 252 (1979) 351. A. Lasota and J.A. Yorke, Rend. Sem. Univ. Padova 64 (1981) 141.
[9] G. Pianigiani, J. Math. Anal. Appl., in press.
[10] J.A. Yorke and E.D. Yorke, J. Stat. Phys. 21 (1979) 263.
[11] J. Guckenheimer, Comm. Math. Phys. 70 (1979) 133. I. Gumowski and C. Mira, Recurrences and Discrete Dynamic Systems (Springer-Verlag, New York, 1980).
[12] Y. Pomeau and P. Manneville, Comm. Math. Phys. 74 (1980) 189.
[13] S. Newhouse, Publ. Math. IHES 50 (1980) 101.
[14] S. Smale, Bull. Am. Math. Soc. 73 (1967) 747. Y. Treve, Topics in Nonlinear Dynamics (Am. Inst. Physics, New York, 1978) p. 147.
[15] J. Testa, J. Perez and C. Jeffries, Phys. Rev. Lett. 48 (1982) 714; C. Jeffries and J. Perez, Phys. Rev. A27 (1983) 601.
[16] E.N. Lorenz, J. Atmos. Sci. 20 (1963) 130.
[17] D.A. Russell and E. Ott, Phys. Fluids 24 (1981) 1976.
[18] B.A. Huberman and J.P. Crutchfield, Phys. Rev. Lett. 43 (1979) 1743.
[19] See also the paper of C. Tresser, P. Coullet and A. Arneodo [J. Physique-Lett. 41 (1980) L-243] for a discussion of discontinuous transitions to chaos involving hysteresis as in figs. 27 and 28.
[20] C. Grebogi, E. Ott and J.A. Yorke, to be published.
[21] J.L. Kaplan, J. Mallet-Paret and J.A. Yorke, to be published.

SYMBOLIC DYNAMICS OF NOISY CHAOS

J.P. CRUTCHFIELD and N.H. PACKARD
Physics Board of Studies, University of California, Santa Cruz, California, USA

One model of randomness observed in physical systems is that low-dimensional deterministic chaotic attractors underly the observations. A phenomenological theory of chaotic dynamics requires an accounting of the information flow from the observed system to the observer, the amount of information available in observations, and just how this information affects predictions of the system's future behavior. In an effort to develop such a description, we discuss the information theory of highly discretized observations of random behavior. Metric entropy and topological entropy are well-defined invariant measures of such an attractor's "level of chaos", and are computable using symbolic dynamics. Real physical systems that display low dimensional dynamics are, however, inevitably coupled to high-dimensional randomness, e.g. thermal noise. We investigate the effects of such fluctuations coupled to deterministic chaotic systems, in particular, the metric entropy's response to the fluctuations. We find that the entropy increases with a power law in the noise level, and that the convergence of the entropy and the effect of fluctuations can be cast as a scaling theory. We also argue that in addition to the metric entropy, there is a second scaling invariant quantity that characterizes a deterministic system with added fluctuations: I_0, the maximum average information obtainable about the initial condition that produces a particular sequence of measurements (or symbols).

1. The role of fluctuations in dynamical systems modeling

The work of Lorenz [1] and Ruelle and Takens [2] has led to the idea that randomness observed in physical systems may in some cases be modeled by low-dimensional chaotic attractors. A growing body of experimental evidence now supports this view [3]. This data also demonstrates that any purely deterministic model is incomplete, since the dynamics of physical systems is inevitably coupled to some source of fluctuations. We shall refer to these fluctuations as *external fluctuations**. An-

other attribute that must be incorporated into an accurate model for observed randomness is fluctuations of the measuring instrument, these we will call *observational noise*. Observational noise differs markedly from heat bath fluctuations in that it does not affect the temporal evolution of the system being observed (assuming a classical measurement process); rather, it directly limits what may be inferred about the system under study. We will be concerned only with the effects of external fluctuations here; for further discussion of this classification of noise types see ref. 4.

Incorporation of any kind of fluctuation into a dynamical description implies that observables become average quantities, the average being taken over all possible fluctuations†. For the case of a chaotic deterministic dynamical system, we are led to the idea that observables are average quantities, where the average is taken with respect to the asymptotic probability distribution. When fluctuations are added to such a system, they produce a new asymptotic probability distribution. A formal expression of how this distribution arises will be presented below. In referring to a probability distribution $P(x)$ we will find it convenient to

---
* We can give an unambiguous definition of this in terms of the ideas presented in this paper: External fluctuations may be regarded as a second dynamical system (coupled to the system of interest) with sufficiently high entropy h_μ so that all the information I from a measurement is lost after the typical time τ used for sampling the first system. In other words, $I/\tau \ll h_\mu$, where I/τ is the information acquisition rate. This allows for an operational definition of a non-deterministic source of random behavior as a deterministic system whose entropy is sufficiently large to preclude an observer's geometric reconstruction of the source's dynamics. All of our information quantities will be measured in bits and so, in particular, all logarithms will be taken to the base 2.
† We will assume fluctuations to be drawn from a stationary ensemble at each time.

also speak of the associated *measure* μ defined by

$$\mu(A) = \int_A P(x)\,\mathrm{d}x,$$

where A is some set.

We will be concerned with the effect of fluctuations on measurements of randomness, in particular, their effect on the metric entropy. As we shall see from numerical computations, the metric entropy is relatively insensitive to observational noise, but is strongly dependent on external fluctuations coupled to the dynamics; and so we will concentrate mostly on this latter case. The prototypical chaotic systems we shall use are iterated maps of the unit interval I onto itself: $x_{n+1} = f(x_n)$, where f is some nonlinear function. These will also be referred to as one-dimensional maps. We will model the effects of external fluctuations with a stochastic difference equation of the form

$$x_{n+1} = f(x_n) + \xi_n, \tag{1}$$

where ξ_n is a delta-correlated random variable. Numerical experiments [5] indicate that the response of the metric entropy to the added fluctuations ξ is insensitive to the details of their probability distribution. We will assume ξ to have zero mean, and to be evenly distributed over some finite interval, with a standard deviation, or *noise level*, σ.

For dynamical systems with added fluctuations there are not many rigorous results. Kifer [6] has proven that for hyperbolic attractors* the invariant measure converges weakly to the correct zero noise limit. Boyarsky [7] proved that for one-dimensional maps that have slope everywhere greater than one, there exists some noise level for which the invariant measure of the system with fluctuations approximates the zero noise invariant measure with arbitrary accuracy (strong convergence), and that *all* initial conditions have time averages that correspond to averages with respect to the invariant measure.

We begin by reviewing entropy measurement techniques for deterministic systems. We will then investigate the effects of external noise on the symbolic dynamics, and discover that the amount of information $I(n)$ about the initial condition that produces a symbol sequence of length n reaches a limit I_0 at some particular length n_c that is dependent on and scales with the noise level. Furthermore, the added noise produces entropy convergence features that also obey scaling laws. After describing the scaling features of the entropy, we will discuss numerical experiments in which we compute the scaling exponents for many different systems. We then describe an alternate entropy-like quantity similar in spirit to the Lyapunov characteristic exponent, conjecturing equality with the symbolic dynamics metric entropy. We conclude with an overview and a brief discussion of some experimental applications.

2. Symbolic dynamics and entropy for deterministic systems

We must first review the case of observing a deterministic dynamical system. We will consider time to be discrete, and the dynamical system to be a map f from a space of states M into itself, $f: M \to M$. We assume f has some ergodic invariant measure $\bar{\mu}$. If f has an attractor, we will restrict our attention to the attractor, and assume that almost all (with respect to Lebesque measure m) initial conditions approach the attractor and have trajectories that are asymptotically described by the measure $\bar{\mu}$ on the attractor; i.e. that for almost all points the measure

$$\mu_N(x) = \frac{1}{N}\sum_{n=1}^{N}\delta_{f^n x}$$

---

* In the context of one-dimensional maps, this means that the absolute value of the slope of the map must be greater than one everywhere on the attractor.

converges weakly to $\bar{\mu}$. Oono and Osikawa [8] refer to this assumption as the "condition for observable chaos".

This may seem like an amazing assumption from a mathematical viewpoint, but it is proven rigorously for axiom-A systems, where $\bar{\mu}$ is the Bowen–Ruelle measure. For maps of the unit interval this assumption will hold for all maps that have an invariant measure that is absolutely continuous with respect to Lebesque measure. Consideration of noise added to the dynamics also makes this assumption plausible for most physical contexts, as will be discussed in following section.

The behavior of a dynamical system $f: M \to M$ can have many symbolic representations, each obtained by using a *measurement partition*, $P = \{P_1, \ldots, P_q\}$, to divide the state space M into a finite number of sets each of which is labeled with a symbol $s_i \in \{1, \ldots, q\} \equiv S$. The time evolution $(x_0, x_1, x_2, \ldots)$ of the dynamical system $f: M \to M$ is then translated into a sequence of symbols labeling the partition elements visited by an orbit

$$s = \{s_0, s_1, s_2, \ldots\}$$

and f itself is replaced by a *shift* operator σ which re-indexes a symbol sequence; that is,

$$\sigma(s) = s',$$

where for each symbol in the sequence s',

$$s'_i = (\sigma(s))_i = s_{i+1}.$$

Thus the shift operator σ merely moves the time origin of a symbol sequence one place to the right.

In the space of all possible symbol sequences

$$\Sigma \equiv \{s = (s_0, s_1, \ldots)\},$$

the *observed* or *admissable sequences* are those which satisfy

$$x_i = f^i(x_0) \in P_{s_i}.$$

The set of admissable sequences Σ_f along with the shift σ is called a *subshift*. (Σ_f, σ) is the symbolic dynamical system induced by f using the measurement partition P.

The symbol sequences of Σ_f are a coding for the orbits of $f: M \to M$. A finite sequence of symbols $(s_0^n, \ldots, s_{n-1}^n)$ defines an *n-cylinder* $s^n \equiv \{s : s_i = s_i^n, i = 0, \ldots, n-1\}$ which is a subset of Σ_f consisting of all sequences whose first n elements match with those of s_i^n. An n-cylinder s^n corresponds to a set of orbits that are "close" to one another in that their initial conditions and first $n - 1$ iterates fall in the same respective partition elements. Since these orbits must follow each other for at least $n - 1$ iterations, they must all have initial conditions that are close, belonging to some set $U \subset M$. We thus have a map Δ from n-cylinders to subsets of M:

$$\Delta(s^n) = \{x \mid f^i(x) \in P_{s_i}, \text{ for } i = 0, \ldots, n - 1\}.$$

To a different n-cylinder will correspond a different set of orbits whose initial conditions are contained in some other set $U' \subset M$. M will become partitioned into as many subsets as there are n-cylinders. As n is increased, this *n-cylinder partition* will become increasingly refined. The refinement caused by taking an increasing number of symbols is illustrated in fig. 1, where M is the unit interval

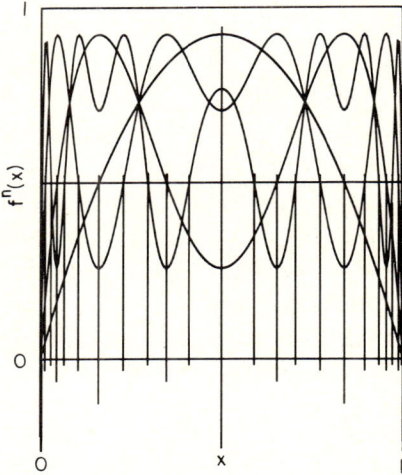

Fig. 1. Construction of the partition induced by taking n symbols (i.e. specifying an n-cylinder) with the measurement partition $\{[0, 0.5], (0.5, 0]\}$. The 1-cylinder, 2-cylinder, 3-cylinder, and 4-cylinder partitions are shown with successively shorter tic marks below the x-axis.

[0, 1], and f is the quadratic logistic equation, $f(x) = rx(1-x)$, with $r = 3.7$. We have used the measurement partition formed by cutting the interval in half at $d = 0.5$, the *critical point* of f where the slope vanishes. We will label the left subinterval "0" and the right "1". We see from the figure that the dividing points for the n-cylinder induced partition are simply the collection

$$\{d, f^{-1}(d), f^{-2}(d), \ldots, f^{-(n-1)}(d) \ldots\}$$

whenever the specified inverse images exist. If the map is not everywhere two onto one (i.e. $r < 4$), some of the inverse images will not exist, corresponding to the fact that some n-cylinders are non-admissable. Changing the measurement partition clearly generates a different set of admissable sequences, just as it generates different n-cylinder partitions.

The usefulness of symbolic dynamics as a representation for the orbits of f can be captured in the following commutative diagram:

$$\begin{array}{ccc} \Sigma_f & \xrightarrow{\sigma} & \Sigma_f \\ \pi \downarrow & & \downarrow \pi \\ M & \xrightarrow{f} & M \end{array}$$

with the projection operator

$$\pi(s_0, s_1, \ldots) = \bigcap_{i=0}^{\infty} f^{-i}(P_{s_i}).$$

One can then study the simpler, albeit abstract, symbolic dynamical system in order to answer various questions about the original dynamical system. Within this construction, every point on the attractor will have at least one symbol sequence representation. There are a few ambiguities in the labeling of orbits by symbol sequences that prevent π from being invertible, but our discussion of the entropy will prove to be insensitive to the ambiguities*.

The space of one-sided symbol sequences can easily be metrized by mapping each symbol sequence to a power series

$$\Phi(x) = \sum_{i=1}^{\infty} \frac{S(f^i x)}{q^i},$$

where $s(x)$ is the symbol labeling the measurement partition element containing x (the denominator is 2^i only if the partition has two elements). For the case of a binary partition, this map identifies every sequence with a binary fraction whose value lies in [0, 1]. We will conveniently confuse s^n with its binary fraction representation unless the distinction is necessary†.

A Cantor set structure in the symbol sequences of the chaotic logistic equation is revealed in fig. 2 by a sequence of probability distributions for n-cylinder binary fractions: with the increase in length of the n-cylinder the distributions show successively more, although narrower, peaks. Another demonstration of the Cantor set structure of

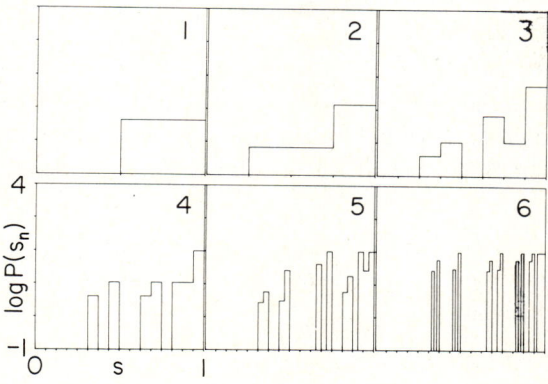

Fig. 2. The Cantor set structure of the subshift (Σ_f, σ) is shown in this sequence of probability distributions for n-cylinders: $n = 1, 2, 3, 4, 5,$ and 6. Each n-cylinder has been mapped onto the unit interval by using its binary fraction. In this example f is the quadratic logistic equation with $r = 3.7$.

---

* An example of one such ambituity is that there can be two symbol sequences that are nowhere the same, but label the same point on the interval: e.g. 100000... and 011111... both label the same point $x = 0.5$ in the limit of infinite length.

† Milnor and Thurston [9] show how to form a slightly more sophisticated "invariant coordinate" which is monotonic. Our entropy calculations do not require this feature, so we use the computationally simpler binary fraction.

Σ_f is the graph of the distribution of symbols s (truncated to a finite n-cylinder with $n = 12$ and mapped onto the unit interval using its binary fraction) versus position x, illustrated in fig. 3.

We will now embark on the task of characterizing the chaotic behavior in a dynamical system using topological and metric entropies, in that order. After giving their definitions, we will show how these quantities may be computed numerically using the symbol sequence representation of orbits. Our analysis follows Shannon [10].

Heuristically, the topological entropy of a dynamical system measures the asymptotic growth rate of the number of resolvable orbits (using a given measurement partition) whose initial conditions are all close. Equivalently, the topological entropy quantifies the average time-rate h of spreading a subset over nearby subsets. This process is most easily illustrated by considering a collection of subsets which form a "cover" of the state space M. The dynamic f spreads a single cover element over other elements after some time t. The number of new cover elements $N(t)$ visited by points in the original cover element can be written,

$$N(t) \sim e^{ht},$$

where $h > 0$ for chaotic dynamical systems. With this geometric motivation, we will now consider a more formal definition of the topological entropy h [11].

For a compact topological space M, with an open cover U, let $N(U)$ be the number of sets in a subcover of minimal cardinality. Two covers U and V may be "combined" to form a refinement W by

$$W = U \vee V$$
$$= \{A \cap B \mid A \in U \text{ and } B \in V\}$$

Now if $f: M \to M$ is a continuous map, the *topological entropy* of f with respect to the cover U is defined as

$$h(f, U) = \lim_{n \to \infty} \log \frac{N(U^n)}{n},$$

where

$$U^n = U \vee f^{-1} U \vee \ldots \vee f^{1-n} U.$$

The topological entropy $h(f)$ of the map itself is then the supremum of $h(f, U)$ over all open covers U.

The supremum is obtained only if the measurement partition is "good" in that there is an unambiguous correspondence between orbits of f and symbol sequences. Only with such a good partition is the topological entropy of Σ_f obtained using partition P exactly $h(f)$, the topological entropy of f. There is no general procedure for finding such a good partition, but we will give numerical evidence that such partitions are easily found for simple piecewise monotone maps of the unit interval. Given such a partition, however, we have a readily computable algorithm for $h(f)$:

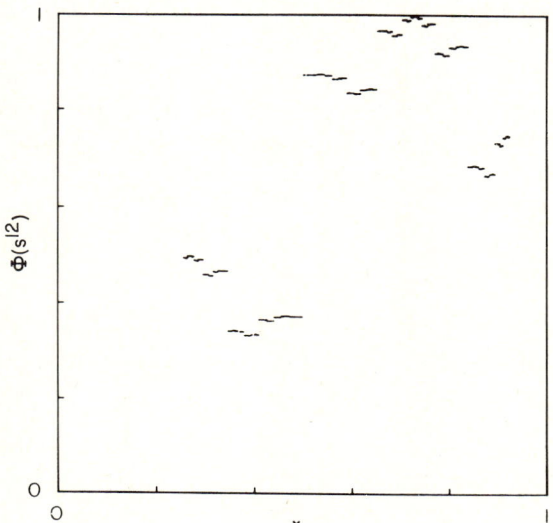

Fig. 3. 2000 iterations of the 'logistic equation' with $r = 3.7$, showing the Cantor set structure of the distribution of sequences in Σ_f. Graphed is $\Phi(s^{12})$ against position x, where s^{12} is the sequence obtained from the initial condition x. The density of points on the x-axis is the asymptotic distribution of f on the unit interval; the density of points on the y-axis is the Cantor distribution illustrated in the previous figure.

simply counting the number of n-cylinders. Note that in the space of symbol sequences Σ_f, each n-cylinder s^n is an open set, and the class of all n-cylinders is an open cover. Thus the topological entropy of the system (Σ_f, σ_f) is given by

$$\lim_{n\to\infty} \frac{\log N(n)}{n} \to h(\sigma_f),$$

where $N(n)$ is the number of admissable n-cylinders*. $N(n)$ is readily obtainable numerically, so this formula presents us with a computable algorithm for the topological entropy†.

In presenting the topological entropy before the metric entropy we have purposely reversed their historical order because there is a sense in which the metric entropy is a generalization of the topological entropy: the metric entropy also measures the asymptotic growth rate of the number of resolvable orbits (using a given measurement partition) having close initial conditions, but weights each orbit with its probability of occurrence.

The definition of metric entropy for the dynamical system (M, f) requires an invariant measure $\bar{\mu}$ and a sigma-algebra of measurable subsets of M: more structure than needed for the definition of topological entropy.

If $P = \{P_i\}$ is a finite measurable partition of M with p elements, we define the entropy of P as

$$H_{\bar{\mu}}(P) = -\sum_{i=1}^{p} \bar{\mu}(P_i)\log(\bar{\mu}(P_i)).$$

---

* For the case of symbolic dynamics, this formula for the topological entropy was first introduced by Parry [12], but is essentially the same as the "channel capacity" introduced by Shannon [10].

† Crutchfield and Shaw [13] have developed other algorithms to compute the topological entropy of a map f based on representing the dynamics as a branching process with a deterministic transition matrix. For certain cases, these techniques allow one to analytically calculate the topological entropy and so to study, for example, the convergence of the topological entropy directly (c.f. ref. 14). These techniques are related to the kneading calculus of Milnor and Thurston [9].

‡ This theorem as well as the original definition of metric entropy are presented in Kolmogorov [15].

Given two partitions P and Q, their refinement is

$$P \vee Q = \{P_i \cap Q_j | P_i \in P \quad \text{and} \quad Q_j \in Q\}.$$

The metric entropy of f with respect to the partition P is defined by

$$h_{\bar{\mu}}(f, P) = \lim_{n\to\infty} \frac{1}{n} H_{\bar{\mu}}(P^n),$$

where

$$P^n = P \vee f^{-1}P \vee \ldots \vee f^{1-n}P.$$

Finally, the metric entropy of f itself is

$$h_{\bar{\mu}} = \sup_{P} h(f, P),$$

where the supremum is taken over all partitions P.

As for the topological entropy, the supremum is obtained only for special partitions; Kolmogorov‡ proved that the desired requirement is that the partition be *generating*. This is the case if the smallest sigma-algebra containing $\Delta(s^n)$ for all $n > 0$ coincides with the sigma-algebra of measurable subsets in M. In simpler terms, a partition is generating if, as the length of all sequences becomes large, the sequences label individual points. Thus, only if P is a generating partition we have

$$h_{\bar{\mu}}(f) = h_{\bar{\mu}}(f, P).$$

Again, if we label the elements of the partition P with symbols, the entropy of $h_\mu(\sigma_f)$ is exactly $h_{\bar{\mu}}(f, P)$, with

$$\mu(s^n) = \int_{\Delta(s^n)} d\bar{\mu} = \bar{\mu}(\Delta(s^n)).$$

Note that the entropy h_μ of (Σ_f, σ) is equal to $h_{\bar{\mu}}(f)$ *only* if the measurement partition is generating. For arbitrary measurement partitions,

$$h_\mu(\sigma_f) \leq h_{\bar{\mu}}(f).$$

Assuming a generating measurement partition, the identification between n-cylinders and elements of the refinement P^n allows us to estimate the measure of each element of P^n by accumulating a frequency histogram for the observed n-cylinders. (Note that P^n is exactly the n-cylinder partition illustrated in fig. 1 for $n = 4$.) We may then obtain an n-symbol estimate for the topological entropy from either

$$h(n) = \frac{\log N(n)}{n}$$

or

$$h(n) = \log N(n) - \log N(n-1),$$

and estimates for the metric entropy from

$$h_\mu = \frac{H_\mu(n)}{n}$$

or

$$h_\mu(n) = H_\mu(n) - H_\mu(n-1).$$

It is easily shown that the latter estimate for h_μ converges more quickly than the former [10], so all of our numerical computations of $h_\mu(n)$ will use this expression. Fig. 4 illustrates an example computation of $h(n)$ and $h_\mu(n)$ for the logistic equation, $f(x) = rx(1-x)$ at a typical parameter value, $r = 3.7$.

In order to illustrate the dependence of the entropy on the measurement partition used, we have computed $h(13)$ and $h_\mu(13)$ for a range of binary (two-element) measurement partitions; the results are illustrated in fig. 5. We call the location $x = d$ at which we decide whether a point x on an orbit is either a "0" or a "1" the *decision point*. For two values of the decision point, $d = 0.5$ and $d = 0.839\ldots$ (an inverse image of the critical point), $h_\mu(13)$ is maximized, giving evidence that these values of d yield a generating partition. Note that $h_\mu(13)$ is greater than the Lyapunov character-

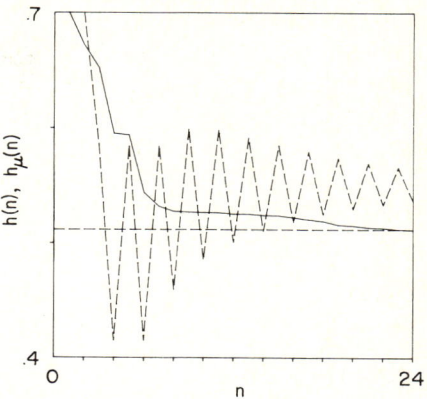

Fig. 4. Entropy convergence for the logistic equation $f(x) = rx(1-x)$, with $r = 3.7$; the solid line represents $h_\mu(n)$ and the oscillating dashed line represents $h(n)$. 2×10^8 iterations were used. The horizontal dashed line is the Lyapunov characteristic exponent.

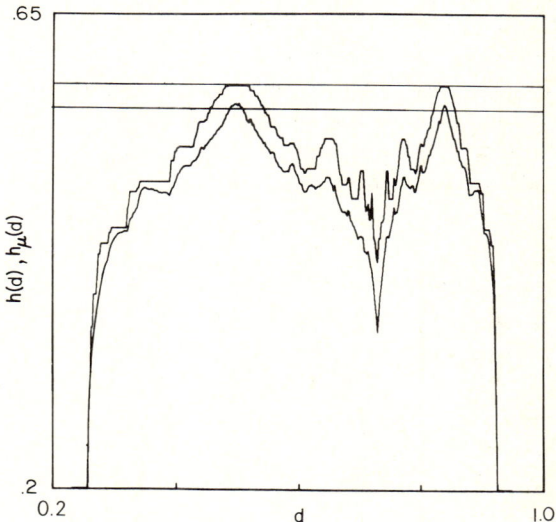

Fig. 5. $h(13)$ (upper curve) and $h_\mu(13)$ for the logistic equation with $r = 3.7$, using different measurement partitions obtained by varying the decision point d. $h(13)$ is actually an average of $h(6),\ldots, h(13)$ to eliminate the oscillitory effects. The upper horizontal line is the topological entropy calculated to one part in 10^6 with the kneading determinant [13, 14]. The lower horizontal line is the Lyapunov characteristic exponent calculated to within 0.1%.

istic exponent (to be discussed in more detail shortly) because the metric entropy has not converged by thirteen symbols (cf. fig. 4).

From the above definition of the metric entropy,

it is easy to see that $h \geq h_\mu$, since $h(f, P^n)$ is maximized when each element of P^n is equally probable (i.e. $\mu(P_i^n) = 1/N(n)$ for all i). In this case, the formula for the metric entropy reduces to that for the topological entropy. This is also evident from a theorem due to Goodwyn [16] and Dinaburg [17], which states that

$$h = \sup_\mu h_\mu,$$

where the supremum is taken over all invariant measures μ.

One of the primary roles of entropy in dynamical systems theory is that it is an invariant [15], which is to say that any two dynamical systems (M, f, μ) and (M', f', μ') have the same metric entropy if they are related by a isomorphism that preserves measure. We will not use this fact at all in our entropy calculations for deterministic systems, but when noise is added to the dynamics, we will address the question of how the invariance of the entropy is affected.

We now introduce Lyapunov characteristic exponents as another measure of chaos, and discuss their relationship to the entropies described above. The Lyapunov characteristic exponents measure the average asymptotic divergence rate of nearby trajectories in different directions of a system's state space [18, 19]. For our one dimensional examples, $f: I \to I$, there is only one characteristic exponent λ. It can be easily calculated since the divergence of nearby trajectories is simply proportional to the derivative of f [19]:

$$\lambda = \lim_{N \to \infty} \frac{1}{N} \sum_{n=1}^{N} \log|f'(x_n)|.$$

Or equivalently, if a continuous ergodic invariant measure $\bar\mu$ exists, then the characteristic exponent is given by

$$\lambda = \int_0^1 \log|f'(x)| \, d\bar\mu.$$

If M is an axiom-A attractor, there is a prescription for constructing a partition which is generating, and the equality of the metric entropy h_μ and the sum of the positive Lyapunov characteristic exponents can be proven [20]. In fact, whenever an absolutely continuous invariant measure exists, a theorem due to Piesin [21] shows that the metric entropy of a diffeomorphism is equal to the sum of the positive exponents*. Ruelle [22] proved that for any C^2 map that has an absolutely continuous invariant measure

$$h_\mu \leq \sum_i \lambda_i^+,$$

where the λ_i^+ are all the positive Lyapunov characteristic exponents, and he has conjectured that equality holds. For a wide class of maps of the unit interval, Ledrappier [23] has shown that an ergodic measure having positive metric entropy is absolutely continuous with respect to Lebesque measure if and only if the metric entropy is equal to the Lyapunov exponent. Shimada [24] obtained good numerical agreement between the characteristic exponent and the metric entropy for the Lorenz attractor and its induced symbolic dynamics using only 9 symbols, and Curry [25] has computed a metric entropy slightly lower than the positive characteristic exponent for a two-dimensional diffeomorphism (Hénon's map)†. Our numerical results for several maps of the unit interval (including the logistic equation) indicate that the metric entropy is indeed equal to the Lyapunov exponent, supporting Ruelle's conjecture and indicating the existence of an absolutely continuous invariant measure whose probability distribution is well approximated by a frequency histogram.

---

* In the general case, the exponents are a function of initial condition, so the sum must be integrated over the attractor, but we will consider only the case of an ergodic attractor where the exponents are constant almost everywhere with respect to the asymptotic invariant measure.

† Curry's underestimate of the entropy is probably due to the fact that the partition he chose was not generating.

3. Symbolic dynamics and entropy in the presence of noise

One of the reasons that there are so few results on the response of the metric entropy to added fluctuations is that there are problems with the definition of metric entropy (as well as its computation) in the presence of fluctuations. There are also problems with the definition and computation of Lyapunov characteristic exponents for systems with added fluctuations. Some of the problems associated with the metric entropy are:

(1) There is no clear definition of a generating partition for a deterministic system with added noise. Increasingly long sequences of measurements can no longer isolate the system into an arbitrarily fine partition element (where for fineness we mean to use Lebesque measure on the unit interval).

(2) A related problem is that the entropy with respect to a particular partition diverges as the partition is made increasingly fine [26], rendering problematic the definition of a "true" entropy that is independent of partition.

(3) Even using a coarse (e.g. binary) partition, a fixed point with added noise will have nonzero entropy if a partition divider is placed on the fixed point*. (This entropy would then give an estimate of the external noise in the system.)

(4) The effect of adding noise will depend on what coordinate system the noise is added to. One

---

* This example is due to Doyne Farmer.

† The fact that the observed asymptotic probability distributions will depend on the coordinate system used suggests that if one has some a priori reason for believing the noise to have a particular distribution (e.g. Gaussian), one should, in principle, be able to adjust the coordinate system used to observe the system until the noise displays the correct distribution. An experimentalist's model would thus include the specification of a *physically preferred coordinate system* in which the noise was added. Most systems may be too complicated to give any clue about the "correct" noise distribution, however. For example, in fluid systems with some underlying low-dimensional chaotic attractor, even if we assume that the fluid is being driven by thermal noise, it is not a priori clear what form will be taken by the noise terms added to the equations of motion on the attractor, since the thermal noise will undoubtedly be filtered by many dynamical effects.

might hope that the response of the metric entropy to noise should be independent in the limit of small noise, but this point is not yet clear from the theory†.

In spite of these problems, we may take a well-defined operational approach to the measurement of metric entropy in the presence of noise: the algorithm embodied in the definitions and estimates yields an unambiguous value of the metric entropy with respect to a particular measurement partition. Any sequence of measurements on a physical system will produce a string of observed symbols; our operational approach will give a measure of the predictability of this string. The measurement partition we will use will be of the same form as that used for the deterministic one-dimensional maps, namely a binary partition of the form $\{[0, d), [d, 1]\}$ where $0 < d < 1$. Given this kind of binary partition, one may again ask if there is a value of d that maximizes the entropy, and we find empirically in fig. 6 that $d = 0.5$ gives a maximum value just as it does for the deterministic case illustrated in fig. 5. This is partial

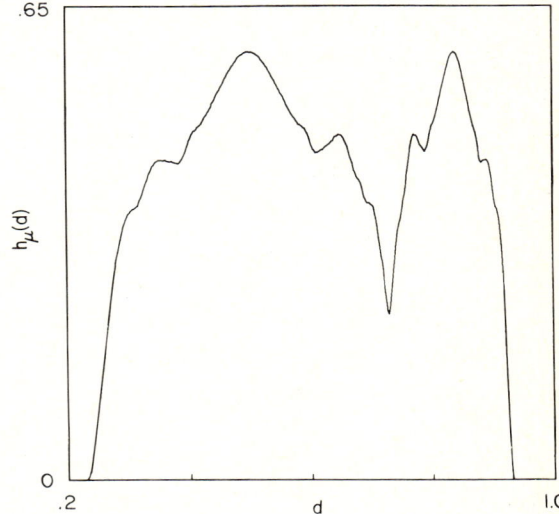

Fig. 6. Entropy h_μ with respect to a binary measurement partition $\{[0, d), [d, 1]\}$ as a function of the decision point d, for the logistic equation with $r = 3.7$ and added noise of width $\sigma = 2^{-7}$. Compare fig. 5.

justification for our use of this particular measurement partition, but because we are considering only binary partitions, we have not escaped points (1) and (2) above. We must again stress that the metric entropy of a noisy process like eq. (1) depends on the measurement partition used, but we will take the liberty of referring to the "metric entropy" of such a system as that computed using the measurement partition $\{[0, 0.5), [0.5, 1]\}$ unless otherwise noted. Though the entropy h_μ diverges as the measurement partition becomes fine [26], we may still conjecture that for measurement partitions with coarser resolution than the noise level*, our computations give an invariant, well-defined value for h_μ†. To begin the discussion of our numerical computations of the entropy in the presence of noise we will first examine a few properties of the asymptotic probability distributions both of the noisy map on the interval and of the shift on the space of observed sequences Σ_f.

Before considering entropy computation, we will first remark on a few features of the invariant measure, which will in turn have certain implications for entropy measurements. In the deterministic case $f: M \to M$, the asymptotic invariant distribution function $\bar{P}(x)$ is the fixed point of the Frobenius–Perron operator L_f given by

$$(L_f P)(x) = \sum_{y = f^{-1}(x)} \frac{P(y)}{|f'(y)|}.$$

This operator may be written as a Fredholm equation

$$(L_f P)(x) = \int \delta(f(y) - x) P(y) \, dy, \qquad (2)$$

where the equivalence is established by integrating

---

* We mean here that the size of the smallest partition element must be larger than the induced noise level. The induced noise level is obtained from the width of the distribution of the added noise by multiplying this width by the map's maximum slope.

† The well-defined value must still be obtained using a supremum over partitions of a given resolution similar to the supremum illustrated in fig. 6.

the right-hand side using a change of variables $y' = f(y)$.

If noise ξ (with a distribution $P_\sigma(\xi)$ having zero mean and width σ) is added to the deterministic map, forming the noisy map

$$x_{n+1} = f_\xi(x_n) = f(x_n) + \xi_n,$$

an additional average must take place with respect to the noise:

$$(L_{f_\xi} P)(x) = \int \delta(f(y) + \xi - x) P_\sigma(\xi) P(y) \, dy$$
$$= \int P_\sigma(f(y) - x) P(y) \, dy. \qquad (3)$$

Thus we see that the deterministic Frobenius–Perron operator is generalized to include the effects of fluctuations by simply replacing the delta function in eq. (2) by the noise distribution function. This formalism has been used by Schraiman, Wayne, and Martin [27], as well as Haken and Meyer-Kress [28], Takahashi [29], and Feigenbaum and Hasslacher [30]. The asymptotic probability distribution for the noisy map is in principle numerically computable using eq. (3). We have not used this expression to compute the distribution (our entropy computations are based on frequency histograms instead), but we may use eq. (3) to infer at least one qualitative property of the asymptotic distribution $\bar{P}(x)$ on the unit interval: Since the distribution must be invariant under the noisy Frobenius–Perron operator, which includes a convolution of the noise distribution, the asymptotic distribution $\bar{P}(x)$ will have no structure on length scales less than the noise level σ.

The primary difference between the symbolic dynamics of a purely deterministic system and that of a deterministic system with added noise is the nature of the identification between a particular symbol sequence and the set of initial conditions that might have produced that sequence. For the deterministic case there is a direct correspondence between symbol sequences and sub-intervals of the unit interval, with the sub-intervals becoming increasingly small as the symbol sequences get longer (cf. the construction of the n-cylinder partition illustrated in fig. 1). When noise is added, instead of there being a sub-interval, every point of which produces a particular sequence, there is a set of points which have some probability of producing a particular sequence. We will label the probability distribution of finding the sequence s^n for an initial condition x as $P_{s^n}(x)$.

For the deterministic case, we have

$$P_{s^n}(x) = X_{\Delta(s^n)},$$

where $X_{[a,b]}$ is the characteristic function over the interval $[a, b]$:

$$X_{\Delta(s^n)}(x) = \begin{cases} 1, & \text{for } x \in \Delta(s^n), \\ 0, & \text{for } x \notin \Delta(s^n), \end{cases}$$

and where $\Delta(s^n)$ is the set of initial conditions that can produce s^n for the deterministic case. We may then use the Frobenius–Perron equation to find $P_{s^{n+1}}(x)$ from $P_{s^n}(x)$. First, for the deterministic case, this gives

$$X_{\Delta(s^{n+1})}(x) = \int_{y \in \Delta(s^n)} \delta(f(y) - x)\bar{P}(y)\,dy.$$

When noise is added, $P_{s^n}(x) = X_{\Delta(s^n)}$ becomes smeared because we must use the noisy Frobenius–Perron operator, which includes a convolution of the noise distribution P_σ:

$$P_{s^{n+1}}(x) = \int P_\sigma(f(y) - x) P_{s^n}(y)\,dy. \qquad (4)$$

The smearing of the partition boundaries, or dividers, that takes place with each application of this operator decreases with successive applications*. The effective width of a partition element increases by $\sigma/|f'(y_i)|^i$, where the y_i are the appropriate inverse images of the deterministic divider. Another way of phrasing this observation is that averaging over fluctuations of width σ at each of n iterations is equivalent, for the purposes of constructing $P_{s^n}(x)$, to averaging over n sets of fluctuations of the initial condition each having a magnitude $\sigma \approx |f'(y_i)|^{-i}$. The convergence of the $P_{s^n}(x)$ to a distribution of a fixed width is illustrated for the logistic equation ($r = 3.7$) in fig. 7a. Note that $\log P_{s^n}(x)$ appears parabolic for large n in the semi-log plots of fig. 7, indicating that $P_{s^n}(x)$ is Gaussian, as might be expected from the repeated convolution of eq. (4).

We see, then, that the picture of bins (elements of an n-cylinder partition) being split into sub-bins (elements of an $(n + 1)$-cylinder partition) for the purely deterministic map (cf. fig. 1) is replaced by probability distributions splitting into daughter probability distributions for a deterministic map with added noise. Consider the situation when the width of the distribution $P_{s^n}(x)$ is large compared to the size of the deterministic bin (i.e. the length of $\Delta(s^n)$): Because $P_{s^n}(x)$ converges to a distribution of fixed width for large enough n, daughter distributions have nearly the same width (and in fact nearly coincide), as illustrated in fig. 8. Since the probability of s^n is given by

$$\mu(s^n) = \int P_{s^n}(x) \bar{P}(x)\,dx,$$

we see that for large enough n†

$$\mu(s^n 1) \approx \mu(s^n 0). \qquad (5)$$

---

* Actually, the width decreases only when the slope evaluated at the appropriate inverse image of the deterministic divider is greater than one. It appears, however (as illustrated in the following figure) that even when there are occasional contributions of slopes less than one, as for the logistic equation, the fact that the "average asymptotic slope" is greater than one causes the width of the distributions $P_{s^n}(x)$ to approach a limit. If the "average asymptotic slope" (this quantity is really well defined only for purely deterministic systems) is less than one (i.e. when the attractor is a periodic orbit) $P_{s^n}(x)$ diverges to cover the entire interval, since in this case all initial conditions end up giving the same periodic symbol sequence.

† We are also using the fact that $\bar{P}(x)$ does not change much over the width of $P_{s^n}(x)$.

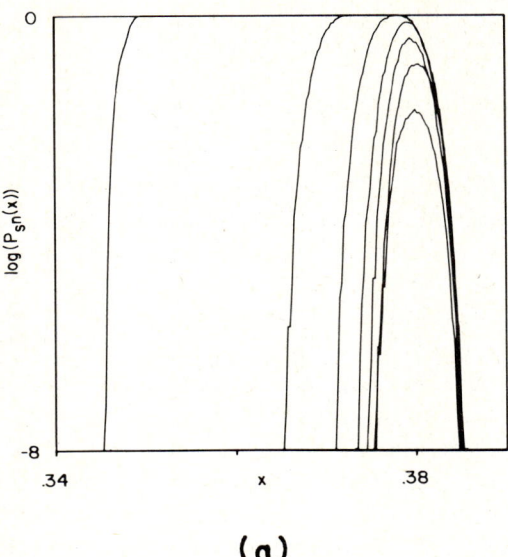
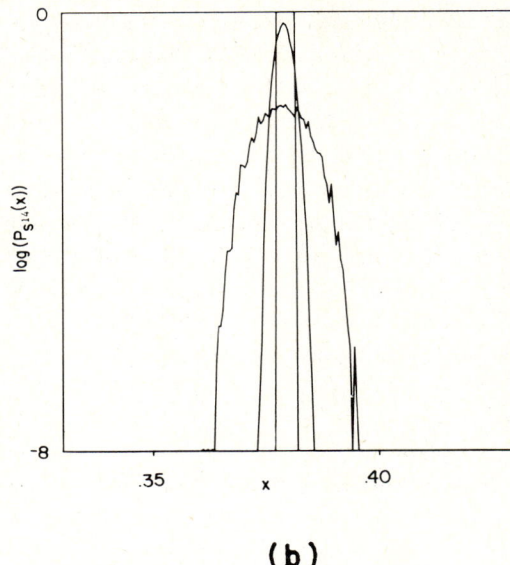

Fig. 7. (a) For a fixed noise level, shown is $P_{s^n}(x)$ for $n = 8, 9, 11, 13, 15, 16$, and 18 (the values of n corresponding to splitting of this particular series of bins in the deterministic case). (b) Fixing $n = 14$, shown is $P_{s^{14}}(x)$ for the deterministic case and for two noise levels $\sigma = 2^{-10}$ and $\sigma = 2^{-7}$. The sequence used was $s^{18} = (010101110111111010)$ (the shorter sequences are truncations: $s^8 = (01010111)$, etc.).

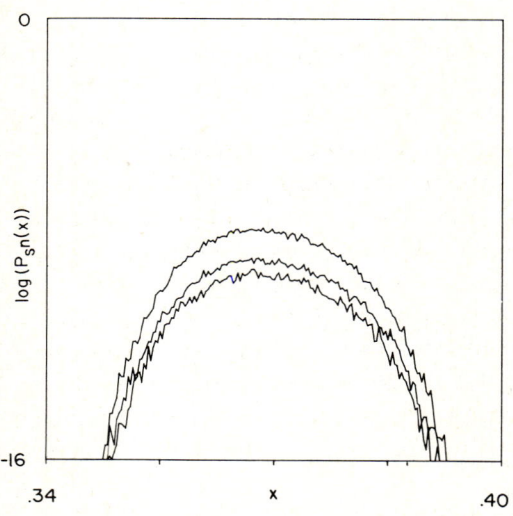

Fig. 8. Splitting of $P_{s^{17}}(x)$ into two daughter distributions $P_{s^{17}1}$ and $P_{s^{17}0}$. The top distribution is $P_{s^{17}}(x)$, the second distribution is $P_{s^{17}1}$, and the third is $P_{s^{17}0}$. s^{17} is the same as used in the previous figure.

This condition has some interesting implications that we will now discuss.

$P_{s^n}(x)$ is the distribution of initial conditions that produce the sequence s^n. We may then ask how much information about the initial condition is obtained by observing the sequence s^n, given the asymptotic distribution on the unit interval $\bar{P}(x)$. The appropriate informational measure turns out to be [31, 32]

$$I(s^n) = \int P_{s^n}(x) \log \frac{P_{s^n}(x)}{\bar{P}(x)} \, dx. \tag{6}$$

Then the average information obtained by specifying n symbols is

$$I(n) = \sum_{s^n} \mu(s^n) I(s^n). \tag{7}$$

For n large enough so that the width of $P_{s^n}(x)$ has reached its noisy asymptotic value we may use the conditions $\mu(s^n 1) \approx \mu(s^n 0)$ and $P_{s^n 0}(x) \approx P_{s^n 1}(x)$ to deduce that $I(n) \approx I(n+1)$, which means that for large enough n, observation of additional symbols gives *no additional information* about the initial condition. Stated another way, in the presence of noise, the *attainable information* about the initial condition reaches some maximum value I_0, which clearly depends on the noise level. The situation is illustrated schematically in fig. 9. For any given noise level, we may augment our conjecture that a well-defined metric entropy exists, and conjecture the existence of *two* well-defined invariant quantities that characterize a deterministic system with noise: h_μ and I_0*.

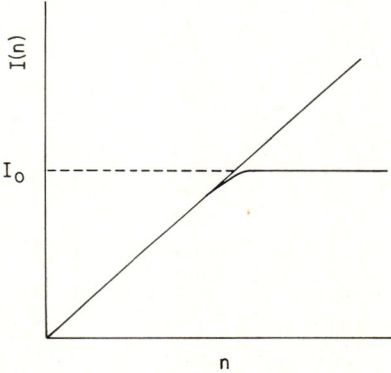

Fig. 9. Schematic illustration of the effect of noise on the attainable information, given by eq. (7), from a sequence of n measurements, or symbols. The straight line corresponds to the deterministic case.

We may also use the condition $\mu(s^n 0) \approx \mu(s^n 1)$ for n large enough to deduce entropy convergence properties. Convergence of the entropy for finite length n symbol sequences is exactly the condition that the symbolic dynamics be equivalent to an m-state Markov process, where m is the least integer that produces a distribution $P_{s^n}(x)$ satisfying

$$\text{Var } P_{s^n}(x) > \text{Var } X_{\Delta(s^m)}(x).$$

The finite state Markov property insures that the entropy reaches its converged value for $n \geq m$; we will call this phenomenon the *noise floor*, and say that the *convergence knee* occurs at $n = m$. Fig. 10 shows that as the noise level is increased, the convergence knee occurs for smaller values of m. The following section shows how these effects may be described in terms of a scaling theory.

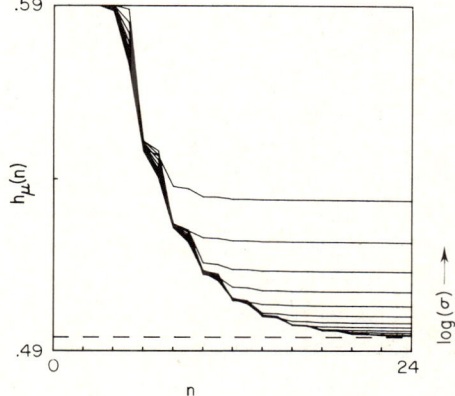

Fig. 10. Entropy convergence of the logistic equation at the parameter value where two bands join to one, $r = 3.67857\ldots$, for increasing noise levels $\sigma = 2^{-18}, \ldots, 2^{-7}$. The Lyapunov characteristic exponent is shown by the dashed line.

4. Scaling properties of entropy measurement

Considering the entropy as a function of both $N = 2^n$ and σ, we may define the normalized *excess entropy*† as

$$\bar{h}_\mu(N, \sigma) = \frac{h_\mu(N, \sigma) - h_\mu(\infty, 0)}{h_\mu(\infty, 0)}.$$

We then find that the data illustrated in fig. 10 displays power law behavior in N:

$$\bar{h}_\mu(N, 0) \sim N^{-\gamma},$$

and power law behavior in σ:

$$\bar{h}_\mu(\infty, \sigma) \sim \sigma^\beta,$$

---

* Of course we must still include the proviso that the measurement partition has coarser resolution than the noise level as long as we use our algorithm to compute h_μ. Rob Shaw [3] makes a similar conjecture for an entropy-like quantity computed using a measurement partition with resolution finer than the noise. We will discuss the relationship between these ideas in section 6.

† This can be considered as the *topological pressure* for finite symbol sequences in the presence of noise because we may assume that $h_\mu(\infty, 0) = \lambda$ (cf. discussion above), and in our numerical computations we actually use λ for the value of $h_\mu(\infty, 0)$.

(where ∞ has been approximated by $N = 2^{24}$) [5]. The least squares fits used to estimate the *convergence exponent* γ and the *noise exponent* β are quite good (cf. table I). The scaling in N is visible in the zero noise curve of fig. 10, and the power law increase with noise is illustrated in fig. 11.

Our observation that the metric entropy increases with a power law in response to added fluctuations is reminiscent of some results concerning the response of the Lyapunov characteristic exponent to added fluctuations. It has been shown for maps with a quadratic maximum that at the asymptotic limit of a band merging cascade (i.e. at the onset of chaos) noise added to the dynamics cause a power law increase in the Lyapunov characteristic exponent [4, 27, 33]. The cause of this power law must, however, be fundamentally different from the cause of the power law reported here. Their derivation of power law behavior of the Lyapunov characteristic exponent at band merging cascades (as well as similar results near tangent bifurcations) relies on the change in the attractor's geometry (i.e. the structure of the attractor on the unit interval) as noise is added. Furthermore, only the nearness to crucial bifurcation parameter values allows the change in the attractor's geometry

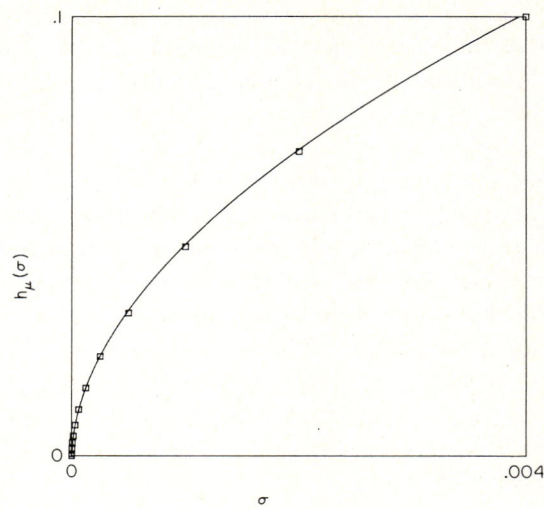

Fig. 11. Power law increase of $\bar{h}_\mu(n = 24, \sigma)$ for the logistic map at $r = 3.67857\ldots$ where two bands join to one.

to be systematically described using renormalization group techniques. The power law behavior we describe here appears to hold more generally, including parameter values away from bifurcation cascades, where the geometry of the attractor changes very little with added noise.

Table I
Numerical calculations of scaling exponents

System	γ	β	ω
Logistic $r = 3.9$	0.48 ± 0.2	0.56 ± 0.05 (1.0)	0.86
Logistic $r = 3.7$	0.4 ± 0.2	0.53 ± 0.05 (1.3)	0.76
Logistic $2 \to 1$ bands	0.38 ± 0.02	0.52 ± 0.01 (0.9)	0.73
Logistic $4 \to 2$ bands	—	0.51 ± 0.02 (1.5)	—
Logistic r_c	—	0.345 ± 0.01 (1.0)	—
Logistic $2 \to 1$ bands (function space perturbation)	0.38 ± 0.02	0.53 ± 0.01 (1.0)	0.72
Collet and Eckmann map ($2 \to 1$ band)	0.41 ± 0.1 (0.80)	0.62 ± 0.1	0.66
Tent, $s = 1.43$	0.55 ± 0.1 (0.86)	1.01 ± 0.01 (0.95)	0.55
Tent, $2 \to 1$ bands	0.50 ± 0.02 (0.82)	1.06 ± 0.08 (35)	0.47
Tent, $2 \to 1$ bands (function space perturbation)	0.51 ± 0.1	1.05 ± 0.08	0.49
Cusp map	—	1.04 ± 0.05 (4.3)	—
Random walk	—	0.92 ± 0.05 (0.05)	—
Toral automorphism	—	0.9 ± 0.02 (0.5)	—

Note: numbers in parentheses are constants of proportionality.

The two numerically observed power laws in N and σ lead us to posit the *scaling hypothesis* that $\bar{h}_\mu(N, \sigma)$ is a homogeneous function of N and σ, namely, that

$$\bar{h}_\mu(\lambda^\gamma N, \lambda^{-\beta}\sigma) = \lambda \bar{h}_\mu(N, \sigma),$$

where λ is an arbitrary change in scale. This sort of scaling hypothesis has been studied extensively in critical phenomena [34], and it is easily shown that the homogeneity of $\bar{h}_\mu(N, \sigma)$ in both variables implies that $\bar{h}_\mu(N, \sigma)$ may be written as a function of a single scaling variable multiplied by a power law. This reduction may be accomplished in two different ways:

$$\bar{h}_\mu(N, \sigma) = \sigma^\beta H(N\sigma^{\beta/\gamma}) \tag{8}$$

or

$$\bar{h}_\mu(N, \sigma) = N^{-\gamma} H'(\sigma N^{-\gamma/\beta}).$$

Since we are interested primarily in the response of $\bar{h}_\mu(N, \sigma)$ to noise, we will concentrate on the first scaling representation of $\bar{h}_\mu(N, \sigma)$.

The scaling hypothesis may be tested empirically (i.e. using data from a numerical simulation) by graphing $\sigma^{-\beta}\bar{h}_\mu(N, \sigma)$ as a function of the scaling variable $N\sigma^{\beta/\gamma}$, and observing whether or not the data lie on a well-defined function $H(N\sigma^{\beta/\gamma})$. The results of this procedure applied to the data shown in fig. 10 are displayed in fig. 12, where we see a convincing numerical verification of the scaling hypothesis. It is interesting to note that while in critical phenomena, scaling often occurs only asymptotically ("asymptotically" means $\sigma \to 0$ or $N \to \infty$ in this context) for this dynamical system we see scaling for *all* σ and N.

We see in fig. 12 that all of the convergence knees are mapped to a single knee of $H(N\sigma^{\beta/\gamma})$. This signals another scaling relation describing the convergence knee, since this implies that the set of (N, σ) for which a convergence knee occurs must satisfy

$N\sigma^{-\gamma/\beta} =$ constant.

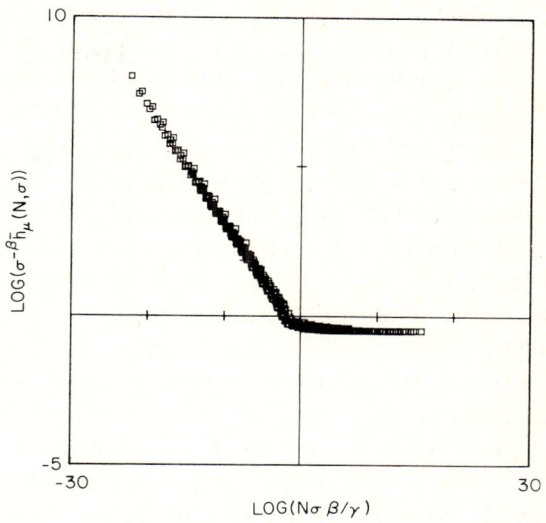

Fig. 12. All the data shown in fig. 10 are replotted here using the homogeneous function representation of eq. (8). The fact that all the data points lie on a well-defined function is verification of the scaling hypothesis.

Either N or σ may be regarded as dependent variables, $N_c(\sigma)$ or $\sigma_c(N)$, in this relation that defines the condition for the occurrence of a convergence knee. And at the convergence knee we may write either

$$N_c(\sigma) \sim \sigma^\omega \quad \text{or} \quad \sigma_c(N) \sim N^{1/\omega},$$

where we find the convergence knee exponent is given by

$$\omega = \gamma/\beta. \tag{9}$$

The same result may be obtained from an eigenvalue equation

$$\frac{\sigma_c(N)}{\sigma_c(2N))} \sim \frac{N_c(2\sigma)}{N_c(\sigma)} \sim \kappa,$$

where $\omega = \log \kappa$.

We have, then, two equivalent interpretations of the eigenvalue κ: First, it is the factor by which the converence knee noise level σ_c must decrease if we are to observe convergence using symbol sequences of length $n+1 = \log(2N)$ rather than of length $n = \log(N)$. Second, if we decrease the noise level by a factor of 2, then κ gives the relative increase in the length of symbol sequence at which we will find convergence. κ will probably be easier to directly measure than the exponents and, at least, it provides a simple way to summarize the net effect of noise on entropy convergence. Values of ω derived from β and γ are tabulated in table I.

The type of relationship between different scaling exponents exemplified by eq. (9) is quite common in the study of critical phenomena. The scaling exponents may be viewed as parameters that describe a surface $\bar{h}_\mu(N, \sigma)$ over the (N, σ) plane. This surface is shown in fig. 13, which illustrates the geometrical significance of the scaling exponents γ, β, and ω.

5. Further numerical experiments

We will now discuss the results obtained from simulating several different systems. For each system we have computed the convergence and noise exponents; the results are tabulated in table I. We will first describe each of the systems studied, then discuss the numerical results.

We study four different one-dimensional maps of the unit interval:

(i) the logistic equation, $f(x) = rx(1-x)$ for five parameter values, $r = 3.9$, $r = 3.7$ (i.e. two "typical" chaotic parameter values), $r = 3.67857\ldots$ (where two bands merge into one), $r = 3.59257\ldots$ (where four bands merge into two), and $r = 3.5699456\ldots$ (the onset of chaos);

(ii) the tent map:

$$f(x) = \begin{cases} sx, & \text{for } 0 < x \leq 0.5, \\ s(1-x), & \text{for } 0.5 < x < 1, \end{cases}$$

with $s = 1.43$ (a "typical" chaotic parameter value with topological entropy approximately equal to that of the logistic equation at $r = 3.7$) and $s = \sqrt{2}$ (the parameter value where two bands join to one);

(iii) Collet and Eckmann's map:

$$f(x) = \begin{cases} 2x, & \text{for } 0 < x < 0.5 - \delta, \\ 1 - \dfrac{(x-0.5)^2}{\delta}, & \text{for } 0.5 - \delta \leq x \\ & \leq 0.5 + \delta \\ 2(1-x), & \text{for } 0.5 + \delta < x < 1, \end{cases}$$

with $\delta = 1/6$ (the parameter value where two bands join to one); and

(iv) the cusp map:

$$f(x) = a(1 - |2x - 1|^{1+\epsilon}),$$

with $\epsilon = -0.05$ and $a = 0.66445776\ldots$ (one of the parameter values where two bands join to one for this ϵ).

We include one two-dimensional system, the linear toral automorphism whose matrix is

$$T = \begin{bmatrix} 0 & 1 \\ 1 & 1 \end{bmatrix}.$$

There is a general procedure for constructing a Markov partition (that is generating) for such a map [35], and we use this as the measurement partition.

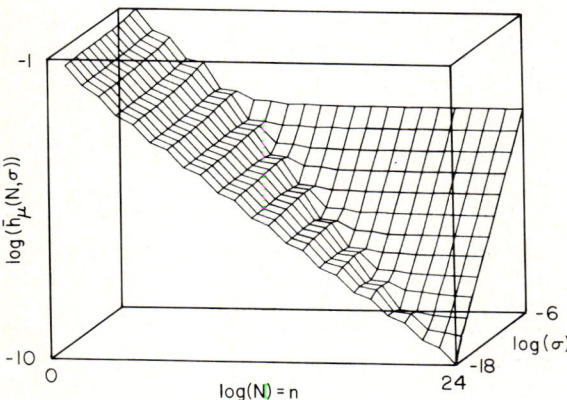

Fig. 13. All the data shown in fig. 10 are replotted as a three dimensional surface. The slope of the line along the front face is $-\gamma$, and the slope of the line along the right face is β. The intersection of these two surfaces defines a line of slope ω in the proper projection onto the scaling variable $N\sigma^\gamma$.

Another system we study has one of the simplest possible deterministic parts; a random walk on the circle (we have identified the ends of the unit interval to prevent escape of the orbit):

$$f_\xi(x) = x + \xi \; [\text{mod } 1],$$

where the circle is coordinatized by the unit interval (with 0 and 1 identified, and the usual measurement partition $\{[0, 0.5), [0.5, 1)\}$ is used to produce the symbol seqquences. The deterministic part of this map is simply the identity, $f(x) = x$, which has no attractor, and which clearly has zero entropy (the symbol sequence is periodic). As noise is added, however, every sequence becomes possible, through for small noise levels, long sequences of 0's and 1's will be most probable. As we have done for maps of the unit interval, we may, for this system, numerically accumulate probability histograms for n-cylinders and compute h_μ as before*. We find that $h_\mu(n)$ converges almost immediately to h_μ, i.e. $h_\mu(2) \approx h_\mu(\infty)$†.

Before discussing the other numerical results contained in table I, we will consider a question concerning the nature of the fluctuations, and how they are coupled to the deterministic system. Eq. (1) represents a very specific model for external fluctuations, namely additive noise. There are, however, many alternatives to perturbing the deterministic function by simply adding noise; one example is multiplicative noise (for the logistic equation this is equivalent to adding noise to the parameter‡). A natural question then arises: which method of adding noise correctly models external fluctuations in a physical system?

Perturbations of a physical system may best be thought of as a perturbation of the dynamics, and not simply a perturbation of the trajectory. A "correct" model for perturbations of a deterministic function $f \in F(M) = \{f: M \to M\}$ would choose a function at each time step from an ensemble of functions, with the ensemble centered about the deterministic zero noise limit f. Additive noise simply represents a choice from an ensemble that extends along a one-parameter family of functions $q \to f_q: M \to M: x \to f(x) + q$. We have modeled the more general case by expanding the function f in a Taylor series (for convenience we will now consider a map on the unit interval: $M = [0, 1]$), and perturb each coefficient separately:

$$f_\xi(x) = (a_0 + \xi_0) + (a_1 + \xi_1)x + (a_2 + \xi_2)x^2 \ldots, \quad (10)$$

where each ξ_i is an independent random variable with zero mean, and where $\{a_i\}$ represent the Taylor coefficients of the deterministic function. For example, when we take the deterministic function to be the logistic map, $f(x) = rx(1 - x)$, the deterministic coefficients are $\{a_0 = 0, \; a_1 = r, \; a_2 = -r, \; a_i = 0 \text{ for all } i > 2\}$.

The entries in table I labeled "function space perturbation" represent noise added as in eq. (10) up to sixth order. Comparing the noise exponents for these systems with the noise exponents obtained from simple additive fluctuations, we see agreement to within numerical error. This result gives some confidence that models using additive fluctuations may reflect behavior of physical systems with external fluctuations quite well.

We will now summarize a few interesting aspects of the results listed in table I. Some of the results may be coincidentally similar and lead to erroneous extrapolations. Conjectures based on these results must be verified with further numerical work as well as theoretical progress. The largest error in most of these computations is due to inaccuracy in the estimation of h_μ in the absence of noise; we have assumed the conjecture $h_\mu = \lambda$ (supported by our numerical evidence) and so

---

* The probabilities of the n-cylinders (and hence h_μ itself) are analytically computable using techniques from the theory of random walks; this calculation will be presented in a future paper.

† This result is similar to the entropy convergence of the toral automorphism.

‡ Crutchfield, Farmer, and Huberman [4] have shown for the logistic equation that for any ensemble of additive fluctuations $\{\xi\}$ there is an equivalent ensemble $\{\xi'\}$ of parametric fluctuations (with a different distribution than that for ξ, in general) that will yield the same time averages over trajectories.

estimate h_μ by the Lyapunov characteristic exponent λ computed using 10^7 iterations (giving an accuracy of $\approx 0.1\%$).

The convergence exponents show no discernable features. For the logistic equation, the convergence exponent decreases from $\gamma = 0.48$ at $r = 3.9$ as the parameter is lowered; at r_c there is no power law convergence. In fact, it is easy to show that at r_c

$$h_\mu(n) \approx \frac{\log n}{n}.$$

The fact that all of the maps at $2 \to 1$ band joining parameter values do not agree in their convergence exponents reveals that the convergence exponent is not constant under topological conjugacy (for all such maps $h = 0.5$). There is no known general technique to compute the convergence exponents, but for special cases (e.g. tent maps at band joinings) γ can be computed exactly to be $\gamma = 0.5$ for $2 \to 1$ band merging [26]. This value agrees extremely well with the numerical value quoted in table I.

Both the random walk and the toral automorphism have $\beta \approx 0.9$ (we see no particular theoretical reason for such a close match). For the logistic map, β decreases from ≈ 0.56 at $r = 3.9$ to ≈ 0.34 at the onset of chaos*, $r_c = 3.5699456\ldots$. For Collet and Eckmann's map at band joinings, we find $\beta \approx 0.6$, indicating that the noise exponent is neither a topological invariant nor universal for quadratic maps.

For all the tent maps simulated, we have $\beta \approx 1.0$, the same value of β as obtained for the cusp map. This leads us to the conjecture: everywhere expanding maps† have a noise exponent $\beta = 1$. There are other reasons for such a conjecture besides the numerical results listed in table I, for instance, the structure of the asymptotic probability distribution on the unit interval. Maps with a critical point

---

* This value for β agrees with the power law increase of the Lyapunov characteristic exponent at r_c for the logistic equation [4, 27, 33].

† A map $f: I \to I$ is everywhere expanding if $|f'| > 1$ for all points on the attractor.

(where the slope vanishes) have distributions with infinite singularities, expanding maps do not.

For maps with critical points, these singularities lead to a very non-uniform probability distribution of symbol sequences. In this case, the highly probable sequences are less affected by noise, and do not readily yield new observable sequences. Consequently, the entropy increases more slowly with noise level for maps with critical points. The first class of seven examples in table I with low noise exponents consists of maps with critical points; whereas the maps in the second class of six examples listed in the table have relatively high noise exponents, but no critical points.

6. Lyapunov characteristic exponents and other measures of chaos in the presence of fluctuations

In the context of deterministic systems, we have seen that for an attractor on the unit interval with one positive Lyapunov characteristic exponent λ and an absolutely continuous invariant measure μ [22],

$$h_\mu \leq \lambda,$$

and we have presented numerical evidence for equality. The purpose of this section is to see how this kind of result may be generalized to include systems with added fluctuations.

Just as the definition of metric entropy is problematic for systems with added fluctuations, so is the definition of Lyapunov characteristic exponents. For one-dimensional maps, the Lyapunov characteristic exponent can no longer be defined as the average slope of the map because the derivative of the noisy map is not defined. Two approaches to this problem have appeared in the literature. The first technique is to compute Lyapunov characteristic exponents numerically by using the deterministic slope of the map along a noisy trajectory [4, 33, 36]. These computations give quite good results at the asymptotic limit of band merging cascades, where the numerical results can be checked against theoretical predictions [27, 33]. This may seem surprising, but the numerical results are probably good for the same reason that the theoretical predictions can be made: at band merging cascades, the response of the Lyapunov characteristic exponent is dominated by the change in the geometrical structure of the attractor* when noise is added.

The second definition of Lyapunov characteristic exponents in the presence of noise is due to Schraiman, Wayne, and Martin [27]:

$$\lambda = \lim_{n\to\infty,\epsilon\to 0} \frac{1}{n} \cdot \log \left| \frac{\langle f_\zeta^n(x) \rangle - \langle f_\zeta^n(x+\epsilon) \rangle}{\epsilon} \right|,$$

where iteration of the noisy map is given by eq. (1), and where $\langle \ldots \rangle$ denotes an average over the ensemble of noise fluctuations. The noise amplitude must be small enough, and the limits taken carefully for this definition to make sense. When thought of as a measure of the initial spreading rate of two noise distributions whose means are separated by ϵ, this expression for λ is close to a third formulation of Lyapunov characteristic exponents in the presence of noise which we will now discuss (equivalence may eventually be proven).

We have defined h_μ in terms of symbolic dynamics (with a generating measurement partition), but there is another important alternate measure of a system's information generation in terms of the average initial spreading rate of narrow probability distributions†. This formulation has been discussed by Shaw [38]; Farmer, Crutchfield, Froehling, Packard, and Shaw [39]; and Farmer [32]. The spreading rate of sharp distributions is close in spirit to the definition of Lyapunov characteristic

---

* By "geometrical structure," we mean the band like structure of the attractor near r_c [37].

† Here we are identifying a narrow probability distribution with the ensemble of states the system may be in after a (precise but finite) measurement. The time evolution of a sharp distribution is obtained by application of the Frobenius–Perron operator as in eq. (2) (see eq. (3) for the case of added noise). The evolution of a sharp probability distribution is illustrated quite graphically in the movie "Mixing Properties of Strange Attractors," made by Doyne Farmer.

exponents (since the spreading of very sharp distributions is governed by the slope of the map) and the correspondence can be made exact for sufficiently simple maps (e.g. piecewise expanding maps). The main reason for discussing this spreading rate here is that it generalizes quite naturally to systems with added fluctuations, and such a measure may in fact be the most appropriate generalization of Lyapunov characteristic exponents for such systems. We will now define the spreading rate and discuss a few qualitative features for different examples, then outline some conjectures relating this picture to the symbolic dynamics quantities already discussed.

As we have noted previously, a one-dimensional map $f: I \to I$ has an associated Frobenius–Perron operator on the space of probability distributions on I given by

$$(L_f P)(x) = \int \delta(f(y) - x) P(y) \, \mathrm{d}y.$$

If f has an asymptotic ergodic invariant measure $\bar{\mu}$, then its distribution function $\bar{P}(x)$ must be a fixed point of the operator L_f. Non-equilibrium distribution functions $P(x)$ approach $\bar{P}(x)$ under successive iterations of L_f. The essential idea is to formulate an informational measure of the *rate* that $P(x)$ approaches $\bar{P}(x)$.

To begin, the measure of the amount of information contained in $P(x)$ relative to $\bar{P}(x)$ is

$$\int P(x) \log \frac{P(x)}{\bar{P}(x)} \, \mathrm{d}x.$$

Now consider how much information is obtained by making a measurement using a measurement partition $A = \{A_i\}$. If the system is found in the ith partition element, the amount of information obtained is

$$I_i = \int X_{A_i}(x) \log \frac{X_{A_i}(x)}{\bar{P}(x)} \, \mathrm{d}x$$

$$= -\log \bar{\mu}(A_i). \tag{11}$$

As time passes, if the system is chaotic, the information obtained by the measurement is lost because the distribution X_{A_i} spreads:

$$I_i(t) = \int L_f^t X_{A_i}(x) \log \frac{L_f^t X_{A_i}(x)}{\bar{P}(x)} dx.$$

Note that for $t = 0$, this equation reduces to eq. (11). We may now ask for the average information loss after a measurement, where the average is to be taken over all possible initial measurements*

$$\bar{I}(t) = \sum_i \bar{\mu}(A_i) I_i(t).$$

Farmer [32] has given an alternate (equivalent) expression for this quantity:

$$I(t) = \sum_{i,j} \bar{\mu}(f(A_i) \cap A_j) \log \frac{\bar{\mu}(f(A_i) \cap A_j)}{\bar{\mu}(f(A_i)) \bar{\mu}(A_j)}.$$

For a deterministic system we then have the situation illustrated in fig. 14a. A sharp distribution containing a significant amount of information $\bar{I}(0)$ gradually relaxes to the asymptotic distribution, at which point $\bar{I}(t) = 0$ for large enough t†. The slope of $\bar{I}(t)$ well before it goes to zero is then a measure of the loss rate of initial information, which we shall call k_μ. k_μ has been conjectured to be equal to h_μ (Shaw [38]; Farmer, Crutchfield, Froehling, Packard, Shaw [39]; Farmer [32]) for deterministic systems‡.

For contrast, consider the case when the measurement partition is used simply to sample a white noise process. In this case, the probability of any measurement outcome is independent of all previous outcomes, so $\bar{I}(t)$ goes to zero after the first time step, as illustrated in fig. 14b.

When noise is added to a deterministic system, and a measurement partition finer than the noise level is used, we expect $\bar{I}(t)$ to behave something like fig. 14c. Much of the information obtained from an initial measurement using a measurement partition with a typical partition element size smaller than the noise level σ is immediately lost as the sharp probability distribution X_{A_i} spreads out on the first time step into a distribution of width $\approx \sigma$. $\bar{I}_0 \equiv \bar{I}(1)$ then represents the true amount of information that can be obtained from a measurement; using any finer measurement partition can give no more information about the future behavior of the system.

We are now in a position to phrase the conjectures relating this picture to the measurements of chaos using symbolic dynamics: (i) $k_\mu = h_\mu$ and (ii) $\bar{I}_0 = I_0$. The noise level must, of course be small enough so that there is some time interval for which $\bar{I}(t)$ displays a well defined constant slope. Numerical experiments are underway to check these conjectures.

7. Concluding comments

The effects of fluctuations added to chaotic deterministic dynamical systems reveal the concept of "infinitely precise points" as invalid in many contexts. A new mathematical foundation of classical mechanics is needed; one that uses primitives derived from noise processes. Ruelle [42] has made significant progress in this direction. Though the inclusion of fluctuations in a dynamical model adds many analytical complications to a subject already incompletely understood, there is hope, based on physical observations and numerical computations, that there may be several rewarding

---

* Note that $\bar{I}(t)$ must be distinguished from $I(n)$ defined in eq. (7); $I(n)$ is the rate that information (with respect to the previous $n - 1$ symbols) is acquired with the observation of new symbols, and $\bar{I}(t)$ is the average rate that information contained in an initial condition (using a particular measurement partition) is lost.

† This is actually a crude picture with details which may change for different systems; e.g.: (i) Phase coherent attractors have $\bar{I}(t) > 0$ as $t \to \infty$ (cf. Farmer et al. [39]); (ii) Rob Shaw [3] has pointed out that for maps with a critical point the initial slope of $\bar{I}(t)$ will be larger than h_μ, and then decrease to h_μ.

‡ Goldstein and Penrose [40] have introduced a similar information loss rate which, for certain systems, Goldstein [41] proved to be equal to the metric entropy.

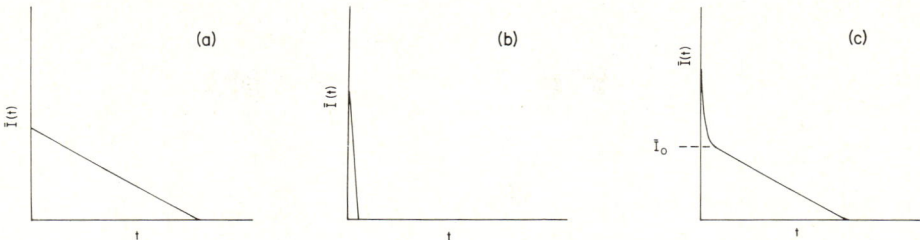

Fig. 14. (a) Schematic representation of $\bar{I}(t)$ for a deterministic system. For certain systems (e.g. one dimensional maps with a critical point) the slope of $I(t)$ will be greater than k_μ at $t = 1$. (b) Schematic representation of $\bar{I}(t)$ for measurements of a white noise process. (c) Schematic representation of $I(t)$ for a deterministic system with added noise, where a measurement partition finer than the noise level has been assumed.

simplifications lurking in the theory. Assuming such a theory may be formulated, most of the numerical results presented here should be consequences of the theory, so they will hopefully point the direction for some future theoretical developments. We will now review our results in this light.

For a chaotic deterministic system, successive measurements (using a "good" measurement partition) pinpoint the initial condition whose orbit produced the observations with arbitrary accuracy (i.e. an arbitrarily large amount of information about the initial condition may be obtained from an arbitrarily long sequence of measurements). When noise is added to the deterministic dynamics, we have observed that the initial condition may be specified only to within some uncertainty, even with an arbitrarily long sequence of measurements. This has led to the proposal that a chaotic system with added fluctuations is characterized by *two* invariant quantities: (i) I_0: the maximum average information (about the initial condition) obtainable from a sequence of measurements; and (ii) h_μ: the average information generation rate (simply the metric entropy in the case of a deterministic system)*.

---

* We have also conjectured these two quantities to be equal to $\bar{I}_0$ and k_μ, the maximum amount of information that can be stored in an initial condition, and the average loss rate of information after a measurement, respectively.

I_0 has not been computed numerically yet, but the information production rate h_μ (with respect to a given measurement partition) is easily computed using the same algorithms used to compute h_μ for deterministic systems. Upon pursuing the question of how h_μ depends on the fluctuations added to the deterministic dynamics, we find that h_μ increases with a power law in the noise level σ: $h_\mu \approx \sigma^\beta$. We have found that this power law increases seems to happen very generally (for all systems studied here). The exact value of the noise exponent β varies with the system under study, though our numerical experiments have led to the conjecture that a wide class of systems (those reducible to a one-dimensional map $f: I \to I$ with $|f'| > 1$) has a noise exponent $\beta = 1$. We have combined the power law response of h_μ with the power law convergence of the entropy as a function of the number of symbols observed, to form a homogeneous function description of entropy measurement. In this context, a scaling hypothesis has been verified numerically.

The power law increase in the metric entropy may be regarded as the discovery of a new phenomenon, an observable feature of the information production properties of any physical system that can successfully be modeled by a low dimensional chaotic dynamical system coupled to external fluctuations. There is a growing body of very good experimental evidence that supports such a model; convincing one-dimensional return maps have been obtained for fluid systems and for chemical systems*.

The noise exponent should be measurable, given reasonable experimental accuracy, though we have no prediction for its value if the one-dimensional map that underlies the observed behavior has a critical point. So far, all the return maps constructed from experimental data appear to have a critical point (or several critical points). There are, however, many physical systems that should be describable by a one-dimensional return map whose slope (absolute value) is always greater than one. One example would be a Benard convection fluid system constrained to excite only those modes described by the Lorenz equations, which have a cusp-like one-dimensional return map. For these systems, we might expect a noise exponent of $\beta = 1$.

Fluctuations are now generally recognized as the source of much of the diverse complexity we see in the world around us (especially in the biosphere). It has been hypothesized (by R. Shaw [46], for example) that what we call "diverse complexity" is a result of intrinsic dynamical properties of some (complicated) dynamical system, in particular, of the system's information generating properties. The informational properties of most of the dynamical systems underlying and producing this complexity are, however, poorly understood. One example of how the current picture of information generation in chaotic dynamical systems must be generalized, is that unlike the chaotic systems studied here, the information generated by the dynamics of complicated evolving systems like the biosphere is stored in physical structures, which then serve as the base for even more complicated evolution. There are many other similar problems to be faced, but the results presented here will hopefully serve as a starting point for the study of the role fluctuations will play in the context of these more complicated systems.

Acknowledgements

The authors wish to thank Doyne Farmer, David Fried, John Gukenheimer, and Rob Shaw, for helpful discussions, and the Center for Nonlinear Studies, Los Alamos National Laboratory, for the warm hospitality offered during the time this work was written up. This work was supported in part by NSF grant 443150-21299.

References

[1] E.N. Lorenz, J. Atmos. Sci. 20 (1963) 130.
[2] D. Ruelle and F. Takens, Comm. Math. Phys. 20 (1974) 167.
[3] For example: J.L. Hudson and J.C. Manken, J. Chem. Phys. 74 (1981) 6171. J.C. Roux, A. Rossi, S. Bachelart and C. Vidal, Phys. Lett. 77A (1980) 391; R. Shaw "The Dripping Faucet as a Model Chaotic System", UCSC Preprint (1982). A. Libchaber (this volume); and H. Hauke and Y. Maeno (this volume).
[4] J.P. Crutchfield and B.A. Huberman, Phys. Lett. 77A (1980) 407; J.P. Crutchfield, J.D. Farmer and B.A. Huberman, "Fluctuations and Simple Chaotic Dynamics", Physics Reports 92 (1982) 45.
[5] J.P. Crutchfield and N.H. Packard, Int. J. Theo. Phys. **21** (1982) 433; and J.P. Crutchfield and N.H. Packard, in Evolution of Ordered and Chaotic Patterns, H. Haken, ed. (Springer, Berlin, 1982).
[6] Ju. I. Kifer, USSR Izvestija 8 (1974) 1083.
[7] M. Boyarsky, Trans. Am. Math. Soc. 257 (1980) 350.
[8] Y. Oono and M. Osikawa, Prog. Theo. Phys. 64 (1980) 54.
[9] J. Milnor and W. Thurston, "On Iterated Maps of the Interval, I and II", Princeton University preprint (1977).
[10] C.E. Shannon and W.E. Weaver, The Mathematical Theory of Communication, (Univ. of Illinois Press, Urbana, Illinois, 1949).
[11] R. L. Adler, A.G. Konheim and M.H. McAndrew, Trans. Am. Math. Soc. 114 (1965) 309.
[12] W. Parry, Trans. Am. Math. Soc. 112 (1964) 55.
[13] J.P. Crutchfield and R. Shaw (unpublished). See J.P. Crutchfield, UCSC Ph.D. dissertation (1983).
[14] P. Collet, J.P. Crutchfield, J.-P. Eckmann, "On Computing the Topological Entropy of Maps", to appear Comm. Math. Phys. (1983).

---

* Cited here are "non-trivial" physical systems in which one might not naively expect to see low-dimensional chaos because of the many degrees of freedom that could potentially participate in the dynamics. Return maps have, of course, been successfully constructed for much simpler physical systems (e.g. electrical oscillator circuits) in which low-dimensional chaos is expected (cf. Crutchfield [43]; Packard, Crutchfield, Farmer, Shaw [44]; Gollub, Romer, and Socolar [45]) because of the few degrees of freedom involved.

[15] A.N. Kolmogorov, Dokl. Akad. Nauk. SSSR 119 (1958) 861; Dokl. Akad. Nauk. SSSR 124 (1959) 754.
[16] L.W. Goodwyn, Proc. Am. Math. Soc. 23 (1969) 697.
[17] E.I. Dinaburg, Sov. Math. Dokl. 11 (1970) 13.
[18] G. Benettin, L. Galgani, A. Giorgilli and J.-M. Strelcyn, Meccanica (1980) 21, and C.R. Acad. Sc. Paris 286 (1978) 431. I. Shimada and T. Nagashima, Prog. Theo. Phys. 61 (1979) 1605.
[19] R. Shaw, Zeit. fur Naturfor. 36a (1981) 80.
[20] R. Bowen and D. Ruelle, Inven. Math. 29 (1975) 181.
[21] Ya. B. Piesin, Uspek. Math. Nauk. 32 (1977) 55.
[22] D. Ruelle, Boll. Soc. Brasil. Math. **9** (1978) 331; see also D. Ruelle, Comm. Math. Phys. 55 (1977) 47.
[23] F. Ledrappier, Ergod. Theo. Dyn. Sys. 1 (1981) 77.
[24] I. Shimada, Prog. Theo. Phys. 62 (1979) 61.
[25] J.H. Curry, "On Computing the Entropy of the Henon Attractor", Institut des Hautes Etudes Scientifique, Burr-sur-Yvette, preprint (1981).
[26] N.H. Packard, UCSC Ph.D. Dissertation (1982).
[27] B. Schraiman, C.E. Wayne and P.C. Martin, Phys. Rev. Lett. 46 (1981) 935.
[28] H. Haken and G. Mayer-Kress, Z. Physik B **43** (1981) 185; Phys. Lett. 84A (1981) 159.
[29] Y. Takahashi, "Observable Chaos and Variational Principle for One-Dimensional Maps", Dept. Math., Coll. Gen. Educ., University of Tokyo, preprint (1981).
[30] M. Feigenbaum and B. Hasslacher, "Irrational Decimations and Path Integrals for External Noise", Phys. Rev. Lett. 49 (1982) 605.
[31] F. Schlogl, Z. Physik 243 (1971) 303.
[32] J.D. Farmer, "Information Dimension and the Probabalistic Nature of Chaos", Zeit. fur Naturfor. 37a (1982) 1304. See also J.D. Farmer, UCSC Ph.D. Dissertation (1981).
[33] J.P. Crutchfield, M. Nauenberg and J. Rudnick, Phys. Rev. Lett. 46 (1981) 933.
[34] H.E. Stanley, Introduction to Phase Transitions and Critical Phenomena (Oxford Univ. Press, New York, 1971).
[35] R.L. Adler and B. Weiss, Mem. Am. Math. Soc. 98 (1970) 1.
[36] H. Haken and G. Mayer-Kress, J. Stat. Phys. 26 (1981) 149.
[37] E.N. Lorenz, New York Acad. Sci. Annals 357 (1980) 282. J.P. Crutchfield, J.D. Farmer, N.H. Packard, R. Shaw, R.J. Donnelly and G. Jones, Phys. Lett. 76A (1980) 1.
[38] R. Shaw, UCSC Ph.D. Dissertation (1980).
[39] J.D. Farmer, J.P. Crutchfield, H. Froehling, N.H. Packard and R. Shaw, New York Acad. Sci. Annals 357 (1980) 453.
[40] S. Goldstein and O. Penrose, J. Stat. Phys. 24 (1981) 325.
[41] S. Goldstein, Israel. J. Math. 38 (1981) 241.
[42] D. Ruelle, Comm. Math. Phys. 82 (1981) 137.
[43] J.P. Crutchfield, UCSC Senior Thesis (1979).
[44] N.H. Packard, J.P. Crutchfield, J.D. Farmer and R. Shaw, Phys. Rev. Lett. 45 (1980) 712.
[45] J.P. Gollub, E.J. Romer and J.E. Socolar, J. Stat. Phys. 23 (1980) 321.
[46] R. Shaw, cover letter for Louis Jacot Prize Competition, Paris (1977).

ON THE QUADRATIC MAPPING $z \to z^2 - \mu$ FOR COMPLEX μ AND z: THE FRACTAL STRUCTURE OF ITS $\mathcal{M}$ SET, AND SCALING

Benoit B. MANDELBROT

IBM Thomas J. Watson Research Center, Yorktown Heights, New York 10598, USA

For each complex μ, denote by $\mathcal{F}(\mu)$ the largest bounded set in the complex plane that is invariant under the action of the mapping $z \to z^2 - \mu$. Mandelbrot 1980, 1982 (Chap. 19) reported various remarkable properties of the $\mathcal{M}$ set (the set of those values of the complex μ for which $\mathcal{F}(\mu)$ contains domains) and of the closure $\mathcal{M}^*$ of $\mathcal{M}$. The goals of the present work are as follows. A) To restate some previously reported properties of $\mathcal{F}(\mu)$, $\mathcal{M}$ and $\mathcal{M}^*$ in new ways, and to report new observations. B) To deduce some known properties of the mapping f for real μ and z, with $\mu \epsilon]-\frac{1}{4}, 2[$ and $z \epsilon]-\frac{1}{2} - \frac{1}{2}\sqrt{1 + 4\mu}, \frac{1}{2} + \frac{1}{2}\sqrt{1 + 4\mu}[$. In many ways, the properties of the transformation f are easier to grasp in the complex plane than in an interval. (This exemplifies the saying that "when one wishes to simplify a theory, one should complexify the variables".) C) To serve as introduction to some recent pure mathematical work triggered by Mandelbrot 1980. Further pure mathematical work is strongly urged.

Introduction. The illustrations are the focus of this paper, and the text is organized around the illustrations, in the form of extended comments. Additional illustrations are found in Mandelbrot 1980, 1982, 1983a, b.

1. Discussion of figs. 1a to 1e. Illustration of the action of $z \to f(z, \mu) = z^2 - \mu$ on a large complex circle. Sequences of algebraic curves approximating the repeller (Julia) sets $\mathcal{F}^*(\mu)$

A transformation becomes easier to study when one has a concrete visual feeling for its action. In the case of $z \to f(z, \mu) = z^2 - \mu$, it is known that the point at infinity is a stable fixed point of f. Hence it is an attractive point. In order for a circle of sufficiently large radius r and center O to be in the domain of attraction of ∞, a sufficient condition is $r > r_W = \frac{1}{2} + \frac{1}{2}\sqrt{1 + 4|\mu|}$. The circle of radius r_W and center O will be denoted by W^O and called "whirlpool circle", and r_W will be called the "whirlpool radius", because the orbits of all the points outside W^O "whirl away" from O.

On the other hand, some complex z are not attracted to ∞. Examples are the bounded fixed points $z' = \frac{1}{2} + \frac{1}{2}\sqrt{1 + 4\mu}$ and $z'' = \frac{1}{2} - \frac{1}{2}\sqrt{1 + 4\mu}$, and their successive pre-images. Let the maximal bounded set invariant under $f(z, \mu)$ be denoted by $\mathcal{F}(\mu)$. By definition of the whirlpool circles, $\mathcal{F}(\mu)$ is contained within W^O. Also, for every value of k, $\mathcal{F}(\mu)$ is contained in the kth pre-image of W^O under $f(z)$, i.e., the pre-image of W^O under $f_k(z)$. This last set is defined by $|f_k(z, \mu)| = r_W$ and will be denoted by W^{-k}. It is an algebraic curve called "lemniscate" (Walsh 1956). The lemniscates corresponding to increasing values of k are non-overlapping and are monotonically imbedded in sequence. They can be called "parallel under $f(z, \mu)$". The set $\mathcal{F}(\mu)$ is the limit of these curves plus their interiors.

Denote by $\mathcal{F}^*$ the boundary of $\mathcal{F}$. The set $\mathcal{F}^*$ is the limit of W^{-k} for $k \to \infty$. It is the repeller set of $f(z, \mu)$ and is also called Julia set. (History: the earliest basic facts about iteration were described in Fatou 1906, and the bulk of the original theory was described near simultaneously in Julia 1918 and Fatou 1919. Since the term "Julia set" has become entrenched, I chose to honor Fatou by denoting this set by $\mathcal{F}^*$.)

Now for some illuminating illustrations. Figs. 1a to 1d represent the interiors of W^O and of several curves W^{-k} in superposition, for four selected values of μ. The goal is to demonstrate intuitively that the topology of $\mathcal{F}$ and of the Julia set $\mathcal{F}^*$

Fig. 1a. Action of $z \to z^2 - \mu$ when the Julia set is a loop-free closed fractal curve.

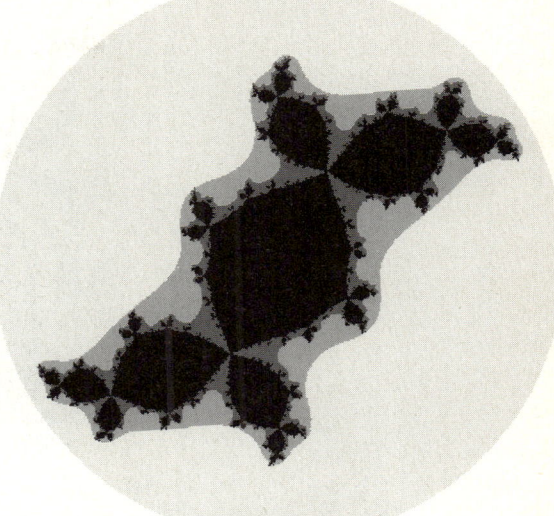

Fig. 1b. Action of $z \to z^2 - \mu$ when the Julia set is a closed fractal curve with loops.

greatly depends on the value of μ: in particular, $\mathscr{F}^*$ can be (a) a loop-free ("simple") curve that bounds a domain, (b) a curve with multiple points that bounds an infinite number of domains, (c) a tree ("branching curve without loop", "dendrite") that does not surround a domain, or (d) a totally disconnected dust. Fig. 1e represents an attractive example of $\mathscr{F}(\mu)$.

2. Discussions of figs. 2a to 2f. Classification of the values of μ by the topology of $\mathscr{F}(\mu)$. The sets $\mathscr{M}$ and $\mathscr{M}^*$. Sequences of algebraic curves approximating $\mathscr{M}^*$. The continent, islands, stellate structures, devil's causeways

This series of figures investigates in detail the set of those values of μ for which $\mathscr{F}(\mu)$ is connected.

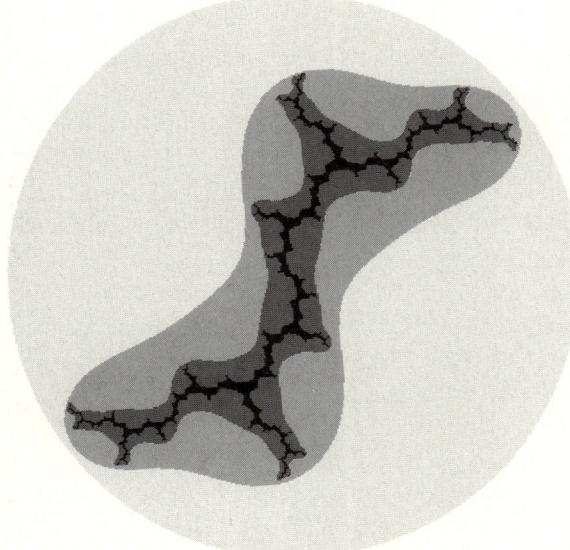

Fig. 1c. Action of $z \to z^2 - \mu$ when the Julia set is a fractal tree.

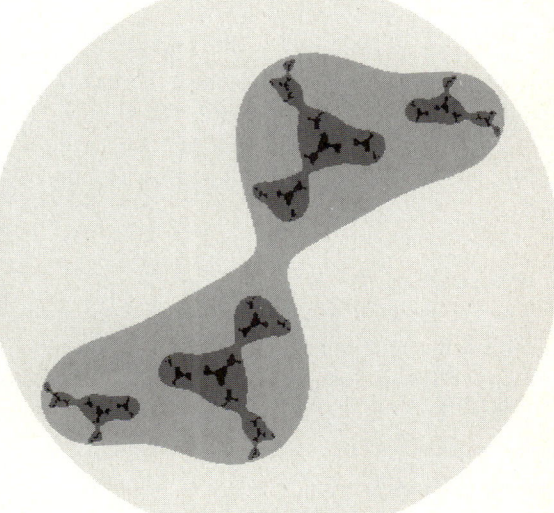

Fig. 1d. Action of $z \to z^2 - \mu$ when the Julia set is a fractal dust ("Cantor set").

Fig. 1e. Interior of a Julia set (repeller set) of $z \to z^2 - \mu$ after two successive sixfold bifurcations.

This set is to be denoted by $\mathcal{M}^*$, and $\mathcal{M}$ will denote the set of values of μ for which $\mathcal{F}(\mu)$ has interior points, that is, includes domains. On a graph, e.g., on fig. 2a, the $\mathcal{M}$ set and the $\mathcal{M}^*$ set cannot be distinguished, but they turn out to be significantly different in structure.

Construction of the $\mathcal{M}$ set and of the $\mathcal{M}^*$ set. To follow the method used in figs. 1a to 1d would be cumbersome and unreliable, but is not necessary because Gaston Julia gave the following direct criterion. The set $\mathcal{F}(\mu)$ is disconnected if and only if the sequence of iterates of $z = 0$, beginning with $-\mu, \mu^2 - \mu$ and $((\mu^2 - \mu)^2 - \mu$, converges to infinity. For this to be the case, a necessary and sufficient condition is that $|f_k(0, \mu)|$ must exceed for some value of k the whirlpool radius $r_w = \frac{1}{2} + \frac{1}{2}\sqrt{1 + 4|\mu|}$ derived in section 1. If $|\mu| > 2$, this condition is satisfied for $k = 1$. Hence the $\mathcal{M}^*$ set is entirely contained within the closed disc $|\mu| \leqslant 2$. Furthermore, the program is simplified (though the runs become a bit longer) if r_w is replaced by a uniform threshold equal to 2. For each k, one can draw in the μ-plane the set defined by $|f_k(0, \mu)| = 2$. As we know from section 1, such a set is an algebraic curve called "lemniscate". Here, all the lemniscates include the point $\mu = 2$, but otherwise they are non-overlapping and monotonically imbedded in sequence. The $\mathcal{M}^*$ set is the limit of these curves plus their interiors.

Fig. 2a. Overall view of the $\mathscr{M}$-set of $z \to z^2 - \mu$.

Fig. 2b. Detail of an $\mathscr{M}$-island: the "speck" to the right of fig. 2a.

Fig. 2c. Detail of an $\mathscr{M}$-island: the "speck" at the bottom of fig. 2a.

As is known, μ is called superstable of minimal period k if $f_k(0, \mu) = 0$ but $f_h(0, \mu) \neq 0$ for every $h < k$. It follows that $f_{nk}(0, \mu) = 0$ for every integer n, hence all superstable μ's fail to iterate to infinity, meaning that they belong to the $\mathscr{M}^*$ set. Walsh 1956 reports that a lemniscate cannot contain a loop within a loop: it is necessarily either a single loop, or a finite union of loops with non-overlapping interiors. It turns out that in the present case, all the lemniscates are single loops for all values of k, hence $\mathscr{M}^*$ is a connected set. But before we tackle this point, other features of the $\mathscr{M}^*$ set must be considered.

The continental subset of $\mathscr{M}$. For reasons that will transpire momentarily, the structure of fig. 2a is clarified by positioning the grid of μ's so that real valued μ's are not tested. A first glance reveals that the great bulk of the black points lie in a large and very highly structured "continent". It has a striking "cactus tree" structure, which I propose to describe as a "molecule" made of an infinity of "atoms". At the center is a "seed atom", which has the shape of a cardioid, and contains all the μ's for which $f(z, \mu)$ has a single stable limit point besides ∞. The exactly circular atom straight to the right from the cardioid contains all the μ's for which $f(z, \mu)$ has a stable cycle of period 2, and the near circular atoms that follow to the right correspond to stable cycles of periods 4, 8, etc. The points where these atoms join are the μ's corresponding to the basic real μ bifurcations. Other near circular atoms that touch the cardioid correspond to cycles of order $k > 2$.

The shape of the $\mathscr{M}^*$ set near the value μ_∞ at the rightmost tip of the continental subset. Scaling property in the plane, and its use to rederive (as corollary) the known scaling property of bifurcations on the real line. Consider the sequence of atoms that converge to the tip of the continent. They seem essentially alike, and seem tangent to two straight half-lines that are symmetric with respect to the real axis and converge to the value μ_∞ defined as the accumulation point of bifurcations. This is a geometric property of scaling, more precisely, of asymptotic geometric scaling.

An inferred consequence is that these atoms' horizontal intercepts decrease geometrically at each bifurcation. This inference is of course well-known to be true, having been discovered by Grossman and Thomae and by Feigenbaum.

To verify the identity of the atom shapes by a more exacting test, $\mathscr{M}^*$ was redrawn by replacing the parameter μ by $v = \log(\mu - \mu_\infty)$. The cardioid shaped seed atom is thereby made much smaller, and the other atoms indeed become near identical.

The big island to the right of the continent. Other islands. In addition to the continent, the $\mathscr{M}^*$ set contains a number of scattered specks. It is hoped that these specks escaped the watchful eye of the editors and the printers of the present Proceedings. The reason why I am concerned is that their counterparts on page 250 of Mandelbrot 1980 came to be erased, on the firm assumption that they could only be dirt!

In fact they are very real, and it may be useful to devote a few lines to telling how I discovered them. Examining my first rough graph of $\mathscr{M}^*$, I too took most of them to be dirt. But the biggest one, positioned to the right of the continent, looked too big to be spurious, and it was easy to verify that it intersects the axis of real μ's along the interval, discovered by Myrberg and Metropolis, Stein and Stein (see Collet and Eckmann 1980), for which $f(z, \mu)$ has a stable cycle of period 3. I had this speck examined in closeup, fig. 2b and it was revealed to be essentially a downsized version of the continent. Other Myrberg intervals that I examined in closeup were also revealed to intersect very small downsized versions of the continent.

Thus, the rightmost tip of the continent continues along the real axis by a peculiar causeway. Because of analogy with the Devil's Staircase (Mandelbrot 1982, page 83), I propose to call it the "Devil's Stepstones". The metaphor starts with large stones set in a stream to accommodate ordinary super giants, then smaller stones are set to accommodate ordinary giants, small giants, super-

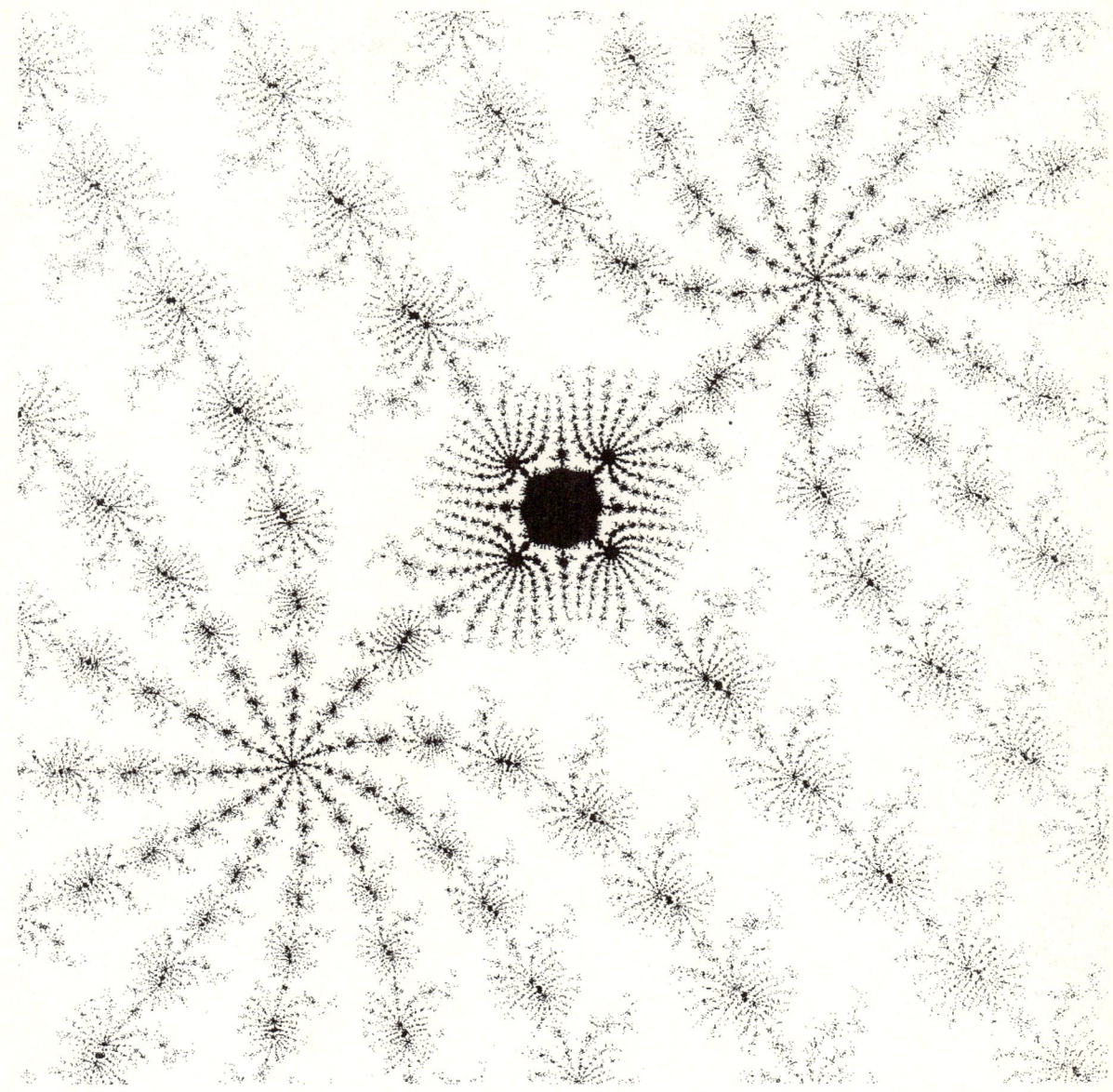

Fig. 2d. Detail of the ℳ-web offshore the continent on fig. 2a.

men, and so on down to devilishly tiny beasts. Ultimately the stones leave no gap of positive width, however small. The real axis runs along the center of this causeway, in a way that is familiar to students of the real transform $f(z, \mu)$.

At this point, I traced several puzzling observations to the same source. The first observation was that for periods 1, 2 and 3, each superstable μ is the "nucleus" of an atom known to belong to the continent or an island off the real axis. However, two superstable μ's of period 4 remained "unattached", and for higher periods the number of "unattached" superstable μ's kept increasing rapidly. One may have argued that some atoms con-

Fig. 2e. The interior of the Julia set when μ is near the nucleus of an offshore "island" of $\mathcal{M}$.

tain multiple-root nuclei, or several distinct nuclei, but these atoms should have looked different from atoms of smallest period 1, 2 or 3, while in fact all atom shapes fell into either of the two patterns exemplified by the seed cardioid and the circle to the right of it.

The second puzzling observation was that except for the point μ_∞, the tips of the continent gave no evidence of being followed by Devil's Stepstones.

The third puzzling observation was already mentioned; when the $\mathcal{M}$-set was traced with low precision on a medium-tight lattice, it seemed surrounded by unattached specks of dirt.

A close-up view of the big speck to the right of $\mathcal{M}^*$ (fig. 2c) settled the three puzzles together; most specks did not vanish but turned out to be islands identical to the continent in their topology and overall form. It soon became clear that these

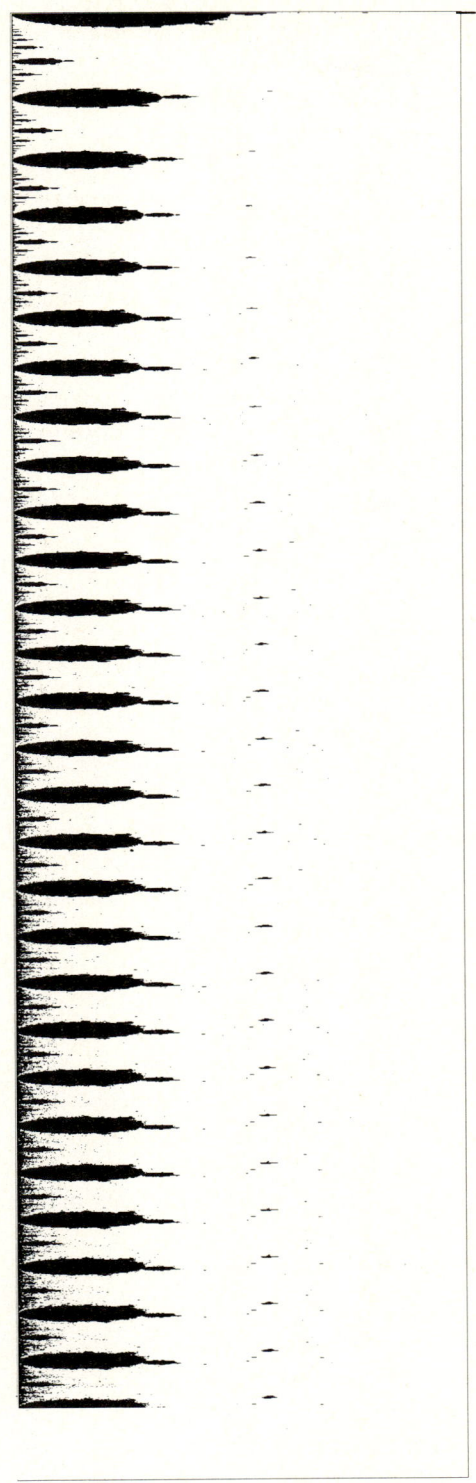

Fig. 2f. Detail of the $\mathscr{M}$ set of $z \to \lambda z(1-z)$ after inversion and compression.

islands do not scatter around haphazardly, but form "stellate" arrays (see Mandelbrot 1980, 1982 for details). An array close to a point of bifurcation of order 11 is seen on fig. 2d. Further closeups revealed increasing numbers of increasingly small islands between larger islands along each "ray". This was reminiscent of the above-mentioned fact that increasingly small islands are "pierced through" by the real axis. At this point, the growing analogy with the real axis suggested that the islands in every ray are linked together by curves that are counterparts of the real axis, but could not be seen because curves other than the axis nearly always fall between the lattice points used in computation. This implies that the stellate structure reflects an underlying tree structure, and that the $\mathscr{M}^*$ set is connected. It may be recalled that $\mathscr{M}^*$ is approximated by a sequence of lemniscates, which (according to general results) might have split into separate loops. The fact that $\mathscr{M}^*$ remains connected implies that the approximating lemniscates are single loops. This was verified to be the case until high order pre-images of the circle $|\mu| = 2$, parts of which started falling between the lattice points.

Needless to say, computer based observations do not provide a substitute for actual proofs. In some cases, full proofs bring rigor without additional insight, but mathematical study of my observations on the $\mathscr{M}^*$ set turned out to be fruitful and useful, witness Douady and Hubbard 1982 and forthcoming works. Needless to say (again) this mathematical study could not have been undertaken without my computer based observations.

The inverse of bifurcation: the notion of confluence. The literature of bifurcation never seems to refer to the opposite effect that is observed, say, when μ starts with a value in an atom other than a seed atom, and changes continuously, without leaving the island, until it reaches a seed atom. Mandelbrot 1980 gave to this inverse operation the name *confluence*. The point is that the continent is the domain of confluence to a stable limit point, and each island is a domain of confluence to a periodic cycle, but not of confluence to a limit point.

The atoms' and the islands' intrinsic coordinates. Homologous points. Given an atom of minimal period k, denote by z_μ any point in the stable cycle corresponding to the parameter value μ. We know that the complex number $f'_k(z_\mu, \mu)$ is less than 1 in modulus. Its real and imaginary parts form intrinsic coordinates for the point μ within the atom to which it belongs. Two points having identical intrinsic coordinates can be called homologous within their atoms. The set of μ's for which $f'_n(z_\mu, \mu)$ is real will be called the atom's "spine". It runs from a point where $f'_k(z_\mu, \mu) = 1$ (which is a cusp in the case of seed atoms), to a point of bifurcation of order 2, where $f'_k(z_\mu, \mu) = -1$.

Furthermore, each atom's position in its island can be identified by an "address", namely the sequence of integers that identify the sequence of bifurcations that lead to this atom starting from the seed cardioid. Each bifurcation is indeed marked by a rational number n_i/m_i, with $m_i \geq 2$ and $0 < n_i < m_i$. Thus, it suffices to write these n_i and m_i in sequence separated by commas. One can agree that the seed cardioid's address is 0 (and other addresses may, but need not, start by 0). The combination of the address of the atom and of the value of $f'_k(z_\mu, \mu)$ forms an intrinsic coordinate for a point μ within the island to which μ belongs. Two points having identical intrinsic coordinates can be called "homologous" within their islands.

An island's spine combines its seed cardioid's spine with the spines of atoms corresponding to bifurcation into $m_i = 2$. Every island spine's endpoint is homologous to the tip μ_∞ of the continental subset of the $\mathcal{M}^*$ set.

"Universality class" argument to explain why the islands are alike. Assume that μ is near a superstable value μ^* of minimum period k. We wish to determine the shape of the atom nucleated by μ^*.

The lowest order terms in the expansion of $f_k(z, \mu)$ near $z = 0$ and $\mu = \mu^*$ can be written as $\beta_k z^2 + \gamma_k(\mu - \mu^*)$. Now let us state and test out a brutal "universality class" argument, then a milder version of it.

The brutal argument claims that the shape of the atom nucleated by μ^* depends only on the lowest terms in the expansion of $f_k(z, \mu)$ near $z = 0$ and $\mu = \mu^*$. If this were the case, μ^* would nucleate a cardioid-shaped seed atom. This atom and the molecule grown upon it would be identical to the continent, except for its size being reduced in the ratio $1/\beta_k \gamma_k$. The milder argument agrees to take account of a few higher order terms near $z = 0$ and $\mu = \mu^*$, while continuing to disregard the behavior of $f_k(z, \mu)$ far from $z = 0$ and $\mu = \mu^*$. This milder argument suggest the following properties. A) The atom nucleated by μ^* is the seed atom of a molecule, and its shape resembles the continental cardioid except for some mild non-linear deformation. B) Other atoms obtained by bifurcation are arrayed around this seed as on the continent, except again for a mild deformation.

Inspection of the actual $\mathcal{M}$ set indicates that when the prediction A) is correct, B) is also correct. Moreover, A) can only fail by A) and B) being replaced by the following properties. A') The atom is not a seed and its shape is near circular. B') Other atoms obtained by bifurcation are arrayed around the atom nucleated by μ^*, in the same way as their counterparts are arrayed around the continent, except for a transformation that straightens out the cusp.

Example: the superstable values for $n = 2$ are the roots of $\mu^2 - \mu = 0$, that is $\mu^* = 0$ and $\mu^* = 1$. Near $\mu^* = 1$, $f_z(z, \mu) = (z^2 - \mu)^2 - \mu = z^4 - 2\mu z^2 + \mu^2 - \mu \sim -2z^2 + (\mu - 1)$. This suggests an atom equal to the basic cardioid downsized in the ratio of $\frac{1}{2}$ and translated to the right by 1. But the actual atom is bigger in every direction and happens to be precisely a disc.

One may expect to find that the condition of validity of the milder universality class predictions A) and B) is that $\beta_k \gamma_k$ be large.

A further universality class argument (not well developed as yet) suggests that atoms increasingly removed from the seed of their islands tend to the universal shape.

"Universality class" argument to explain the shape of $\mathcal{F}^*(\mu)$ when μ^* lies in an island. The brutal universality class argument also makes a prediction concerning $\mathcal{F}^*(\mu)$: C) The Julia set $\mathcal{F}^*(\mu^*)$, call it

a "little dragon", obtains by reducing in the ratio β_k the Julia set which the full $\mathcal{F}(z, \mu)$ predicts for the point that lies in the continent and is homologous to μ, namely $\mu' = \beta_k \gamma_k (\mu - \mu^*)$. The milder universality class argument makes the prediction C') The portion of the Julia set $\mathcal{F}^*(\mu)$ near $z > 0$ obtains by reducing in the ratio β_k the Julia set which the full $\mathcal{F}(z, \mu)$ predicts for the point that lies in the continent and is homologous to μ.

Inspection of actual sets $\mathcal{F}^*(\mu)$ indicates that the portion of $\mathcal{F}^*(\mu)$ near $z = 0$ is indeed the little dragon predicted by the milder C'). But the brutal prediction C) gives a quite incorrectly pallid idea of the structure of the whole of $\mathcal{F}^*(\mu)$. This set does *not* reduce to the little dragon near $z = 0$, but is made up of an infinity of mildly deformed replicas of this little dragon.

These replicas from Devil's Stepstones with the same structure we already encountered in the shape of $\mathcal{M}^*$. (Fig. 2e is relative to a case where μ is very close to the nucleus of the cardioid on fig. 2c.) That is, these replicas are strung along a tree. As to this tree's shape, it brings in something entirely foreign to the universality argument. Indeed, this shape is determined by μ, and not merely by the point in the continent that is homologous to μ. This shape varies fairly slowly with μ, and is approximately determined by μ^*.

To introduce an even rougher but useful approximation, let us begin by bringing in the parameter value μ'' corresponding to the tip of the island containing μ^*. This point is homologous to the classical real point μ_∞ at the tip of the continent. In the next section's discussion of the Julia sets $\mathcal{F}^*(\mu)$, we shall see that μ_∞ is among the values of μ for which $\mathcal{F}^*(\mu)$ is a tree having a real interval as spine, and other ribs but no flesh. When μ'' is the tip of an island other than the continent, $\mathcal{F}^*(\mu'')$ is also a tree (though it contains no straight interval). Now we come back to $\mathcal{F}^*(\mu)$: it is found that the replica dragons belonging to this set string along a tree approximated by $\mathcal{F}^*(\mu'')$.

Rough estimates of the counterparts of the ratio δ for bifurcations of order > 2. For the purpose of this subsection, it is best to change the coordinates by replacing the parameter μ by the parameter $\lambda = 1 \pm \sqrt{1 + 4\mu}$. This corresponds to the mapping $z \to f^*(z, \lambda) = \lambda z (1 - z)$. The corresponding transform of the $\mathcal{M}$ set is shown on page 250 of Mandelbrot 1982. The continent is no longer of the same shape as the islands, since, instead of being seeded by a cardioid, it is now seeded by two discs. But this transformed shape has its own assets. A first advantage, which had been ascertained with pen and paper, is that the bifurcation from a stable fixed point to a cycle of period m recurs at the points where either λ or $2 - \lambda$ is of the form $\exp(2\pi i n / m)$, with n an integer less than m.

A second advantage only transpired after the $\mathcal{M}$ set had been computed and could be examined. It was observed that the tips of the "major" sprouts around the circle $|\lambda - 2| = 1$, defined as the sprouts rooted at the points $\exp(2\pi i / m)$, appear to be placed along a larger circle whose diameter begins at $\lambda = 1$ and ends somewhere beyond $\lambda = 3$. This suggests that one perform an inversion of the $\mathcal{M}$ set with respect to $\lambda = 1$, with $\lambda = 3$ remaining fixed. This inversion should yield sprouts placed between parallel lines. Furthermore, the transform of the root of the mth major sprout should lie at a vertical distance from $\lambda = 3$ equal to $2 \tan(\frac{1}{2}\pi - \frac{1}{2} 2\pi/m) = 2 \cotan(\pi/m)$. For large m, this yields $2m/\pi$, i.e., a series of equally spaced points.

The inverted $\mathcal{M}$ set shown in fig. 2f is plotted using very different units along the two axes, so that the graph remains legible yet covers many values of m. The above hunch is confirmed, except for $m = 2$ and 3. That is, an extrapolation from sprouts with a higher value of m would yield a smaller sprout for $m = 2$. Denote by A the height of the inverted sprouts for larger m. Assuming circular atoms, these properties of inversion yield the result that the relative linear size of the sprout of order m is $A \sin^2(\pi/m)[2 - A \sin^2(\pi/m)]^{-1}$. This is roughly the ratio of successive absolute changes in μ between bifurcations of order m, that is, the mth counterpart of the $1/\delta$ ratio of Grossman and Thomae and of Feigenbaum. In fact, this ratio is for all m close to m^{-2}.

3. Discussion of figs. 3a to 3f. The repeller stack for real μ and complex z. Illustrations of the influence of the value of μ on the shape and the topology of the repeller (Julia) set $\mathscr{F}^*(\mu)$

In figs. 3a to 3d, the horizontal coordinates x and y are the real and imaginary parts of z, and the vertical coordinate is μ. The figures represent a stack of Julia sets $\mathscr{F}^*(\mu)$ for μ ranging from $-\frac{1}{4}$ to 2. The goal is to show that the shape of $\mathscr{F}^*(\mu)$ varies continuously, while the topology of $\mathscr{F}(\mu)$ moves around discontinuously. The stack was sliced along the plane $xO\mu$, and the two halves have been separated.

Fig. 3a. Perspective view of the top portion of the repeller stack of Julia sets for real μ (vertical) and complex z (horizontal).

Fig. 3b. Top portion of the "veils" within fig. 3a.

Fig. 3c. Perspective view of the repeller stack minus the web: outside view.

Fig. 3d. Perspective view of the repeller stack: inside cut showing bifurcations.

Fig. 3e. Detail of the top portion of fig. 3c: cut along the plane $y = 10^{-3}$ with x and μ as coordinates.

Computation of the stack. The theory of $\mathscr{F}^*(\mu)$ involves two "proof-of-existence" constructions. The first is used in figs. 1a to 1d. The second consists in tracing the pre-images of the unstable fixed point $z' = \frac{1}{2} + \frac{1}{2}\sqrt{1 + 4\mu}$. This second construction is efficient only if μ is near 0, i.e., when $\mathscr{F}^*(\mu)$ is an uncomplicated loop. In general, either construction requires prohibitively long computer runs to yield an acceptable approximation. For the sake of efficient computation, it was found best to devise several alternative constructions and to use them in combination. After the fact, these programs turned out to help in understanding the facts. Fig. 3a combines some "veils" and a "shell", while fig. 3b represents the veils alone (with fewer stages for the sake of clarity), and figs. 3c and 3d represent the shell alone.

The ribs and veils. For each μ, the backbone of the horizontal section of the stack is the real interval from $]-z', z'[$. The other ribs are the pre-images of $]-z', z'[$ under $f(z)$; fig. 3b, shows them up to order 8. Since $]-z', z'[$ is well known to fail to converge to ∞ under $f(z)$, the backbone and the ribs belong to the set $\mathscr{F}(\mu)$ and can be said to form its "skeleton". The ribs corresponding to different

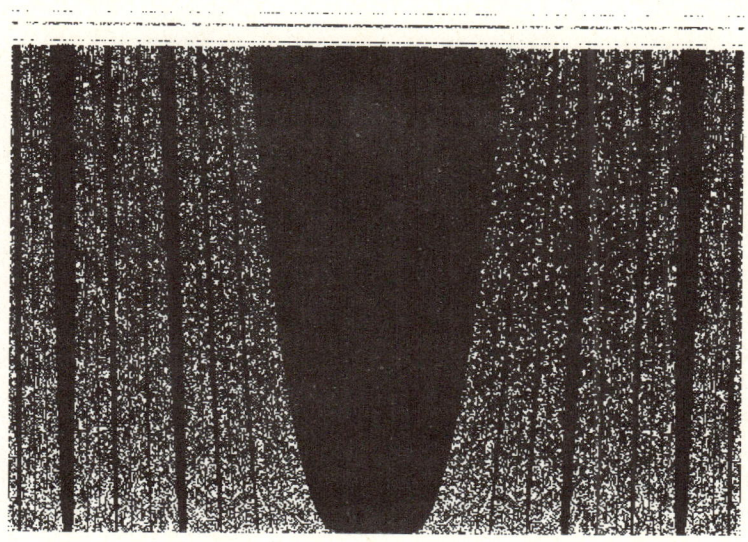

Fig. 3f. Even finer detail of the top portion of fig. 3c: cut along the plane $y = 10^{-10}$ with x and μ as coordinates.

μ's merge together to form a series of "veils". They include a square wall in the plane $y = 0$, and a rounded wall in the plane $x = 0$. Moving through a superstable μ, the veils change from hanging on the rounded wall to hanging on the square wall, or from hanging on a high-order veil to hanging on one of lower order. The pre-images of the unstable fixed point z' are the rib tips. The precise relationship between the ribs' closure and the set $\mathscr{F}(\mu)$ depends on the value of μ.

Superstable μ's. For superstable values of μ, the ribs' closure is a domain, and is identical to $\mathscr{F}(\mu)$. Hence, (by the same anatomical analogy) $\mathscr{F}(\mu)$ can be said to include no proper flesh. The obvious example is $\mu = 0$, when $\mathscr{F}(\mu)$ is the disc of unit radius, and the kth order ribs are segments joining 0 to the points of the form $\exp(2^{-k}\pi i/n)$, with n an integer and $0 < n < 2^{k+1}$.

Chaotic μ's. For the chaotic values of μ, the closure of the ribs is a tree, and is again identical to $\mathscr{F}(\mu)$. The obvious example (though a degenerate one) is $\mu = 2$. Indeed, the set $\mathscr{F}(2)$ and its ribs both reduce to the backbone $[-2, 2]$. To obtain $\mathscr{F}(2)$, it is obviously faster to draw the backbone than to use the proof of existence construction that dots $\mathscr{F}(2)$ with the dense pre-images of $z = 2$.

Whenever μ is close to either a superstable or a chaotic value, the maximal invariant set $\mathscr{F}(\mu)$ is rapidly approximated by only a few levels of ribs. Since $\mathscr{F}(0)$ is simply a disc and all the other superstable or chaotic μ's fall in]1, 2[, it was found best to draw a few levels of ribs for every μ in]1, 2[(anyhow the cost of "unnecessary" computation is less than the cost of determining whether or not the computation was worth performing).

Stable but not superstable μ's. The remaining real values of $\mu \epsilon\,]-\frac{1}{4}, 2[$ are the μ's for which a stable fixed point or a finite period exists but is not superstable. For these values of μ, the set of ribs is not dense in a domain and does not remain a tree even in the limit. To describe the resulting structure of $\mathscr{F}^*(\mu)$, let us mix the previous anatomical metaphor with a botanical one: we can say that for these μ's, the trees' branches join asymptotically to form a "canopy". Clearly, an inspection of the pre-images of the unstable fixed point z' could not distinguish between the cases when the branch tips are disconnected and form dusts, and cases where the branch tips are connected.

The shell. As already mentioned, the proof of existence construction of $\mathscr{F}^*(\mu)$ via the pre-images of z' is efficient when $\mathscr{F}^*(\mu)$ is an uncomplicated loop, that is, for μ near $\mu = 0$. Whenever $\mathscr{F}^*(\mu)$ is even moderately kinky, the cusp shaped kinks remain unfilled even after other portions of $\mathscr{F}^*(\mu)$ have been covered many times over. (For the cognoscendi: the reason is that this method reconstitutes the invariant measure on the Julia set, and this measure can be extraordinarily uneven.)

An efficient graphic method is one that spends roughly equal times on each portion of $\mathscr{F}^*(\mu)$. The shell in fig. 1 was drawn by the following shell generator (Norton, 1982). Each horizontal plane was covered with a square lattice and the position of a lattice point was saved in computer memory whenever, a), its kth iterate falls within a circle of radius 2, and, b), the kth iterate of at least one of its neighbors falls outside of that circle. These points are identified by a search method that starts with the unstable fixed point z', and is very efficient, because the number of wasted points (tested but not saved) is only a small multiple of the number of points that are saved.

Unfortunately, whenever $\mathscr{F}^*(\mu)$ is *very* kinky, as is the case for $\mu \epsilon]1, 2[$, the shell generator misses many points in $\mathscr{F}^*(\mu)$. It misses "A-pieces" that (by definition) are so thin that they squeeze between the lattice points. And it misses "B-pieces" that (by definition) are large but connect to the unstable fixed point z' through A-pieces.

When the shell is examined from the inside, fig. 2d, these A- and B-pieces above do not matter, because they would be hidden anyhow. And inclusion of the ribs would hide the evidence discussed below. When, to the contrary, the shell is examined from the outside, Figs. 3a, b, c, A- and B-pieces do matter. Luckily, many of the points missed by the shell generator are picked up by the rib generator, and the combination of the two yields a sensible idea of the outside shape of the stack. (Note that I had not originally planned to split this figure open, but an inside view was computed by mistake, a *felix culpa*.)

Basic observations. As μ increases, the repeller set varies continuously, but its topological characteristics change back and forth. The largest invariant set varies continuously within each Myrberg interval of μ. However, let μ decrease through the value $-\frac{1}{4}$, or through a value homologous to $-\frac{1}{4}$ within an island. The result is that a continuous canopy becomes punctured, leaving a dust and allowing the flesh to "evaporate". On opposite sides of a chaotic μ, the tree tips combine into canopies in different fashions.

Near $\mu = 0$, the shell is extremely smooth. More generally, as μ moves away from $\mu = 0$, the unsmoothness of $\mathscr{F}^*(\mu)$ increases very slowly and gradually. This led me to conjecture that the fractal dimension of $\mathscr{F}^*(\mu)$ is a very regular function of μ: infinitely differentiable and perhaps analytic. This hunch was proven true in Ruelle 1982.

Interpretation of the bottom portion of the inside view of the shell. Bifurcations. Aside from a clearly visible circle for $\mu = 0$, the most striking feature of this view resides in rows upon rows of protuberances. The lowest row lies at height $\mu = \frac{3}{4}$. For real μ below $\frac{3}{4}$, the mapping $f(z)$, whether in real or complex z, has a stable limit point. For $\mu = \frac{3}{4}$, this stable limit bifurcates into a stable limit cycle of period 2. At the same time, one sees that $\mathscr{F}^*(\mu)$ changes from being a simple loop to being an infinitely knotted one. (The protuberances are denumerable, and have only two limit points: the fixed point z' and $-z'$.) The next highest row of protuberances marks the second bifurcation, and so on.

The reader is surely acquainted with Robert May's tree diagram, May 1976, which maps the variation with μ of the values of x for all the points in the corresponding stable cycle. Were it superposed on fig 2d, this tree would be rooted at the sharp tip to the bottom left, and each branch point would hang on a suitable protuberance. One may extend May's diagram to map the variation with μ

of the "real preorbit of the cycle", defined as the set of real z that eventually fall exactly in the cycle. The resulting preorbit map diagram would be made of many trees, with a branch hanging on every protuberance of fig. 2d.

Interpretation of the top portions of the inside and outside views of the shell. The Myrberg intervals of μ. The top of the stack is characterized by a nearly blank wall, which we know from the veil generating construction. However, this wall is interrupted by mysterious hanging "knobs" forming horizontal strips. Each strip corresponds to a Myrberg interval of values of μ.

Interpretation of some horizontal or vertical sections of the stack. When μ lives in a Myrberg interval, a horizontal section of the stack is a tree formed of Devil's Stepstones. Since the stepstones vary continuously with μ, those which intersect the plane $x = 0$ would form a kind of Devil's Corduroy.

Now take vertical sections. The $y = 0$ section of the whole stack is bounded to the side by the half parabola $\mu = x^2 - 2x$ for $x > \frac{1}{2}$, the half parabola $\mu = x^2 + 2x$ for $x < -\frac{1}{2}$, and the segments {from $(\mu = -\frac{1}{4}, x = -\frac{1}{2})$ to $(\mu = -\frac{1}{4}, x = \frac{1}{2})$} and {from $(\mu = 2, x = -2)$ to $(\mu = 2, x = 2)$}. The bottom of this vase-shaped outline is filled solid and the top is surmounted by Myrberg strips.

Now consider analogous vertical sections of the top of the shell for $y = 10^{-3}$ (fig. 3e) and of a detail for $y = 10^{-10}$ (fig. 3f). Here, one sees a large number of black shapes, each of them a deformed version of the vase shape of the overall stack. When μ is such that the iterates of $f(z, \mu)$ are not chaotic, the intersection of $\mathcal{F}^*(\mu)$ and the wall $y = 0$ is a denumerable set. Its points of accumulation number 2 when μ lies in the continent, and are themselves denumerable when μ lies in an island. When μ is chaotic, the intersection of $\mathcal{F}^*(\mu)$ and the wall is an interval.

Acknowledgment

The programs to generate the illustrations in this paper, many of them elaborate, are due to my colleague V.A. Norton.

References

P. Collet and J.P. Eckman, Iterated Maps on the Interval as Dynamical Systems, Birkhauser (Boston, 1980).

A. Douady and J. Hubbard, Comptes Rendus (Paris) 294-I (1982) 123–126.

P. Fatou, Sur les solutions uniformes de certaines équations fonctionnelles, Comptes Rendus (Paris) 143 (1906) 546–548.

P. Fatou, Sur les équations fonctionnelles, Bull. Société Mathématique de France 47 (1919) 161–271; 48 (1920) 33–94; 48 (1920) 208–314.

G. Julia, Mémoire sur l'itération des fonctions rationnelles, J. de Mathématiques Pures et Appliquées 4 (1918) 47–245. Reprinted (with related texts) in: Julia Oeuvres (1968) I 121–319.

B.B. Mandelbrot, Fractal aspects of the iteration of $z \to \lambda z(1-z)$ for complex λ and z. Nonlinear Dynamics, R.H.G. Helleman, ed., Annals of the New York Academy of Sciences 357 (1980) 249–259.

B.B. Mandelbrot, The Fractal Geometry of Nature (Freeman, San Francisco, 1982).

B.B. Mandelbrot, Forthcoming article in Scientific American (1983a).

B.B. Mandelbrot, Forthcoming monograph (1983b).

R. May, Simple mathematical models with very complicated dynamics, Nature 261 (1976) 459.

V.A. Norton, Generation and display of geometric fractals in 3-D, Computer Graphics 16 (1982) 61–67.

D. Ruelle, Repellers for real analytic maps, Ergodic theory and dynamical systems 2 (1982) 99–107.

J.L. Walsh, Interpolation and approximation by rational functions in the complex domain, American Mathematical Society Colloquium Publication, number 20 (1956).

THE TWIST MAP, THE EXTENDED FRENKEL–KONTOROVA MODEL AND THE DEVIL'S STAIRCASE

Serge AUBRY*

Center for Nonlinear Studies, Los Alamos National Laboratory, Los Alamos, New Mexico 87545, USA

This paper reviews exact results which we obtained on the discrete Frenkel–Kontorova (FK) model and its extensions, during the past few years. These models are associated with area preserving twist maps of the cylinder (or a part of it) onto itself. The theorems obtained for the FK model thus yields new theorems for the twist maps. We describe the exact structure of the ground-states which are either commensurate or incommensurate and assert the existence of elementary discommensurations under certain necessary and sufficient conditions. Necessary conditions for the trajectories to represent metastable configurations, which can be chaotic, are given. The existence of a finite Peierls–Nabarro barrier for elementary discommensurations is connected with a property of non-integrability of the twist map. We next prove that the existence of KAM tori corresponds to "undefectible" incommensurate ground-states and give a theorem which asserts that when the phonon spectrum of an incommensurate ground-state exhibits a finite gap, then the corresponding trajectory is dense on a Cantor set with zero measure length. These theorems, when applied to the initial FK model, allow one to prove the existence of the transition by "breaking of analyticity" for the incommensurate structures when the parameter which describes the discrepancy of the model from the integrable limit varies. These theorems also allow one to obtain a series of rigorous upper bounds for the stochasticity threshold of the standard map which for the fifth order approximation already approaches within 25% the value which is numerically known. Finally, we describe a theorem proving the existence of a devil's staircase for the variation curve of the atomic mean distance versus a chemical potential, for certain properties of the twist map which are generally satisfied.

1. Introduction – Description of the models

Up to now, applications of the properties of nonintegrable maps and particularly the possibility that they have to exhibit chaotic behavior, have been mostly devoted to physical systems which are really dynamical. However, they also have interesting applications for understanding *static* structural properties of condensed matter. The aim of this paper is to describe some of these applications. Instead of giving a detailed report of our talk (which would be too long), we mostly focus on the rigorous results which we obtained. The reader can refer to [16] where the physical applications of this work have been focused at the expense of a precise mathematical description which as a counterpart is given here.

* On leave of "Laboratoire Léon Brillouin" Orme des Merisiers 91191-Gif-sur-Yvette Cedex, France and D.R.P. Université Pierre et Marie Curie, Paris France.

We initially studied the Frenkel–Kontorova model [2] (noted hereafter FK model). However, due to difficulties in the publication of these early works, these results have only been published in parts and with incomplete proofs in journals of limited audience. We take the opportunity of this paper to recall, to clarify and to emphasize some particularly important points which apparently have been ignored or misunderstood in the literature, but which already gave answers to certain presently controversial questions (for example on the existence of chaotic ground states in the FK model). The exact results which we obtained on its ground states and on its metastable states, also turned out to have important applications for the standard map. We recently improved and extended these results to a larger class of models corresponding to twist maps and for which we obtained interesting new theorems. In this paper, we describe them in the most recently improved form, but we do not include their proofs which are generally long and complicated. However, we detail some corollaries which have immediate applications with their proofs when they are simple. The first parts of the most important proofs are submitted to publication (Refs 6 and 7). The second part (ref. 8) is still in preparation.

This study is essentially analytical and yields only qualitative results of topological nature. However, explicit rigorous calculations can be carried out on a particular but pathological model with the form [1] where $V(x)$ is replaced by a piecewise parabola periodic potential [2, 3, 7]. We also performed few numerical calculations mostly for the illustration of the theory (figs. 1 and 4). Some recent numerical calculations [14] have also been performed on the transition by breaking of analyticity in order to explicit critical quantities and critical exponents.

Let us describe now, the Frenkel–Kontorova [1] model, in its original version. It corresponds to a chain of elastically coupled atoms submitted to a periodic potential

$$\phi(\{u_i\}) = \sum_i [\lambda V(u_i) + W(u_{i+1} - u_i) - \mu(\mu_{i+1} - \mu_i)]. \quad (1a)$$

The atom i is at abscissa u_i. The coupling potential W is harmonic,

$$W(u_{i+1} - u_i) = \tfrac{1}{2}(u_{i+1} - u_i)^2 \quad (1b)$$

(The energy unit is chosen such that the coupling constant in (1b) be one). The periodic potential V with period $2a$ is sinusoidal.

$$V(u_i) = \tfrac{1}{2}\left(1 - \cos\frac{\pi u_i}{a}\right). \quad (1c)$$

λ, the amplitude of this potential, is an adjustable parameter. The chain is subject to a tensile force μ (or a chemical potential) which allows one to change the distance between neighboring atoms in the absence of periodic potential ($\lambda = 0$). The configurations $\{u_i\}$ of model (1) which have the most physical interest are those which correspond to the ground-states for various boundary conditions or with free ends and those which correspond to metastable configurations. All these configurations are solutions of the equation

$$\frac{\partial \phi}{\partial u_i} = (-u_{i+1} - u_{i-1} + 2u_i) + \frac{\lambda \pi}{2a}\sin\frac{\pi u_i}{a} = 0, \quad (2)$$

but this equation also exhibits many other unphysical solutions (in our physical context) which correspond to unstable configurations. (Note that the parameter μ disappears when writing eq. (2)).

This equation can be recursively solved [2] by iterating the area-preserving two-dimensional map T_T which maps the point P_i with coordinates (u_i, u_{i-1}) onto the point P_{i+1} with coordinates (u_{i+1}, u_i). From eq. (2), we get

$$P_{i+1} = T_T(P_i) = \left(2u_i + \frac{\lambda \pi}{2a}\sin\frac{\pi u_i}{a} - u_{i-1}, u_i\right). \quad (3)$$

This map can be folded onto a torus $[0, 2a[\times [0, 2a[$ by defining

$$\theta_i = u_i (\mathrm{mod}\, 2a). \quad (4)$$

It is now well known that such a map exhibits many kinds of trajectories which are either chaotic or not. Figs. 1 shows some trajectories for $\lambda = 0.15$ (fig. 1a), $\lambda = 0.20$ (fig. 1b) and $\lambda = 0.25$ (fig. 1c). About 1000 iterated points have been plotted from each initial point. These figures exhibit trajectories which are either rotating on one or several smooth closed curves or are chaotic. The behavior of two-dimensional area-preserving maps has been intensively studied particularly during the past few years and we refer for example to the important work of Greene [9] on this subject.

By the change of variables

$$p_i = u_{i+1} - u_i \quad (5)$$

this map becomes the well-known standard map

which have been studied as a model for certain dynamical systems (for example the motion of an ion in a plasma)

$$(p_{i+1}, \theta_{i+1}) = T_s(p_i, \theta_i)$$
$$= \left(p_i + \frac{\lambda\pi}{2a}\sin\frac{\pi\theta_i}{a}, \; p_{i+1} + \theta_i\right). \quad (6)$$

This standard map, which maps the cylinder $[0, 2a[$ onto itself, is a prototype for the twist maps of the annulus onto itself (see ref. 5). An annulus is defined as the part of the cylinder (p, θ) which is limited by two circular sections $p = \rho_0$ and $p = \rho_1$. A twist map is a map $P_{i+1} = T(P_i)$ of the annulus onto itself which satisfies

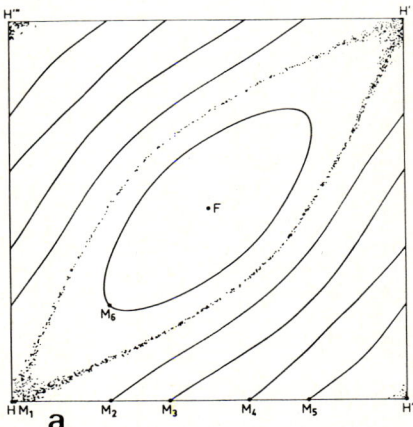

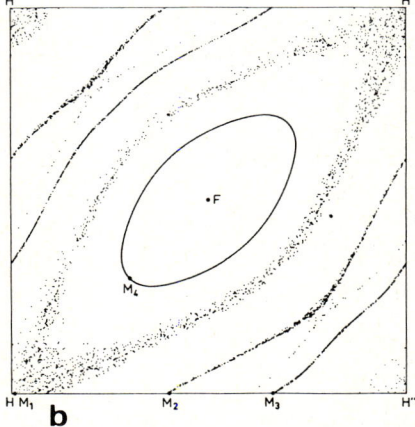

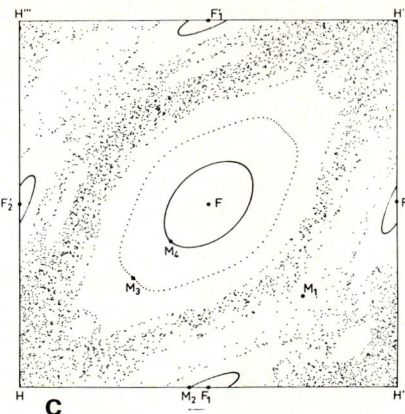

Fig. 1. Map of the transformation T_T in (6) showing the trajectories of the initial points M plotted on the figures for $\lambda = 0.15$ (a); $\lambda = 0.2$ (b); and $\lambda = 0.25$ (c). For each initial M_i, about 1000 points of the trajectory $T_s^n(M_i)$ $0 \leq n \leq 1000$ have been plotted. For $\lambda = 0.15$, most trajectories lie on smooth closed curves (KAM tori) except the trajectory generated by M_1 which maps a chaotic cloud of points in a narrow area. For $\lambda = 0.2$ this chaotic area becomes much wider while for $\lambda = 0.25$ this chaotic area fills most of the map except in some isolated islands.

$$\begin{pmatrix} p_{i+1} \\ \theta_{i+1} \end{pmatrix} = T\begin{pmatrix} p_i \\ \theta_i \end{pmatrix} = \begin{pmatrix} T_1(p_i, \theta_i) \\ T_2(p_i, \theta_i) \end{pmatrix}, \quad (7)$$

where

1) T_1 and T_2 are differentiable in p and θ with continuous derivatives. T is area preserving and invertible;

2) T_1 and T_2 have period 2π with respect to the variable θ,

3) for any fixed value of θ, $T_2(p, \theta)$ is a strictly monotonic function of p;

4) The two boundaries of the annulus are invariant by T which also preserves their orientation.

This standard map T_s in (6) allows one to represent any stationary configuration of model (1) modulo $2a$ (which can be either physically stable or unstable) by a trajectory in the dynamical system with the discrete time i and the evolution operator T_s. But let us emphasize again, that our specific problem is not to find the properties of arbitrary trajectories, but to find those which correspond to physically stable configurations. Let us also emphasize that the physical stability of a configuration must not be confused with the stability in the map of the associated trajectory.

Although our theory was initially developed for a slightly generalized form of model (1), we recently found that the method which we used can be extended with few changes to a wider class of one dimensional models with first neighbor interactions. The map associated with these models by extremalizing their energy turns out to include the class of twist maps above defined in (7) but our map T is not necessarily restricted to an annulus. The energy of this class of model (or variational form) which contains model (1) as a particular case is

$$\phi(\{u_i\}) = \sum_i L(u_{i+1}, u_i), \tag{8a}$$

where $L(x, y)$ is an arbitrary function of the two variables x and y which has the following properties:

1) $L(x, y)$ is continuous with a lower bound;
2) $L(x, y)$ is diagonally periodic with period $2a$, that is for any x and y

$$L(x + 2a, y + 2a) = L(x, y); \tag{8b}$$

3) the crossed second derivative of $L(x, y)$ is strictly negative: that is, there exists a positive constant C such that for any x and y

$$-\frac{\partial^2 L}{\partial x \partial y}(x, y) > C > 0. \tag{8c}$$

By setting $p_i = \partial L(u_i, u_{i-1})/\partial u_i$, the conjugate variable of u_i, the equation $\partial \phi/\partial u_i = 0$ generates an area-preserving map $(p_{i+1}, \theta_{i+1}) = T(p_i, \theta_i)$ with the same properties as the twist map (7) except that it maps the cylinder (or a part of it) onto itself and not necessarily an annulus onto itself.

Our theory [6] introduces a distinction between the concept of minimum energy configuration (m.e. configuration) and the concept of ground states. The reason for this distinction is that under certain boundary conditions, for example the constraint

$$\lim_{N-N' \to \infty} u_N - u_{N'} = 2a, \tag{10}$$

the configuration of model (1) which satisfies this condition and which has the minimum energy is in fact a defect (a soliton in the continuum limit) and is not usually considered as a ground state. The set of minimum energy configurations is defined as the set of all possible limits of ground states of finite systems with arbitrary boundary condition at u_N and $u_{N'}$ when N goes to $+\infty$ and N' goes to $-\infty$. This set of m.e. configurations is called $\mathcal{Q}$. We keep the name ground state for m.e. configurations which are represented by recurrent trajectories in the associated map. (A recurrent trajectory returns to any neighborhood of any point of the trajectory.) This definition turns out to correspond to the usual intuition of a ground state (see ref. 6 for more details). This set is called $\mathcal{G}$ and is included in $\mathcal{Q}$.

We found the topological structure of the sets $\mathcal{Q}$ and $\mathcal{G}$ without any explicit calculation of m.e. configuration. These results are described in the following section 2. Before the description of these results let us briefly explain the general ideas which allow one to find a method which works when some topological and symmetry properties are satisfied.

1) We note that the set $\mathcal{Q}$ is closed for the weak topology: that is, the limit of a convergent sequences of m.e. configurations is a m.e. configuration. This property is only a consequence of the fact that the energy of the model depends continuously on the atomic positions.

2) We note the existence of a group of transformation G' which transforms a configuration into other configurations with the same energy. Particularly $\mathcal{Q}$ and $\mathcal{G}$ are invariant by G'. This group G' is defined by the transformations $g_{n,p}$ which transforms a configuration $\{u_i\}$ into

$$g_{n,p}(\{u_i\}) = \{u_{i+n} - 2pa\}, \tag{11}$$

where n and p are two arbitrary integers. This

property is a consequence of the homogeneity of the model (all the atoms play an identical role) and of the periodicity condition (10b).

3) Condition (10c) allows one to prove the fundamental lemma which is:

Fundamental lemma. Let $\{u_i\}$ and $\{v_i\}$ be two m.e. configurations, then the sequence $(u_i - v_i)$ has at most one node for $-\infty < i < \infty$ (i.e. one change of sign). If the two configurations $\{u_i\}$ and $\{v_i\}$ are asymptotic for $i \to \pm \infty$, the point at infinity must be considered as a node.

Considering a m.e. configuration $\{u_n\}$, the group G' allows one to construct an infinite number of m.e. configurations from which the limits are also m.e. configurations. These m.e. configurations can be compared one with another with the above fundamental lemma, which yields inequalities. By combining these methods in a sequence of proofs which is quite long and complicated [6], one finds the exact topological structure of the set $\mathcal{Q}$ and $\mathcal{G}$ now described.

2. Topological structure of the set of m.e. configurations and of ground state in the extended FK model (proofs in ref. 6b)

We first found

Theorem 1. For any m.e. configuration in $\mathcal{Q}$, the limit

$$l = \lim_{N-N' \to \infty} \frac{1}{N-N'}(u_N - u_{N'}) \quad (12)$$

is defined and does not depend on the way by which $(N - N')$ goes to infinity.

Conversely, for any value of l, there exists a m.e. configuration $\{u_i\}$ in $\mathcal{Q}$ such that the limit (11) is l.

The corresponding trajectory in the twist map has the winding number $l/2a$ which is its mean number of revolutions around the cylinder per iteration of the map. Because of this theorem, we can split the set $\mathcal{Q}$ (and $\mathcal{G}$) into subsets $\mathcal{Q}_l$ (and $\mathcal{G}_l$) which are defined as the configurations in $\mathcal{Q}$ (and $\mathcal{G}$) with winding number $l/2a$ and such that

$$\mathcal{Q} = \bigcup_l \mathcal{Q}_l \quad \text{and} \quad \mathcal{G} = \bigcup_l \mathcal{G}_l, \quad (13)$$

with for any $l \neq l'$

$$\mathcal{Q}_l \cup \mathcal{Q}_{l'} = \emptyset \quad \text{and} \quad \mathcal{G}_l \cup \mathcal{G}_{l'} = \emptyset. \quad (14)$$

The two following theorems describe the structure of $\mathcal{Q}_l$ and $\mathcal{G}_l$, first for $l/2a$ an irrational number and next for $l/2a$ a rational number:

Theorem 2. Let $l/2a$ be an irrational number then:

1) The set $\mathcal{Q}_l$ of m.e. configurations of the above defined extended FK models, is nonempty and is totally ordered: that is, if $\{u_i\} \neq \{v_i\}$ both belonging to $\mathcal{Q}_l$, then for all n either

$$u_n < v_n, \quad (15a)$$

or

$$u_n > v_n. \quad (15b)$$

2) The whole set $\mathcal{G}_l$ of ground state configurations of model (8) ($\mathcal{G}_l \subset \mathcal{Q}_l$) is nonempty and can be parametrized with one or two hull functions $f(x)$ which are strictly increasing. a) When $f(x)$ is continuous, a unique function allows one to parametrize the full set $\mathcal{G}_l$. b) When $f(x)$ is discontinuous, two determinations $f^+(x)$ and $f^-(x)$ are necessary to parametrize $\mathcal{G}_l$. $f^+(x)$ and $f^-(x)$ correspond the right continuous and the left continuous determination of the same discontinuous, strictly increasing function. In other words, we have

$$\lim_{\substack{\delta \to 0 \\ \delta > 0}} f^-(x + \delta) = f^+(x) \quad (16a)$$

and

$$\lim_{\substack{\delta \to 0 \\ \delta < 0}} f^+(x + \delta) = f^-(x). \quad (16b)$$

c) When $f^\pm(x)$ is discontinuous at x_0, it is also discontinuous at the points $x_0 + hl + 2ka$, where h and k are arbitrary integers. As a result, the set of discontinuity points of $f^\pm(x)$ is dense on the real axis.

d) The functions $g^\pm(x) = f^\pm(x) - x$ are periodic with the period $2a$ of $L(x, y)$.

e) Finally, for any ground state which belongs to $\mathcal{G}_l$, there exists a phase α and a determination of $f: f^+$ or f^- when f is discontinuous (the determination is unique when f is continuous) such that

$$u_n = f^\pm(nl + \alpha) = nl + \alpha + g^\pm(nl + \alpha). \tag{17}$$

Conversely, any configuration $\{u_n\}$ defined by (17) for an arbitrary phase and one of the two determinations f^+ or f^- when f is discontinuous, is a ground state in $\mathcal{G}_l$.

This hull function $f(x)$ obviously depends on $(l/2a)$. A configuration $\{u_n\}$ as defined by (17) is called incommensurate. It describes a crystal structure of atoms at distance l which is modulated by the function g with the period $2a$ incommensurate with l. Let us now describe the structure of $\mathcal{Q}_l$ and $\mathcal{G}_l$ for $(l/2a)$ rational.

Theorem 3. Let $l/2a = r/s$ be a rational number (r and s are two irreducible integers). Then

1) The set $\mathcal{G}_l$ is nonempty and is totally ordered. (i.e. for $\{u_i\} \neq \{v_i\}$ in $\mathcal{G}_l$ then for all n we have either (15a) or (15b)).

2) For any $\{u_i\}$ in $\mathcal{G}_l$, we have for all n

$$u_{n+s} = u_n + 2ra. \tag{20}$$

(This ground-state is called commensurate. It has a unit cell of s atoms with length $2ra$.)

3) When the set $\mathcal{G}_l$ is continuous, which means that it can be parametrized by continuous functions $\{u_n(\alpha)\}$ where α is a continuous parameter which varies from $-\infty$ to $+\infty$ (for example u_0), then $u_n(\alpha)$ is a continuous strictly increasing function of α and we have

$$\mathcal{Q}_l = \mathcal{G}_l. \tag{21}$$

4) When $\mathcal{G}_l$ is a discontinuous set, it is closed and there exists for each discontinuity a couple of ground states $\{v_n^-\}$ and $\{v_n^+\}$ in $\mathcal{G}_l$ such that there exist no ground states in $\mathcal{G}_l$, $\{v_n\}$ which satisfy for all n

$$v_n^- < v_n < v_n^+. \tag{22}$$

Then, there exists a m.e. configuration $\{u_n\}$ in $\mathcal{Q}_l$ such that for all n

$$v_n^- < u_n < v_n^+ \tag{23a}$$

and

$$\lim_{n \to +\infty} (v_n^+ - u_n) = 0, \tag{23b}$$

$$\lim_{n \to -\infty} (u_n - v_n^-) = 0. \tag{23c}$$

Such a configuration $\{u_n\}$ is called an "advanced elementary discommensuration". There also exist m.e. configurations $\{u_n\}$ in $\mathcal{Q}_l$ called "delayed elementary discommensuration" such that for all n

$$v_n^- < u_n < v_n^+ \tag{24a}$$

and

$$\lim_{n \to -\infty} (v_n^+ - u_n) = 0, \tag{24b}$$

$$\lim_{n \to \infty} (u_n - v_n^-) = 0. \tag{24c}$$

5) The union of $\mathcal{G}_l$ and of the set of advanced elementary discommensurations in $\mathcal{Q}_l$ is called $\mathcal{Q}_l^+$. Similarly, the union of $\mathcal{G}_l$ with the set of delayed elementary discommensurations is called $\mathcal{Q}_l^-$. Then $\mathcal{Q}_l^+$ and $\mathcal{Q}_l^-$ are totally ordered sets (with the definition given in (15)) and we have

$$\mathcal{Q}_l = \mathcal{Q}_l^+ \cup \mathcal{Q}_l^-, \tag{25a}$$

$$\mathcal{G}_l = \mathcal{G}_l^+ \cup \mathcal{G}_l^-. \tag{25b}$$

This theorem proves that when the boundary condition (11) is satisfied with $l/2a$ a rational number, then the ground state is indeed commensurate and satisfies (20). It can be obtained by finding the absolute minimum of the energy per unit cell with this condition (20). There generally exist s minima (modula $2a$) (r and s are irreducible integers) because of the invariance of the energy per unit cell under the s cyclic permutations $\{u_n\} \to \{u_{n+p}\}\, p = 1, 2, \ldots s$. There may also exist ks minima (for example $k = 2$ is possible if the model has a symmetry by reflexion) or also a continuum of minima but these two situations are exceptional.

In this theorem, we distinguish two different situations. The situation where $\mathcal{G}_l$ is a continuous set is found for example in the case of integrable maps. It corresponds to the absence of locking of the commensurate configuration by the lattice and can be considered as exceptional. The situation where $\mathcal{G}_l$ is discontinuous turns out to be the most general case. Then, the lattice locking does not vanish. This is a necessary and sufficient condition to have discommensurations (see fig. 2). These are called elementary discommensurations because they correspond to the minimum energy of the system for certain boundary conditions similar to (10). They were already known as solitons in continuous models for incommensurate structures [10]. Thus, we also prove their existence (under certain conditions) in a discrete model for any commensurability ratio r/s.

Since any twist map (7) corresponds to a variational form (8) for some choice of $L(x, y)$, these theorems predict the existence of certain trajectories in the twist map with particular properties as a corollary of theorems 1, 2 and 3:

Let ω_0 be the winding number of T (defined by (7)) on the invariant circle $p = \rho_0$ and ω_1, its winding number on the invariant circle $p = \rho_1$. In order to fix the ideas, we assume that $\omega_0 < \omega_1$. Then for any $\omega_0 \leq \omega \leq \omega_1$, there exists a trajectory with winding number ω. If ω is an irrational number, this trajectory is quasi-periodic (in an extended sense because function f in (17) is not necessarily continuous) and is dense either on a continuous closed loop or on a Cantor set which is parametrized by the function $f^\pm$ in (17). (This result has also been recently proved by Mather [24].) If ω is rational number r/s, it is a periodic cycle $\{F_i\} = 1, \ldots, s$ with period s ($T^s(F_i) = F_i$). When the set of periodic cycles with period s of T does not form a closed continuous loop around the cylinder, (unlike certain integrable twist maps) there exist initial points h which by applying the transformations T^{sn} are asymptotic to one of the points F_i of the periodic cycle for $n \to -\infty$ and to another point F_j of the same periodic cycle for $n \to +\infty$. (These points F_i and F_j are in consecutive

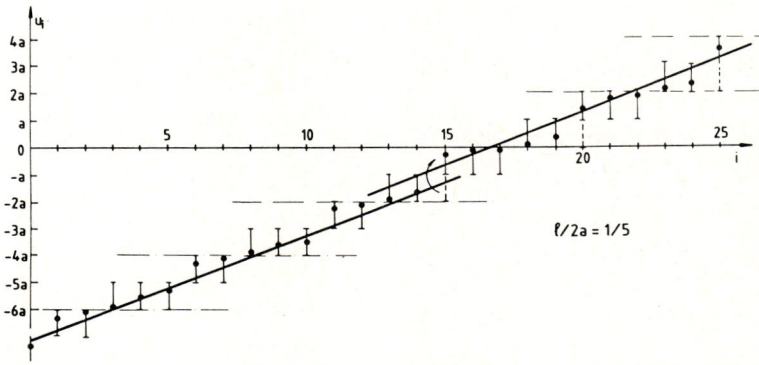

Fig. 2. Scheme of an advanced elementary discommensuration $\{u_i\}$ for $l/2a = 1/5$. u_i is plotted as a function of i. The phase shift, $2a/5$, occurs in the region $14 < i < 15$. Far from this region the configuration is commensurate.

order with the order relation given in (23)). Such points are called in mathematics "heteroclinic points". This point h belongs to the intersection of two curves (see fig. 3): the dilating sheet W_i^+ of F_i which is the set of points which converge to F_i by iterating the transformation T^s and the contracting sheet W_j^- of F_j which is the set of points which converge to F_j by T^s. Let us note that the point F_i must be linearly unstable with respect to T^s (that is, the Jacobian matrix of T^s at F_i has a real eigenvalue with modulus larger than one) in order to be allowed to apply a theorem which predicts the existence of a dilating sheet W_i^+ (ref. 11). It may happen, although F_i is unstable with respect to the operator T^s, that its Jacobian matrix has an eigenvalue with modulus one. Then, a proof of the existence of a continuous dilating or contracting sheet is necessary. (We have not yet performed this proof).

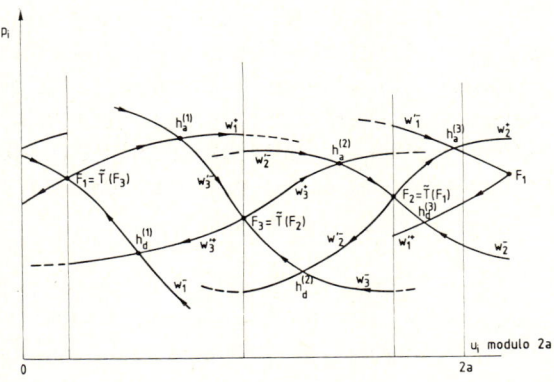

Fig. 3. Scheme showing the initial points of the trajectories in the twist map which represent the commensurate ground-state for $l/2a = 2/3$: $F_1 = T(F_3)$, $F_2 = T(F_1)$, $F_3 = T(F_2)$. (These points from a periodic cycle with period 3). The beginning of the dilating sheet of F_1, F_2 and F_3 are also represented with only one intersection point one with each other. The arrow indicates the direction of the motion of a point of the sheet by the twist map. Thus it indicates if the sheet is dilating or contracting. The trajectories generated by the points $h_a^{(1)}$, $h_a^{(2)}$ and $h_a^{(3)}$ correspond to advanced elementary discommensurations. Those generated by $h_d^{(1)}$, $h_d^{(2)}$ and $h_d^{(3)}$ corresponds to delayed elementary discommensurations.

3. Metastable configurations and their corresponding trajectories in the twist map

Theorems 1, 2 and 3 definitely prove that although the equations $\partial \phi / \partial u_i = 0$ exhibit many chaotic solutions, the ground state of model (8) is never chaotic, whatever the boundary condition (12) is, and in particular it has no entropy. Nevertheless, as we already pointed out several years ago in ref. 2 (for the simpler model (1)) model (8) may exhibit for certain boundary conditions (11) metastable configurations which are chaotic but have more energy per atom than the real ground states. The ground state which is obtained for the same boundary conditions is called defectible, while if it is the unique metastable configuration, it is called undefectible.

In this section, we investigate some of the necessary properties of the trajectories in the twist map which correspond to metastable configurations. We also investigate the linear stability of the trajectories in the map and show that this concept of stability is not connected to the physical stability of the corresponding configuration although these two concepts have sometimes been confused in the literature. Let $\{p_i, u_i\}$ be a trajectory of the twist map T. The corresponding configuration $\{u_i\}$ is a solution of the equation $\partial \phi / \partial u_n = 0$,

$$\frac{\partial L}{\partial u_n}(u_{n+1}, u_n) + \frac{\partial L}{\partial u_n}(u_n, u_{n-1}) = 0. \tag{26}$$

By definition, the physical stability (called metastability) of this configuration $\{u_n\}$ means that the second order expansion of the energy (8) with respect to small atomic displacements $\{\delta_n\}$,

$$\delta \phi = \tfrac{1}{2} \sum_n \left[\frac{\partial^2 L(u_{n+1}, u_n)}{\partial u_n^2} + \frac{\partial^2 L(u_n, u_{n-1})}{\partial u_n^2} \right] \delta_n^2 + 2 \frac{\partial^2 L(u_{n+1}, u_n)}{\partial u_{n+1} \partial u_n} \delta_{n+1} \delta_n, \tag{27}$$

is a positive quadratic form in $\{\delta_n\}$. This condition is equivalent to the positivity of the phonon frequencies squared obtained from the time Fourier transform of the small motion equations,

$$\omega^2 \delta_n = -\frac{\partial \delta \phi}{\partial \delta_n}$$

$$= \frac{\partial^2 L(u_{n+1}, u_n)}{\partial u_{n+1} \partial u_n} \delta_{n+1} + \frac{\partial^2 L(u_n, u_{n-1})}{\partial u_n \partial u_{n-1}} \delta_{n-1}$$

$$+ \left[\frac{\partial^2 L(u_{n+1}, u_n)}{\partial u_n^2} + \frac{\partial^2 L(u_n, u_{n-1})}{\partial u_n^2} \right] \delta_n \quad (28)$$

(the atoms have a unit mass and δ_n also denotes the time Fourier transform of $\delta_n(t)$).

For each value of ω, this equation can be recursively solved from the knowledge of δ_0 and δ_1. Then the vector (δ_{n+1}, δ_n) is a linear function of the vector (δ_n, δ_{n-1}). It is convenient to set the new variables

$$\pi_n = \frac{\partial^2 L(u_n, u_{n-1})}{\partial u_n^2} \delta_n + \frac{\partial^2 L(u_n, u_{n-1})}{\partial u_n \partial u_{n-1}} \delta_{n-1} \quad (29)$$

in order to find a linear relation

$$\begin{pmatrix} \pi_{n+1} \\ \delta_{n+1} \end{pmatrix} = (\mathbf{J}(p_n, u_n) - \omega^2 \mathbf{K}(p_n, u_n)) \begin{pmatrix} \pi_n \\ \delta_n \end{pmatrix}, \quad (30a)$$

where $\mathbf{J}(p_n, u_n)$ is the Jacobian matrix of the twist map T at (p_n, u_n) and

$$\mathbf{K}(p_n, u_n) = \frac{-1}{\partial^2 L(u_{n+1}, u_n)/\partial u_{n+1} \partial u_n}$$

$$\times \begin{pmatrix} \frac{\partial^2 L(u_{n+1}, u_n)}{\partial u_n^2} & 0 \\ 0 & 1 \end{pmatrix} \quad (30b)$$

When $\omega = 0$, eq. (28) to be solved, only needs to perform the product of Jacobian matrices:

$$\mathbf{M}_n(p_0, u_o) = \mathbf{J}(p_n, u_n) \mathbf{J}(p_{n-1}, u_{n-1}) \ldots \mathbf{J}(p_0, u_o). \quad (31)$$

Otherwise, the behavior of $\mathbf{M}_n$ for n going to infinity just determines the Lyapounov exponent γ of this trajectory by the definition

$$\gamma = \lim_{n \to \infty} \frac{1}{2n} \ln \|\mathbf{M}_n^t \mathbf{M}_n\|. \quad (32)$$

When γ is zero, the trajectory $\{p_n, u_n\}$ is called linearly stable. Because this matrix product does not diverge (or slowly diverges) we can prove [8] that the zero frequency belongs to the phonon spectrum given by eq. (28).

When γ is not zero, the trajectory $\{p_n, u_n\}$ is unstable with respect to the initial conditions. Then, the zero frequency mode may not belong to the phonon spectrum, but if it does belong, the corresponding eigenstates in the neighborhood of the zero frequency are necessarily exponentially localized.

As a result, one sees that the linear stability of the trajectory $\{p_n, u_n\}$ gives information only on the spectrum of the small motion equation at the frequency zero, but no information on the physical stability of the corresponding configurations. Indeed, our previous papers exhibit examples of trajectories which are either linearly stable or unstable in the twist map and for which the corresponding configurations are either stable or unstable or vice-versa (see for example ref. 21). However, one can use the recursive relation (30) in order to find a necessary condition for the physical stability of the stationary configurations satisfying (26). Because of the condition (8c) the off diagonal terms of the Jacobi matrix (A Jacobi matrix is a symmetric tridiagonal matrix) defined by the eq. (28) or by the quadratic form (27), are all negative. Then, it can be proved [8] that

Theorem 4. A trajectory in a two-dimensional map (associated to a one-dimensional model with first neighbour interactions) corresponds to a metastable configuration if and only if, any sequence δ_n $(-\infty < n < +\infty)$ generated from any arbitrary initial condition (π_0, δ_0) by the product along this trajectory of the Jacobian matrices (3), has at most one change of sign.

Note that this theorem also applies to model (8) when the periodicity condition (8b) is dropped. The map is then on the two-dimensional plane and not on the cylinder. The proof of this theorem is an application of the theory of Jacobi matrices.

(See for example ref. 13) (A well-known corollary of this theory, asserts that the eigenenergies of a one-dimensional Schrödinger equation are in the same order as the number of nodes of the corresponding eigenstates.) This theorem has straightforward applications for predicting the physical instability of the configurations corresponding to certain trajectories. We have with the same hypothesis as in theorem 4.

Corollary of theorem 4. The configurations corresponding
  1) to a periodic cycle which is elliptic;
  2) to a periodic cycle which is hyperbolic (or parabolic) with reflexion; and
  3) to trajectories dense on one or several differentiable tori (KAM tori) which are homotopic to zero are physically unstable.

(A more complicated proof of this corollary was already given in ref. 21, appendices A and B. This result was also given in ref. 2.) For its proof, we first examine the case of a periodic cycle of the twist map with period s. We consider the sequence of matrices $\mathbf{M}_{ks}$ in (31) which is equal to $\mathbf{M}_s^k(p_0, u_0)$. When the periodic cycle is elliptic, the matrix $\mathbf{M}_s$ is by definition equivalent to a rotation (in nonorthogonal axis), then the vector (π_{ks}, δ_{ks}) is rotating on an ellipse around the origin. Therefore, the sequence δ_{ks} (and also δ_n) has infinitely many changes of sign which by theorem 4 proves the first assertion of the corollary. When the periodic cycle is hyperbolic with reflexion, by definition, the matrix $\mathbf{M}_s$ has two real negative eigenvalues with product 1. If (π_0, δ_0) is chosen to be an eigenvector of $\mathbf{M}_s$, the signs of δ_{ks} change for each consecutive k, because the corresponding eigenvalue is negative. The sequence δ_{ks} has then infinitely many changes of sign which proves the second assertion of the corollary.

When the trajectory $\{p_n, u_n\}$ is rotating and dense on a set of s differentiable tori (KAM tori) which are homotopic to zero (which means that they can be shrunk continuously into a point on the manifold of the map), the configuration $\{u_n\}$ can be parametrized with s periodic differentiable functions $g_1, g_2, \ldots, g_s$ with period 2π,

$$u_{ks+p} = g_p(k\theta + \alpha), \tag{33}$$

where α is arbitrary phase and θ is the average of the angle of rotation of T^s on each torus which is incommensurate with 2π. By inserting (33) in (26) and by differentiating with respect ot the phase α, it comes out that

$$\delta_{ks+p} = g'_p(k\theta + \alpha) \tag{34}$$

is a solution of eq. (28) for $\omega = 0$ (This sequence (34) is also generated from δ_0 and δ_1 by a product of Jacobian matrices). Since the derivative of any periodic function has at most two changes of sign per period, and because $\theta/2\pi$ is irrational, the sequence generated by (34) has infinitely many changes of sign. The third assertion of the corollary is thus proved by theorem 4.

There often exist KAM tori of the twist map which are not homotopic to zero (they go around the cylinder). Then the parametrization of the trajectory on this torus takes a form different from (34), which is (as in (17).

$$u_n = nl + \alpha + g(nl + \alpha), \tag{35}$$

where g is a differentiable function with period $2a$. The corresponding configuration is physically stable when

$$\delta_n = 1 + g'(nl + \alpha) \tag{36}$$

is always positive. We will see in the following section that this condition is always satisfied for such a KAM torus.

As a result, the metastable configurations of model (8) are represented by trajectories which do not satisfy the condition of the corollary of theory 4 and thus can be either

  1) hyperbolic or parabolic periodic cycles without reflexion; or
  2) dense on a KAM torus which is *not* homotopic to zero; or

3) Imbedded in the chaotic region (however, this condition does not imply that they are chaotic).

We have examples for these three cases. However, these conditions are not sufficient to have metastable configurations. Using theorem 4, it is particularly easy to check numerically the physical stability of the configuration corresponding to a trajectory. It suffices to perform the Jacobian matrix product (31) along a trajectory which is obtained by iterating the map T. Then we check the changes of sign of an arbitrary sequence δ_n. *All* our numerical experiments [14] for a chaotic trajectory have shown that any sequence δ_n exhibits a great density of change of sign. As a result *all* the observed trajectories which are chaotic in the map correspond to physically unstable configurations. This results confirms the early observation of Shilling and Thomas [15]. But this numerical experiment does not prove that chaotic configurations which are physically stable do not exist. (In fact, we can prove rigorously their existence in model (1) for λ large enough.) It only suggests that the chaotic metastable configurations are represented by a set of trajectories which have *zero measure* in the map, and thus are numerically inaccessible because of the limited accuracy of the computer. By contrast, the KAM tori which are nonhomotopic to zero, (when they exist) have a finite measure and are shown to correspond to undefectible ground-states (see the following section 4). We did not prove this conjecture but ref. 16 gives some other physical arguments which support this assumption.

Consequently, the numerical calculations of the *chaotic metastable* configurations, are *not reliable* when they are simply generated by map iterations. In order to avoid these map problems in the chaotic region, we obtained the metastable configurations by a variational method [14].

Integrating the set of equations

$$\frac{du_i}{ds} = -\frac{\partial \phi}{\partial u_i} \tag{37}$$

with respect to the variable s, yields a solution $\{u_i(s)\}$ which, for any initial configuration $\{u_i(0)\}$, converges to a limit $\{u_i^\infty\}$ which is necessarily a metastable configuration. A special choice of the initial conditions which is given by theorem 1 in refs. 6a or 4, or theorem 2 in ref. 6b (but a symmetry hypothesis is also required to have this theorem) yields a limit which is a ground state. (The solutions shown in fig. 1 of ref. 4 were calculated by this way.) It seems that the problem of studying the physical stability of the configurations generated by map iterations has not been carefully considered in some of the recent publications on this subject (see for example refs. 18 and 19). In the second reference [19] it is particularly obvious, in virtue of the corollary of theorem 4 that the configurations which are represented by KAM tori homotopic to zero, cannot be ground states because they are physically unstable. (See also refs. 16 and 20 for a more detailed comment of these references.)

4. General theorems on the transition by breaking of analyticity and the Peierls–Nabarro barrier

We turn back to the study of the ground states which has been done in section 2. Theorem 2 considered two situations for the incommensurate ground-states of model (8). In the first situation, the hull function is continuous (and generally analytical in analytical models because of the KAM theorem). In the second situation, the hull function becomes discontinuous on a dense set of points. In model (1), the variation of the parameter λ allows one to get a transition from the first situation to the second one. We called this transition: transition by breaking of analyticity [2]. We noted that this transition corresponds to the occurrence of a lattice locking on the incommensurate ground-states: that is, in other words the occurrence of a finite Peierls–Nabarro barrier (noted hereafter PN barrier) which must be passed through for translating continuously the incommensurate ground state. In this section, we describe some of the exact results which we obtained on the PN barrier and the transition by breaking of analyticity. The application of these results to the standard map allows one to easily obtain bounds for the stochasticity threshold.

Let us first examine the case for which $l/2a$ is a rational number and $\mathscr{G}_l$ is a discontinuous set (theorem 3).

It is proven that it is impossible to continuously slide the corresponding commensurate ground states without passing energy barriers. We also recently proved [8] that there necessarily exists another stationary commensurate configuration $\{v_n\}$ which just corresponds to the top in energy of the continuous paths corresponding to the translation of the commensurate ground-state (by keeping it commensurate) which pass the lowest possible barrier for the energy per unit cell. The periodic cycles of the twist map corresponding to the commensurate ground-state $\{u_n\}$ (which are hyperbolic or exceptionally parabolic without reflexion) and the periodic cycles corresponding to this commensurate configuration $\{v_n\}$ (which are either elliptic, or hyperbolic or parabolic with reflexion in both cases) are those which have been considered by Greene [9] for studying the stochasticity threshold of the KAM tori in the standard map. When $\mathscr{G}_l$ is discontinuous, we know that there exist elementary advanced and delayed discommensurations. Let $\{u_n\}$ be for example an advanced discommensuration and $\{v_n^+\}$ and $\{v_n^-\}$ the two commensurate ground-states with the properties described in (23). The configuration $\{v_{n+s} - 2ra\}$ is also an advanced discommensuration which satisfies the same conditions (23). It corresponds to the discommensuration $\{v_n\}$ translated by $-s$ lattice spacings or equivalently by a unit cell of the commensurate ground-state. To define the Peierls–Nabarro barrier of this discommensuration, we consider a continuous path $\mathscr{C}(t) = \{w_n(t)\}$ such that

$$\mathscr{C}(0) = \{w_n(0)\} = \{u_n\} \tag{38a}$$

and

$$\mathscr{C}(1) = \{w_n(1)\} = \{u_{ns} - 2ra\}. \tag{38b}$$

It joins the two translated configurations. The energy difference (which is proved to be finite)

$$E(\mathscr{C}(t)) = \sup_t \phi(\{w_n(t)\}) - \phi(\{u_n\}) \tag{39}$$

is the energy barrier which is passed through for the translation of this discommensuration along the path $\mathscr{C}(t)$. The PN barrier of the discommensuration $\{v_n\}$ is defined as

$$E_{\text{PN}}(\{v_n\}) = \inf_{\mathscr{C}(t)} E(\mathscr{C}(t)), \tag{40}$$

which is the lowest energy barrier which must be passed for a continuous translations of the discommensuration.

We pointed in section 2 that an advanced discommensuration is represented by the trajectory of a heteroclinic point h which belongs to the intersection of the dilating sheet W^+ of the point F_i^-, (which is the initial point of the trajectory corresponding to $\{v_n^-\}$) and to the contracting sheet W^- of the point F_j^+ corresponding to $\{v_n^+\}$ (F_i^- and F_j^+ are fixed points for the twist map). Then we prove

Theorem 5. The Peierls–Nabarro barrier of the elementary discommensuration vanishes if and only if the dilating sheet W^+ of F_i^- and the contracting sheet W^- of F_j^+ merge into a unique continuous curve which joins F_i^- to F_j^+. (The merged curve which corresponds both to W^+ and W^-, is called a separatrix.)

It is the situation which occurs in integrable maps. Thus this theorem proves that if the PN barrier does not vanish, the map cannot be integrable. However, we have not yet completely elucidated the nature of the intersection of W^+ and W^- when this PN barrier does not vanish. We expect that the intersection of W^+ and W^- is always transverse or in other words that the curves W^+ and W^- are not tangent at their intersection.

Now, we turn back to the case of the incommensurate ground-states. We can prove several theorems. The two first ones deal with the case for which there exists in the twist map an invariant continuous and closed curve Γ_l which is non-homotopic to zero and on which the twist map is conjugate to a rotation with winding number $l/2a$. In other words, a trajectory $\{p_n, u_n\}$ on this curve Γ_l can be parametrized by a continuous hull function $f_k(x)$ such that for all n

$$u_n = f(nl + \alpha), \tag{41}$$

with $f_k(x)-x$ periodic with period $2a$ (α is some arbitrary phase). Then we proved the following theorem [4, 8]:

Theorem 6. Let us assume the existence of an invariant continuous curve Γ_l on which the twist map is conjugate to a rotation with winding number $l/2a$, then this set Γ_l is identical to the set of trajectories representing the ground-state of $\mathcal{G}_l$. (This theorem also applies when $l/2a$ is rational).

Particularly, this curve Γ_l can be a KAM torus with an irrational winding number $l/2a$. When this KAM torus exists, it necessarily represents the set of ground state $\mathcal{G}_l$. Since we know that when KAM tori exists, they have a finite measure on the cylinder, most of them (that is with probability 1) can be approached by sequences of KAM tori with winding numbers $l_i/2a$ such that l_i goes to l either with $l_i > l$ or $l_i < l$. Let us call these tori "true" KAM tori. Most KAM tori are "true". Then, we have the theorem.

Theorem 7. When the set of incommensurate ground states $\mathcal{G}_l$ is represented by a "true" KAM torus Γ_l, then the incommensurate ground states of $\mathcal{G}_l$ are undefectible (by definition a ground-state is called undefectible, when, apart a phase shift, it is the only metastable configuration of the system with the same boundary conditions (12)).

In the situations, considered by theorems 6 and 7, the PN barrier which corresponds to the translation of the incommensurate structure is zero. Then the gap in the phonon spectrum of the incommensurate ground-state $\{u_n\}$ given by eq. (28) is proven to vanish. (The gap is the smallest phonon frequency given by (28).) Conversely a finite PN barrier does not imply a finite gap although generally they are both finite (or both zero). However when the gap is finite, we obtained the following theorem which has a quite complicated proof [8]:

Theorem 8. Let $\{u_n\}$ be an incommensurate ground-state of model (8). Let us assume that the gap in frequency of the small motion eq. (28) is strictly positive. Then, the hull function f describing the incommensurate ground-state is discrete (see theorem 2). In other words, $f^\pm(x)$ can be written as a sum of step functions,

$$f^\pm(x) = \sum_i f_i Y^\pm(x - x_i), \tag{42}$$

where f_i is the amplitude of the step function located at x_i (By definition $Y^\pm(x) = 0$ for $x < 0$, $Y^\pm(x) = 1$ for $x > 0$ and $Y^+(0) = 1$, $Y^-(0) = 0$). Then, the Lyapounov exponent γ given by (32) for this incommensurate ground state is strictly positive.

For reasonably differential models 8, we conjectured in ref. 6a that the hull function f of an incommensurate ground state should be either

1) absolutely continuous: that is, $f(x)$ is differentiable almost everywhere and for arbitrary x_0, x_1:

$$f(x_1) - f(x_0) = \int_{x_0}^{x_1} f'(\zeta) \, \mathrm{d}\zeta; \tag{43}$$

or 2) singular continuous ($f(x)$ is continuous with a zero derivative almost everywhere) or 3) discrete ($f(x)$ is discontinuous and can be written with the form (42)).

We have not rigorously proven this conjecture

but we have shown in ref. 4 that model (1) exhibits situations for which the hull function f is either analytical or discrete. The following section 5 reports these proofs with more details which yield incidently a series of exact upper bounds for the transition by breaking of analyticity or equivalently for the stochasticity threshold of the standard map.

5. Existence proof and exact bounds for the transition by breaking of analyticity in the standard map

In the standard map (6) associated with model (1), the Kolmogorov–Arnold–Moser theorem [5, 23] predicts that for almost any irrational $l/2a$, there exists $\lambda_2(l)$ such that for $|\lambda| < \lambda_2(l)$, there exists an invariant torus on which the map is conjugate to a rotation with winding number $l/2a$. Then applying theorem 6 yields that the trajectories of this KAM torus represent the ground states of $\mathscr{G}_l$ and that their hull function is differentiable. Conversely, when λ becomes large enough the intuitive image of the problem, suggests that the atoms locate in the bottoms of the periodic potential and thus that the function f becomes discrete.

This hull function satisfies the functional equation

$$F(x) = f(x+l) + f(x-l) - 2f(x)$$
$$= \frac{\lambda \pi}{2a} \sin\left(\frac{\pi}{a} f(x)\right), \quad (44)$$

which is obtained by inserting (17) in (21). Because of the periodicity property of this model we can restrict our study to the case $0 < l < 2a$. Since $f(x)$ is monotonous increasing it comes out that for any x

$$f(x+l-2a) < f(x) < f(x+l). \quad (45a)$$

These inequalities (45) in (44) yields

$$|F(x)| < 2a. \quad (45b)$$

As a result, when

$$\lambda > 4a^2/\pi > \hat{\lambda}_c, \quad (46)$$

eq. (44) and inequality (45b) shows that the hull function $x + g(x) = f(x)$ cannot take any value $(2n+1)/a$ where n is an integer. Consequently, $f(x)$ must be discontinuous and because of theorem 6 there exist no invariant continuous curves which are nonhomotopic to zero on which the standard map is conjugate to a rotation. As a result, there exist no KAM tori for any winding number. Then inequality (46) gives a rigorous upper bound for the stochasticity threshold $\hat{\lambda}_c$ for the standard map which has been calculated by Greene [9] and which is in our units

$$\hat{\lambda}_c \# 0.9716 \times \frac{2a^2}{\pi^2} = \sup_l \lambda_c(l). \quad (47)$$

In fact this bound can be improved by only considering the positivity of the quadratic form (27)

$$\delta\phi = \frac{1}{2} \sum_n \left[\left(2 + \frac{\lambda \pi^2}{2a^2} \cos\frac{\pi}{a} u_n\right) \delta_n^2 - 2\delta_{n+1}\delta_n \right] \quad (48)$$

for any ground state $\{u_n\}$. Assuming that the hull function $f(x)$ be continuous, it is possible to choose the phase α such that

$$u_0 = f(\alpha) = a \quad (49)$$

(where we expect that the discontinuity of f should first appear) be on the top of the periodic potential. Next, we prove that in certain range of λ, all stationary configuration $\{u_n\}$ with $u_0 = a$, are such that their quadratic form (48) is not positive. No ground state can exist with a continuous hull function whatever the atomic mean distance and consequently no KAM torus nonhomotopic to zero can exist. For this proof we set

$$u_1 = x. \tag{50a}$$

The stationary eq. (2) yields

$$u_{-1} = 2a - u, \qquad u_2 = 2x - a + \frac{\lambda \pi}{2a} \sin\left(\frac{\pi}{a} x\right),$$

$$u_{-2} = 2a - u_2. \tag{50b}$$

For convenience, we also set

$$A = 2 - \frac{\lambda \pi^2}{2a^2}, \tag{51a}$$

$$X = 2 + \frac{\lambda \pi^2}{2a^2} \cos\frac{\pi}{a} x, \tag{51b}$$

$$Y = 2 - \frac{\lambda \pi^2}{2a} \cos\frac{\pi}{a}\left(2x + \frac{\lambda \pi}{2a} \sin\frac{\pi}{a} x\right). \tag{51c}$$

First, we consider the minor of order 1, $\Delta_1 = A$. When it is negative, the quadratic form (48) cannot be positive. For

$$\lambda > \frac{4a^2}{\pi^2} > \lambda_c, \tag{52}$$

A is negative and there exist no KAM tori. Second, we consider the minor of order 2,

$$\Delta_2 = \begin{vmatrix} A & -1 \\ -1 & X \end{vmatrix} = AX - 1. \tag{53a}$$

When $\lambda < 4a^2/\pi^2$, it is smaller for any x than

$$\left(2 - \frac{\lambda \pi^2}{2a^2}\right)\left(2 + \frac{\lambda \pi^2}{2a^2}\right) - 1 = 3 - \left(\frac{\lambda \pi^2}{2a^2}\right)^2.$$

Consequently, Δ_2 is always negative when

$$\frac{4a^2}{\pi^2} > \lambda > \sqrt{3}\frac{2a^2}{\pi^2} > \hat{\lambda}_c. \tag{53b}$$

When this condition is satisfied, there also exist no KAM tori which are nonhomotopic to zero. A third order bound is obtained by considering the minor of order 3,

$$\Delta_3 = \begin{vmatrix} X & -1 & 0 \\ -1 & A & -1 \\ 0 & -1 & X \end{vmatrix} = X(AX - 2), \tag{54a}$$

which is negative for any x when

$$\frac{4a^2}{\pi^2} > \lambda > \sqrt{2}\frac{2a^2}{\pi^2} > \hat{\lambda}_c. \tag{54b}$$

This inequality improves the upper bounds (46), (52) and (53a) for the stochasticity threshold $\hat{\lambda}_c$. By considering higher order minors of the quadratic form, we obtain better bounds for λ_c. For example, we consider the order five:

$$\Delta_5 = \begin{vmatrix} Y & -1 & 0 & 0 & 0 \\ -1 & X & -1 & 0 & 0 \\ 0 & -1 & A & -1 & 0 \\ 0 & 0 & -1 & X & -1 \\ 0 & 0 & 0 & -1 & Y \end{vmatrix}$$

$$= (XY - 1)(AXY - 2Y - A). \tag{55a}$$

In order to avoid cumbersome calculations, which in principle are possible, we only checked numerically the sign of $(AXY - 2Y - A)$ for $0 \le x < 2\pi$ with A, X, and Y given by (51). Then, we found that for

$$\sqrt{2} \ge \frac{\lambda \pi^2}{2a^2} > 1.230 \pm 0.005 > \hat{\lambda}_c \tag{55b}$$

$(AXY - 2Y - A)$ is negative for any x. As a result, either Δ_5 or $(XY - 1)$ (which also is a minor of the quadratic form (48)) is negative. Consequently, when (55b) is satisfied, there exist no KAM tori nonhomotopic to zero, which still improves the upper bound of $\hat{\lambda}_c$. Note that this bound 1.230 ± 0.005 is now only 25% above the value (47) calculated by Greene and that this result is a strict bound obtained with a very short numerical calculation. (Note that J. Mather also obtained the bound 4/3 with a method which is apparently different [24].) We conjecture that the sequence of bounds obtained by writing the positivity for $u_0 = a$ of the sequence of minors Δ_n which follows Δ_1, Δ_2, Δ_3 (Δ_4) and Δ_5 converges to the exact value of $\hat{\lambda}_c$, but we have not checked numerically this assertion.

Let us turn back to the study of the functional eq. (44). We reproduce here, for the model (1) the proof of ref. 4 (which we hope more clear) which shows that for λ large enough the hull function f becomes discrete. When (46) is satisfied, we have

$$\left|\sin\frac{\pi}{a}f(x)\right| \leq \frac{4a^2}{\lambda\pi} \tag{56}$$

which implies that for any x there exists an integer n such that

$$-f_0 + na \leq f(x) \leq f_0 + na, \tag{57a}$$

with

$$f_0 = \frac{a}{\pi}\arcsin\frac{4a^2}{\lambda\pi}. \tag{57b}$$

We now write that the diagonal terms of the quadratic form (48) is positive which yields another inequality for all x,

$$-\frac{4a^2}{\lambda\pi^2} \leq \cos\frac{\pi}{a}f(x). \tag{58}$$

Otherwise, inequality (57a) for n odd implies by using (57b)

$$\cos\frac{\pi}{a}f(x) \geq \cos\frac{\pi}{a}(f_0 + na)$$

$$= -\left(1 - \left(\frac{4a^2}{\lambda\pi}\right)^2\right)^{1/2}. \tag{59a}$$

When

$$\lambda > \frac{4a^2}{\pi^2}\sqrt{\pi^2 + 1}, \tag{59b}$$

inequalities (58) and (57a) are incompatible for n odd, thus the integer n which appears in (57a) must be even. As a result, when (59b) is satisfied, we obtain for all x

$$\cos\frac{\pi}{a}f(x) > \cos\frac{\pi}{a}f_0 = 1 - \left(\frac{4a^2}{\lambda\pi}\right)^2 > 0. \tag{59c}$$

Now, we can apply theorem (8) for proving that function f is discrete by checking that (59c) implies that the gap of the phonon spectrum is larger or equal to $\sqrt{1 - (4a^2/\lambda)^2}$ and thus strictly positive. But, a direct proof is also quite simple. For that, we prove that the continuous part $F_c(x)$ of $F(x)$ in (44) is a constant by proving that it is both periodic and monotonically increasing.

$F_c(x)$ is periodic because it is the variation $h_c(x+l) - h_c(x)$ from x to $(x+l)$ of the continuous part of the periodic function $h(x) = f(x) - f(x-l) = l + g(x) - g(x-l)$. (Note however that the continuous part of a periodic function is not necessarily periodic.)

$F_c(x)$ is monotonically increasing because in the last member of (44), 1) $f(x)$ is monotonically increasing; 2) $\sin(\pi/a f(x))$ is strictly increasing in the vicinity of each value taken by $f(x)$ because of the inequality (59c). As a result, $f(x)$ obtained from $F(x)$ by (44) is also discrete.

The results described in this section rigorously prove the existence of a breaking of analyticity in the standard map although we have not proved that it exactly occurs at a well defined λ_c. Anyway we obtained explicit bounds of $\hat{\lambda}_c$. This transition is numerically found to be well defined on the figs. 4 which show the trajectories corresponding to the ground states for $l/2a = 441/997$ (which is practically an irrational number) and for $\lambda = 0.167$, $\lambda = 0.20$ and $\lambda = 0.212$. (These ground-states have not been calculated by iterating the standard map because, as we know, it is an unstable process for $\lambda > \lambda_c(l)$ but by using the gradient method described by eq. (37)).

6. Final remarks on the devil's staircase and the order without periodicity

The above theorems have an application for the theory of the devil's staircase which we briefly describe now.

Let us consider model (8) to which we add a tensile force μ (or chemical potential)

$$\phi_\mu(\{u_i\}) = \sum_i [L(u_{i+1}, u_i) - \mu(u_{i+1} - u_i)] \qquad (60)$$

(as for model (1), the addition of this tensile force does not change the twist map associated to this model). The ends of the chain are let free for finding the ground-state of this model, we first

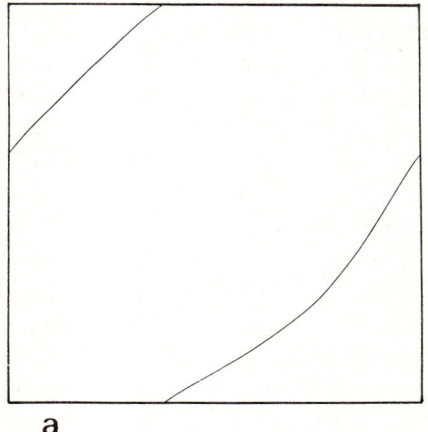

a

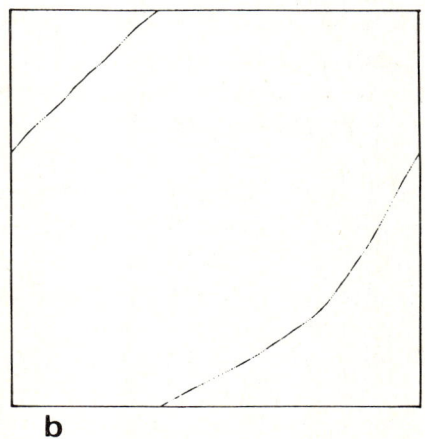

b

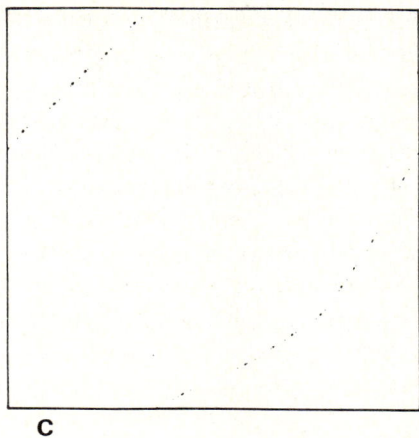

c

Fig. 4. From Aubry and André (1980), ref. 4. Trajectories of the map (6) representing the ground-state of the FK model for $l/2a = 441/997$ (which is practically an irrational number) for $\lambda = 0.167$ (a); $\lambda = 0.2$ (b); and $\lambda = 0.212$ (c). Note the sharp change in the aspect of the trajectory which signals the transition by breaking of analyticity at $\lambda_c \# 0.2$. For $\lambda < \lambda_c$ (fig. 4a) the trajectory is dense on a KAM torus. At $\lambda \approx \lambda_c$ (fig. 4b) the density of a point on the torus exhibits critical fluctuations; while for $\lambda > \lambda_c$ (fig. 4c) the trajectory is dense and quasi-periodic on a Cantor set which survives to the KAM torus. Compare with Figs. 1 which exhibits arbitrary trajectories. In fact, all of them corresponds to unstable configurations except those generated by the points M_2, M_3, M_4 and M_5 of fig. 1a, the trajectory of which generate a KAM torus nonhomotopic to zero.

consider the average energy per atom (for $\mu = 0$)

$$\psi(l) = \lim_{N \to \infty} \frac{1}{N} \sum_{i=1}^{N} L(u_{i+1}, u_i) \qquad (61)$$

for the ground-state(s) with atomic mean distance l (which we proved to be a well-defined function) and we minimize the energy per atom $\psi(l) - \mu l$. Then, we prove that the atomic mean distance l varies as a devil's staircase versus μ. We have

Theorem 9 The variation curve $l(\mu)$ of the atomic mean distance l of the ground state of model (60) with free ends versus the tensile force μ has the following properties.

1) the curve $l(\mu)$ is monotonically increasing and is continuous.

2) for each rational $l/2a = r/s$, $l(\mu)$ is constant

on a finite interval δ if and only if the corresponding set $\mathcal{G}_l$ (described in theorem 3) is discontinuous.

In general, when the twist map is not integrable $\mathcal{G}_l$ is not continuous for all rationals l. As a result $l(\mu)$ has a constant step at each rational $l/2a$.

This curve is called a devil's staircase [25]. In this book, B. Mandelbrot also shows other physical examples which involves such pathological curves. On the basis of solid physical arguments, we conjectured [2, 3] that this curve $l(\mu)$ is a complete devil's staircase for $l_1 < l < l_2$ when for all irrational $l/2a$ in this interval, the set $\mathcal{G}_l$ are discontinuous. (By definition, a devil's staircase is called complete [2] when it is entirely composed of steps, or equivalently when $l(\mu)$ has a zero derivative almost everywhere, or equivalently when the Stieltjes measure $l(\mu)$ has no absolutely continuous part.) We also conjectured that it becomes incomplete (that is its derivative becomes finite on a finite measure set) when for some $l/2a$ irrational (which have finite measure) the sets $\mathcal{G}_l$ are represented by KAM tori. (Let us mention that our theory would become rigorous, if a *uniform* bound of the exponential interactions between the discommensurations could be obtained.) Anyway, we can exhibit exact models (which however have some pathologies) in which a complete devil's staircase [7] can be proved to exist and also explicitly calculated. As we explained in refs. 2, 3 and 16 a complete devil's staircase physically corresponds to an irreversible but *continuous* transformation which is a quite unusual behavior. But, indeed similar features been observed in certain experiments.

It has also been experimentally observed structures which are neither periodic or quasi-periodic (incommensurate). Are they chaotic? We generalized some aspects of this theory on the twist maps, to all structures in any dimensions which are obtained from the minimization of an energy (i.e. a variational form). We introduced an abstract dynamical system in which the usual time group is replaced by the translation group of the space in which the structure is imbedded. Using this representation, we proved that *there always exists* a "minimal invariant closed set" (by definition, it does not contain any smaller closed set invariant under the action of the group) which corresponds to a ground-state. Translated in physical terms, this property implies the existence of ground states with a new kind of long range order which could be neither periodic nor incommensurate. We called this new kind of long range order "weak periodicity". It also corresponds physically to a "local order at all scales". In ref. 16, we briefly describe this theory but with some more details than here. Particularly, surprising examples of "undecidable structures" obtained by tiling the plane are given, proving that such strange structures do exist in theoretical models. Moreover they have no entropy. Let us emphasize that our assertions are not in contradiction with those of Ruelle [27] on the existence of "turbulent ground state", although they seem to disagree. Indeed for Ruelle, "turbulent" means nonperiodic and "non-quasi-periodic". With this definition, we agree with his assertion on the existence of turbulent ground state. However, our definition of turbulent is more restrictive because we require that the structure has a finite entropy.

Although, we have no proof, we believe that except in exceptional models with accidental degeneracy, the ground state of most models obtained by minimizing a free energy has no entropy although it can be neither periodic nor quasi-periodic. It is necessarily "weakly periodic" (but this property is still quite physically imprecise). Of course, we do not exclude defectible ground-states for which there may exist many other metastable configurations. Although they have more energy than the ground-state these configurations should play an important role for the thermodynamical properties of the structure [16].

References

[1] T. Kontorova and Y.I. Frenkel, Zh. Eksp. & Teor. Fiz. 8 (1938) 89, 1340, 1349. F.C. Frank and J.M. Van der

Merwe, Proc. Roy. Soc. (London A198 (1949) 205. S.C. Ying, Phys. Rev. B3 (1971) 4160.
[2] S. Aubry, in Solitons and Condensed matter physics, A.R. Bishop and T. Schneider, eds. Solid State Sciences 8 (Springer, Berlin, 1978), p. 264.
[3] S. Aubry, Ferroelectrics 24, (1980). For a detailed version of this paper see S. Aubry, Lect. Notes in Math. 925 (1980) 221 (Springer).
[4] S. Aubry and G. André, in Colloquium on group theoretical methods in physics, Ed. L. P. Horwitz and Y. Ne'eman, Annals of the Israel Phys. Soc. 3 (1980) 133.
[5] J. Moser, Stable and Random motions in Dynamical Systems (Princeton Univ. Press, Princeton, NJ, 1973).
[6] S. Aubry, On modulated crystallographic structures, exact results on the classical ground-states of a one-dimensional model, (1978) unpublished; S. Aubry and P.Y. Le Daeron, The discrete Frenkel Kontorova model and its extension. I. Exact results for the ground-states, preprint submitted to Physica D.
[7] S. Aubry, Exact models with a complete devil's staircase, preprint, to be published in J. of Phys. C.
[8] S. Aubry, P. Y. Le Daeron and G. André, in preparation.
[9] J. Greene, J. Math. Phys. 20 (1979) 1183.
[10] W.L. McMillan, Phys. Rev. B14 (1976) 1496.
[11] Marsden and McKracken, The Hopf Bifurcation and its Application, Applied Math. Sci., vol. 19 (Springer, Berlin, 1976).
[12] S. Smale, Bull. of AMS 73 (1967) 747.
[13] N.I. Akhiezer, The Classical Moment Problem and some Related Questions in Analysis, (Oliver & Boyd, Edinburgh and London, 1965).
[14] M. Peyrard and S. Aubry, to be published in J. Phys. C (1983).
[15] R. Shilling, private communication (1982).
[16] S. Aubry, Journal de Physique (Paris) 44 (1983) 147.
[17] S. Aubry, in Physics of Defects, R. Balian et al., Les Houches 35 (1981) 431 (North-Holland, Amsterdam).
[18] P. Bak, Phys. Rev. Lett. 46 (1981) 791.
[19] P. Bak and V.L. Pokrovsky, Phys. Rev. Lett. 47 (1981) 958.
[20] P.Y. Le Daeron and S. Aubry, Metal insulator transition in the Peierls chain, submitted to J. of Phys. C.
[21] S. Aubry, On the dynamic of structural phase transition. Lattice locking and Ergodic theory, unpublished (1977).
[22] F. Nabarro, Theory of Crystal Dislocations (Clarendon, Oxford, 1967).
[23] S. Aubry, in Intrinsic Stochasticity in Plasmas, G. Laval and D. Gresillon, eds. (Edition de Physique, Orsay, 1979).
[24] J. MacKay, this conference.
[25] B. Mandelbrot, Fractals, (W.F. Freeman, San Francisco, 1977); see also The fractal geometry of nature, (1982) to appear, same editor. See also this conference.
[26] F. Riesz and B. Nagy, Functional analysis (Frederik Ungar, New York, 1965).
[27] D. Ruelle, this conference.

INFLUENCE OF SOLITONS IN THE INITIAL STATE ON CHAOS IN THE DRIVEN DAMPED SINE-GORDON SYSTEM

A.R. BISHOP, K. FESSER and P.S. LOMDAHL
Center for Nonlinear Studies and Theoretical Division, Los Alamos National Laboratory, Los Alamos, New Mexico 87545, USA

and

S.E. TRULLINGER
Department of Physics, University of Southern California, Los Angeles, California 90089-0484, USA

The appearance of chaos in the a.c. driven, damped sine-Gordon equation is studied numerically. Several transitions from periodic to chaotic behaviour are investigated in detail for flat initial conditions. Spatial structures (breather, kink) in the initial conditions smooth out many of these transitions and give rise to an interesting symbiosis of time and spatial intermittency. This symbiosis appears to be due to the competition between the background tendency towards chaos and the system's preference to maintain a spatial pattern. The way that this competition is relieved is also found to depend very strongly on symmetry in the initial conditions.

1. Introduction

Recently, there has been a great deal of interest in the nature of transitions to chaotic behavior in simple nonlinear systems as a parameter (or set of parameters) in the governing deterministic equation(s) of motion is varied. In particular, single classical particles moving in one of a wide class of nonlinear potentials and subjected to damping and a sinusoidal driving force can exhibit [1–3] a great variety of deterministic motions ranging from simply-periodic to fully chaotic. An especially useful example, because of its relevance to many physical systems, is the sinusoidal potential [2, 3] (e.g., $mgl(1 - \cos\phi)$ for a simple pendulum of mass m and length l in a gravitational field, g). The single-particle dynamics for this model has been studied [1–3] carefully, by numerical (analog or digital) integration of the nonlinear equation of motion, as a function of damping ϵ, the frequency ω_d and strength Γ of the oscillatory external driving force, $\Gamma \sin \omega_d t$. Using primarily power spectral density diagnostics, a remarkably rich variety of particle responses has been (partially) mapped out in this 3-parameter space. Feigenbaum sequences [4] of period doubling are limited to a very small region of this parameter space (unless a d.c. driving force is also added [3]), but chaos and high-order harmonic and sub-harmonic flows form a very complex array of dynamical responses. Phase space diagnostics familiar from dynamical systems theory [5] (Poincaré sections, limit cycles, attractors, strange attractors, Lyapunov exponents, etc.) are useful in characterizing these responses. For example, in the large amplitude (Γ) chaotic regime Poincaré sections reveal [2] a characteristic strange attractor with periodic (in ϕ) structure due to the periodicity of the potential.

The model single-particle studies cited above are very provocative in view of their many potential physical applications to solid state systems, electronic devices, etc. For example, the recently suggested explanation [3] of anomalous noise rise in SUPARAMPS afforded by nonperturbative, non-linear but deterministic dynamical studies is quite appealing. However, most potential applications, especially in low-dimensional materials, demand the consideration of *spatial* degrees of freedom (rather than a single, spatially *averaged* quantity). This raises several new and challenging questions. As has been shown recently in independent investigations [6, 7], coupled particles moving in nonlinear local potentials (e.g., the sine-Gordon chain) are susceptible to "chaotic" responses, but they can also exhibit spatially *coherent* structures (e.g., solitons) in the presence of temporal chaos. Thus, the question of whether such spatially coherent structures can co-exist with temporal chaos has been answered in the affirmative. Both of these numerical simulation studies [6, 7] were performed on the driven, damped sine-Gordon chain governed by the nonlinear equation of motion

$$\phi_{tt} - \phi_{xx} + \sin \phi = F(x, t) - \epsilon \phi_t, \qquad (1)$$

where the subscripts t and x denote partial derivatives with respect to time and distance along the chain, respectively, ϵ is a damping parameter and $F(x, t)$ is an external force. In ref. 6, this force was chosen to be spatially inhomogeneous with sinusoidal time dependence and uniform initial conditions; it was found that, for some parameter ranges, kink–antikink pairs were created chaotically. In ref. 7, on the other hand, $F(x, t)$ was taken to be spatially homogeneous (uniform) with the form

$$F = \Gamma \sin \omega_d t, \qquad (2)$$

where Γ is the driving amplitude and ω_d the driving frequency, and a brief survey of possible responses (chaotic and periodic) was presented for a variety of parameter values and choices of initial conditions (flat as well as spatially inhomogeneous).

In this paper, we focus our attention on the specific question of the influence of spatially inhomogeneous, nonlinear structures (breather and kink solutions of the unperturbed sine-Gordon equation) in the initial state on the nature of nonchaotic ↔ chaotic transitions as compared to the responses when flat initial conditions ($\phi_t = \phi_x = 0$) are imposed. Thus, the parameters ϵ and ω_d are held fixed (at values giving interesting results) and Γ and the type of initial condition are varied. The particular values chosen are $\epsilon = 0.2$ and $\omega_d = 0.6$, for which a rather rich sequence of chaotic ↔ nonchaotic transitions are found [2, 3] in the single-particle case as Γ is varied (see section 2 below). Only periodic (or periodic modulo 2π) boundary-condition results are reported here.

Our motivations for such a study are several. Firstly, the question of possible persistence of initial spatial structures for very long times having been settled in ref. 7, we wish to refine our understanding of their influence on the transitions into and out of chaotic regimes. This is of special importance when a variety of initial conditions are available in a particular system or when we do not have the ability to control them precisely (e.g., when thermal fluctuations are present). Secondly, even in the absence of thermal fluctuations, one may wish to develop statistical descriptions [8] of the chaotic behavior involving averages over initial conditions. In such cases, information regarding the various "basins of attraction" (i.e., sets of initial conditions) for different responses may prove essential. For example, in section 3 below, we find that two different breather initial states lead to essentially the same dynamical behavior in chaotic regimes after transients have died away, while kink initial states (see section 4) lead to much different behavior. Forthermore, both breathers and kinks lead to much different responses than obtained with flat initial conditions. It appears then, that perhaps a very large number of basins of attraction exist in the multi-dimensional phase space of the (homogeneously) driven sine-Gordon chain. Our results regarding this point are preliminary and far from complete; the complex variety of responses necessitates a methodical approach to their explanation and we report here our first results on this topic. More detailed and thorough findings will be published subsequently. Finally, the motivations for our choice of the driven sine-Gordon chain for study are its many physical applications (e.g., Josephson transmission lines [9] and charge-density-wave systems [10]) as well as our having the hope of carrying out analytic studies in at least the quiescent regimes since the coherent soliton structures, and their evolution under weak perturbations, are so well studied [11], and susceptible to analysis by inverse spectral transform theory or related "soliton" techniques.

Using a versatile predictor-corrected scheme, we integrate the coupled differential equations for the chain of 120 particles used in this study for times equalling several thousand characteristic time units (single-pendulum small-oscillation periods)– typically 50,000–70,000 integration steps. Our simulation is of course a discrete version of eq. (1)–we have used here a sixth-order finite difference scheme to approximate the 2nd-order spatial derivative. Some discreteness effects were shown in ref. 7, but here we concentrate on a near continuum regime with the characteristic spatial scale (unity) in eq. (1) covering 5 grid points. Note that these discretizations are *not* those necessary to maintain integrability in the undriven, undamped chain, but they are relevant to real solid state lattice contexts. We have improved our power spectra over those in ref. 7 by incorporating data smoothing techniques to remove the spurious skirts of the "delta-function" peaks occurring at the fundamental driving frequency and harmonics. In addition to the primary diagnostic tools represented by the position-dependent power spectra $S(x, \omega)$ and plots of ϕ and ϕ_t vs. x at certain times as employed in ref. 7, we have added phase-plane trajectory plots, Poincaré sections, and plots of the spatial averages, $\langle \phi \rangle$ and $\langle \phi_t \rangle$, of $\phi(x, t)$ and $\phi_t(x, t)$ vs. time. In the phase-plane trajectory plots shown below, we plot ϕ_t vs ϕ for particles at particular sites along the chain for 5000 closely-spaced time values which are usually chosen after initial transients have died away. The Poincaré sections shown below are computed using the data shown in the corresponding phase-plane plots, by plotting ϕ_t vs. ϕ for a given particle at equally spaced time intervals corresponding to the positive zero-crossings of the applied force [eq. (2)], as in ref. 2. Since the particular time step used in the integration routine does not (except by coincidence) yield values of ϕ and ϕ_t at times corresponding precisely to the zero-crossings, we have added a simple 4-point interpolation algorithm to approximate the values ϕ and ϕ_t at these crossing times. These interpolated values are in general quite accurate enough. However, in one case shown in section 3, the scatter of points and fineness of the ϕ_t scale in the Poincaré plot indicates that a more accurate interpolation can be necessary in unusual cases. In some of the Poincaré sections for chaotic situations, we have folded the ϕ values onto a 2π interval, since for long times the attractor will fill in to eventually become periodic in ϕ, as in ref. 2, and we can take advantage of this to obtain a higher density of points in one "unit cell" of the attractor "lattice".

With these tools added we are able to give convincing evidence for the periodic or chaotic nature of the various responses observed. In all cases we have thus been able to verify that "noisy" power spectra correlate with scatter in the phase plots and characteristic "folded" structures (strange attractors) in the Poincaré sections.

The outline of the remainder of the paper is as follows. In section 2 we describe the main results we have obtained when flat initial conditions are imposed on the chain. The responses in these cases are equivalent to single-particle responses since the applied force is homogeneous and thus each particle responds in exactly the same way, i.e., $\phi(x, t)$ and $\phi_t(x, t)$ are independent of x. We thus have a valuable calibration for studying the influence of spatially *inhomogeneous* initial conditions.

In section 3, we present results for two cases of spatially inhomogensous initial conditions where the chain is started with the profile of a breather solution of the unperturbed sine-Gordon equation (right-hand side of eq. (1)). The two cases differ in that the internal vibration (or "breathing") frequency, ω_b, of the breathers is set equal to $\omega_b = 0.2$ and 0.7, respectively. These give a narrow, high amplitude profile and a wider, lower amplitude profile, respectively. In this section we also examine the influence of asymmetry in the initial conditions; namely, whether or not the initial breather profile is symmetric around the chain center. We find that this has a strong effect on the time evolution, and in particular the competition between overall flow in ϕ-space and maintaining spatial structure. In general we find that this symmetry strongly inhibits tendencies towards chaotic motion or large amplitude uniform motions in ϕ-space (see section 2). However, the symmetry can be broken very easily and then an interesting intermittency (described in section 3) is one way in which the above competition is resolved.

In section 4 we present results for the case when a single sine-Gordon kink is present at rest in the initial state and the boundary condition, $\phi(L, t) - \phi(0, t) = 2\pi$, is imposed. We find that, in some regimes, certain features of the dynamics of arbitrarily chosen particles on the chain can be interpreted in terms of (distorted) kink oscillation back and forth along the chain, or of uniform drifting of the kink center-of-mass.

Finally, in section 5 we summarize our results, compare the gross features of the dynamical response of the chain with different initial conditions, and discuss several remaining questions which require further study.

2. Flat initial condition

In this section we begin our presentation of the results of numerical integration of eq. (1), in this case for flat initial conditions $[\phi(x, 0) = \phi_t(x, 0) = 0]$. As mentioned above, the results serve as a calibration for comparing the response of the chain when inhomogeneous initial conditions are imposed (sections 3 and 4). The only parameter which we vary in this section is the strength, Γ, of the periodic, homogeneous driving force [eq. (2)] applied to the chain.

In table 1a, we indicate the main types of responses encountered as Γ increased from zero. We have located the threshold value of Γ for the first onset of chaos to lie in the range 0.6158–0.6159. In fig. 1a we show the power spectrum $S(0, \omega)$ at $\Gamma = 0.6158$ for a particular particle ($x = 0$, the chain center) on the chain where $S(x, \omega)$ is defined [7] as the square of the Fourier time transform of $\phi_t(x, t)$. One observes peaks in the power spectrum at the fundamental driving frequency $\omega_d = 0.6$ and its odd harmonics, indicating simple (but anharmonic) periodic response with period $T = 2\pi/\omega_d$. An examination of the phase-plane plot (fig. 1b) for the same particle confirms this conclusion since the trajectory is a closed orbit with the fundamental period T. Note that ϕ executes its motion within a *single* (dis-

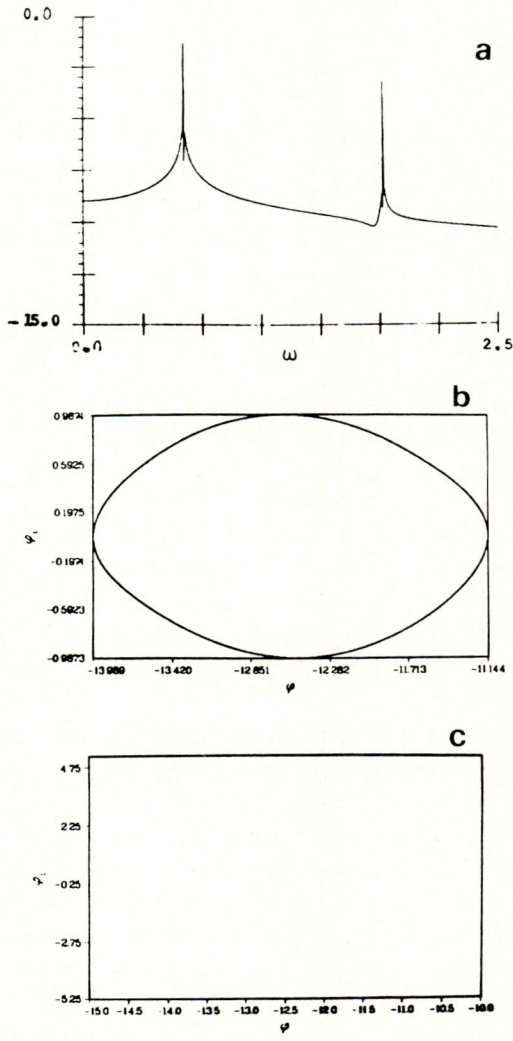

Fig. 1. Flat initial conditions, $\Gamma = 0.6158$, (a) power spectrum, (b) phase-plane, and (c) Poincaré section.

Table 1

Γ	(a) Flat	Γ	(b) Breather (broken symmetry)	Γ	(c) Breather (symmetric)	Γ	(d) Kink
0.94	Chaotic (odd harmonics)	0.89	Chaotic with intermittent periodicity	0.89	Chaotic	0.75	Chaotic initial transient followed by simple periodicity
0.9015	Period 3 (odd 1/3 subharmonics)						
0.87	Chaotic (odd 1/3 subharmonics)		Symbiotic intermittency between translation and locking in x, with large and small amplitude motions in ϕ				
0.75	Chaotic (odd harmonics)				Simply periodic		Quasi-periodic
0.7425	Chaotic (all harmonics) Increasing dc component with decreasing Γ		Power spectrum sensitive to sampling period		Spatial period 1/2 structure		Kink oscillates and uniformly drifts
0.731	"Running" chaotic (all harmonics)				Increasing chaotic initial transients with increasing Γ		
0.729	Period 5, (all 1/5 subharmonics) No dc component						
0.724	"Running" (all 1/2 subharmonics) Period 2, dc component		Spatial period 1/2 structure in locked state				
0.68075	"Running" (all harmonics) dc component					0.64	Periodic kink oscillations dressed by random kink-breather collisions
0.67	Chaotic (all harmonics) Increasing dc component with increasing Γ	0.6197	Breather synchronization to driving frequency	0.6197			
	Chaotic (odd harmonics)	0.57	Spatial smoothing Simply periodic	0.57	As 1b	0.59	Simply periodic Entrained kink oscillation
0.61585	Simply periodic (odd harmonics)				As 1b		

torted) well of the cos ϕ potential and the orbit is nearly elliptical. The corresponding Poincaré section (described above) shown in fig. 1c is a single point, again indicating simply-periodic motion.

As Γ is incresaed to 0.6159, the response changes dramatically, as can be seen in fig. 2. The power spectrum (fig. 2a) now has a very "noisy" character, with the fundamental driving frequency riding on a broad chaotic background (note the change in logarithmic vertical scale between figs. 1a and 2a). The phase-plane plot in fig. 2b exhibits a scatter of points distributed over a *large* number of potential wells. The Poincaré section in fig. 2c shows a strange attractor with characteristic folded structure. There is a remnant of the periodic motion, seen for $\Gamma = 0.6158$, evident in the phase-plane plot as coarse elliptical figures superimposed on the scatter background. For certain intervals of time the system locks into a periodic motion; this *intermittency* (in time) does not necessarily occur regularly thus giving rise to the chaotic background in the power spectrum. The spacing of the ellipses in the phase-plane reflects the 2π-periodicity of the cos ϕ potential. We have taken advantage of this eventual periodicity to obtain a higher density of points in the Poincaré section [fig. 2c] by "folding" the ϕ values into the range $0 < \phi \leqslant 2\pi$, i.e., one "unit cell" of the strange attractor *lattice* [2]. This intermittent periodic behavior becomes increasingly pronounced as Γ is increased further into the chaotic regime. In fig. 3 we show the corresponding phase-plane plots

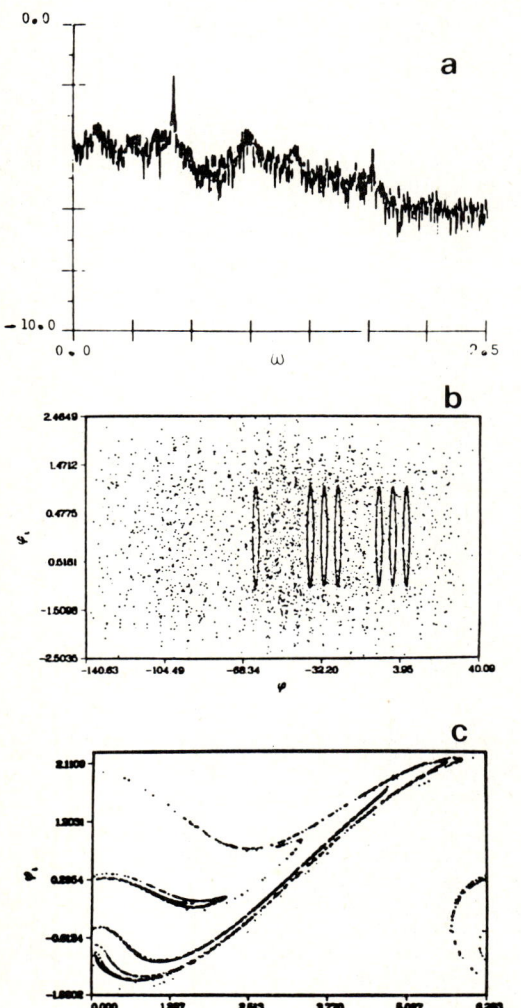

Fig. 2. Flat initial conditions, $\Gamma = 0.6159$, (a) power spectrum, (b) phase-plane, and (c) Poincaré section.

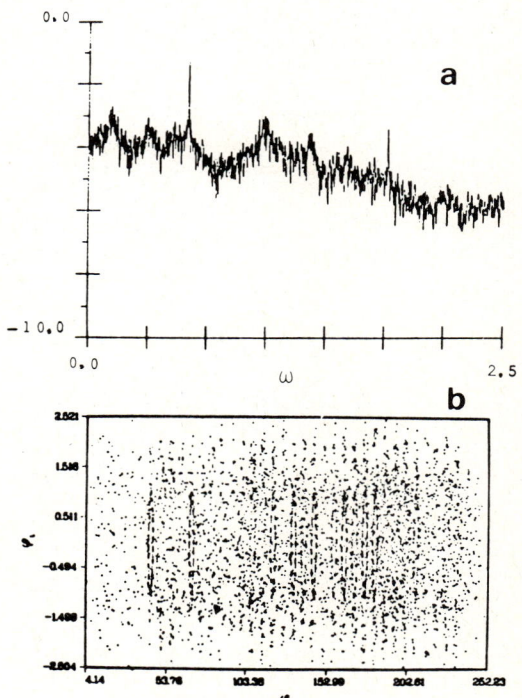

Fig. 3. Flat initial conditions, $\Gamma = 0.6160$, (a) power spectrum, (b) phase-plane.

and power spectra for $\Gamma = 0.6160$. Beyond $\Gamma = 0.6162$, the locking into periodic motion disappears (i.e., becomes indiscernable) from the phase-plane plots. As Γ is increased further into the first chaotic regime, the phase-plane plots exhibit increasingly diffuse scatter and the strange attractor lattice exhibits a more complicated folded structure.

The next threshold value is at $\Gamma = 0.6807\text{--}8$, where a transition from chaotic to "running periodic" response occurs: compare fig. 4 ($\Gamma = 0.6807$) and fig. 5 ($\Gamma = 0.6808$). In the power spectrum (fig. 5a) we observe, in addition to the fundamental driving frequency and *all* of its harmonics (contrast with fig. 1a where only odd harmonics are present), a strong component at zero frequency (d.c.). This results from a net *nonzero* time average of ϕ_t, which in turn results from a *net* advancement of the chain over the still-present potential barriers. Hence the term "running periodic" is used to describe such a response. The pattern of the phase-plane plot, fig. 5b, is the result of two features of this running mode: after moving over one barrier the particle oscillates once in this well before advancing to the next one. This unusual motion is illustrated more clearly in fig. 5c, where ϕ and ϕ_t are shown as a function of time, as they are in fig. 4d. The folded Poincaré section (not shown) for this periodic motion is a single point as it should be (contrast with fig. 4c). Notice the precursor to the uniform running in fig. 4 (compare figs. 4b, 5b, 6b and figs. 4d, 5c).

The next transition encountered as Γ is increased occurs at a value somewhat less than 0.724. The new state is still "running-periodic" (fig. 6b; $\Gamma = 0.725$) but subharmonic peaks at half-integral multiples of the driving frequency have appeared (fig. 6a), indicating additional "rocking" in the motion as the chain passes over the wells. Note that the Poincaré section (fig. 6c) shows two points (broadened by our interpolation procedure).

As Γ is increased further a narrow window

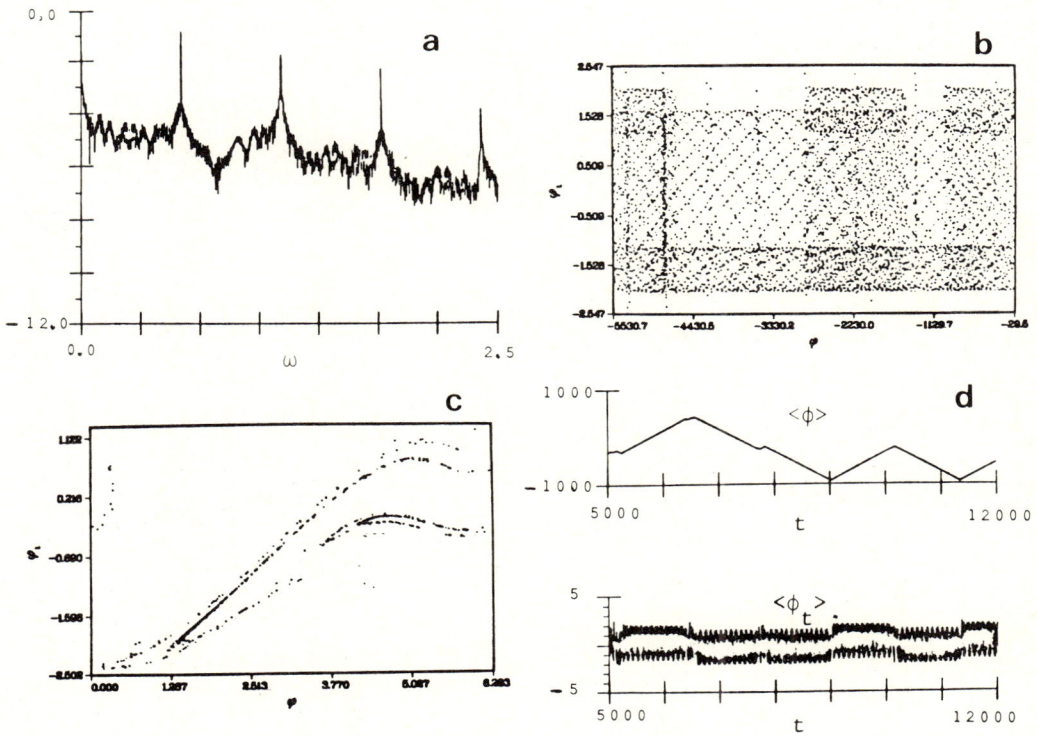

Fig. 4. Flat initial conditions, $\Gamma = 0.6807$, (a) power spectrum, (b) phase-plane, (c) Poincaré section, and (d) $\langle \phi \rangle$ and $\langle \phi_t \rangle$ vs. t.

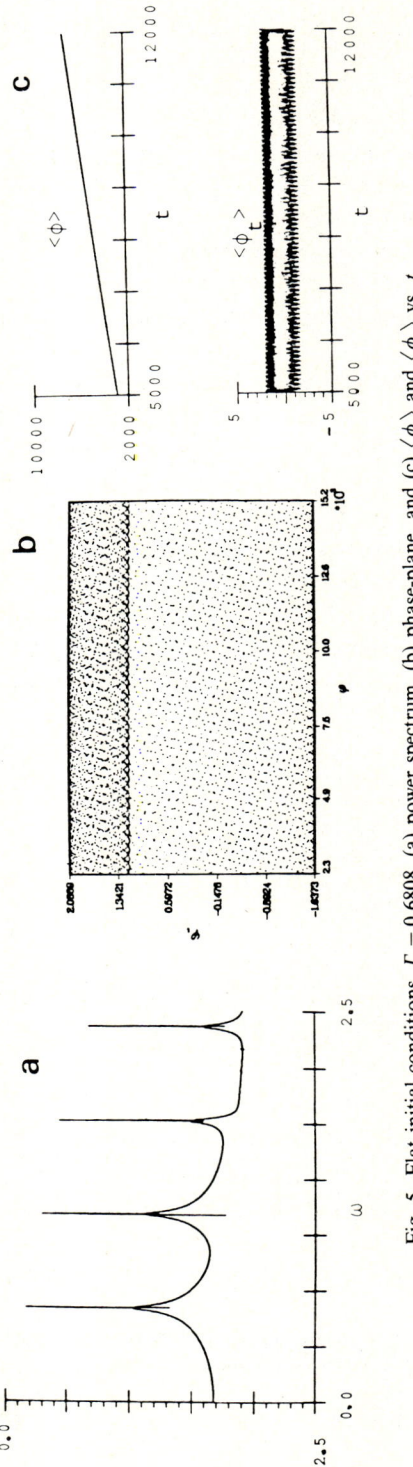

Fig. 5. Flat initial conditions, $\Gamma = 0.6808$, (a) power spectrum, (b) phase-plane, and (c) $\langle\phi\rangle$ and $\langle\phi_t\rangle$ vs. t.

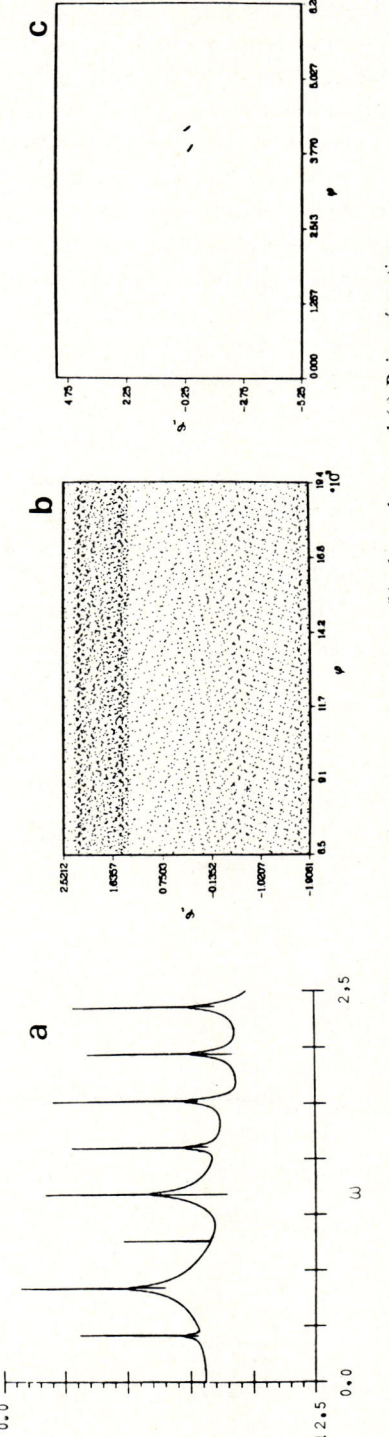

Fig. 6. Flat initial conditions, $\Gamma = 0.7250$, (a) power spectrum, (b) phase-plane, and (c) Poincaré section.

$(0.729 \leq \Gamma \leq 0.731)$ of period 5 behavior, without a d.c. component, is encountered before this running-oscillating mode becomes "running-chaotic", where first the small oscillation within the well goes chaotic ($\Gamma = 0.742$; fig. 7) and then subsequently the advancement over the barriers, where the system is confined in one well intermittently ($\Gamma = 0.743$, fig. 8): the direction of "running" is intermittently reversed, as shown in fig. 8b. Beyond $\Gamma = 0.75$ the d.c. component in the power

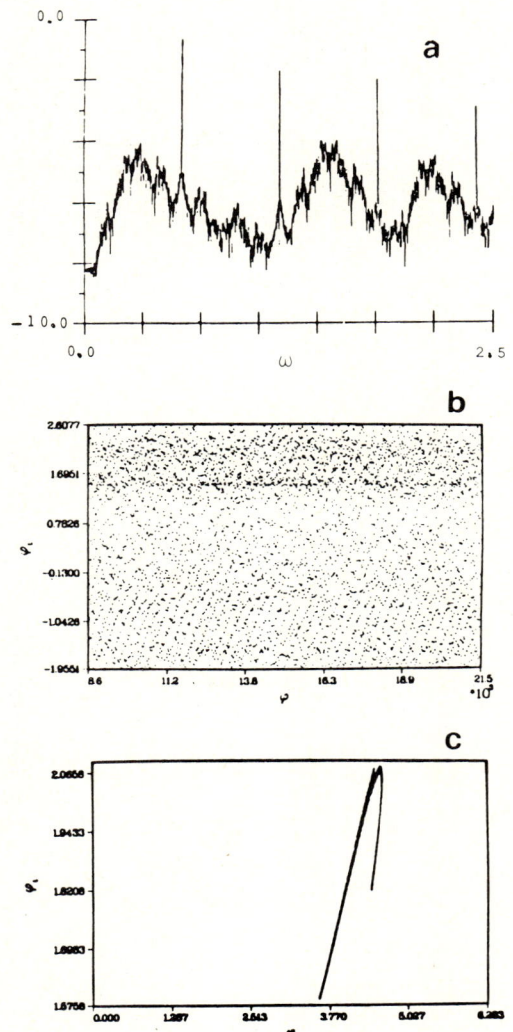

Fig. 7. Flat initial conditions, $\Gamma = 0.7420$, (a) power spectrum, (b) phase-plane, and (c) Poincaré section.

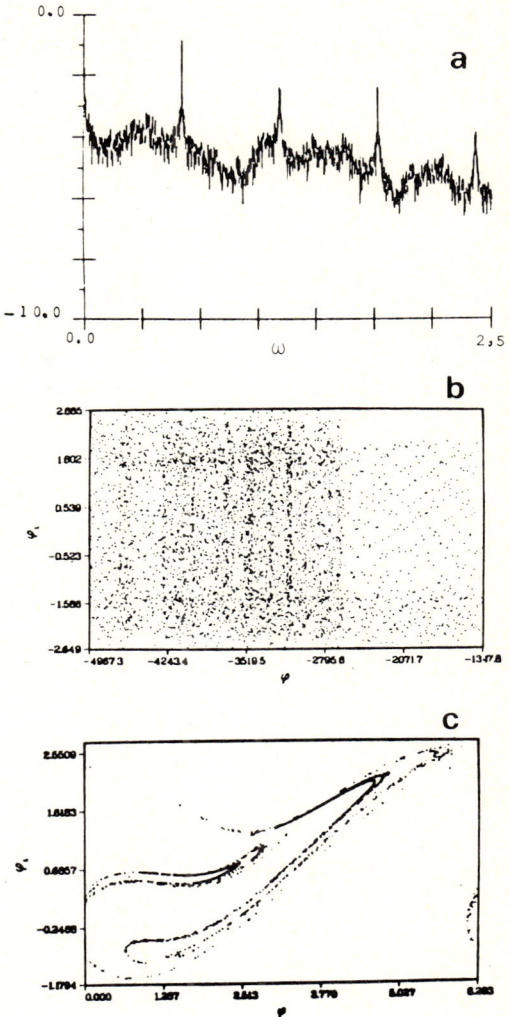

Fig. 8. Flat initial conditions, $\Gamma = 0.7430$, (a) power spectrum, (b) phase-plane, and (c) Poincaré section.

spectrum vanishes thus indicating the disappearance of the overall advancement of ϕ, while the system still shows chaotic behavior. In addition the even harmonics of the driving frequency have died out. For $\Gamma > 0.87$ the third subharmonic starts to rise from the still noisy (chaotic) power spectrum. For $\Gamma > 0.900$ the system moves out of this chaotic regime into a simple period-3 behavior (compare fig. 9 ($\Gamma = 0.900$) and fig. 10 ($\Gamma = 0.901$)). At $\Gamma = 0.94$ we again encounter a

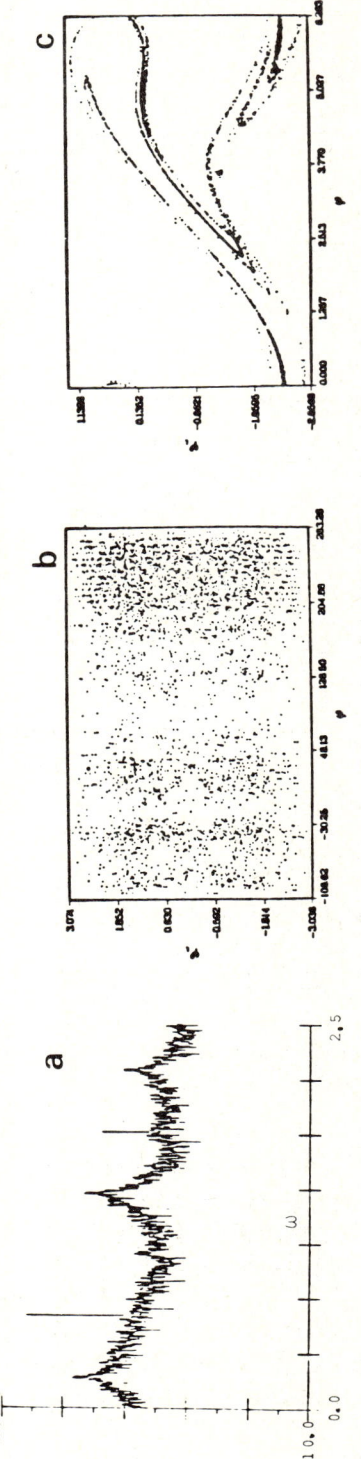

Fig. 9. Flat initial conditions, $\Gamma = 0.9000$, (a) power spectrum, (b) phase-plane, and (c) Poincaré section.

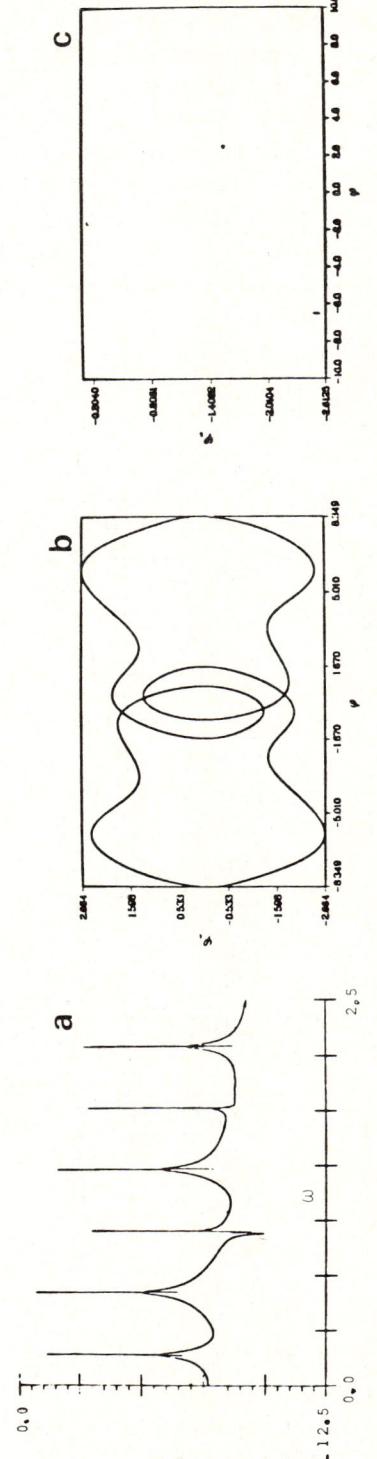

Fig. 10. Flat initial conditions, $\Gamma = 0.9010$, (a) power spectrum, (b) phase-plane, and (c) Poincaré section.

chaotic region with only odd harmonics present in the power spectrum.

There exist additional chaotic ↔ nonchaotic transitions for larger values of Γ, but we do not describe them in detail here.

3. Breather initial conditions

We now turn to a discussion of our results when breather initial conditions are imposed on the system. The non-translating breather solution of the unperturbed sine-Gordon equation is given by

$$\phi(x, t) = 4 \tan^{-1}\left\{\frac{(1-\omega_b^2)^{1/2}\cos(\omega_b t)}{\omega_b \cosh[x(1-\omega_b^2)^{1/2}]}\right\}, \quad (3)$$

where $\omega_b (0 < \omega_b < 1)$ is the internal vibration or "breathing" frequency. We have chosen to study two cases, namely $\omega_b = 0.2$ and 0.7, which are of a narrow and a wide initial profile, respectively. As for all the results in this paper, the frequency of the driving force is held fixed at $\omega_d = 0.6$, and we see that one initial breather has $\omega_b < \omega_d$ while the other has $\omega_d > \omega_d$. Surprisingly, we find that both of these inhomogeneous initial conditions evolve with time onto the same basic attractor, although with sensitivity to initial conditions in the chaotic regimes. In tables Ib, and Ic we have indicated the different responses as Γ is varied, for $\omega_b = 0.2$. In table Ic we show results in the case where a strict spatial symmetry was imposed on the time evolution through the symmetry in the initial condition (3). Table Ib shows the significant differences that occur when this symmetry is even slightly broken (as can be expected physically) in parameter regimes where chaotic tendencies exist. The symmetry can be broken in many ways but the results here used the device of slightly asymmetrizing the location of certain grid points. This asymmetry was made very small, close to machine accuracy, to test the stability against the spatial symmetry. Not surprisingly we find a strong dependence on this change of initial condition when there is an underlying tendency towards chaos. More interestingly, however, there is not a simple divergence of orbits (from which a Lyapunov exponent might be estimated). Rather there is an intermittent recurrence phenomenon in which the initial near-symmetry is almost completely restored during (apparently) random time intervals or the center of symmetry simply flows in x-space. This unusual behavior is intimately related to large amplitude motions in ϕ and chaotic flow. Only at large values of $\Gamma (\geq 0.9)$ does the very weak seed of asymmetry lead to a divergent loss of spatial symmetry.

Most of the scenarios we find can be appreciated qualitatively in terms of *competitions* between "background" (i.e., similar to the flat initial conditions case) and spatial structure tendencies. As we saw in section 2, the former motions are very sensitive to Γ and even for $\Gamma \leq 1$ there are regimes of chaotic or large-amplitude ϕ motions. On the other hand, the presence of spatial structure in general inhibits this tendency. We describe below the ways in which the competition is resolved.

Consider first the case in which the spatial symmetry was slightly broken (table Ib). For $\Gamma \leq 0.575$ we find a spatial smoothing of the initial breather structure and the final evolution is the same non-chaotic, entrained motion as for a single particle (i.e., flat initial conditions, section 2). In the range $0.575 \leq \Gamma \leq 0.6197$ we observe an interesting symbiosis of spatial and temporal patterns. For Γ just above 0.575 we find an evolution into a synchronized breather-like state [11] oscillating with a frequency ω_d – an example is shown in fig. 11a, $t = 555$. [Figs. 11a-c, 12a-c, 15c, 17a show plots of ϕ (upper figure) and ϕ_t (lower figure) versus x at various times t.] For $\Gamma \geq 0.616$ – where the "background" would like to run through large ϕ variations (cf. chaos sets in for flat initial conditions) – the breather begins to increasingly distort into a kink–antikink pattern for part of its cycle (see fig. 11b, $t = 500$). At $\Gamma = 0.6197$–8 a transition to a permanent spatial period $1/2$ structure occurs (fig. 11c, $t = 500$), as the kink and antikink interact through the periodic boundary conditions and bind [this threshold value of Γ and the resulting spatial pattern can depend on the

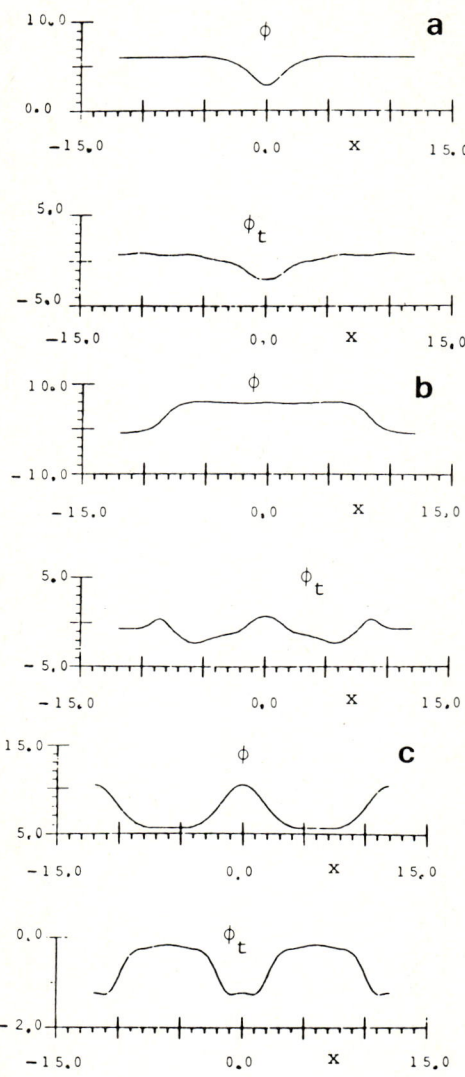

Fig. 11. Breather initial conditions (broken symmetry), ϕ and ϕ_t for (a) $\Gamma = 0.5750$, $t = 555$, (b) $\Gamma = 0.6160$, $t = 500$, and (c) $\Gamma = 0.61975$, $t = 500$.

system length]. The spatial pattern is close to that of two breathers but is probably best described in general as a period 1/2 wave, similar to the Jacobi elliptic function solutions known [12] for the finite length *unperturbed* sine-Gordon system – the observed structure is periodic on *half* the system's length. It is worth noting that the spatial structure has *removed* the tendency towards time

chaos – however the competing forces (the background tendency towards flow in ϕ space vs. the systems preference to maintain the initial spatial structure) have resulted in a *change* in the spatial pattern. For $0.6198 < \Gamma \leqslant 0.885$ this same competition results in *intermittent* periods of evolution where the spatial period 1/2 structure either translates in x (not necessarily uniformly in time) or is broken before becoming locked in the period 1/2 pattern again [13]. An example is shown in fig. 12 where $\Gamma = 0.6805$; for a while the period 1/2 pattern is locked (see fig. 12a for $t = 510$), then it is de-locked and translates (fig. 12b for $t = 1500$), then the pattern becomes re-locked (fig. 12c for $t = 1816$). The corresponding power spectrum $S(x = 8.2, \omega)$, phase-plane and Poincaré section are shown in figs. 12d, e, f. The frequency and length of these intermittent bursts vary non-uniformly as Γ is increased in the above range, presumably related to the variety of background flow tendencies (cf. table Ia). Interestingly, the intermittent spatial pattern de-locking is accompanied by *large* random fluctuations in ϕ (i.e., the background). During the spatially locked periods the time-evolution of ϕ is confined to very few wells and is nonchaotic. Thus this spatial and temporal intermittency is the way the competing tendencies are accomodated by the system (i.e., the way the "frustration" is relieved [13]). The intermittency is evidenced in the phase-plane and Poincaré sections. In fig. 13a we show a phase-plane plot at a characteristic point for $\Gamma = 0.6808$. The dark structure to the right (enlarged in fig. 13b) corresponds to motion in the spatially locked periodic regime, while the structure at the left and the scattered points correspond to the spatially intermittent periods of de-locking. The intermittency also shows up in the power spectra – if taken in a spatially locked period 1/2 time interval (fig. 13c, $3000 < t < 5800$) no chaos is apparent, but if it includes times with spatially de-locked patterns it is chaotic in the sense of broadened peaks and higher, non-uniform background (fig. 13d, $1000 < t < 3900$ and fig. 13e, $2000 < t < 5600$). Indeed in some cases the de-locked time is so great

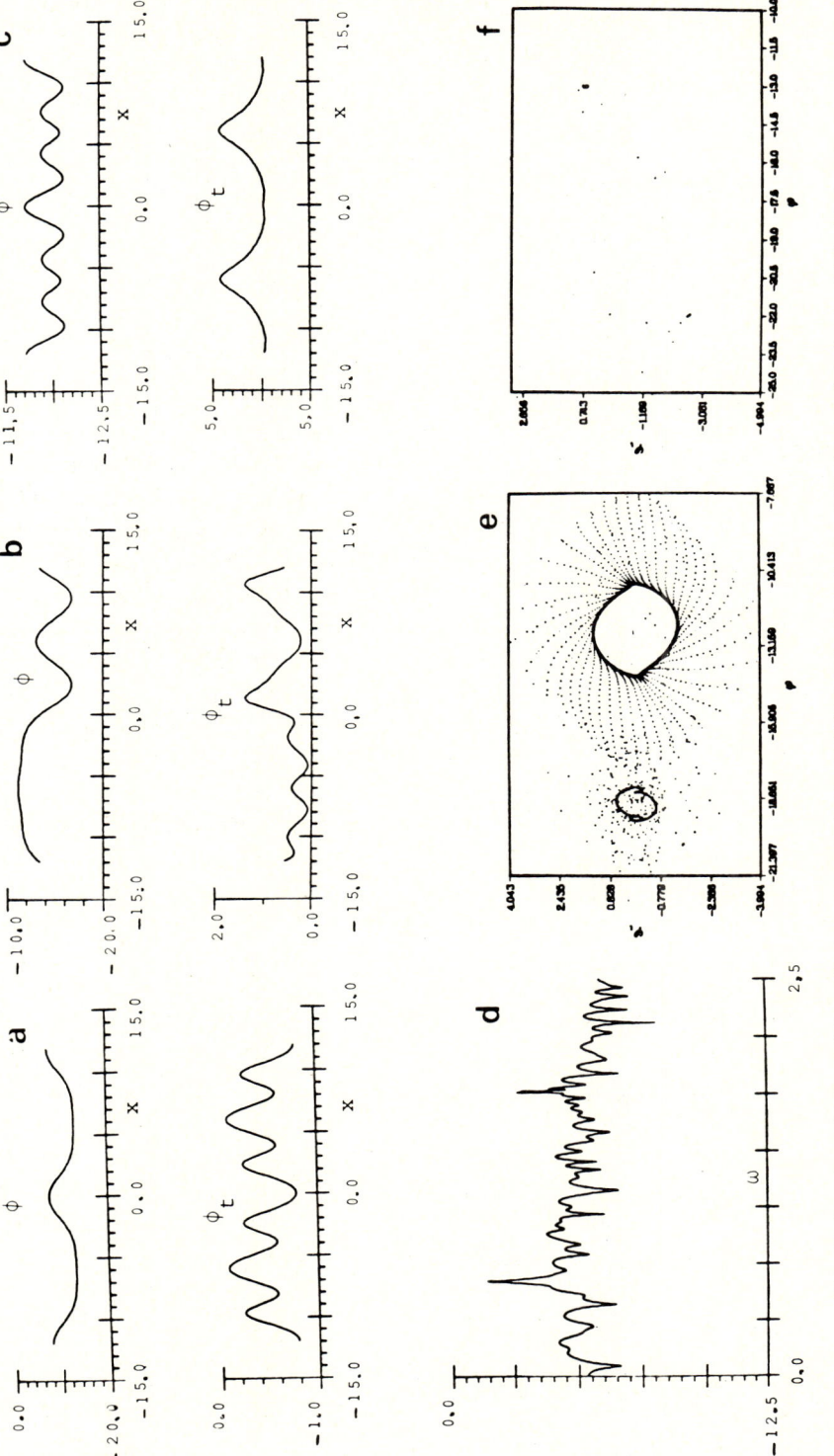

Fig. 12. Breather initial conditions (broken symmetry), $\Gamma = 0.6805$, (a) ϕ and ϕ_t for $t = 510$, (b) ϕ and ϕ_t for $t = 1500$, (c) ϕ and ϕ_t for $t = 1816$, (d) power spectrum $S(x = 8.2, \omega)$, (e) phase-plane ($x = 8.2$), and (f) Poincaré section ($x = 8.2$).

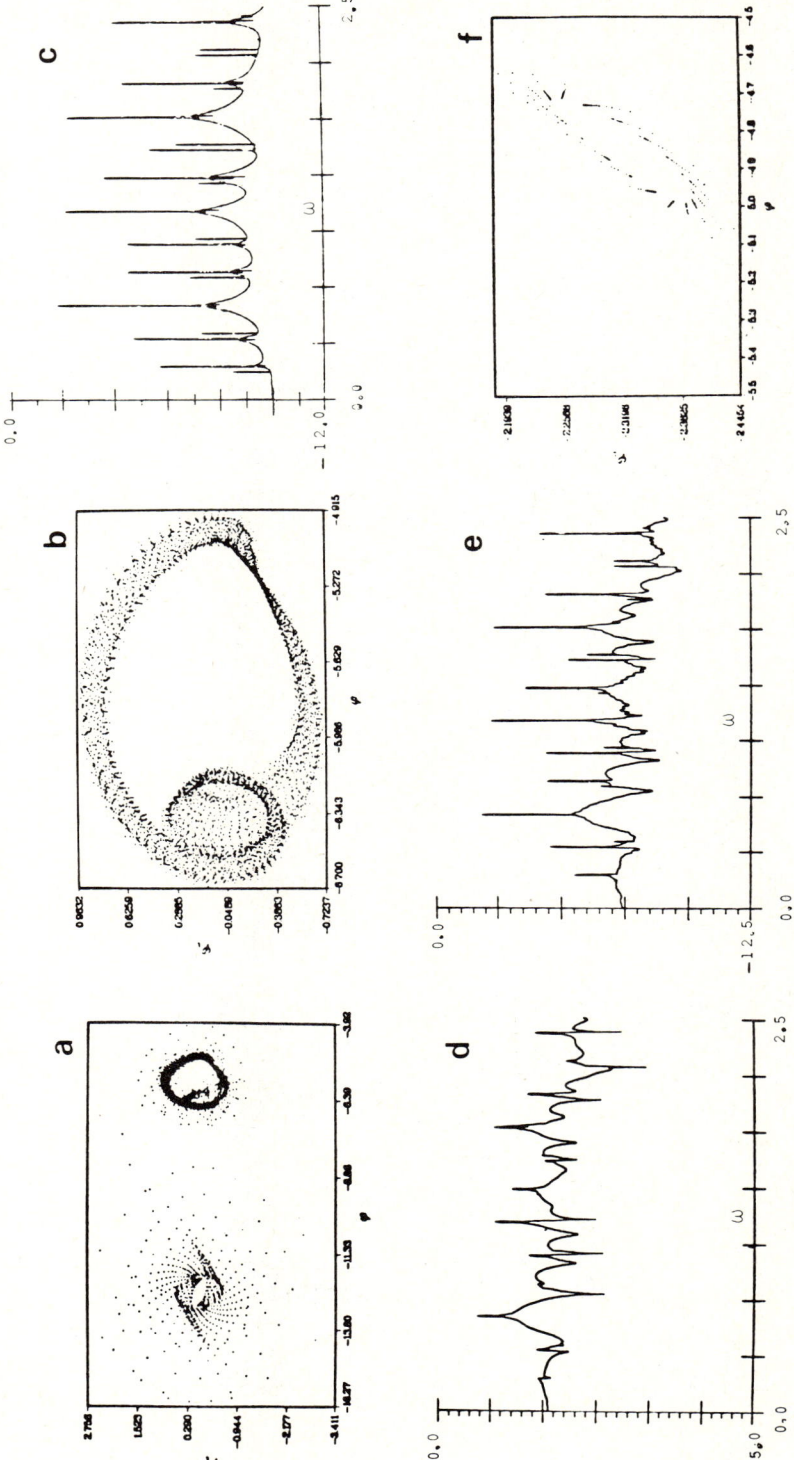

Fig. 13. Breather initial conditions (broken symmetry), $\Gamma = 0.6808$, (a) phase-plane, (b) phase-plane (enlarged part of (a)), (c) power spectrum (taken for $3000 < t < 5800$), (d) power spectrum (taken for $1000 < t < 3900$), (e) power spectrum (taken for $2000 < t < 5600$), and (f) Poincaré section (taken for $3000 < t < 5800$).

that the spectra can look broad in the same sense as flat initial condition time chaos. This is shown in fig. 14a for $\Gamma = 0.6810$ (figs. 14b, c show the phase-plane and Poincaré section, respectively). In the non-chaotic regimes, the periodic motions can be quite highly structured – e.g. period 17 at $\Gamma = 0.6808$ (fig. 13f) which appears close to a quasi-periodic motion on a torus – compare the amplitudes of the harmonics in the corresponding power spectrum (fig. 13c). The lowest-frequency peak occurs at $\omega = (5/17)\,\omega_d$ and peaks occur also at $(11/17)\,\omega_d$, $(17/17)$, $(23/17)\,\omega_d$, etc. The fact that we have 17 "sticks" in the Poincaré section rather than strict points is an indication of the accuracy of our interpolation scheme for determining ϕ and ϕ_t at the zero-crossings of the applied force. There is a quite sharp change of character at $\Gamma \approx 0.885$, but even for $\Gamma > 0.885$ there are intermittent bursts of spatially locked period 1/2 motion, which is shown in the plase-plane plot for $\Gamma = 0.890$ (fig. 15a – note the barely visible trace of a periodic structure around $\phi \approx 10$. The power spectrum however now shows a very broad noisy band (fig. 15b) – compare with the intermittency observed in the flat initial condition case (fig. 2a). Also the spatial period 1/2 structure is now completely broken for most of the time, see fig. 15c which shows the waveform for $t = 1550$. The Poincaré section (fig. 15d) shows no structure at all. We stress that in view of the intermittency occuring over long time intervals, caution must be exercised in taking and interpreting power spectra in the whole parameter range $\Gamma \gtrsim 0.62$.

We turn now to the breather initial condition (3) imposed with perfect symmetry (in our case around the chain center) (table Ic). Here we find that the competition described above cannot be relieved by developing intermittent flows or asymmetries in x-space. In fact for $\Gamma \lesssim 0.9$ the motion is simply periodic (with the same thresholds at $\Gamma \approx 0.57$ and 0.6197, as above), although as $\Gamma \to 0.9$ the spatial patterns become increasingly complex and there are increasingly long chaotic transients before a periodic motion sets in (indicating the increasing ease with which the periodic motion can be disturbed). In this sense, for $0.57 \lesssim \Gamma \lesssim 0.9$, the presence of spatial structure dominates and *totally* inhibits chaotic or large amplitude ϕ variations. However, for $\Gamma \gtrsim 0.9$, the converse holds and background motions dominate, producing chaos and, judging from the Poincaré sections, the motion is on a strange attractor of high dimension, with many modes participating. Compare results for $\Gamma = 0.89$ and 0.92, figs. 16a–c and d–f respectively, and recall the typical low-dimensional

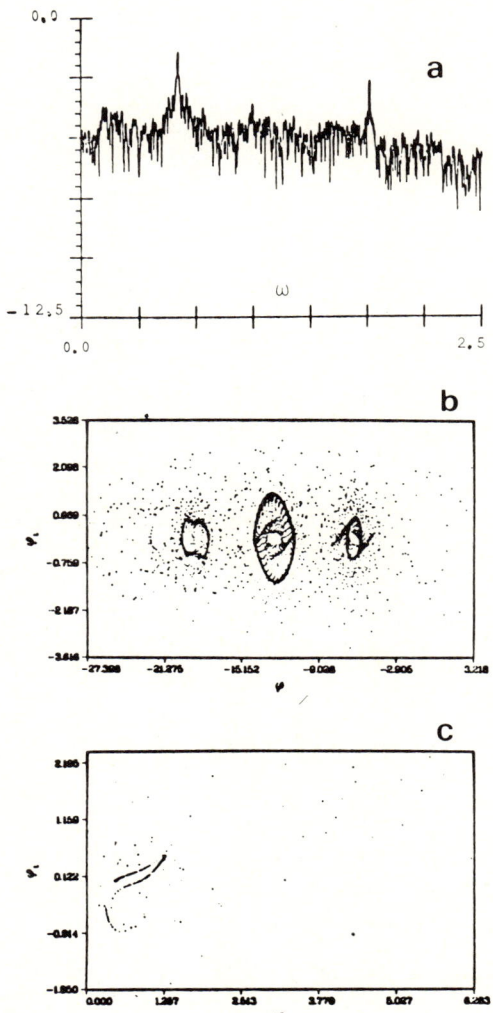

Fig. 14. Breather initial conditions (broken symmetry), $\Gamma = 0.6810$, (a) power spectrum, (b) phase-plane, and (c) Poincaré section.

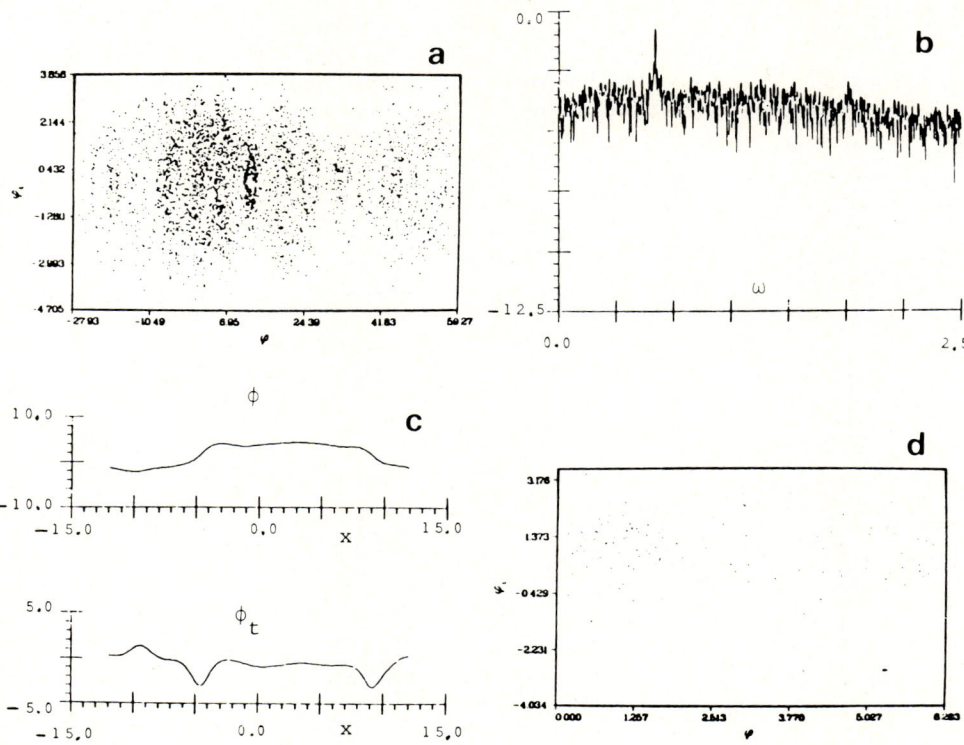

Fig. 15. Breather initial conditions (broken symmetry), $\Gamma = 0.8900$, (a) phase-plane, (b) power spectrum, (c) ϕ and ϕ_t for $t = 1550$, and (d) Poincaré section.

attractors in section 2 (e.g. fig. 2c). Figs. 16a,d show power spectra; figs. 16c,f show spatially averaged ϕ and ϕ_t; fig. 16b is a phase-plane plot at $x = 0$; fig. 16e is a Poincaré section.

We have also investigated behavior at higher Γ values. Details will be described elsewhere, but we note that the competitions and influences of symmetry described above have additional novel effects. For example, with flat initial conditions there is another band of Γ values for which "running" motion occurs: $1.745 \lesssim \Gamma \lesssim 2.005$. Interestingly, for the breather condition with seeded asymmetry, the breather profile uses the possibility of developing a small asymmetry to deform to a sufficiently *low* amplitude, extended breather profile which then "runs" in ϕ just as for the flat case. In contrast, if strict symmetry is imposed this compromise does *not* occur and the running mode is inhibited. However, Γ is sufficiently large that large amplitude ϕ variations are still imposed (partly through kink–antikink production [6]) and result in chaotic evolution, preceded at smaller Γ by quasi-periodicity. Thus, strict symmetry encourages *chaotic* relief of the characteristic competition in this case.

4. Kink initial condition

In this section, summarized in table Id, we describe some of our results for the case when a static kink-soliton solution

$$\phi(x) = 4\tan^{-1}[\exp(x)], \qquad (4)$$

is placed at the center of the chain as an initial condition and the boundary conditions $\phi(L, t) - \phi(0, t) = 2\pi$ are maintained rather than the periodic boundary conditions employed in all

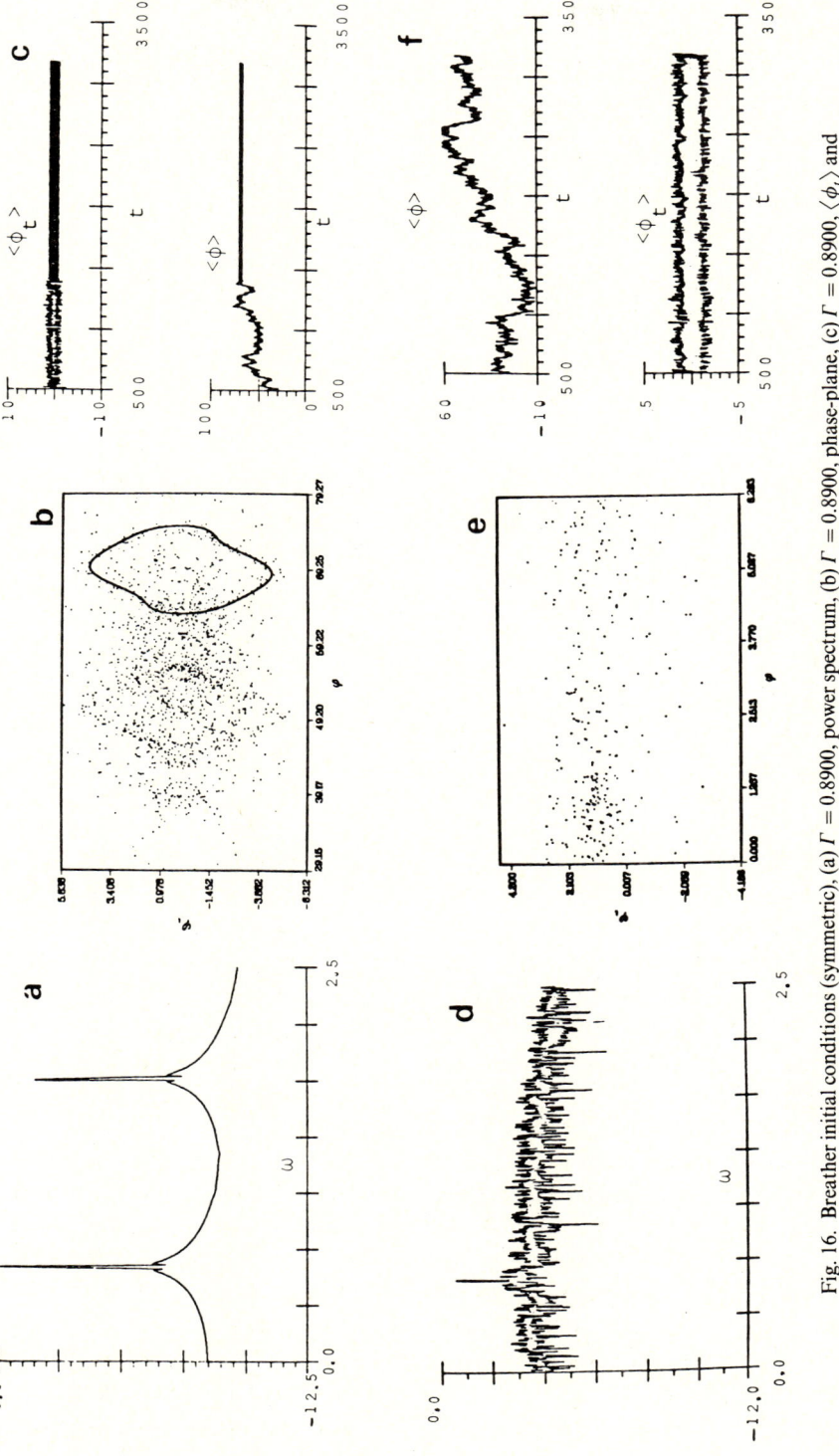

Fig. 16. Breather initial conditions (symmetric), (a) $\Gamma = 0.8900$, power spectrum, (b) $\Gamma = 0.8900$, phase-plane, (c) $\Gamma = 0.8900$, $\langle \phi_t \rangle$ and $\langle \phi \rangle$ vs. t, (d) $\Gamma = 0.9200$, power spectrum, (e) $\Gamma = 0.9200$, Poincaré section, and (f) $\Gamma = 0.9200$, $\langle \phi \rangle$ and $\langle \phi_t \rangle$ vs. t.

the cases above. These boundary conditions are consistent with the topological nature of the kink in an infinitely long system.

In the range $\Gamma \lesssim 0.59$ the motion of the chain is characterized by a small amplitude oscillation of the background around $\phi \approx 0$ or 2π, and of the kink center-of-mass around $x \approx 0$. These oscillations are synchronized with the driving frequency ω_d and can be described for small Γ with linear perturbation theory [14]. For $0.59 \lesssim \Gamma \lesssim 0.64$, similar to the breather case, the background tendency towards flow in ϕ space competes with the spatial structure: the system typically sets up a breather oscillation in addition to the kink (see fig. 17a where $\Gamma = 0.6$). This breather collides with the kink in a random way, leading to a noisy dressing of the power spectrum (shown in fig. 17b). The kink undergoes large displacements during the collisions, but oscillates periodically around fixed locations on the chain between collisions (see fig. 18c). The phase-plane plot (taken at the midpoint of the chain) is shown in fig. 17c. These intermittent collisions of the breather and kink increase with Γ and eventually lead to the same degree of broad band chaotic background found for flat initial conditions – see the power spectrum for $\Gamma = 0.63$ (fig. 18a) taken for a time interval where these collisions take place. The corresponding plase-plane plot is shown in fig. 18b. Fig. 18c shows the spatial average of ϕ and ϕ_t as functions of time. Clearly, as for the breather case we need to use caution in interpreting the power spectra. With the driving strength in the range $0.64 \lesssim \Gamma \lesssim 0.75$ we observe the appearance of an incommensurate *low* frequency component giving rise to side-bands and thereby apparent broadening of the peaks at the harmonics of the driver. Fig. 19 gives an example for $\Gamma = 0.65$. The Poincaré section shows periodicity in ϕ (fig. 19c). In this regime the kink center-of-mass drifts slowly but uniformly in x, superimposed on an oscillation at frequency ω_d (see fig. 19b, which shows the spatial averages of ϕ and ϕ_t). An accompanying breather also persists as for small Γ but it now drifts with the kink as a composite pattern.

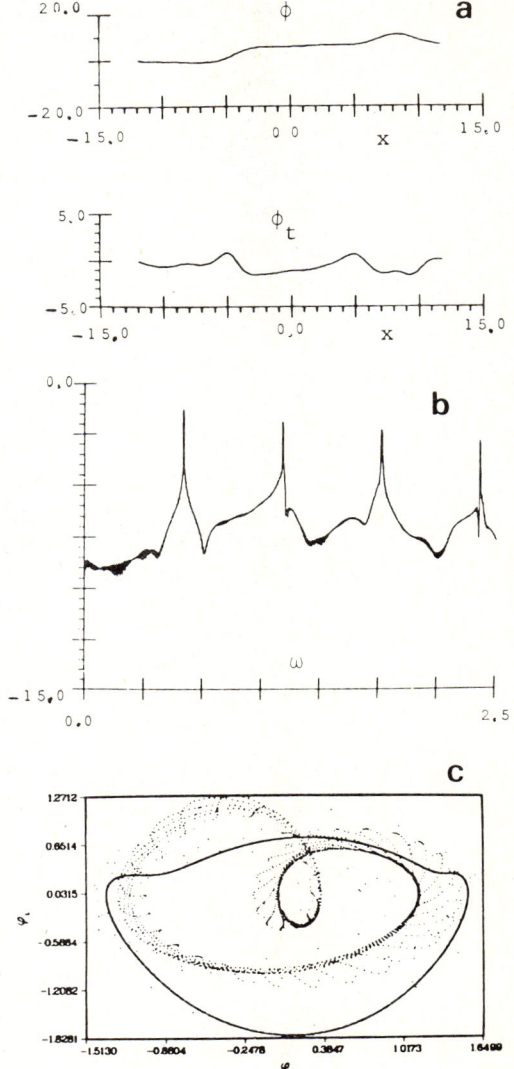

Fig. 17. Kink initial conditions, $\Gamma = 0.6000$ (a) ϕ and ϕ_t for $t = 500$, (b) power spectrum, and (c) phase-plane.

Finally when Γ is increased beyond 0.75 the low frequency component disappears (the kink no longer drifts) while the initial time the system spends in a chaotic state before it locks into a simple periodic motion increases with Γ. This is illustrated in fig. 20 for $\Gamma = 0.9$, figs. 20a and 20b are the power spectrum and phase-plane plots taken for $500 < t < 2500$, whereas figs. 20c and 20d are taken for $2000 < t < 6500$. The periodic cycle

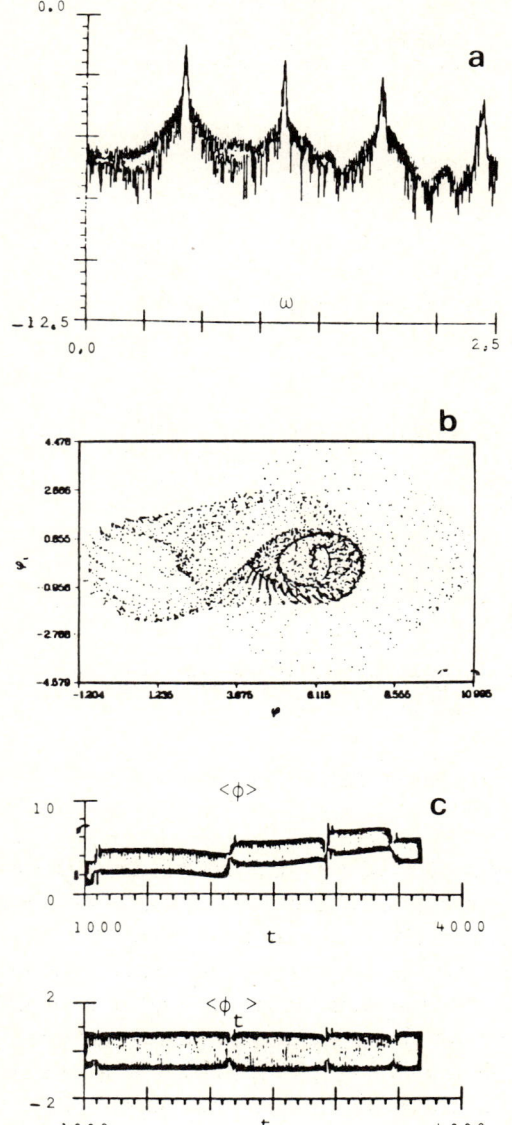

Fig. 18. Kink initial conditions, $\Gamma = 0.6300$, (a) power spectrum, (b) phase-plane, and (c) $\langle\phi\rangle$ and $\langle\phi_t\rangle$ vs. t.

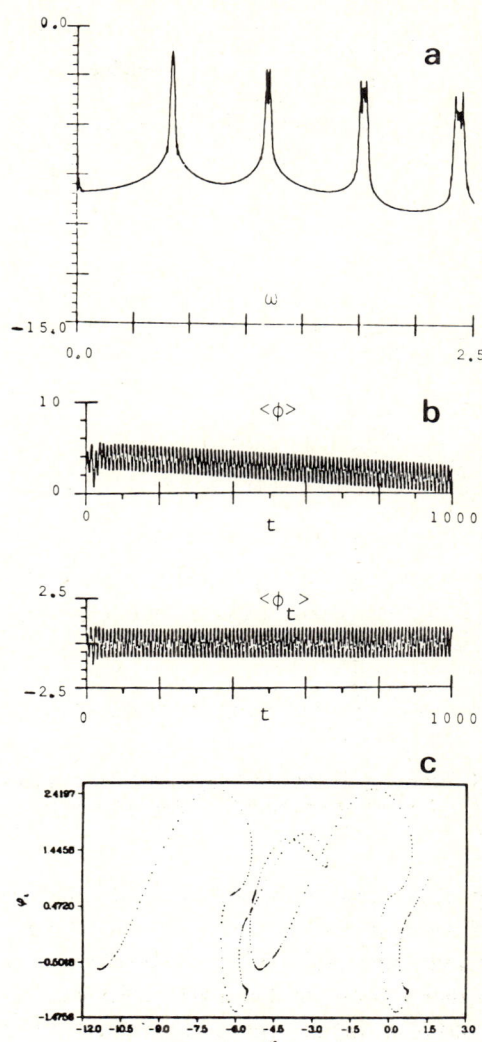

Fig. 19. Kink initial conditions, $\Gamma = 0.6500$, (a) power spectrum, (b) $\langle\phi\rangle$ and $\langle\phi_t\rangle$ vs. t, and (c) Poincaré section.

of fig. 20d is also visible in fig. 20b. This example clearly shows how spectra and phase-plane plots can change considerably over long periods of time.

In this section, we have seen again that an inhomogeneity has inhibited true chaos – this is only established at larger Γ values, as we will describe elsewhere.

5. Concluding remarks

We have studied the influence of spatial structure in the initial conditions on the appearance and character of time chaos in the a.c. driven, damped sine-Gordon equation. We have found an interesting symbiosis of time and spatial structures in the response to the external force. Spatial patterns tend to *inhibit* time chaos of the sort seen with flat (no spatial structure) initial conditions – but *com-*

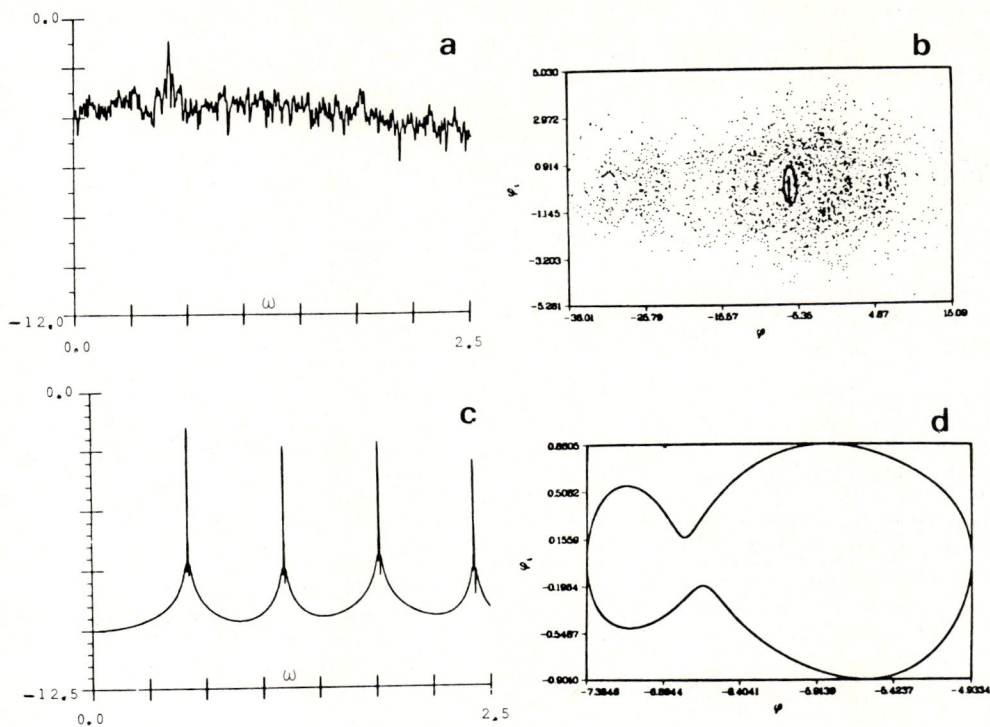

Fig. 20. Kink initial conditions,, $\Gamma = 0.9000$, (a) power spectrum (taken for $500 < t < 2500$), (b) phase-plane (taken for $500 < t < 2500$), (c) power spectrum (taken for $2000 < t < 6500$), and (d) phase-plane (taken for $2000 < t < 6500$).

peting tendencies of background motion versus spatial structure can lead to a rich *intermittency* in space and time [13]. This is a different kind of chaos than is found in the single particle case [1–3]. New basins of attractions are found – interestingly breathers with different internal frequencies seem to lie in the same basin for the cases we have studied. We have also found that spatial *symmetry* imposes even more stringent restrictions on available dynamics, but that such a symmetry is broken very easily.

Presently the *spatially* Fourier transformed quantities, $S(q, t)$ and $S(q, \omega)$, are being analyzed [15] to give more insight into the spatial structures described above. Further extensions will include time-series analysis to extract the geometry of the attractors and to clarify the possible existence of low dimensional maps. It seems clear from the raw Poincaré sections that in some cases (e.g. fig. 16) many modes are involved and that the dimension of the attractor is larger than for flat initial conditions. A collective coordinate analysis is being investigated in order to distinguish, e.g., the kink center-of-mass from the background motion information which we are also extracting numerically. Finally, we reiterate [7] the wider setting for studying the coexistence of temporal chaos in the presence of persistent spatial structure – this is a generic phenomenon with additional examples including dislocations in Couette flow, clumps in turbulent plasmas or large-scale structures in turbulent fluids.

Acknowledgements

We would like to thank Doyne Farmer for valuable discussions. One of us (SET) wishes to acknowledge the support of an Alfred P. Sloan Research Fellowship. The work at Los Alamos was supported by the USDOE.

References

[1] B.A. Huberman and J.P. Crutchfield, Phys. Rev. Lett. 43 (1979) 1743;. C. Herring and B.A. Huberman, Appl. Phys. Lett. 36 (1980) 976.
[2] B.A. Huberman, J.P. Crutchfield and N.H. Packard, Appl. Phys. Lett. 37 (1980) 750. D. D'Humieres, M.R. Beasley, B.A. Huberman and A. Libchaber, Phys. Rev. A 26 (1982) 3483.
[3] N.F. Pedersen and A. Davidson, Appl. Phys. Lett. 39 (1981) 830. M. Cirillo and N.F. Pedersen, Phys. Lett. 90A (1982) 150.
[4] M. Feigenbaum, J. Stat. Phys. 19 (1978) 25. 21 (1979) 669. Los Alamos Science 1 (1980) 4.
[5] e.g., Nonlinear Dynamics, R.H.G. Helleman, ed., Annal. N.Y. Academy of Sciences 357 (1980).
[6] J.C. Eilbeck, P.S. Lomdahl and A.C. Newell, Phys. Lett. 87A (1981) 1.
[7] D. Bennett, A.R. Bishop and S.E. Trullinger, Z. Phys. B 47 (1982) 265.
[8] e.g., R.V. Jenson and C.R. Oberman, Physica 4D (1982) 183.
[9] e.g., A.C. Scott, F.Y.F. Chu and S.A. Reible, J. Appl. Phys. 47 (1976) 3272. P.S. Lomdahl, O.H. Soerensen and P.L. Christiansen, Phys. Rev. B 25 (1982) 5737.
[10] e.g., M.J. Rice, A.R. Bishop, J.A. Krunhansl and S.E. Trullinger, Phys. Rev. Lett. 36 (1976) 432.
[11] e.g., D.J. Kaup and A.C. Newell, Proc. Roy. Soc. (London). A361 (1978) 413.
[12] R.M. DeLeonardis, S.E. Trullinger and R.F. Wallis, J. Appl. Phys. 51 (1980) 1211; and references therein.
[13] cf. frustrated Hamiltonian systems relieved by discommensurations. e.g. S. Aubry, these proceedings.
[14] M.B. Fogel, S.E. Trullinger and A.R. Bishop, Phys. Lett. 59A (1976) 81.
[15] A.R. Bishop, K. Fesser, P.S. Lomdahl, W.C. Kerr, M.B. Williams and S.E. Trullinger, Phys. Rev. Lett. 50 (1983) 1095.

CHAPTER 5

TRANSITION TO CHAOS
IN MAPS OF THE CIRCLE

A RENORMALISATION APPROACH TO INVARIANT CIRCLES IN AREA-PRESERVING MAPS

R.S. MACKAY[†]

Plasma Physics Laboratory, Princeton University, Princeton NJ 08544, USA

Kadanoff and Shenker introduced a renormalisation approach to invariant circles in area-preserving maps. This paper makes more precise the connection between invariant circles and the renormalisation operator. Restricting attention to noble rotation numbers, the stability of a simple fixed point of the renormalisation is analysed, corresponding to a linear twist map. It is found to be essentially attracting, so that noble circles persist under perturbation, giving a new view on KAM theory. Shenker and Kadanoff found evidence for another fixed point, corresponding to a map with a non-smooth noble circle. Further evidence is given in this paper. It has essentially only one unstable direction, and its stable manifold is believed to give the boundary of the set of twist maps with a noble circle. Finally, noble circles are shown to be locally most robust, in an important sense.

Contents

1. Area-preserving twist maps 283
2. Invariant circles 285
3. Action representation 286
4. Renormalisation 286
5. Commuting pairs 287
6. Simple fixed point 288
7. Compactness of DN_1 at the simple fixed point . 290
8. Significance of the simple fixed point 291
9. Connections between invariant circles and nearby periodic orbits 291
10. Critical noble circles 293
11. Critical fixed point 295
12. Golden curves for pairs converging to any fixed point 296
13. Robustness of noble circles 296
14. Conclusion 298

1. Area-preserving twist maps

Nonlinear stability in many conservative systems is equivalent to existence of invariant circles for related area-preserving maps. I will motivate this paper with a simple but significant example from plasma physics, namely, flow along magnetic field lines. To a first approximation, charged particles in a magnetic field follow the field lines in tight helices.

[†] Present address: Department of Applied Mathematics, Queen Mary College, Mile End Road, London E1 4NS, England.

The idea of fusion devices such as the tokamak is to confine them by a magnetic field which is largely toroidal. If the field lines remain confined within the device, there is a chance that the particles will too.

Field line flow can be reduced to iteration of a *return map F* on a poloidal section, given by following the field lines once around the device. Fig. 1 shows some orbits of such a return map for a real magnetic field (Sinclair et al. [1]) (see also White et al. [2]). Since magnetic flux is conserved, this return map preserves the area form $\mathbf{B} \cdot (\xi \times \eta)$, where ξ, η are tangent vectors to the poloidal section. By Dar-

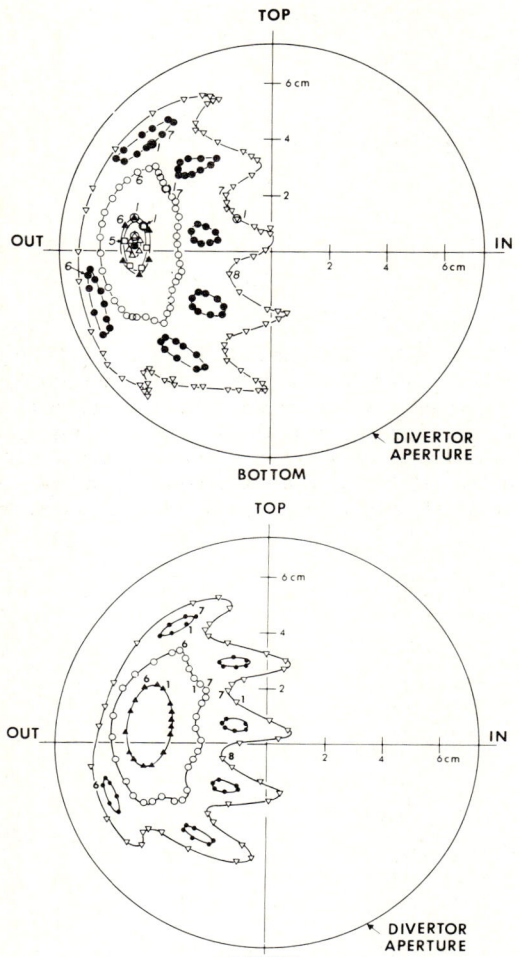

Fig. 1. Some orbits of a return map for a stellarator field (from Sinclair et al. [1]). The upper figure was produced by following a low energy electron beam injected paralel to the field. The lower figure was produced by integrating the field computationally.

boux's theorem [3], coordinates can be chosen to make it preserve the usual area.

Typically, a tokamak has a magnetic axis, that is, a field line which closes after one revolution. This corresponds to a fixed point of the return map. Also, the field is designed to have *rotational transform*, i.e. the other field lines twist around this axis. So in the poloidal section orbits rotate around the fixed point. Finally, tokamaks generally have *mag-netic shear*, that is, the rotation rate varies with distance from the magnetic axis.

Thus we are led to consider *area-preserving twist maps*, that is, maps

$$(\theta', z') = F(\theta, z). \quad (1.1)$$

with θ an angle variable, det $DF = 1$, and $\partial \theta'/\partial z$ of constant sign.

For the field line problem (and many others), (θ, z) are some sort of polar coordinates centred on a fixed point, and there is a coordinate singularity there. Removing the fixed point gives a map which can be regarded as acting on a cylinder. Maps of the cylinder derived in this way, however, have some special properties. Firstly, they are *end-preserving*. Secondly, given an area-preserving map of a cylinder, and a set U containing all points below some level z_1, and no points above some other level z_2, the difference between the areas of U and $F(U)$ is independent of U, and is called the *Calabi invariant* of F. For maps of the cylinder derived from a map of the plane by removing a fixed point, the Calabi invariant is clearly zero. Thus I will restrict attention to the *class A* of end-preserving, area-preserving twist maps of a cylinder with zero Calabi invariant.

Often I will want to consider a *lift* of F, rather than the map F itself. This means θ is regarded as a coordinate on a line rather than a circle, so we get a *periodic* map (of period 1, say), i.e.

$$\theta'(\theta + 1, z) = \theta'(\theta, z) + 1. \quad (1.2)$$

Maps of class A are relevant to many other conservative problems, for instance, all Hamiltonian systems with two degrees of freedom, or with one degree of freedom and periodic time dependence. This includes other examples from plasma physics, such as the motion of a charged particle in a 2-D field, guiding centre motion in 3-D, and ray tracing for waves in 2-D. They also have applications in other fields such as celestial mechanics, and solid-state physics (e.g. Aubry, this volume).

2. Invariant circles

If a map of class A has an invariant circle encircling the cylinder, then the circle traps everything below it. Conversely, Birkhoff (1932) [4] showed that an encircling invariant circle is necessary for confinement of any connected open set containing all points below some level. In all that follows I will restrict attention to this class of invariant circles. Note that zero Calabi invariant is necessary for existence of any such circle.

Next I introduce an important quantity for an invariant circle. Poincaré (Nitecki [5]) showed that for a homeomorphism g of a circle (or really, for a lift of g to a periodic homeomorphism of the line), the limit:

$$\omega = \lim_{q\to\infty} \frac{g^q(\theta_0)}{q} \tag{2.1}$$

exists and is the same for all θ_0. It is called the rotation number of g. In the case that g is the restriction of an area-preserving map to an invariant circle, ω is called the *rotation number* of the circle.

There are systems which have an invariant circle for every rotation number is some range. A map of a surface is said to be *integrable* if it possesses a differentiable invariant function which is not constant on any open set. For example, an axisymmetric magnetic field has a *flux function*. An extremum of the invariant is typically surrounded by many circles on which the invariant is constant. They are invariant under some power of the map, and so is the region between any two. Liouville (Arnold, [3]) showed that if the derivative of the invariant is non-zero on a compact connected invariant set, then there exist *angle–action coordinates* (θ, z), in which the map takes the standard integrable form

$$(\theta', z') = (\theta + \omega(z), z). \tag{2.2}$$

So the set is foliated by invariant circles.

Integrable maps are very special. For example, (2.2) has an invariant circle for every rotation number in a range, including rationals, but generically there are no rational circles. Nevertheless, conditions close to integrable are very common. For example, a map is arbitrarily close to integrable near enough to any typical elliptic point. A remarkable theory, due to Kolmogorov, Arnold and Moser, shows that systems close enough to an integrable one possess an arbitrarily large fraction of the invariant circles of the integrable system. The particular result most relevant to this paper is a corollary (Mather, private communication) of the Moser twist theorem (Moser, [6]). First, let us introduce some terminology. ω is called a *Diophantine number* (Niven, [7]), if

$$\exists C > 0,\ \tau \text{ such that } \left|\omega - \frac{p}{q}\right| \geq \frac{C}{q^\tau}\ \forall p, q \in \mathbb{Z}, q > 0. \tag{2.3}$$

An invariant circle is called *smooth* if the motion on it is sufficiently differentiably conjugate to rotation (the number of derivatives depending on τ for a Diophantine rotation number), i.e. if there is a sufficiently differentiable coordinate function ψ on the circle, with differentiable inverse, such that the map sends ψ to $\psi + \omega$. Then the result is that:

Smooth Diophantine circles persist, for small enough perturbation in class A.

On the other hand, here are maps of class A with no encircling invariant circles. For example, the *standard map*,

$$z' = z - \frac{k}{2\pi}\sin 2\pi\theta,$$
$$\theta' = \theta + z', \tag{2.4}$$

has no invariant circles for $|k| \geq 2\pi$, because then it has an accelerator mode (Chirikov [8]), and even for $|k| > 4/3$ (Mather [9]).

The size of the perturbations allowed by the Moser twist theorem depends only on C and τ in the Diophantine condition, and the local twist. It is largest for τ small and C large. In any interval, the number(s) for which τ can be taken smallest and C largest (excluding a finite set of q) is always a *noble* number (terminology due to Percival [10]). These are the numbers whose *continued fraction expansion*,

$$\omega = m_0 + \cfrac{1}{m_1 + \cfrac{1}{m_2 + \cdots}} \equiv [m_0, m_1, m_2, \ldots],$$

$$m_i \in \mathbb{Z}, m_i \geqslant 1 \text{ for } i \geqslant 1, \tag{2.5}$$

has $m_i = 1$ for all large enough i. They satisfy a Diophantine condition with $\tau = 2$, the smallest possible. The noblest of them all is the *golden ratio*:

$$\gamma = [(1,)^\infty] = \frac{\sqrt{5} + 1}{2}, \tag{2.6}$$

which has the largest possible value for C (for $\tau = 2$) of $1/\gamma^2$ (Prasad, [33]). This leads one to suspect that typically noble circles may be the most robust, in the sense that the last circle to break up in any region, as a parameter varies, will be a noble. For this reason I will concentrate almost entirely on nobles.

The proofs in KAM theory generally give unrealistically low estimates of the perturbation sizes sufficient for persistence of invariant circles. In this paper, I develop a new approach to KAM theory which, I believe, gives the boundary of the set of twist maps with an invariant circle of given rotation number.

3. Action representation

Before I describe the renormalisation approach to invariant circles, I will need an important representation for area preserving twist maps. As this representation does not require periodicity, I use coordinates (x, y) in place of (θ, z). Given a function $\tau(x, x')$, with $\tau_{12}(x, x')$ of constant sign, the relations

$$y' = \tau_2(x, x'),$$
$$y = -\tau_1(x, x'), \tag{3.1}$$

generate an area-preserving map $T: (x, y) \to (x', y')$ (where subscript i refers to the derivative with respect to the ith argument). It satisfies the twist condition (section 1), since

$$\frac{\partial x'}{\partial y} = -\frac{1}{\tau_{12}(x, x')}. \tag{3.2}$$

Conversely, every area-preserving twist map can be generated in the above fashion. In a given coordinate system, the generating function is unique up to addition of a constant. I call it the *action generating function*.

The rule for composition of action generating functions (where defined) is that of *stationarity*, i.e. the generating function for the composition TU of two maps T, U with generating functions τ, v, is

$$v \oplus \tau(x, x'') = v(x, x') + \tau(x', x''), \tag{3.3}$$

where $x'(x, x'')$ is chosen to make the sum stationary with respect to variations in x', i.e.

$$0 = v_2(x, x') + \tau_1(x', x''). \tag{3.4}$$

That $v \oplus \tau$ generates TU can be seen immediately from (3.1).

Note for a periodic map, that since $\tau(\theta + 1, \theta' + 1)$ generates the same map as $\tau(\theta, \theta')$, they can differ only by a constant. This constant can easily be shown to be the Calabi invariant.

4. Renormalisation

Now I will motivate the renormalisation. Rotation number can be generalised to other orbits than those on an invariant circle. I say that (the orbit of) (θ, z) has rotation number

$$\lim_{q \to \infty} \frac{\pi_1 F^q(\theta, z)}{q} \tag{4.1}$$

if the limit exists (which it need not), where π_1 is the projection onto the first coordinate. Without loss of generality, consider the orbit of the origin **0**. If it has rotation number ω, then

$$\pi_1 F^q R^p(\mathbf{0}) = q\omega - p + o(q), \quad \text{as } q \to \infty, \quad (4.2)$$

where

$$R(\theta, z) = (\theta - 1, z). \quad (4.3)$$

If **0** belongs to a circle on which F is topologically conjugate to rotation then we have the stronger statement

$$\pi_1 F^{q_n} R^{p_n}(\mathbf{0}) \to 0, \quad \text{if } q_n\omega - p_n \to 0. \quad (4.4)$$

In the case of differentiable conjugacy to rotation, one can say even more:

$$\pi_1 F^{q_n} R^{p_n}(\mathbf{0}) \sim K(q_n\omega - p_n) \quad \text{as } q_n\omega - p_n \to 0. \quad (4.5)$$

This suggests that we consider the sequence of maps

$$B_n F^{q_n} R^{p_n} B_n^{-1}, \quad (4.6)$$

where the B_n are coordinate changes, looking on successively smaller scales.

A choice of p_n, q_n for which $q_n\omega - p_n$ is particularly small is given by the *convergents* of ω. They are the successive truncations

$$\frac{p_n}{q_n} = m_0 + \cfrac{1}{m_1 + \cfrac{1}{\cdots + \cfrac{1}{m_n}}} \equiv [m_0, m_1, \ldots, m_n] \quad (4.7)$$

of its continued fraction expansion. For this choice, there is a systematic way to generate the sequence (4.6). Define the *renormalisation operator* N_m (for $m \in \mathbb{Z}$), acting on the pairs (U, T) of maps, by

$$N_m : \begin{cases} U' = BTB^{-1}, \\ T' = BT^m U B^{-1}. \end{cases} \quad (4.8)$$

B is a coordinate change, chosen to *renormalise* the pair $(T, T^m U)$ in some sense, for which I will not make a specific choice here. Then it follows from the definition of convergents that

$$N_{m_n} \ldots N_{m_0}(F, R)$$
$$= (B_n F^{q_{n-1}} R^{p_{n-1}} B_n^{-1}, B_n F^{q_n} R^{p_n} B_n^{-1}), \quad (4.9)$$

where B_n is the composition of the successive coordinate changes B. This is essentially the same renormalisation as that introduced by Kadanoff [11] and Shenker [12], and applied to the dissipative case by Feigenbaum et al. [13] and Rand et al. [14] (see also the articles by Shenker and by Siggia, in this volume). It is also closely related to the approximate renormalisation of Escande and Doveil [15] for Hamiltonians.

An apparent problem with the renormalisation is that in looking on successively smaller scales one loses the periodicity of the map in θ. In the next section, I will show how the essence of the periodicity can be saved, by generalising the class of periodic maps to that of commuting pairs of maps.

5. Commuting pairs

To say that F is periodic in θ is equivalent to saying that F commutes with R (4.3). So let us generalise the important concepts for periodic maps, in particular, rotation number and Calabi invariant, to commuting pairs (U, T). I use coordinates $\mathbf{x} = (x, y)$ in place of (θ, z) to indicate that there is not necessarily any periodicity in x.

Firstly I generalise orbits and invariant circles.

Definitions. The *orbit* of $\mathbf{x}$ under a commuting pair (U, T) is $\{U^q T^p \mathbf{x} : p, q \in \mathbb{Z}\}$.

A point $\mathbf{x}$ is *periodic* if $\exists\, (p, q) \in \mathbb{Z}^2 \setminus \{\mathbf{0}\}$ so that $U^q T^p \mathbf{x} = \mathbf{x}$. It has *type* (p, q) if these are the smallest such integers ($q \geq 0$).

An *invariant curve* is a curve from $x = -\infty$ to $+\infty$, invariant under both U and T.

Next, I generalise rotation number. If (θ, z) has rotation number ω under a periodic map F, then

$$\frac{\pi_1 F^q R^p(\theta, z)}{q} = \omega - \frac{p}{q} + o(1), \quad \text{as } q \to \infty, \quad (5.1)$$

This tends to zero for a sequence p_n, q_n iff $p_n/q_n \to \omega$. So generalise (and also allow $\omega = \infty$, i.e. consider rotation number as belonging to the projective line):

x has *rotation number* $\omega \in \mathbb{R}P$ under (U, T) if for all sequences $p_n, q_n \in \mathbb{Z}$ so that $r_n = \max(|p_n|, |q_n|) \to \infty$, then

$$\frac{\pi_1 U^{q_n} T^{p_n} x}{r_n} \to 0 \quad \text{iff} \quad \frac{p_n}{q_n} \to \omega. \quad (5.2)$$

Poincaré's theorem (section 2) generalises to invariant curves of a commuting pair, under the condition that

$$\exists m, n \in \mathbb{Z}, K > 0 \text{ so that } \pi_1 U^m T^n x \leqslant x - K \quad (5.3)$$

for all points x on the curve (cf. $\pi_1 R(\theta, z) = \theta - 1$). So an invariant curve has a rotation number [19].

For the *twist condition*, I want both U and T to have action generating functions, i.e. $\partial x'/\partial y$ should have constant sign. This is probably more restrictive than necessary, and slightly unfortunately so, as R does not satisfy the twist condition. If U, T commute and have generating functions v, τ, then the generating functions for UT and TU can differ by only a constant, so I call it the *Calabi invariant* $C(v, \tau)$.

I call the extension of class A to commuting pairs *class AA*. Presumably Moser twist and other results like Mather's theorem (section 9) would generalise under suitable conditions.

I can now make some nice connections between rotation number and the renormalisation:

x has rotation number ω under (U, T) iff Bx has rotation number ω' under $N_m(U, T)$, where ω, ω' are related by $\omega = m + 1/\omega'$.

x has rotation number $\omega_0 = [m_0, m_1, \ldots]$ under (U, T) iff $B_{n-1} \ldots B_0 x$ has rotation number $\omega_n = [m_n, \ldots]$ under $N_{m_{n-1}} \ldots N_{m_0}(U, T)$, where the B_j are the successive coordinate changes.

(U, T) has an invariant curve of rotation number ω_0 iff $N_{m_{n-1}} \ldots N_{m_0}(U, T)$ has an invariant curve of rotation number ω_n.

In particular, a periodic map F has an invariant circle of rotation number ω_0 iff $N_{m_{n-1}} \ldots N_{m_0}(F, R)$ has an invariant curve of rotation number ω_n.

Restricting B to linear diagonal scale changes $B(x, y) = (\alpha x, \beta y)$, N_m induces the following renormalisation on action generating functions:

$$\begin{aligned} v'(x, x') &= \alpha\beta\, \tau\left(\frac{x}{\alpha}, \frac{x'}{\alpha}\right), \\ \tau'(x, x') &= \alpha\beta\, v \oplus \tau \oplus \cdots \oplus \tau\left(\frac{x}{\alpha}, \frac{x'}{\alpha}\right). \end{aligned} \quad (5.4)$$

Note that it preserves zero Calabi invariant.

The idea of renormalisation is not new to KAM theory. Most proofs consist in finding successive coordinate changes to make the system look more like a linear twist, restricting attention each time to a narrower annulus (see, for example, Moser [6], Herman [16], Rüssmann [17], Gallavotti [18]). So scale changes are made in the z-direction. I believe that the freedom we have to make scale changes in the θ direction too will make this renormalisation more powerful. The only expense is that at each step one has to change the generators for the group $\{F^q R^p : p, q \in \mathbb{Z}\}$. The main benefit will be that we will probably get the boundary in class A of the maps with an invariant circle of given rotation number.

6. Simple fixed point

Quadratic irrationals have eventually periodic continued fraction expansion. So for maps with an orbit of quadratic irrational rotation number, $[b_0, \ldots b_j, (c_1, \ldots c_k,)^\infty]$, this suggests that one might find asymptotic behaviour under $N_{c_k} \ldots N_{c_1}$, after removing the aperiodic head by applying $N_{b_j} \ldots N_{b_0}$.

I will look at the simplest case, namely, nobles, for which the repeat pattern $[c] = [1]$. N_1 (4.8) has two important fixed points, the main objects of discussion in this paper. I begin with the *simple fixed point*:

$$T: \begin{cases} x' = x + y + 1, \\ y' = y, \end{cases} \quad U: \begin{cases} x' = x + \dfrac{y}{\gamma} - \gamma, \\ y' = y, \end{cases}$$

$$B: \begin{cases} x' = -\gamma x, \\ y' = -\gamma^2 y. \end{cases} \tag{6.1}$$

It corresponds to a linear shear, with $y = 0$ as a golden curve.

One can analyse its stability under N_1,

$$DN_1: \begin{cases} \delta U' = B\delta T B^{-1} \\ \delta T' = B\delta T U B^{-1} + BDT_{UB^{-1}} \cdot \delta UB^{-1}. \end{cases} \tag{6.2}$$

Here we are using linearity of B to identify DB with B, for simplicity of notation. Also we are ignoring contributions to $\delta U'$, $\delta T'$ due to variation of B with (U, T), which would depend on the particular prescription for renormalising (T, TU). These contributions are only in the direction of coordinate changes, so they will have no essential effect.

At the simple fixed point, DN_1 is

$$\delta U'_x(x, y) = -\gamma \delta T_x\left(-\frac{x}{\gamma}, -\frac{y}{\gamma^2}\right),$$

$$\delta T'_x(x, y) = -\gamma \delta U_x\left(-\frac{x}{\gamma}, -\frac{y}{\gamma^2}\right)$$

$$-\gamma \delta T_x\left(-\frac{x}{\gamma} - \frac{y}{\gamma^3} - \gamma, -\frac{y}{\gamma^2}\right)$$

$$-\gamma \delta U_y\left(-\frac{x}{\gamma}, -\frac{y}{\gamma^2}\right),$$

$$\delta U'_y(x, y) = -\gamma^2 \delta T_y\left(-\frac{x}{\gamma}, -\frac{y}{\gamma^2}\right), \tag{6.3}$$

$$\delta T'_y(x, y) = -\gamma^2 \delta U_y\left(-\frac{x}{\gamma}, -\frac{y}{\gamma^2}\right)$$

$$-\gamma^2 \delta T_y\left(-\frac{x}{\gamma} - \frac{y}{\gamma^3} - \gamma, -\frac{y}{\gamma^2}\right).$$

Let us define an order of monomials:

$$1 < y < x < y^2 < xy < x^2 < \cdots \tag{6.4}$$

and say a polynomial is of *rank* p, q is its largest monomial, with respect to this order, is $x^p y^q$. Then observe that at the simple fixed point, DN_1 never increases the rank of a polynomial perturbation. Thus, we can put DN_1 into Jordan normal form on the space of polynomial perturbations, with polynomial eigenvectors or generalised eigenvectors. What happens on the rest of the space is discussed in the next section. Expanding δU_x, δT_x, δU_y, δT_y in power series, and ordering the coefficients by rank, and in the above order within rank, we see that DN_1 is block upper triangular, with 4×4 diagonal blocks:

$$\begin{vmatrix} 0 & -\gamma & 0 & 0 \\ -\gamma & -\gamma & -\gamma & 0 \\ 0 & 0 & 0 & -\gamma^2 \\ 0 & 0 & -\gamma^2 & -\gamma^2 \end{vmatrix} \times (-\gamma)^{-p}(-\gamma^2)^{-q} \tag{6.5}$$

for rank p, q. This diagonal block has eigenvalues

$$-\gamma^2, 1, -\gamma^3, \gamma \times (-\gamma)^{-p}(-\gamma^2)^{-q} \tag{6.6}$$

with respective eigenvectors

$$\begin{vmatrix} 1 \\ \gamma \\ 0 \\ 0 \end{vmatrix}, \begin{vmatrix} \gamma \\ -1 \\ 0 \\ 0 \end{vmatrix}, \begin{vmatrix} 1/2\gamma \\ \gamma/2 \\ 1 \\ 1 \end{vmatrix}, \begin{vmatrix} \gamma \\ -\gamma \\ \gamma \\ -1 \end{vmatrix}. \tag{6.7}$$

$$B_{pq}, \quad D_{pq}, \quad A_{pq}, \quad C_{pq}.$$

For each of these eigenvectors for the block, one can determine coefficients of lower rank to make eigenvectors or generalised eigenvectors for DN_1. They span the space of polynomial perturbations. As there is a lot of degeneracy, the coefficients need not be uniquely determined.

N_1 leaves invariant several important spaces, namely:

i) commuting pairs;
ii) area-preserving pairs;
iii) commuting area-preserving pairs with zero Calabi invariant (class AA);
iv) symmetric commuting pairs;

v) coordinate transforms of the fixed point;
vi) all intersections of the above.

One would like to use the freedom (due to degeneracy) in determining the eigenvectors or generalised eigenvectors, to choose them to respect these invariant subspaces. This can be done, and the result is shown in Table I, where they are labelled by their terms of maximal rank, as in (6.7). The details are described in MacKay [19].

Note with regard to iv) above, that we say T is *symmetric* if $(TS)^2$ is the identity, where

$$S(x, y) = (-x, y). \tag{6.8}$$

Symmetry alone, however, is not preserved by N_1. This symmetry property corresponds to the important class of *reversible* systems (Devaney [20]), but unfortunately there is not room here to discuss them.

7. Compactness of DN_1 at the simple fixed point

Next I show that DN_1 is a compact operator at the simple fixed point, in a suitable norm, and that the error in truncating it at finite degree goes to zero as the degree goes to infinity. Thus, standard results in functional analysis (e.g. Krasnosel'skii et al., [21] section 18) imply that the diagonalisation of table I is complete, apart from a component with eigenvalue 0. It is clear, incidentally, that table I does not cover the whole space, as there are arbitrarily small perturbations of the simple fixed point, in class AA, which are not coordinate transforms of the simple fixed point. For example, for $k = 0$ the standard map is equivalent to the simple fixed point, and has a whole circle of points of type $(0, 1)$, but for $k \neq 0$ there are only two.

To show compactness of DN_1, I show that N_1 is analyticity improving in a neighbourhood of the simple fixed point, on suitable domains. Specifically, if T is analytic on the product of discs $|x| \leq X$, $|y| \leq Y$, for some $X, Y > 0$ with

$$X > \gamma^3 + \frac{Y}{\gamma} \tag{7.1}$$

and U is analytic on $|x| \leq X'$, $|y| \leq Y'$, with

$$\frac{X}{\gamma} < X' < \gamma X, \quad \frac{Y}{\gamma^2} < Y' < \gamma^2 Y, \tag{7.2}$$

then (U', T') is analytic on larger discs, for (U, T) close enough to the simple fixed point. Close enough means with respect to the l_1 norm for power series expansions in the discs (7.1, 7.2).

Table I
Decomposition of the spectrum of DN_1 at the simple fixed point, according to area preservation (a.p.), commutativity (comm), coordinate transforms of the fixed point (c.c), non-zero Calabi invariant (C.I.), and symmetry (s). Prefix n- stands for non-

				Classification		Eigenvector	Eigenvalue	Eigenvalues greater than or equal to 1 in modulus
a.p.	comm	c.c.		s		B_{pq}, p even	$-\gamma^2(-\gamma)^{-p}(-\gamma^2)^{-q}$	$-\gamma^2, 1, -1$
					D_{0q}	$(-\gamma^2)^{-q}$	1
				ns	B_{pq}, p odd	$-\gamma^2(-\gamma)^{-p}(-\gamma^2)^{-q}$	γ
		C.I.		ns	A_{00}	$-\gamma^3$	$-\gamma^3$
	n-comm			ns	D_{pq}, $p \geq 1$	$(-\gamma)^{-p}(-\gamma^2)^{-q}$	
					C_{p0}	$\gamma(-\gamma)^{-p}$	$\gamma, -1$
n-a.p.	comm	c.c.			A_{p0}, $p \geq 1$	$-\gamma^3(-\gamma)^{-p}$	$\gamma^2, -\gamma, 1$
				s	A_{pq}, $q \geq 2$, p odd	$-\gamma^3(-\gamma)^{-p}(-\gamma^2)^{-q}$	
				ns	A_{pq}, $q \geq 2$, p even	$-\gamma^3(-\gamma)^{-p}(-\gamma^2)^{-q}$	
		n-c.c.		ns	A_{p1}	$\gamma(-\gamma)^{-p}$	$\gamma, -1$
	n-comm			ns	C_{pq}, $q \geq 1$	$\gamma(-\gamma)^{-p}(-\gamma^2)^{-q}$	

The error in truncating DN_1 at degree d is less than $C\lambda^d$, where

$$\lambda = \max\left(\frac{X}{\gamma X'}, \frac{X'}{\gamma X}, \frac{Y}{\gamma Y'}, \frac{Y'}{\gamma Y}, \frac{\frac{X}{\gamma} + \frac{Y}{\gamma^3} + \gamma}{X}\right) < 1. \quad (7.3)$$

8. Significance of the simple fixed point

If one restricts attention to class AA, table I shows that all polynomial directions from the simple fixed point are coordinate changes. Taking section 7 into consideration, this implies that, modulo coordinate changes, the simple fixed point attracts a neighbourhood, in fact faster than exponentially. This is, of course, what one should expect from KAM theory. Note also that the simple fixed point is attracting in the space of symmetric commuting pairs, as one expects from the reversible version of Moser's twist theorem [6].

$N_1(U, T)$ possesses a golden curve iff (U, T) does, but I want to show that convergence of $N_1^n(U, T)$ to a pair with a golden curve implies that (U, T) has a golden curve. For convergence to the simple fixed point, this follows from Moser's twist theorem (assuming it generalises to commuting pairs). It would give a $C^{3+\epsilon}$-neighbourhood of the simple fixed point (for any $\epsilon > 0$), in which all commuting pairs have a (smooth) golden curve. Convergence of $N_1^n(U, T)$ to the simple fixed point (in the $C^{3+\epsilon}$ topology) implies that $N_1^{n_0}(U, T)$ is in this neighbourhood for some n_0, and so (U, T) has a (smooth) golden curve (cf. Escande and Doveil [15]).

It may not be necessary, however, to use Moser's twist theorem. Mather [22] proved a necessary and sufficient condition for existence of an invariant circle, to be discussed in the next section. In section 12, I will show how one can probably use this theorem to prove that convergence of $N_1^n(F, R)$ to any fixed point, not just the simple one, implies that F has a golden circle. I suspect that even use of this theorem is not necessary at the simple fixed point.

Most proofs of results in KAM theory restrict one to pretty small perturbations (although Herman (private communication) is obtaining much more realistic results) determining a reasonably large neighbourhood of attraction for the simple fixed point. The results of sections 6 and 7 can be extended to other rotation numbers than nobles. The simple fixed point of N_1 belongs to a simple line, invariant and attracting under all of the N_m. Presumably, the size of the basin of attraction diminishes to zero as $m \to \infty$, thus setting a restriction on the growth rate of m_i for convergence to the simple line.

9. Connections between invariant circles and nearby periodic orbits

Now I discuss some connections between invariant circles and nearby periodic orbits. The first is a conjecture of Greene [23]. The other is Mather's theorem referred to in the previous section. This section will lead us to another fixed point of N_1.

I return to the setting of periodic maps. In the action representation a periodic orbit of type (p, q) corresponds to a sequence

$$\boldsymbol{\theta} = \theta_0, \theta_1, \ldots \theta_q = \theta_0 + p$$

for which the *action*

$$W(\boldsymbol{\theta}) = \sum_{i=0}^{q-1} \tau(\theta_i, \theta_{i+1}) \quad (9.1)$$

is stationary with respect to variations in $\boldsymbol{\theta}$. Birkhoff (1927) [24] showed that a map in class A has at least two periodic orbits of type (p, q) for each rational p/q in lowest terms, in an appropriate interval. If $\tau_{12} < 0$, the periodic orbits respectively minimize and minimaximize the action $W(\boldsymbol{\theta})$ over an appropriate set of $\boldsymbol{\theta}$. For $\tau_{12} > 0$, interchange "max" and "min". I shall restrict attention without loss of generality to the former case.

The linear stability of a periodic orbit can be measured by its *residue* (Greene [25]):

$$R = (2 - \text{Tr } DF^q)/4, \quad (9.2)$$

where DF^q is the derivative of F^q at any of its points. In the action representation, considering without loss of generality the case $q = 1$, the residue of a fixed point is

$$R = \frac{\frac{d^2}{d\theta^2}\tau(\theta, \theta)}{4\tau_{12}(\theta, \theta)}. \quad (9.3)$$

Thus, the minimizing periodic orbits found by Birkhoff have non-positive residue R^-, and the minimaximizing orbits have non-negative residue R^+. They give rise to *island chains*. Fig. 2 shows some island chains for the *quadratic map*.

$$\begin{cases} x' = p - y - x^2, \\ y' = x, \end{cases} \quad (9.4)$$

for parameter $p = 2.38216325159$. For purposes of orientation, thin island chains, as when they are born by bifurcation from a periodic orbit, have residue close to zero.

Greene [23, 26, 34] suggested a connection between existence of invariant circles and the stability of nearby periodic orbits. This has also been followed up by Schmidt [27] and Bialek [28]. Given ω irrational, let us restrict attention to the Birkhoff periodic orbits of type (p_n, q_n) with p_n/q_n convergents of ω. Calling their residues $R_n^\pm$, one finds numerically one of three cases:

i) *Subcritical*. $R_n^\pm \to 0$, and it looks as if the island chains converge to a smooth invariant circle of rotation number ω.

ii) *Critical*. $R_n^\pm$ are eventually bounded away from 0 and $\pm\infty$, and it looks as if the island chains converge to a non-smooth invariant circle of rotation number ω.

iii) *Supercritical*. $R_n^\pm \to \pm\infty$, and it looks as if there is no circle of rotation number ω.

The critical case is shown in fig. 2 for $\omega = 1/\gamma^2 = [0, 2, (1,)^\infty]$ for the quadratic map. The convergents are $2/5, 3/8, 5/13, \ldots$, and these island chains can be seen to be converging to the outermost invariant circle. It is non-smooth in the sense that it has thin spots (smoothness refers to the conjugacy, not the graph).

The conjecture is that under suitable conditions, one could replace "and it looks as if" in the above by "which implies that". A partial result in this direction follows from the Moser twist theorem, namely, if there is a smooth Diophantine circle, then $R_n^\pm \to 0$ (in fact, faster than exponentially) (Mather, private communication), so in the critical and supercritical cases there is no smooth circle (for Diophantine rotation number). The converse, however, is not known.

A necessary and sufficient condition for existence of an invariant circle with irrational rotation number has been proved by Mather [22, 29], based on another property of nearby island chains. If one defines $\Delta W_{p/q}$ to be the difference in actions between the two Birkhoff orbits of type (p, q):

$$\Delta W = W_{\text{minimax}} - W_{\text{min}}, \quad (9.5)$$

then Mather's result states that

There exists an invariant circle of irrational rotation number ω iff $\Delta W_{p/q} \to 0$ as $p/q \to \omega$. In the case that there is no invariant circle then $\Delta W_{p/q}$ has a positive limit, and there is an invariant Cantor set of rotation number ω.

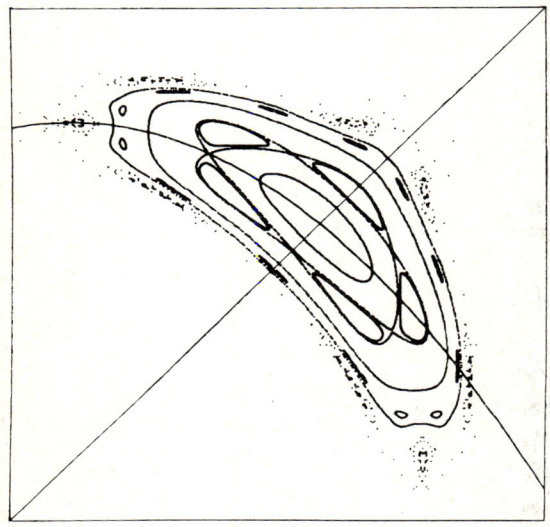

Fig. 2. Some orbits of the quadratic map, for $p = 2.38216325159$, and two symmetry lines.

Similar results in this direction were found by Aubry (e.g., this volume).

In the subcritical and critical cases above I find numerically that $\Delta W_{p_n/q_n} \to 0$ (faster than exponentially in the subcritical case, as Moser twist implies for a smooth circle), so there is an invariant circle, and in the supercritical case $\Delta W_{p_n/q_n}$ tends to a positive limit, so there is no circle, but there is a Cantor set. This is shown in fig. 3 for $\omega = 1/\gamma^2$, and four parameter values in the quadratic map, one subcritical, one critical, and two supercritical.

Note that existence of a smooth Diophantine circle is stable to perturbation, by Moser twist. Also extensions of Mather's theorem [22], show that non-existence of an invariant circle of given rotation number is stable to perturbation. So one expects, and finds numerically, the subcritical and supercritical cases to be open sets in class A.

These approaches to finding where there are invariant circles have the advantage that they generalise directly to continuous time systems. There is no need to choose a surface of section or evaluate a return map. Finding periodic orbits and evaluating their residues and actions is a relatively straightforward procedure, especially if one takes advantage of symmetries the system may possess (see, for example, MacKay [19]).

10. Critical noble circles

I now wish to concentrate on the critical case of the previous section. For reversible maps F in class A with a critical quadratic irrational circle, Shenker and Kadanoff [12] found scaling behaviour in the neighbourhood of certain points. The behaviour appears to be the same for all quadratic irrationals with the same repeat pattern, in most maps.

In particular, for nobles one obtains fig. 4 in critical cases, if one looks on a small enough scale and in appropriate coordinates in the neighbourhood of the point where the critical noble circle crosses a "dominant symmetry line". In the picture, the symmetry line has been transformed to the X-axis, and the noble circle crosses it at the origin. Note that everything in the picture repeats itself on a smaller scale and turned over, in the smaller box. Asymptotically, the scaling factors are

$$\begin{aligned}\alpha &= -1.4148360 \quad \text{in } Y, \\ \beta &= -3.0668882 \quad \text{in } X.\end{aligned} \qquad (10.1)$$

In summary, it looks as if, for a map with a

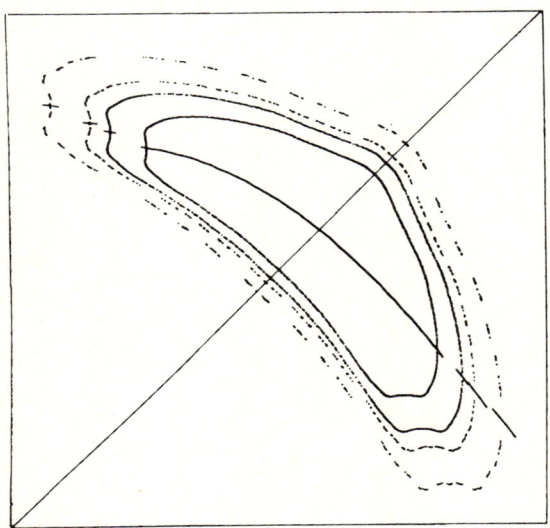

Fig. 3. An orbit of rotation number $1/\gamma^2$ for four parameter values in the quadratic map.

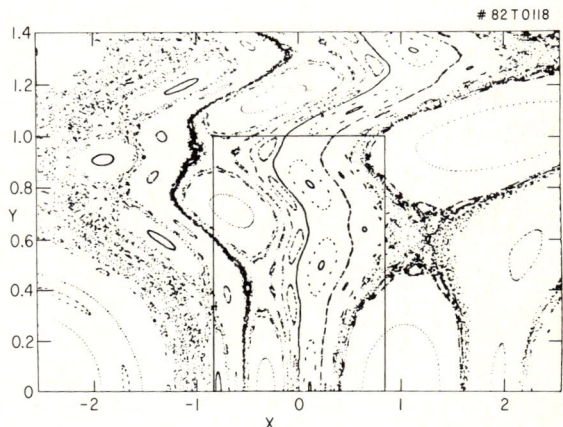

Fig. 4. Some orbits of the universal map F^*.

critical noble there are coordinate changes B_n, such that the maps

$$B_n F^{q_n} R^{p_n} B_n^{-1} \qquad (10.2)$$

converge to some universal map F^*, with

$$B_{n+1} \simeq BB_n \quad \text{as } n \to \infty, \quad B(Y, X) = (\alpha Y, \beta X). \qquad (10.3)$$

In terms of the renormalisation, it looks as if $N_1^n N_{b_j} \ldots N_{b_0}(F, R)$ converges to a fixed point (U^*, T^*), with $T^* = F^*$, $U^* = BF^*B^{-1}$.

Given a one parameter family passing through a critical case, one finds further self-similarity. For example, the parameter values p_n at which $R_n^+ = 1$, converge asymptotically geometrically to the critical value, at rate $1/\delta$

$$\delta = 1.6280. \qquad (10.4)$$

This is the way I located critical parameter values. There is a faster way, however. For a critical noble

$$\begin{aligned} R_n^+ &\to 0.2500888, \\ R_n^- &\to -0.255426. \end{aligned} \qquad (10.5)$$

The convergence is at rate

$$\delta' = -0.6108. \qquad (10.6)$$

Thus the parameter values p_n' where $R_n^+ = 0.2500888$ converge at rate $\delta'/\delta = -0.3752$, faster than $1/\delta$.

The self-similarity can be summarised by saying that it looks as if there is a reparametrisation μ, and (parameter dependent) coordinate changes B_n such that the one-parameter families

$$B_n F^{q_n}_{\mu\delta^{-n}} R^{p_n} B_n^{-1} \qquad (10.7)$$

converge to a universal one parameter family F^*_μ, with B_n scaling as in (10.3). In renormalisation language, the fixed point (U^*, T^*) has an unstable manifold of essentially only one dimension, with eigenvalue δ. The dominating attraction rate on its stable manifold is δ'.

Figs. 5 and 6 show some orbits of F^*_μ for $\mu = -0.3$, $+0.3$, subcritical and supercritical cases. The scale in parameter is chosen to make the minimaximizing point of type (1, 1) (or [1], in continued fraction notation) have residue $R^+_{[1]} = 1$ at $\mu = 1$. Fig. 7 shows how $R^+_{[1]}$ varies with μ. The universal family is the significant object for any renormalisation scheme. Further properties are given in section 13 and MacKay [19], including critical exponents.

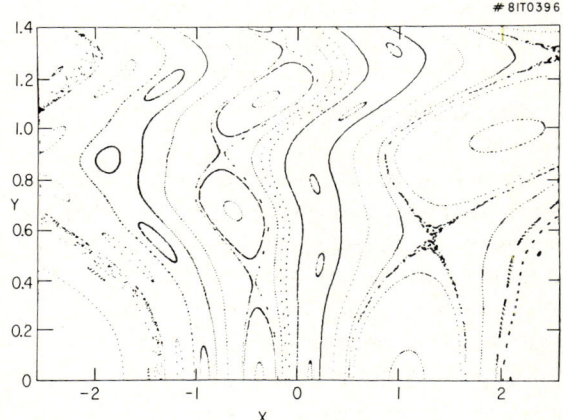

Fig. 5. Some orbits of the universal one parameter family F^*_μ, for $\mu = -0.3$.

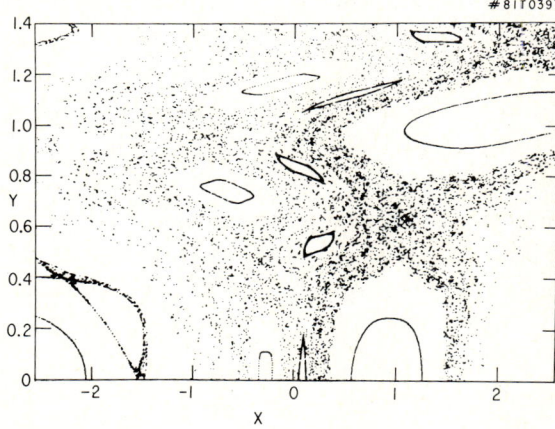

Fig. 6. Some orbits of the universal one parameter family F^*_μ, for $\mu = +0.3$.

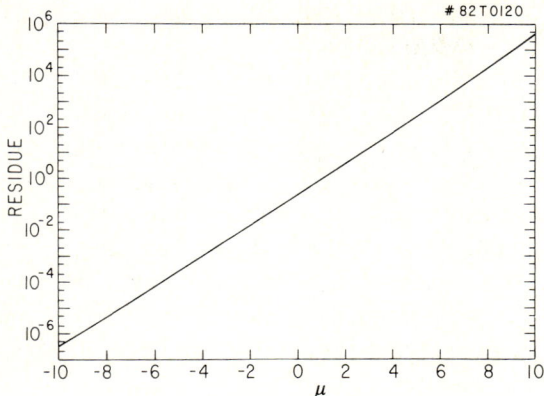

Fig. 7. Dependence of $R_{[1]}^+$ on μ.

11. Critical fixed point

The results of the previous section strongly suggest that there is another fixed point of N_1. Following the pioneering work of Kadanoff [30], I worked in the action representation, using the induced renormalisation (5.4). In order for the truncation at finite degree to have vanishing effect as the degree goes to infinity, it is necessary to find domains of expansion in $\mathbb{C}^2$ such that v, τ analytic in their domains implies v', τ' analytic on the same domains and slightly more, for v, τ close enough to the fixed point in the l_1 norm. However, I couldn't find any domain of the form

$$|a_{11}x + a_{12}x' - c| \leq 1,$$
$$|a_{21}x + a_{22}x' - c'| \leq 1, \tag{11.1}$$

for which this seemed to be satisfied.

From the previous section, we expect the fixed point to be symmetric. So I considered a modified renormalisation:

$$N_{s1}: \begin{cases} v'(x, \dot{x}') = \alpha\beta\, \tau\left(-\dfrac{x'}{\alpha}, -\dfrac{x}{\alpha}\right), \\ \tau'(x, x') = \alpha\beta\, \tau \oplus v\left(\dfrac{x}{\alpha}, \dfrac{x'}{\alpha}\right). \end{cases} \tag{11.2}$$

This is the same as N_1, restricted to the space of symmetric commuting pairs, but permits good domains. I determined α and β by

$$\tau_1\left(0, \frac{1}{\alpha}\right) = 0, \quad \frac{\alpha}{\beta} = \tau_{12}\left(0, \frac{1}{\alpha}\right), \tag{11.3}$$

which forces the normalisation

$$v_1'(0, 1) = 0, \quad v_{12}'(0, 1) = 1. \tag{11.4}$$

The domains I used were

$$|x - c| \leq r, \quad |x' - c'| \leq r', \tag{11.5}$$

with

$$\begin{aligned} c &= 0.050707985, & r &= 0.502060282, \\ c' &= -0.655406307, & r' &= 0.329680205, \end{aligned} \tag{11.6}$$

for τ, and its rescaled and reflected version for v. This choice is close to optimal, if (11.2) is considered as one second order equation, and has an analyticity improvement factor of at least 1.1374. If regarded as two first order equations, the domain for v should be diminished a little.

Newton's method was used to find a fixed point of N_{s1}, truncating at various degrees. The results appear to converge as the degree is increased, and are consistent with the findings of section 10. For example, the values of α, β, δ, δ' for several truncation levels are presented in table II. δ nd δ' were found by diagonalising the derivative DN_{s1}. The eigenvalues of DN_{s1} larger than 0.4 in modulus whose eigenvectors are symmetric are shown and interpreted in table III. There is only one relevant eigenvalue not contained inside the unit circle, namely, δ, as expected. For details on the identification of the eigenvalues, see MacKay [19].

This procedure could in principle be carried to arbitrary precision. Also, existence of the fixed point and bounds on its spectrum could be proved in the same way as Lanford [31] and Eckmann et al. [32] did for period doubling in one-dimensional and area-preserving maps, respectively.

Table II
Values of α, β, δ, δ' for the fixed points of N_{s1}, truncated at several degrees

Degree	α	β	δ	δ'
14	−1.414836085	−3.066888192	1.6279496	−0.61083048
15	−1.414836021	−3.066888344	1.6279506	−0.61083021
16	−1.414836072	−3.066888224	1.6279499	−0.61083040
17	−1.414836052	−3.066888269	1.6279502	−0.61083026
18	−1.414836062	−3.066888246	1.6279500	−0.61083028

Table III
Spectrum of DN_1 at the critical fixed point, showing those eigenvalues greater than 0.4 in modulus which have symmetric eigenvectors

Eigenvalue	Compare with	Value	Interpretation
7.0208826	$\gamma\alpha\beta$	7.0208826	Constant terms in action
−3.0668882	β	−3.0668882	Coordinate change
−2.6817385	$-\alpha\beta/\gamma$	−2.6817385	Constant terms in action
1.6279500	δ	1.6280	Relevant direction (10.4)
−1.5320950	β/α^2	−1.5320951	Coordinate change
1.0000001	α/α	1	Scale change
1.0000000	β/β	1	Scale change
−0.7653736	β/α^4	−0.7653736	Coordinate change
−0.6108303	δ'	−0.6108	Essential convergence rate (10.6)
0.4995593	α/α^3	0.4995601	Coordinate change

12. Golden curves for pairs converging to any fixed point

In this section I show how Mather's theorem probably implies that if $N_1^n(F, R)$ converges to any fixed point with $\alpha\beta > 1$, then F has a golden circle. Let (v^*, τ^*) be the generating functions for the fixed point. Commutation and zero Calabi invariant imply that $\tau^*(x, x)$ has a minimum and maximum, so write $\Delta\tau^*$ for their difference. Then, provided the domains of convergence are large enough to include the relevant Birkhoff periodic points, convergence to the fixed point implies that

$$(\alpha\beta)_n \ldots (\alpha\beta)_1 \Delta W_{[(1, \gamma^n]} \to \Delta\tau^* \qquad (12.1)$$

from (5.4), where the $(\alpha\beta)_j$ are the successive values of $\alpha\beta$. Convergence to the fixed point also implies that

$$(\alpha\beta)_j \to \alpha\beta > 1, \qquad (12.2)$$

so

$$\Delta W_{[(1, \gamma^n]} \to 0 \qquad (12.3)$$

and F has a golden circle, by Mather's theorem. Note that the convergence need only be C^1 for Mather's theorem to apply. The same argument would imply existence of a circle of any frequency if $N_{m_n} \ldots N_{m_0}(F, R)$ remains in a region with $\Delta\tau$ bounded and $\alpha\beta$ bounded above 1.

13. Robustness of noble circles

One of the most significant features of figs. 4 and 6 is that in the critical case, the noble circle appears to be (locally) the only circle, and in the supercritical case there appear to be no circles at all. The dots all belong to one orbit. As further evidence, I measured residues and differences in actions for other periodic orbits than the convergents of the noble. Given the noble $[a, (1,)^\infty]$, where a is a finite sequence of integers, I considered the periodic orbits with rotation number $[a, (1,)^n b]$, for finite sequences b. Figs. 8 and 9 show the residues R^+ in the limit as $n \to \infty$, in the critical case. In these figures I have used the natural ambiguity

$$[b_0, \ldots, b_m + 1] = [b_0, \ldots, b_m, 1] \tag{13.1}$$

to group the points into a tree which branches two ways at each point. The point to notice is that they are all bounded away from 0 (assuming that one can extrapolate the trends). For a smooth Diophantine circle, however, the residues for its convergent periodic orbits must tend to zero. Thus a critical noble has a neighbourhood containing no smooth Diophantine circles. Fig. 10 is the universal "fractal diagram" [27] for the neighbourhood of any noble. Since it shows all residues increasing with the parameter μ, there are no smooth Diophantine circles in the supercritical case either.

In the subcritical case, of course, we expect to have a smooth noble circle. Another corollary (Mather, private communication) of Moser twist is that a smooth Diophantine circle has others arbitrarily close. In fact each smooth Diophantine circle is a density point in the set of smooth Diophantine circles. So there are lots of circles in the subcritical case.

Next we consider differences in actions. Fig. 11 shows $(\alpha \beta)^n \Delta W_{[a, (1,)^n b]}$, in the limit as $n \to \infty$, for various b in the critical case. Apart from the sequence $[(1,)^m]$, converging to γ, we plotted points only for b with $b_0 > 1$, as the self-similarity allows one to fill in for $b_0 = 1$. They are all bounded away from 0, so using Mather's result, there are no invariant circles with irrational rotation number with continued fraction expansion $[a, (1,)^n b_0, \ldots]$, $b_0 > 1$]. Thus there are no irrational circles apart from the noble $[a, (1,)^\infty]$. Assuming a conjectured

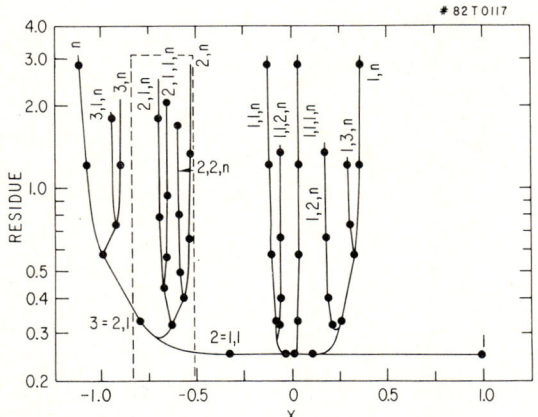

Fig. 8. Residues $R^+_{[b]}$ of periodic orbits of F^*, plotted against position X on the dominant half-line.

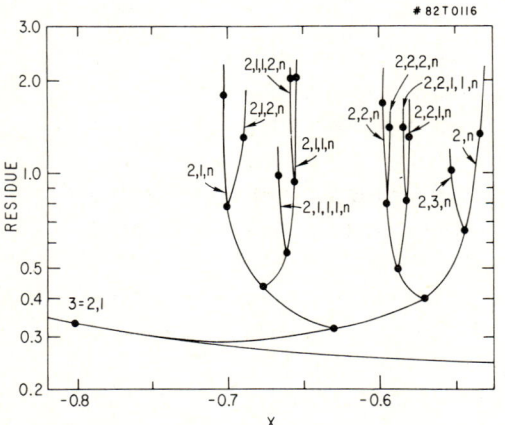

Fig. 9. Inset to fig. 8.

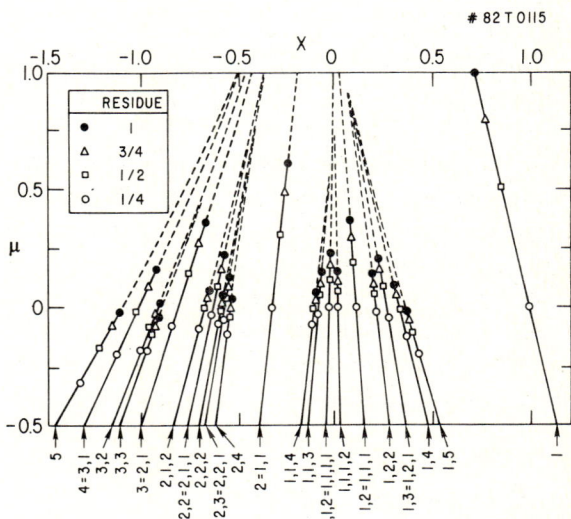

Fig. 10. Position X on the dominant half-line of periodic points of F^*_μ, plotted against μ, and indicating how their residues change.

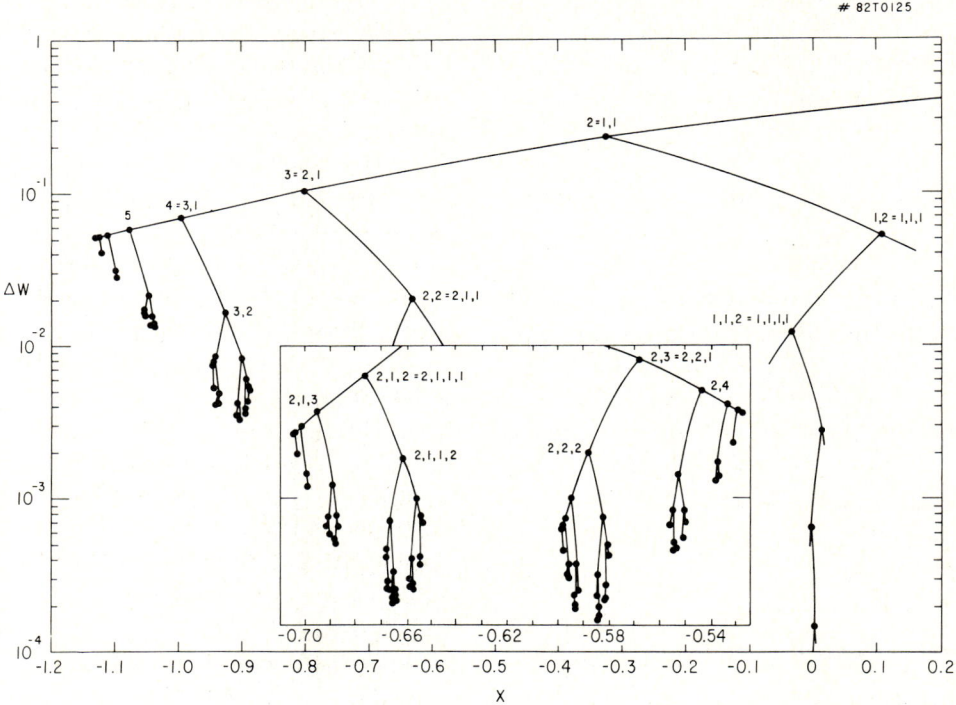

Fig. 11. Values of ΔW for various periodic orbits of F^*, plotted against position X on the dominant half-line.

extension of Mather's work to rational circles, there are no rational circles either.

Thus I conclude that noble circles are robust in an important sense, namely, a critical noble has a neighbourhood containing no other invariant circles, and all narrow enough connected neighbourhoods of a supercritical noble (Cantor set) contain no invariant circles at all. Since nobles are dense, one would like to conjecture a stronger result, namely, that isolated circles are typically nobles, but deducing this from the previous statement would require estimates on the sizes of the neighbourhoods, which I do not have at present.

14. Conclusion

In conclusion, I speculate a picture like fig. 12, in class AA (modulo coordinate changes). The critical fixed point has a codimension 1 stable manifold W^s, which, I believe, separates the space, at least locally, into pairs with a smooth golden curve and those with no golden curve. Any one parameter family crossing W^s transversally (in which I include non-zero speed) will have asymptotically the same behaviour, on a small enough scale in space and parameter, and on a long enough timescale, as the "universal" one parameter family given by a natural parametrisation of the one-dimensional unstable manifold W^u. In some sense there is a fixed point at infinity too, as I find asymptotic behaviour in the supercritical case too. How to express it, however, is not clear, as it has infinite actions and residues, and $\beta = -\infty$ (although $\alpha = -1$).

One could do exactly parallel analysis for any quadratic irrational rotation number, but other irrationals will require a modified treatment. We

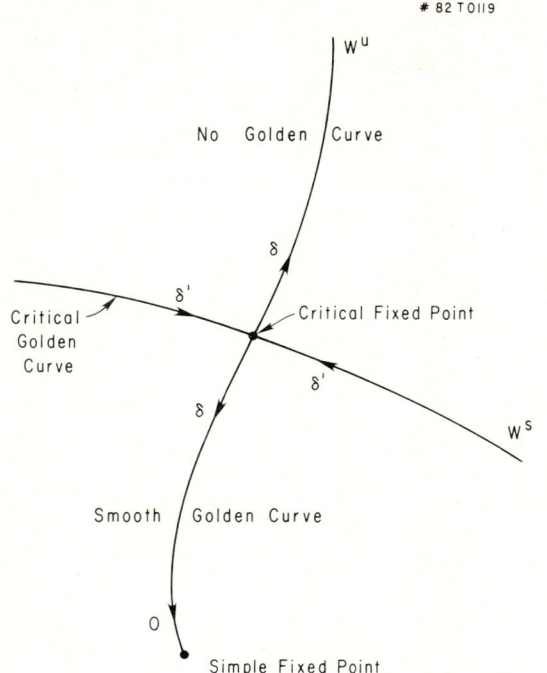

Fig. 12. Schematic of the action of N_1 in the space of commuting pairs in class AA (modulo coordinate changes).

indicated, however, that nobles are probably the most significant.

The ideas of this paper also carry over directly to the problem of existence and breakup of invariant tori in dissipative systems. The simple fixed point is very simple, and analysis of its spectrum much simpler than in the area preserving case (e.g. see Feigenbaum et al., 1982). Construction of a neighbourhood of attraction is also easier [19]. I obtained the critical fixed point to an accuracy of 10^{-7}, working in power series of degree 19 in x^3 on carefully chosen domains [19] (see also Rand et al. [17], and Feigenbaum et al. [13]).

Lastly, the ideas of this paper can be extended to invariant tori of arbitrary dimension. For a $2n$-dimensional symplectic map, for example, one would consider commuting $(n+1)$-tuples of maps. Rotation numbers would lie in $\mathbb{R}P^n$ and the renormalisation would generalise in an obvious way.

Acknowledgements

I wish to thank John Greene very much for his constant advice and encouragement. He has been a great inspiration and help to me. Also I am indebted to John Mather for teaching me the mathematics of this problem, and for explaining many things to me. I am grateful also to Leo Kadanoff and Scott Shenker, Dominique Escande and Fabrice Doveil, and Eric Siggia and David Rand, for freely discussing their work on the same subject. Finally, I wish to acknowledge valuable conversations with Martin Kruskal, Oscar Lanford, Ian Percival, Giovanni Gallovotti, Rafael de la Llave, Philippe Similon, Mitchell Feigenbaum, George Schmidt, Rick Jensen, Serge Aubry, and Franco Vivaldi.

I thank Joel Hosea for use of fig. 1, and I am grateful to the Aspen Center for Physics for their hospitality during part of the time that this paper was in progress. This work was supported by U.S. Department of Energy contract DE-ACO2-76-CHO3073, and U.K. Science Research Council grant B/80/3001.

References

[1] R.M. Sinclair, J.C. Hosea and G.V. Sheffield, Rev. Sci. Instruments 41 (1970) 1552.
[2] R.B. White et al., IAEA conference proceedings, IAEA-CN-41/T-3 (1982).
[3] V.I. Arnold, Mathematical Methods of Classical Mechanics (Springer, New York, 1978).
[4] G.D. Birkhoff, Bull. Soc. Math. de France 60 (1932) 1; reprinted in Collected Mathematical Papers, Vol. II (AMS, New York, 1950), p. 111.
[5] Z. Nitecki, Differentiable Dynamics (MIT press, Cambridge, Mass., 1971).
[6] J.K. Moser, Stable and Random Motions (Princeton Univ. Press, Princeton, 1973).
[7] I. Niven, Irrational Numbers, Carus Math. Monographs, no. 11 (Wiley, New York, 1963).
[8] B.V. Chirikov, Phys. Repts 52 (1979) 263.
[9] J.N. Mather, "Non-existence of invariant circles", preprint, Princeton 1982.
[10] I.C. Percival, "Chaotic boundary of a Hamiltonian map", Physica 6D (1982) 67.
[11] L.P. Kadanoff, Phys. Rev. Lett. 47 (1981) 1641.
[12] S.J. Shenker and L.P. Kadanoff, J. Stat. Phys. 27 (1982) 631.

[13] M.J. Feigenbaum, L.P. Kadanoff and S.J. Shenker, "Quasiperiodicity in dissipative systems: a renormalisation analysis", Physica 5D (1982) 370.
[14] D. Rand, S. Ostlund, J. Sethna and E.D. Siggia, Phys. Rev. Lett. 49 (1982) 132, and Physica D (submitted).
[15] D.F. Escande and F. Doveil, Phys. Lett. 83A (1981) 307; J. Stat. Phys. 26 (1981) 257.
[16] M.R. Herman, "Demonstration du théorème des courbes translatées de nombres de rotation de type constant", manuscript, Paris, and Les Houches notes, 1981.
[17] H. Rüssmann, "On the existence of invariant curves of twist mappings of an annulus", preprint, Mainz, Germany, 1981.
[18] G. Gallavotti, "Perturbation theory of classical Hamiltonian systems", to appear in Lecture notes in Physics, J. Frolich, ed. (Springer, Berlin, 1982).
[19] R.S. MacKay, "Renormalisation in Area Preserving Maps", Thesis, Princeton 1982 (Univ. Microfilms Int., Ann Arbor MI).
[20] R. Devaney, Trans. AMS 218 (1976) 89.
[21] M.A. Krasnosel'skii, G.M. Vainikko, P.P. Zabreiko, Ya.B. Rutitskii and V.Ya. Stetsenko, Approximate solution of operator equations (Wolters-Noordhoff, Groningen, 1972).
[22] J.N. Mather, "A criterion for the non-existence of invariant circles", preprint, Princeton 1982.
[23] J.M. Greene, J. Math. Phys. 20 (1979) 1183.
[24] G.D. Birkhoff, 1927, Dynamical Systems (AMS Colloq. Publ., vol. 9, revised 1966).
[25] J.M. Greene, J. Math. Phys. 9 (1968) 760.
[26] J.M. Greene, Annals of New York Acad. Sci. 357 (1980) 80.
[27] G. Schmidt, Phys Rev. 22A (1980) 2849.
[28] G. Schmidt and J. Bialek, Physica 5D (1982) 397.
[29] J.N. Mather, "Existence of quasiperiodic orbits for twist homeomorphisms of the annulus" Topology 21 (1982) 457.
[30] L.P. Kadanoff, Proceedings of the 9th Midwestern Solid State Theory Seminar (Argonne), in press (1981).
[31] O.E. Lanford III, "A computer assisted proof of the Feigenbaum conjectures", preprint, IHES (1981).
[32] J-P. Eckmann, H. Koch, and P. Wittwer, "A computer-assisted proof of universality for area preserving maps", preprint UGVA-DPT 1982/04-345, 1982.
[33] A.V. Prasad, J. London Math. Soc. 23 (1948) 169.
[34] J.M. Greene, in Nonlinear dynamics and the beam–beam interaction, eds. M. Month and J.C. Herrera, AIP Conf. Proc. 57 (1979) 257.

QUASIPERIODICITY IN DISSIPATIVE SYSTEMS: A RENORMALIZATION GROUP ANALYSIS

Scott J. SHENKER

The James Franck Institute, The University of Chicago, Chicago, Illinois 60637, USA

Abstract

Dynamical systems with quasiperiodic behavior, i.e., two incommensurate frequencies, may be studied via discrete maps which show smooth continuous invariant curves with irrational winding number. In this paper these curves are followed using renormalization group techniques which are applied to a one-dimensional system (circle) and also to an area-contracting map of an annulus. Two fixed points are found representing different types of universal behavior: a trivial fixed point for smooth motion and a nontrivial fixed point. The latter represents the incipient breakup of a quasiperiodic motion with frequency ratio the Golden Mean into a more chaotic flow. Fixed point functions are determined numerically and via an ϵ-expansion and eigenvalues are calculated.

This work was done in collaboration with Leo P. Kadanoff and Mitchell J. Feigenbaum and is described more fully in refs. 1 and 2. Similar work has been done in parallel by D. Rand, S. Ostlund, J. Sethna, and E. Siggia (see Siggia's contribution to this Conference and ref. 3). Related work on Hamiltonian systems can be found in refs. 4–8.

References

[1] S.J. Shenker, Physica 5D (1982) 405.
[2] M.J. Feigenbaum, L.P. Kadanoff and S.J. Shenker, Physica 5D (1982) 370.
[3] D. Rand, S. Ostlund, J. Sethna and E. Siggia, Phys. Rev. Lett. 49 (1982) 132.
[4] J.M. Greene, J. Math. Phys. 9 (1968) 760; 20 (1979) 1183.
[5] S.J. Shenker and L.P. Kadanoff, J. Stat. Phys. 27 (1982) 631.
[6] L.P. Kadanoff, Phys. Rev. Lett. 37 (1981) 1641.
[7] L.P. Kadanoff, Proc. 9th Midwestern Solid State Theory Seminar (Argonne, IL), in press.
[8] R. Mackay, Princeton Preprint (1982).

A UNIVERSAL TRANSITION FROM QUASI-PERIODICITY TO CHAOS

Eric D. SIGGIA
Department of Physics, Cornell University, Ithaca, NY 14853, USA

Abstract

A common route to chaos in dissipative systems proceeds from periodic to quasi-periodic flow (with two independent frequencies). Then, in the absence of rotational symmetry, the system generally mode locks before becoming turbulent. Beyond these qualitative features, the numerous experiments that have examined this regime differ in detail. Dynamical system theory had made the occurrence of the above transitions plausible but has provided no nontrivial quantitative and model independent information.

This situation, on the theoretical side, has recently changed with a proposal on how to modify the experiments so as to make the transition to chaos occur in a quantitatively universal manner [1, 2]. The essence of our proposal follows from K.A.M. theory which is the weak coupling limit of the strong coupling problem relevant to the turbulent transition. In addition to the Rayleigh number, the experimenter must control a second parameter so as to maintain the frequency ratio in the quasi-periodic state at a fixed irrational value. The golden mean, $(\sqrt{5} - 1)/2$, is the optimal ratio experimentally.

The universality, which is restricted to the low frequencies in the time series, is obtained under the above circumstances because the transition to chaos is continuous. In particular the singular low frequency structure in the spectrum develops continuously as $R \to R_T$ from below or as the frequency ratio approached $(\sqrt{5} - 1)/2$ at R_T. These assertions are rigorously established by a renormalization group analysis that resembles the one developed by Feigenbaum to account for the universal features of period doubling.

We again stress that all the low frequency complex amplitudes obtained from either a fluid experiment or a forced nonlinear oscillator at the quasi-periodic to turbulent transition are universal. At present, the theoretical predictions are most easily derived numerically by iterating the map

$$\phi' = \phi + \omega_0 - \frac{a}{2\pi} \sin(2\pi\phi)$$

for $a = 1$ (corresponding to $R = R_T$) and adjusting ω_0 to achieve the desired rotation number.

References

[1] D. Rand, S. Ostlund, J. Sethna and E. Siggia, Phys. Rev. Lett., submitted.
[2] S. Shenker, Physica 5D (1982) 405; M. Feigenbaum, L. Kadanoff and S.J. Shenker, preprint.

CHAPTER 6

FLUIDS AND VORTICES

COADJOINT ORBITS, VORTICES, AND CLEBSCH VARIABLES FOR INCOMPRESSIBLE FLUIDS

Jerrold MARSDEN* and Alan WEINSTEIN*

Department of Mathematics, University of California, Berkeley, California 94720, USA

This paper is a study of incompressible fluids, especially their Clebsch variables and vortices, using symplectic geometry and the Lie–Poisson structure on the dual of a Lie algebra. Following ideas of Arnold and others it is shown that Euler's equations are Lie–Poisson equations associated to the group of volume-preserving diffeomorphisms. The dual of the Lie algebra is seen to be the space of vorticities, and Kelvin's circulation theorem is interpreted as preservation of coadjoint orbits. In this context, Clebsch variables can be understood as momentum maps. The motion of N point vortices is shown to be identifiable with the dynamics on a special coadjoint orbit, and the standard canonical variables for them are a special kind of Clebsch variables. Point vortices with cores, vortex patches, and vortex filaments can be understood in a similar way. This leads to an explanation of the geometry behind the Hald–Beale–Majda convergence theorems for vorticity algorithms. Symplectic structures on the coadjoint orbits of a vortex patch and filament are computed and shown to be closely related to those commonly used for the KdV and the Schrödinger equations respectively.

1. Introduction

The purpose of this paper is to use the methods of symmetry and reduction to study incompressible fluids, especially their Clebsch variables and vortices. The original techniques introduced in Marsden and Weinstein [41] were used in a study of the Hamiltonian structure of plasmas in Marsden and Weinstein [42]. Compressible flow, magnetohydrodynamics, and elasticity require the use of semidirect products, which will be the subject of another publication (Marsden, Ratiu and Weinstein [40]).

Symmetry and Hamiltonian systems are related in the following way to the topic "order in chaos" of this conference:

1) *Order*. Physical systems often exhibit "order" simultaneous with symmetries. For example, soliton-like behavior is frequently linked with symmetry and complete integrability.

2) *Chaos*. When the symmetry of a system is broken, Hamiltonian structures can be useful in detecting chaos by the method of Melnikov (see Holmes and Marsden [28–30]).

Hamiltonian systems written in "non-canonical" variables can be elegantly understood in terms of reduction of the standard canonical variables, such as the Lagrangian configuration map and its conjugate momentum in fluid mechanics. When this is done, one obtains a Poisson manifold which is a union of symplectic leaves. An orbit beginning on a leaf stays on it, so these leaves appear as "constraint manifolds". In fluid mechanics these are the "Lin constraint" manifolds and are exactly the coadjoint orbits for the configuration manifold, which is a group.

'Constraints' are of three types. First of all there are the type above, which correspond to symplectic leaves in a larger phase space and indicate that a reduction has taken place. Second, there are constraints which are imposed on a model for purposes of idealization or simplification, such as incompressibility, rigidity, or the passage from electrofluid dynamics to magnetohydrodynamics. We expect that this second kind can be understood in the context of the first kind using an enlarged Poisson manifold and the framework of Weinstein

---

* Research partially supported by NSF grants MCS 81-07086 and MCS 80-23356, DOE Contract DE-AT03-82ER12097, and the Miller Institute.

[59]. Finally, there are constraints like $\operatorname{div} E = \rho$, which are, from the four-dimensional point of view, some of the Euler–Lagrange equations of the theory. The latter can also be understood in terms of zero sets of momentum maps and reduction, as in Marsden and Weinstein [42].

Clebsch, or canonical, variables for a system can be understood in terms of Poisson maps from symplectic to Poisson manifolds. For many systems these maps are, or are constructed from, momentum maps for symmetry groups. We note that Holm and Kuperschmidt [27] have taken the opposite approach, using Clebsch representations to derive the non-canonical Poisson structures.

In addition to the topics above, this paper contains discussions of point vortices, vortex patches, and vortex filaments. These objects form coadjoint orbits whose symplectic structures are related respectively to those for particles on $\mathbb{R}^1$, the KdV equation, and the Schrödinger equation. Canonical variables for these systems are particular instances of Clebsch variables.

Space does not permit the inclusion of extensive background material. Readers should consult our earlier papers and lectures listed in the bibliography, along with Arnold [6] (especially appendices 2 and 5) and Abraham and Marsden [1] (especially chapter 4).

2. Poisson manifolds, momentum maps, and reduction

A *Poisson manifold* is a manifold P together with a Lie algebra structure $\{\ ,\ \}$ on the space $C^\infty(P)$ of smooth real valued functions on P such that $\{f, g\}$ is a derivation in each argument.

If G is a Lie group and $\mathfrak{G}$ is its Lie algebra, the dual space $\mathfrak{G}^*$ carries a natural Poisson structure defined as follows. For $\mu \in \mathfrak{G}^*$ and $F: \mathfrak{G}^* \to \mathbb{R}$, define $\delta F/\delta \mu \in \mathfrak{G}$ by

$$\mathrm{D}F(\mu) \cdot v = \left\langle v, \frac{\delta F}{\delta \mu} \right\rangle,$$

where $\mathrm{D}F$ is the derivative of F, $v \in \mathfrak{G}^*$, and $\langle\ ,\ \rangle$ is the pairing between $\mathfrak{G}^*$ and $\mathfrak{G}$. For $F, G \in C^\infty(\mathfrak{G}^*)$, define

$$\{F, G\}_-(\mu) = -\left\langle \mu, \left[\frac{\delta F}{\delta \mu}, \frac{\delta G}{\delta \mu}\right]\right\rangle,$$

where $[\ ,\ ]$ is the standard (left) Lie bracket on $\mathfrak{G}$. The bracket $\{\ ,\ \}_-$ is the one induced on $\mathfrak{G}^*$ by identifying $C^\infty(\mathfrak{G}^*)$ with the *left* invariant functions on T^*G. We denote this structure by $\mathfrak{G}^*_-$. The corresponding bracket with the $+$ sign is associated to *right* invariant functions and is denoted $\mathfrak{G}^*_+$. In finite dimensions the formula for the bracket on $\mathfrak{G}^*_\pm$ in terms of a basis e_i and dual basis e^i, with $\mu = \Sigma \mu_i e^i$, is

$$\{F, G\}_\pm(\mu) = \pm \sum c^k_{ij} \frac{\partial F}{\partial \mu_i} \frac{\partial G}{\partial \mu_j} \mu_k,$$

where c^k_{ij} are the structure constants for the Lie algebra, defined by $[e_i, e_j] = \Sigma_k c^k_{ij} e_k$.

This formula for the bracket on $\mathfrak{G}^*_+$ is due to Lie [34], pp. 235 and 294. It was rediscovered by Berezin [9] and is closely related to results obtained by Arnold, Kirillov, Kostant, and Souriau around the same time.

If P is a Poisson manifold, the Hamiltonian system on P corresponding to a function $H: P \to \mathbb{R}$ is the vector field X_H on P such that real-valued functions on P evolve by $\dot F = \{F, H\}$. Since Poisson structures define maps of covectors to vectors, X_H is just the image of $\mathrm{d}H$.

Every Poisson manifold is a union of symplectic manifolds, its "symplectic leaves". Trajectories of X_H starting in a particular leaf necessarily stay there. Thus, these leaves may be viewed as constraint surfaces. For $\mathfrak{G}^*_+$ or $\mathfrak{G}^*_-$, the symplectic leaves are coadjoint orbits. More is known about the structure of Poisson manifolds, extending some of Lie's work [34] on function groups. Namely, Weinstein [59] shows that, at least on a linearized level, a Poisson manifold is near each point the product of a symplectic space and the dual of a Lie algebra. This helps to explain why Lie–Poisson brackets on $\mathfrak{G}^*$ are so fundamental.

Let P be a Poisson manifold and G a Lie group. Assume that G acts on P by a left (resp. right) action by Poisson maps, i.e. maps $\phi: P \to P$ such that $\{F \circ \phi, G \circ \phi\} = \{F, G\} \circ \phi$ for all $F, G \in C^\infty(P)$. By a *Hamiltonian map* for this action we mean a Lie algebra homomorphism (resp. anti-homomorphism) $\hat{J}: \mathfrak{G} \to C^\infty(P)$ such that $X_{\hat{J}(\xi)} = \xi_P$ for each $\xi \in \mathfrak{G}$, where ξ_P denotes the associated infinitesimal generator of the action. Define $J: P \to \mathfrak{G}^*$ by $\langle J(x), \xi \rangle = \hat{J}(\xi)(x)$. We call J the *momentum map*.

Proposition 2.1. Let $J_L: P \to \mathfrak{G}^*$ be a momentum map for a left action of G on P. Then

$J_L: P \to \mathfrak{G}^*_+$ is a Poisson map.

Likewise, if J_R is the momentum map for a right action, then

$J_R: P \to \mathfrak{G}^*_-$ is a Poisson map.

Proof. By definition of the Lie–Poisson bracket,

$$\{F, G\}_+(\mu) = \left\langle J(x), \left[\frac{\delta F}{\delta \mu}, \frac{\delta G}{\delta \mu}\right]\right\rangle$$

$$= \hat{J}\left(\left[\frac{\delta F}{\delta \mu}, \frac{\delta G}{\delta \mu}\right]\right)(x).$$

Since $\hat{J}$ is a Lie algebra homomorphism,

$$\hat{J}\left(\left[\frac{\delta F}{\delta \mu}, \frac{\delta G}{\delta \mu}\right]\right)(x) = \left\{\hat{J}\left(\frac{\delta F}{\delta \mu}\right), \hat{J}\left(\frac{\delta F}{\delta \mu}\right)\right\}(x).$$

The proof will be complete if we can show that

$$d\left(\hat{J}\left(\frac{\delta F}{\delta \mu}\right)\right) = d(F \circ J),$$

where $\delta F / \delta \mu$ is regarded as a constant element of $\mathfrak{G}$ evaluated at $\mu = J(x)$. Indeed, we have

$$d(F \circ J) \cdot v_x = dF(\mu) \cdot dJ(x) \cdot v_x$$

$$= \left\langle dJ(x) \cdot v_x, \frac{\delta F}{\delta \mu}\right\rangle$$

for $v_x \in T_x P$. Also,

$$d\left(\hat{J}\left(\frac{\delta F}{\delta \mu}\right)\right) \cdot v_x = d\left(\left\langle J(x), \frac{\delta F}{\delta \mu}\right\rangle\right) \cdot v_x$$

$$= \left\langle dJ(x) \cdot v_x, \frac{\delta F}{\delta \mu}\right\rangle,$$

since $\delta F / \delta \mu$ is regarded as a constant element of $\mathfrak{G}$. ∎

Note that Ad*-equivariant momentum maps in the usual sense for actions on symplectic manifolds are momentum maps in the present sense. In what follows, most of the momentum maps we consider are standard ones from symplectic geometry (see Abraham and Marsden [1], sec. 4.2).

A consequence of the formula $d(\hat{J}(\delta F/\delta \mu)) = d(F \circ J)$ proved above is the following fact about collective Hamiltonians. (cf. Marle [35] and Guillemin and Sternberg [23]):

Corollary 2.2. Let $F \in C^\infty(\mathfrak{G}^*)$. Then

$$X_{F \circ J}(x) = \left(\frac{\delta F}{\delta \mu}\right)_P(x),$$

where $\delta F/\delta \mu$ is evaluated at $\mu = J(x)$. (This holds for momentum maps associated with either left or right actions).

The distinction between left and right in 2.1 may be clarified by the following remarks. Let G be a Lie group and T^*G be its cotangent bundle. Consider the map $\lambda: T^*G \to \mathfrak{G}^*$ given by

$$\langle \lambda(\alpha_g), \xi \rangle = \langle \alpha_g, TL_g \cdot \xi \rangle,$$

i.e. left translate covectors to the identity; $\mathfrak{G}^* = (T_e G)^*$. This map is a Poisson map of T^*G to $\mathfrak{G}^*_-$ because it is the momentum map associated with *right* translations of G on T^*G. (Abraham and Marsden [1], p. 302). Thus, T^*G/G (quotient by the *left* action) yields $\mathfrak{G}^*_-$ with the Lie–Poisson structure (including the minus sign).

There is a converse to 2.1. Namely, if $J: P \to \mathfrak{G}^*$ is a Poisson map, then J is a momentum map for an action (of the simply connected covering group) of G. See Fong and Meyer [22] for the symplectic case. Thus, when the range space is the dual of a Lie algebra one loses no generality in the search for Poisson maps by looking among momentum maps.

If P is a Poisson manifold and G acts on P by Poisson maps, then P/G is a Poisson manifold. (Assume for present purposes that P/G has no singularities.) Indeed, we may identify functions on P/G with G-invariant functions on P, so that the bracket on P is inherited by P/G. We call P/G the *reduced Poisson manifold*. For example, T^*G reduced by the left action of G is just $\mathfrak{G}^*_-$.

Example 2.4. (The rigid body). Here we take $G = SO(3)$ so that $\mathfrak{G}$, its Lie algebra, is identifiable with $\mathbb{R}^3$ and the Lie bracket with the cross product. A point $m \in \mathfrak{G}^*_-$ represents the angular momentum in "body coordinates". (See Abraham and Marsden [1] or Arnold [6] for the explanation of this terminology.) The Hamiltonian H is the kinetic energy of the body, a positive definite quadratic function of m. By choosing an appropriate orthonormal basis of $\mathbb{R}^3$ (and corresponding orthonormal dual basis of $\mathbb{R}^{3*}$) we can assume that H is diagonal:

$$H(m) = \frac{1}{2}\left(\frac{m_1^2}{I_1} + \frac{m_2^2}{I_2} + \frac{m_3^2}{I_3}\right),$$

where I_1, I_2, I_3 are positive constants, the moments of inertia. Let us work out the Lie–Poisson equations $\dot{F} = \{F, H\}_-$ in this case. Clearly $\delta F/\delta m$ is just the vector in $\mathbb{R}^3$ with components $(\partial F/\partial m_1, \partial F/\partial m_2, \partial F/\partial m_3)$. Thus

$$\{F, H\}_-(m) = -\left\langle m, \frac{\delta F}{\delta m} \times \frac{\delta H}{\delta m}\right\rangle,$$

the triple product. If we choose $F(m) = m_1$, the equation $\dot{F} = \{F, H\}_-$ reads

$$\dot{m}_1 = -\begin{vmatrix} m_1 & m_2 & m_3 \\ 1 & 0 & 0 \\ \frac{m_1}{I_1} & \frac{m_2}{I_2} & \frac{m_3}{I_3} \end{vmatrix} = \frac{I_2 - I_3}{I_2 I_3} m_2 m_3.$$

The equations for $\dot{m}_2$ and $\dot{m}_3$ are obtained by cyclic permutation. These are the famous Euler equations for a force-free rigid body. It is trivial to check that $(d/dt)(m_1^2 + m_2^2 + m_3^2) = 0$; i.e. $\|m\|^2$ is constant in time. The spheres $\|m\| = $ constant are exactly the coadjoint orbits for $SO(3)$. Thus $\mathfrak{so}(3)^*$ is the union of these symplectic manifolds (plus the origin). Their preservation by the Euler equations corresponds to the conservation of angular momentum.

The *heavy* top requires the semi-direct product $E(3) = SO(3) \times \mathbb{R}^3$; see Vinogradov and Kupershmidt [54] p. 236, Guillemin and Sternberg [23], and Marsden, Ratiu, and Weinstein [40].

The reduction of Poisson manifolds is related to reduction of symplectic manifolds with symmetry. Let J be a momentum map for the G-action on P, and assume that P is symplectic. Suppose that $\mathcal{O} \subset \mathfrak{G}^*$ is a coadjoint orbit in $\mathfrak{G}^*$ and that J is transversal to $\mathcal{O}$. Then $J^{-1}(\mathcal{O})/G$ (assuming it is without singularities) is the *reduced symplectic manifold* [41].

Proposition 2.3. The symplectic leaves of P/G are $J^{-1}(\mathcal{O})/G$.

This follows readily from the definitions and the fact that $T_x(J^{-1}(\mathcal{O}))$ splits into $T_x(G \cdot x) + \ker dJ(x)$, whose summands are symplectic orthogonal complements of each other.

3. Symplectic variables and gauge groups

We begin with a Poisson manifold P as the basic space of physical variables for a theory. Suppose that $H: P \to \mathbb{R}$ is a given energy function, so that the trajectories of X_H describe the dynamics of the system.

Definition 3.1. By *symplectic variables* (or "*Clebsch variables*") we mean a symplectic manifold R and a Poisson map $\psi: R \to P$.

Recall that if $P = \mathfrak{G}^*$, then ψ will be a momentum map. If coordinates in R are found which bring the symplectic form into canonical form, they are called *canonical variables*. In many examples R is a cotangent bundle. One can always in principle use Darboux' theorem to find the canonical variables.

If we let $H_\psi = H \circ \psi$, then X_{H_ψ} projects to X_H, and so integral curves for H_ψ project to those for X_H. Thus, by introducing possibly redundant information, one can write the equations in the new R variables in symplectic Hamiltonian form, and using canonical variables, in canonical form.

Example 3.2. For the rigid body, $P = \mathfrak{G}_-^*$ where $G = SO(3)$, as in example 2.4. Now $SO(3)$ has the same Lie algebra as $SU(2)$, which acts symplectically on $\mathbb{C}^2$. The induced momentum map $\psi : \mathbb{C}^2 \to P$ defines Cayley–Klein parameters as special symplectic variables. A related construction for general Lie groups can be found in Weinstein [59].

Definition 3.3. Let $\psi : R \to P$ be symplectic variables for a Poisson manifold P. The associated *gauge transformations* are the symplectic diffeomorphisms $\phi : R \to R$ such that $\psi \circ \phi = \psi$.

Let K be a Lie group of gauge transformations such that K acts on R from the left (resp. right) and is transitive on each fiber $\psi^{-1}(p)$ and has a momentum map $J_K : R \to \mathfrak{K}^*$ where $\mathfrak{K}$ is the Lie algebra of K. By 2.1, J_K is a Poisson map from R to $\mathfrak{G}_+^*$ (resp. $\mathfrak{G}_-^*$). We call K a *gauge group*; J_K is the corresponding conserved quantity.

Example 3.4. In electromagnetism, $R = T^*\mathfrak{A}$ where $\mathfrak{A}$ is the space of vector potentials A, and P is the Poisson space of E's and B's with div $B = 0$ and with bracket

$$\{F, G\} = \int_{\mathbb{R}^3} \left(\frac{\delta F}{\delta E} \cdot \text{curl} \frac{\delta G}{\delta B} - \frac{\delta G}{\delta E} \cdot \text{curl} \frac{\delta F}{\delta B} \right) dx$$

(see Born and Infeld [10] and Marsden and Weinstein [42, §4].) Here we choose K to be the standard group of gauge transformations $A \mapsto A + \nabla \phi$. The symplectic leaves in P are the sets where div $E = \rho$ is given. (These leaves are obtained from R by reduction, as shown in the preceding reference. It is also explained there how to couple electromagnetism with other continuum theories such as fluids and plasmas.) Here the momentum map is given by $J_K(A, Y) = -\text{div } Y$ where $Y = -E$ is the variable conjugate to A, and $\psi(A, Y) = (-Y, \text{curl } A)$.

Example 3.5. A gauge group for Cayley–Klein parameters is $U(1)$, and $\mathbb{C}^2$ reduced by $U(1)$ is $\mathfrak{so}(2)^* \approx \mathfrak{so}(3)^*$.

The reduction result mentioned in the preceding examples holds in general.

Proposition 3.6. Reduced spaces for $J_K : R \to \mathfrak{K}^*$ are symplectic leaves in P.

Proof. Let $\mathcal{O} \subset \mathfrak{K}^*$ be a coadjoint orbit. Let $\psi_\mathcal{O} : J_K^{-1}(\mathcal{O}) \to P$ be ψ restricted to $J_K^{-1}(\mathcal{O})$. By K-equivariance, $\psi_\mathcal{O}$ induces a Poisson map:

$$\Psi_\mathcal{O} : J_K^{-1}(\mathcal{O})/K \to P.$$

Since K acts transitively on fibers, $\Psi_\mathcal{O}$ is one-to-one. Since $J^{-1}(\mathcal{O})/K$ is symplectic, it embeds via $\Psi_\mathcal{O}$ as a symplectic leaf. ∎

The following diagrams summarize the situation:

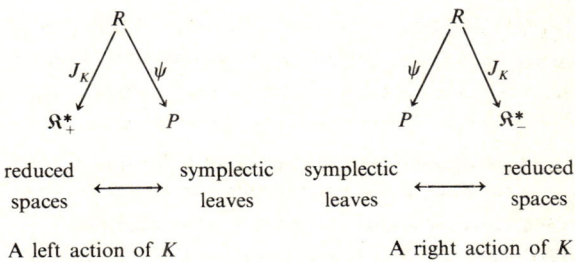

| reduced spaces | ⟷ | symplectic leaves | symplectic leaves | ⟷ | reduced spaces |

A left action of K · · · A right action of K

Notice that if $H : P \to \mathbb{R}$ is our given Hamiltonian and $H_\psi = H \circ \psi$, then J_K is a conserved quantity for

H_ψ, since H_ψ is invariant under gauge transformations. If J_K is constant on ψ-fibers (for example this holds if K is abelian) then it induces a map $\bar{J}_K : P \to \mathfrak{K}^*$ which gives constants of the motion for H. However these give no new information on conserved quantities in view of the preceding proposition and conservation of symplectic leaves in P. (For example, the conservation laws of Levich [33] are of this type and are explained in section 5 below).

In some cases, such as incompressible fluid mechanics, P itself is a Lie–Poisson space $\mathfrak{G}^*_+$ and ψ is a momentum mapping associated to a left action. Then the above diagram becomes

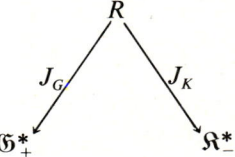

The situation is now symmetric, and G is also a gauge group for R regarded as symplectic variables for the Lie–Poisson space $\mathfrak{G}^*_+$. We say that we have a *dual pair* (see Weinstein [59]). A simple example of a dual pair is obtained by considering the left and right actions of a group G on T^*G:

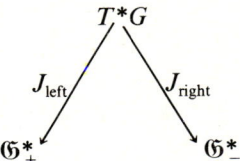

For applications of this idea to semi-direct products, see Marsden, Ratiu and Weinstein [40].

4. Ideal fluid flow as a Lie–Poisson system

The configuration space for ideal (incompressible, homogeneous) fluid flow on a region Ω (a region in $\mathbb{R}^n$ or a compact Riemannian manifold with boundary) is $\mathscr{D}_{\mathrm{vol}}$, the group of volume-preserving diffeomorphisms of Ω to itself. The phase space is $T^*\mathscr{D}_{\mathrm{vol}}$ (where dual spaces are understood in the sense of L^2 pairings). As we shall detail below, the kinetic energy of a fluid is *right* invariant on $T^*\mathscr{D}_{\mathrm{vol}}$ and so induces, by reduction, a Hamiltonian system on the Poisson manifold $\mathscr{X}^*_{\mathrm{vol}}$ where $\mathscr{X}_{\mathrm{vol}}$, the space of divergence-free vector fields on Ω parallel to $\partial\Omega$, is the Lie algebra of $\mathscr{D}_{\mathrm{vol}}$, and the + Lie–Poisson structure (+ because of *right* invariance) is used. The Lie bracket on $\mathscr{X}_{\mathrm{vol}}$ (the left Lie algebra of $\mathscr{D}_{\mathrm{vol}}$) is the negative of the standard commutator bracket of vector fields*.

The picture above, but using $\mathscr{X}_{\mathrm{vol}}$ and $T\mathscr{D}_{\mathrm{vol}}$ rather than the dual spaces was known to Arnold [4]; its functional analytic details were established by Ebin and Marsden [21]. The use of $\mathscr{X}^*_{\mathrm{vol}}$ in this problem has also been emphasized by Morrison [48]. (For compressible flow, see Morrison and Greene [49] and Marsden, Ratiu and Weinstein [40].)

We arrive naturally at the space $\mathscr{X}^*_{\mathrm{vol}}$ starting only with these two assumptions:
1) the phase space is $T^*\mathscr{D}_{\mathrm{vol}}$;
2) the Hamiltonian is right invariant.

Assumption 1 is taken for granted: the deformation and its conjugate momentum describe the state of a fluid (or generally a continuum) in material coordinates. This has been accepted as basic since the time of Euler and Lagrange. Assumption 2 is simply a fact, following from the change of variables theorem (see below or one of the aforementioned references for details).

Now $\mathscr{X}^*_{\mathrm{vol}}$ consists of the linear functionals on $\mathscr{X}_{\mathrm{vol}}$. These we identify with the one-forms α modulo exact one-forms. Note that exact one-forms are L^2-orthogonal to the divergence-free vector fields:

$$\int_\Omega \mathrm{d}f \cdot v \, \mathrm{d}x = 0$$

if div $v = 0$, where $\mathrm{d}x$ is the volume-form on Ω.

One can represent an element $[\alpha]$ of $\mathscr{X}^*_{\mathrm{vol}}$ by the

---

* Arnold [6] uses the left Lie algebra of vector fields so has conventions on brackets of vector fields that is opposite from the standard convention.

two-form $d\alpha$ and the integrals $\Gamma_1, \ldots, \Gamma_l$ of α over a basis $\gamma_1, \ldots, \gamma_l$ of the first homology of Ω; i.e. elements ω of $\mathscr{X}_{\text{vol}}^*$ consist of the *vorticity* $d\alpha$ and circulations around non-contractible loops: $\omega = (d\alpha, \Gamma_1, \ldots, \Gamma_l)$. For simplicity of exposition we shall assume that Ω is simply connected so we may ignore the Γ_i's. As in the abstract theory, for $F : \mathscr{X}_{\text{vol}}^* \to \mathbb{R}$, we define $\delta F / \delta \omega \in \mathscr{X}_{\text{vol}}$ by

$$DF(\omega) \cdot [\sigma] = \int_\Omega \left\langle \frac{\delta F}{\delta \omega}, \sigma \right\rangle dx,$$

where $\langle \, , \, \rangle$ denotes vector–covector pairing and σ stands for any representative of its class $[\sigma]$. The + Lie–Poisson bracket we use is

$$\{F, G\}(\omega) = \int_\Omega \left\langle \omega, \left[\frac{\delta F}{\delta \omega}, \frac{\delta G}{\delta \omega}\right] \right\rangle dx.$$

This bracket was explicitly written down in Kuznetsov and Mikhailov [32], Morrison [48] and Olver [64]. It was implicitly known to Arnold [5]. Its connection with Lie–Poisson structures and reduction is due to the present authors.

Now the Hamiltonian function for incompressible flow is the kinetic energy. This is defined on $T\mathscr{D}_{\text{vol}}$ by

$$H(V_\eta) = \int_\Omega \tfrac{1}{2} \langle V_\eta, V_\eta \rangle \, dx, \qquad (\text{H}_{\text{material}})$$

where V_η, a vector field over $\eta \in \mathscr{D}_{\text{vol}}$, is the material velocity of the fluid (i.e. $V_\eta(X)$ is the velocity at $\eta(X)$ of the particle with material point $X \in \Omega$). By the change of variables theorem, H is right invariant on $T\mathscr{D}_{\text{vol}}$ (see Ebin and Marsden [21] for details). In terms of the Eulerian velocity variable,

$$H(v) = \tfrac{1}{2} \int \|v\|^2 \, dx. \qquad (\text{H}_{\text{Eulerian}})$$

Also, H induces a right-invariant function on $T^*\mathscr{D}_{\text{vol}}$ given at the identity, i.e. on $\mathscr{X}_{\text{vol}}^*$ by

$$H(\omega) = \tfrac{1}{2} \int \langle \Delta^{-1} \omega, \omega \rangle \, dx, \qquad (\text{H}_{\text{vorticity}})$$

where $\langle \, , \, \rangle$ is the metric pairing of two-forms and Δ is the Laplace–DeRham operator, $\Delta = d\delta + \delta d$. (At the identity, ω is identified with $dv^\flat$, where $v \in \mathscr{X}_{\text{vol}}$, $v^\flat$ meaning the corresponding one-form); note that $\int \langle \Delta^{-1} \omega, \omega \rangle \, dx = \int \langle \Delta^{-1} dv^\flat, dv^\flat \rangle \, dx = \int \langle \delta \Delta^{-1} dv^\flat, v^\flat \rangle \, dx = \int \langle v^\flat, v^\flat \rangle \, dx$ since $\delta v^\flat = 0$ (i.e. div $v = 0$).

There are three equivalent ways of representing the following Euler equations for perfect incompressible flow:

$$\left. \begin{aligned} \frac{\partial v}{\partial t} + v \cdot \nabla v &= -\nabla p, \\ \text{div } v &= 0, \\ v \text{ parallel to } \partial \Omega, \end{aligned} \right\} \qquad (\text{E})$$

in which the density is $\rho = 1$*.

First, one views (E) directly on $\mathscr{X}_{\text{vol}}$ via reduction from the corresponding right-invariant system on $T\mathscr{D}_{\text{vol}}$. This is the perspective developed by Arnold [4], and expositions are available in several places (Arnold [6], Marsden [36], and Abraham and Marsden [1]).

Second, one views (E) directly on $T\mathscr{D}_{\text{vol}}$ or $T^*\mathscr{D}_{\text{vol}}$ as an evolution equation for the configuration $\eta \in \mathscr{D}_{\text{vol}}$ of the fluid and its conjugate velocity or momentum. As Arnold showed, *the equations (E) are equivalent to the geodesic equations on $T\mathscr{D}_{\text{vol}}$ (or $T^*\mathscr{D}_{\text{vol}}$) for the right-invariant (weak) Riemannian metric on $T\mathscr{D}_{\text{vol}}$ whose value at the identity is the L^2 inner product*

$$\langle v, w \rangle = \int_\Omega \langle v(x), w(x) \rangle \, dx.$$

It was shown in Ebin and Marsden [21] that the spray of this metric (that is the vector field on a suitable Sobolev completion of $T\mathscr{D}_{\text{vol}}$ whose inte-

---

* For ideal but, inhomogeneous flow on $T\mathscr{D}_{\text{vol}}$ together with the existence theory, see Marsden [37]. The system is not right invariant, but rather yields a Lie–Poisson system on the semidirect product $\mathscr{X}_{\text{vol}} \times$ Functions; see Marsden, Ratiu and Weinstein [40] and Henyey [26].

gral curves give solutions on $T\mathscr{D}_{\text{vol}}$) is C^∞, so the Picard method for ordinary differential equations suffices to prove existence and uniqueness of solutions for short time.

Third, we can view (E) in terms of the vorticity. To do, let us temporarily use traditional fluid mechanics notation and write $\omega = \nabla \times v$, rather than $\omega = dv^\flat$. Taking the curl of (E) gives the *vorticity form* of the equations:

$$\frac{D\omega}{Dt} - \omega \cdot \nabla v = 0, \tag{V}$$

where $D\omega/Dt = \partial \omega/\partial t + v \cdot \nabla \omega$ is the material derivative (see, for instance, Chorin and Marsden [18, p. 32]). Here v is determined by ω through the equations

$$\omega = \nabla \times v,$$

$$\operatorname{div} v = 0, \quad v \parallel \partial \Omega.$$

If we regard ω as a two-form again, then

$$v^\flat = \delta \psi \quad \text{and} \quad \psi = \Delta^{-1}\omega,$$

where ψ is a 2-form, the 'stream function', δ is the codifferential, and $\Delta = d\delta + \delta d$ is the Laplace–DeRham operator. (The solution for v in terms of ω is to be supplemented by specified circulations if Ω contains non-contractible loops.)

Theorem 4.1. *The vorticity equations* (V) *are equivalent to the Lie–Poisson equations* $\dot{F} = \{F, H\}$ *on* $\mathscr{X}_{\text{vol}}^*$ *where H is given by* $(H_{\text{vorticity}})$.

This theorem follows directly from the general facts about reduction already mentioned, but we shall verify it by hand. If we let

$$\frac{\delta H}{\delta \omega} = v \in \mathscr{X}_{\text{vol}},$$

then by definition

$$DH(\omega) \cdot [\sigma] = \int \langle v, \sigma \rangle \, dx,$$

where σ is a 1-form on Ω and $[\sigma]$ is its equivalence class in $\mathscr{X}_{\text{vol}}^*$, identified with the 2-form $d\sigma$. Thus,

$$DH(\omega) \cdot [\sigma] = \int \langle \Delta^{-1}\omega, d\sigma \rangle \, dx = \int \langle v, \sigma \rangle \, dx.$$

Since δ and d are adjoints, we get $v^\flat = \delta \Delta^{-1}\omega$. In other words, $\delta H/\delta \omega$ is nothing other than the corresponding velocity field of the vorticity. Thus,

$$\{F, H\} = \int \left\langle \omega, \left[\frac{\delta F}{\delta \omega}, \frac{\delta H}{\delta \omega}\right] \right\rangle dx$$

$$= \int \left\langle \omega, \left[\frac{\delta F}{\delta \omega}, v\right] \right\rangle dx.$$

Now the Lie algebra bracket $[\delta F/\delta \omega, v]$ is the negative of the usual Lie bracket and so is given by $\mathrm{L}_v \delta F/\delta \omega$, where L_v is Lie differentiation. Integrating by parts, we get

$$\{F, H\} = -\int \left\langle \mathrm{L}_v \omega, \frac{\delta F}{\delta \omega} \right\rangle dx,$$

where, as above, the pairing between $\mathscr{X}_{\text{vol}}^*$ and $\mathscr{X}_{\text{vol}}$ is understood in terms of representatives of the classes of two forms. Thus, by definition of $\delta F/\delta \omega$,

$$\{F, H\} = -DF(\omega) \cdot \mathrm{L}_v \omega.$$

Therefore, the equations $\dot{F} = \{F, H\}$ are, by the chain rule, equivalent to the *Lie form of the vorticity equation*

$$\frac{\partial \omega}{\partial t} + \mathrm{L}_v \omega = 0, \tag{LV}$$

where L_v is Lie differentiation of two-forms. Equation (LV) just says that ω is Lie transported by the flow. Now (LV) is equivalent to the form (V) using some simple vector identities†. This completes the verification of theorem 4.1. (Notice that the use of differential forms enables us to replace $\Omega \subset \mathbb{R}^3$ (or

---

† $\mathrm{L}_v \omega = \mathrm{d}i_v \omega + i_v \mathrm{d}\omega = \mathrm{d}i_v \omega$. If $\hat{\omega}^\flat = (*\omega)$, then $(i_v \omega) = (\hat{\omega} \times v)^\flat$ so $\mathrm{d}i_v \omega = \operatorname{curl}(\hat{\omega} \times v) = -\hat{\omega} \cdot \nabla v + v \cdot \nabla \hat{\omega}$.

$\mathbb{R}^2$) by any Riemannian manifold and reveals the geometric interpretation of the vorticity equation.) ■

The equation (LV) enables us to check directly some other general facts about Lie–Poisson equations. Let us verify that *solution curves to (LV) remain on coadjoint orbits* in $\mathscr{X}^*_{\text{vol}}$.

The (right) coadjoint action of $\mathscr{D}_{\text{vol}}$ on $\mathscr{X}^*_{\text{vol}}$ is readily checked to be the pull-back action $(\eta, \omega) \mapsto \eta^* \omega$. Thus, the coadjoint orbit through $\omega \in \mathscr{X}_{\text{vol}}$ is

$$\mathcal{O}_\omega = \{\eta^* \omega \mid \eta \in \mathscr{D}_{\text{vol}}\},$$

but the solution to (LV) for given initial condition $\omega(0)$ is simply

$$\omega(t) = \eta(t)^* \omega(0),$$

where $\eta(t)$ is the flow of $v(t)$. Hence it is clear that the vorticity stays on $\mathcal{O}_{\omega(0)}$. This transport of vorticity by the flow is nothing other than Kelvin's circulation theorem. Thus, *the preservation of coadjoint orbits, right invariance on $T^*\mathscr{D}_{\text{vol}}$, and Kelvin's circulation theorem are all equivalent*.

We know from the general theory that $\mathcal{O}_\omega$ is a symplectic manifold. Let us compute its symplectic structure. Tangent vectors to $\mathcal{O}_\omega$ at ω are given by elements $L_u \omega$, where $u \in \mathscr{X}_{\text{vol}}$.

Theorem 4.2. The symplectic structure Ω_ω on $T_\omega \mathcal{O}_\omega$ is given by

$$\Omega_\omega(L_{u_1}\omega, L_{u_2}\omega) = \int \omega(u_1, u_2) \, dx.$$

Proof. We use the general Kirillov–Kostant–Souriau formula $\Omega_\mu(\xi_{\mathfrak{G}^*}(\mu), \eta_{\mathfrak{G}^*}(\mu)) = \langle \mu, [\xi, \eta] \rangle$ for the symplectic structure on coadjoint orbits (see Abraham and Marsden [1], p. 303. Here there is a + sign since we are dealing with a *right* invariant system). In our case this reads

$$\Omega_\omega(L_{u_1}\omega, L_{u_2}\omega) = \langle \omega, -[u_1, u_2] \rangle.$$

(Recall that our convention is to always use the left Lie bracket. For $\mathscr{D}$ or $\mathscr{D}_{\text{vol}}$, this is the *negative* of the usual Lie bracket; see Abraham and Marsden [1, Ex. 4.1G]). Let $\omega = d\alpha$. Then the preceding equation gives

$$\Omega_\omega(L_{u_1}\omega, L_{u_2}\omega) = -\int \alpha \cdot [u_1, u_2] \, dx,$$

according to our definition of the pairing. Now write $[u_1, u_2] = L_{u_1} u_2$ and integrate by parts to get

$$\Omega_\omega(L_{u_1}\omega, L_{u_2}\omega) = \int (L_{u_1}\alpha) u_2 \, dx$$

$$= \int (i_{u_1} d\alpha + d i_{u_1} \alpha) \cdot u_2 \, dx.$$

The second term vanishes since $\text{div } u_2 = 0$. Thus we get

$$\Omega_\omega(L_{u_1}\omega, L_{u_2}\omega) = \int (i_{u_1}\omega) \cdot u_2 \, dx = \int \omega(u_1, u_2) \, dx$$

as claimed. ■

5. Clebsch variables for ideal flow

According to our general scheme in section 3, symplectic variables for ideal flow are provided by momentum maps $J: R \to \mathscr{X}^*_{\text{vol}}$. To find such variables it suffices to seek symplectic manifolds on which $\mathscr{D}_{\text{vol}}$ acts and compute their momentum maps. Since $\mathscr{X}^*_{\text{vol}}$ carries the + Lie–Poisson structure, we should seek a *left* action. The classical Clebsch variables, and more, can be readily found by such an approach.

Consider the action of $\mathscr{D}_{\text{vol}}$ on the space $\mathscr{F}$ of real valued functions on Ω by $(\eta, \lambda) \mapsto \eta \cdot \lambda = \lambda \circ \eta^{-1}$. This induces in the usual way a symplectic left action on $T^*\mathscr{F} = \mathscr{F} \times \mathscr{F}^*$. Identify $\mathscr{F}^*$ with $\mathscr{F}$ via the L^2 pairing using the given volume element.

Proposition 5.1. The momentum map of the above action is given by $J: (\lambda, \mu) \mapsto \omega = d\lambda \wedge d\mu$.

Proof. Let $u \in \mathscr{X}_{\text{vol}}$. The corresponding infinitesimal generator on $\mathscr{F}$ is $-L_u \lambda$. Thus, using the formula

$$\langle \xi, J(\alpha_q) \rangle = \langle \alpha_q, \xi_Q(q) \rangle$$

for the momentum map of a lifted action (Abraham and Marsden [1, p. 283]) we get

$$\langle u, J(\lambda, u) \rangle = \int \mu \cdot (-L_u \lambda) \, dx$$
$$= \int (L_u \mu) \lambda \, dx$$
$$= \int \lambda \, d\mu \cdot u \, dx.$$

Thus the one form representing $J(\lambda, \mu)$ is $\lambda \, d\mu$. The corresponding two-form is $d(\lambda \, d\mu) = d\lambda \wedge d\mu$. Thus $J(\lambda, \mu) = d\lambda \wedge d\mu$. ∎

It follows directly that if the Euler equations are expressed in terms of λ and μ they will be in canonical Hamiltonian form, a result already known to Clebsch [19]. We call the canonical variables λ, μ *Clebsch variables*.

The gauge group of Clebsch variables consists of all canonical transformations $\mathscr{S}$ of $\mathbb{R}^2$. Indeed, these are the transformations of the (λ, μ) variables that leave ω invariant. Now we get a dual pair as one can readily check:

```
                    F × F*
                   /      \
J = Clebsch map   /        \ j
                 /          \
          𝒳*_vol(Ω)          ℐ*
```

The momentum map j for the action of $\mathscr{S}$ on $\mathscr{F} \times \mathscr{F}^*$ is given by $(\lambda, \mu) \mapsto (\lambda \times \mu)_* \, d\mu$ where we identify $\mathscr{I}^*$, the dual of the Lie algebra $\mathscr{I}$ of $\mathscr{S}$, with densities on $\mathbb{R}^2$, as in Marsden and Weinstein [42].

The example above is a member of a family of dual pairs. If $(P_{\text{sym}}, \omega_{\text{sym}})$ is a symplectic manifold and $(P_{\text{vol}}, \omega_{\text{vol}})$ is a volume manifold, i.e. a manifold carrying a volume element ω_{vol}, let $\mathscr{M}$ be the space of maps from P_{vol} to P_{sym}. Then the groups G_{vol} and G_{sym} of volume-preserving and symplectic diffeomorphisms of P_{vol} and P_{sym} respectively act on $\mathscr{M}$ by compositions. Their momentum maps are $J_{\text{vol}}(\eta) = \eta^* \omega_{\text{sym}} \in \mathfrak{G}^*_{\text{vol}}$ regarded as two-forms on P_{vol} as before and $J_{\text{sym}}(\eta) = \eta_* \omega_{\text{vol}} \in \mathfrak{G}^*_{\text{sym}}$, regarded as densities on P_{sym}. Then

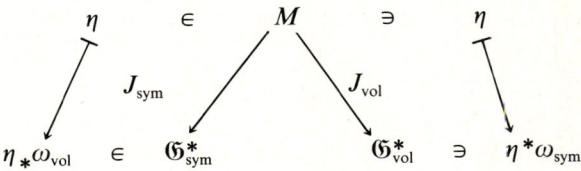

is a dual pair.

Notice that in the case of fluids j is a constant of the motion for the induced Hamiltonian system on $\mathscr{F} \times \mathscr{F}^*$ since $\mathscr{S}$ is, by construction, an invariance group. For example, if we compose j with evaluation against an element ϕ of $\mathscr{I}$, we get the constant of the motion

$$\int_{\mathbb{R}^2} \phi(\lambda, \mu)[(\lambda \times \mu)_* \, dx] = \int_\Omega \phi(\lambda, \mu) \, dx.$$

These invariants were found by Levich [33]. From the general theory, we know that these give no more information than the invariance of coadjoint orbits. In particular, even though one can find infinitely many such invariants in involution, they can never suffice to prove complete integrability. One might call these invariants "kinematic", since their conservation depends only on the Poisson structure and not on the specific Hamiltonian which generates the dynamics. Similar remarks apply to Ebin [63].

Note that if $\omega = d\lambda \wedge d\mu$, then the velocity field $v^\flat$ is such that $v^\flat - \lambda \, d\mu$ is closed; if it is exact, we can write

$$v^\flat = d\alpha + \lambda \, d\mu,$$

which is the classical Clebsch representation for the velocity field. This expression is interpreted in terms of canonical variables in Marsden, Ratiu and Weinstein [40].

Notice next that if $\omega = d\lambda \wedge d\mu$, and we are in $\mathbb{R}^3$, say, then the *helicity* is given by

$$\mathcal{H} \stackrel{\text{def}}{=} \int_{\mathbb{R}^2} v^{\flat} \wedge \omega = \int_{\mathbb{R}^3} v^{\flat} \wedge d(\lambda \wedge d\mu)$$

$$= -\int_{\mathbb{R}^3} dv^{\flat} \wedge \lambda \, d\mu \quad \text{(integration by parts)}$$

$$= -\int_{\mathbb{R}^3} (d\lambda \wedge d\mu) \wedge \lambda \, d\mu = 0.$$

However, in a flow with linked vortex lines, the helicity does not vanish (see Moffatt [46] for a discussion), so *the Clebsch representation leaves out many interesting coadjoint orbits.*

Remark. Invariance of the helicity follows at once from preservation of coadjoint orbits: $\partial \omega / \partial t + L_v \omega = 0$ (cf. Moffatt [46], p. 41).

It has likewise been noted by Bretherton [12] that the variational principle of Seliger and Whitham [52], which is closely related to the Clebsch representation, does not permit knotted vortex lines.

Helicity can be reincorporated if, in the general $(P_{\text{sym}}, P_{\text{vol}})$ dual pair, one replaces $\mathbb{R}^2$ by the two-sphere S^2 and imposes certain boundary conditions on $\mathbb{R}^3$ to allow compactification to S^3. The helicity is then a multiple of the topological Hopf invariant; see Kuznetsov and Mikhailov [32].

6. Two-dimensional flow

The "vorticity-bracket"

$$\{F, G\}(\omega) = \int_{\Omega} \left\langle \omega, \left[\frac{\delta F}{\delta \omega}, \frac{\delta G}{\delta \omega} \right] \right\rangle d\mu$$

introduced in section 4 has a particularly simple formulation in two dimensions. A volume (i.e. area) element in two dimensions in also a symplectic structure, so that each divergence-free vector field v may be thought of as a Hamiltonian vector field X_{ψ}; the Hamiltonian function ψ coincides with the *stream function* for v discussed in section 4. Assuming that the manifold Ω on which the fluid flows is connected, the function ψ is determined by v up to a constant, so we may identify the Lie algebra $\mathcal{X}_{\text{vol}}$ with $C^{\infty}(\Omega)/$constants. The dual space $\mathcal{X}_{\text{vol}}^*$ is then identified with generalized functions ω on Ω with $\int_{\Omega} \omega \, d\mu = 0$, and the reader may check that this correspondence is consistent with the previous identification of $\mathcal{X}_{\text{vol}}^*$ with the vorticities.

Now the Lie algebra bracket $[v_1, v_2]$ in $\mathcal{X}_{\text{vol}}^*$ is the negative of the Lie bracket of vector fields $[v_1, v_2] = -L_{v_1} v_2 = -L_{X_{\psi_1}} X_{\psi_2} = X_{\{\psi_1, \psi_2\}}$ (Abraham and Marsden [1], p. 194). Thus the Lie algebra bracket corresponds to the Poisson bracket $\{\psi_1, \psi_2\}$ of stream functions, (with the standard sign conventions), so we may write the vorticity bracket in the form

$$\{F, G\}(\omega) = \int_{\Omega} \omega \left\{ \frac{\delta F}{\delta \omega}, \frac{\delta G}{\delta \omega} \right\} d\mu,$$

where ω, $\delta F / \delta \omega$, and $\delta G / \delta \omega$ are all thought of as functions on Ω, and the bracket inside the integral is the "ordinary" two-dimensional Poisson bracket* on the symplectic manifold Ω.

In the remainder of this section, we will exhibit two different kinds of Clebsch variables for the vorticity bracket variables on the individual coadjoint orbits.

The first Clebsch representation, introduced by Morrison [47, 48] is to write ω as a bracket $\omega = \{\mu, \lambda\}$. Then if we consider the pair (λ, μ) to

---

* As Morrison [48] has also noted, this same as the Poisson bracket for the Poisson–Vlasov equation in one space dimension, as well as for the guiding center plasma beam equation. It is shown in Weinstein [60] that the same bracket can also be used for geostrophic fluid flow and plasma drift waves.

be canonical variables, the mapping $(\lambda, \mu) \mapsto \omega$ turns out to be a Poisson mapping. This construction is a special case of one proposed in a general Lie group context by Kazhdan, Kostant, and Sternberg [31], who apply it to show the integrability of the Calogero dynamical system. If $\mathfrak{G}$ is any Lie algebra, then G acts on $\mathfrak{G}$ by the adjoint representation, and this action lifts to a symplectic action on $T^*\mathfrak{G} \approx \mathfrak{G} \times \mathfrak{G}^*$. The momentum map is $(\xi, \mu) \mapsto \mathrm{ad}^*_\xi \mu$. If $\mathfrak{G}$ carries an Ad-invariant metric, we may identify $\mathfrak{G}^*$ with $\mathfrak{G}$, and the momentum map of the lifted adjoint action is simply $(\xi, \eta) \mapsto [\eta, \xi]$. (The ad-invariant metric for the vorticity algebra is the L^2 inner product.)

We note that the Lagrangian description in $T^*\mathcal{D}_{\mathrm{vol}}$ is itself a "Clebsch representation". Since $T^*\mathcal{D}_{\mathrm{vol}} \approx \mathcal{D}_{\mathrm{vol}} \times \mathcal{X}^*_{\mathrm{vol}}$, and elements of $\mathcal{D}_{\mathrm{vol}}$ can be parametrized by their generating functions, this representation also involves two functions of two variables. The advantage of Morrison's representation $(\lambda, \mu) \mapsto \{\mu, \lambda\}$ is that canonical variables are readily available, and no generating functions are necessary.

The second Clebsch representation is more complicated to describe, but more "efficient" in that ω is represented by one function of two variables and one function of *one* variable. It will be motivated by a general construction for a Lie group G with a bi-invariant metric, following ideas from group theory (see Weinstein [59].) We seek a symplectic invariant submanifold S of $T^*G \approx TG \approx G \times \mathfrak{G}$. Such a submanifold must necessarily be of the form $S = G \times K \subset G \times \mathfrak{G}$. If K is an open subset in a subspace K_1 of $\mathfrak{G}$, the condition that S be symplectic is that for each $\theta \in K$, the form on $K_1^\perp$ ($\perp$ relative to the bi-invariant metric) given by $(\xi, \eta) \mapsto \langle \theta, [\xi, \eta] \rangle$ be nondegenerate. It is natural to take K_1 as a maximal commuting subalgebra of $\mathcal{X}_{\mathrm{vol}}$, with K the regular elements in K_1. The space $K_1^\perp$ is then the sum of the "root spaces"; i.e. the two-dimensional invariant subspaces for the adjoint action of K on $\mathfrak{G}$.

For *two*-dimensional flow we let $G = \mathcal{D}_{\mathrm{vol}}$; the L^2 pairing is the bi-invariant metric. At this point we can drop our motivating ideas from group theory and concentrate on specific calculations in the algebra $\mathcal{X}_{\mathrm{vol}} = C^\infty(\Omega)$. For simplicity, we shall take Ω to be the (x, y) plane and K_1 to consist of the functions which depend upon y alone. (The case where Ω is the 2-sphere and K_1 consists of functions of the latitude is also an instructive example to work out.)

The orthogonal complement $K_1^\perp$ consists of those ω for which $\int \omega(x, y)\, \mathrm{d}x$ is identically zero in y and can therefore be equally well expressed as those ω for which the horizontal Fourier transform $\hat\omega(k_x, y) = \int e^{-ixk_x} \omega(x, y)\, \mathrm{d}x$ vanishes along the line $x = 0$. In the same representation, ω belongs to K_1 if and only if $\hat\omega(k_x, y)$ is the form $\delta(k_x)f(y)$.

Now if ω_1 and ω_2 belong to $K_1^\perp$ and $f(y)$ belongs to K_1, we have

$$\langle f, \{\omega_1, \omega_2\}\rangle = -\langle \omega_1, \{f, \omega_2\}\rangle$$

$$= \int \omega_1(x, y) f'(y) \frac{\partial \omega_2}{\partial x}(x, y)\, \mathrm{d}x\, \mathrm{d}y$$

$$= \int \hat\omega_1(k_x, y) \hat\omega_2(k_x, y)\, \mathrm{i}k_x f'(y)\, \mathrm{d}x\, \mathrm{d}y.$$

We can expect this to be nondegenerate when $f'(y)$ is nowhere vanishing, i.e. when the flow for which f is the stream function has no stationary points. These, then, are the "regular elements" and define the set K.

Now we can define our Clebsch representation to be in the space $\mathcal{D}_{\mathrm{vol}} \times K = \{(\gamma, f) | \gamma$ is an area-preserving diffeomorphism of $\mathbb{R}^2$ and f is a function of y such that $f'(y) > 0\}$. The Clebsch representation is $\omega = f \circ \gamma$; its image consists of those vorticity functions ω such that whenever $a < b$, the set $\omega_{a,b} = \{(x, y) | a \leq \omega(x, y) \leq b\}$ is a band of infinite area stretching to infinity in both directions. The Poisson structure on $\mathcal{D}_{\mathrm{vol}} \times K$ can be calculated in terms of that on $T^*\mathcal{D}_{\mathrm{vol}}$; although this structure is symplectic, we do not know of any explicit construction of canonical variables on it. (Implicit constructions are guaranteed by the version of Darboux's theorem in Marsden [38], lec. 1.)

7. Point vortices

We now consider point vortices in two dimensions with $\Omega = \mathbb{R}^2$. To do so, we need to allow delta functions as possibilities for the vorticity. Given N vortices in the xy plane with positions (x_i, y_i) and circulations Γ_i, the associated vorticity is

$$\omega = \sum_{i=1}^{N} \Gamma_i \delta_{(x_i, y_i)} \, dx \wedge dy.$$

For fixed $\Gamma_1, \ldots, \Gamma_N$, the set of all such vorticities forms a coadjoint orbit in $\mathcal{X}_{\text{vol}}^*$. The coadjoint action is just the action that moves the points (x_i, y_i) by the diffeomorphism $\eta \in \mathcal{D}_{\text{vol}}$. If $u_1, u_2 \in \mathcal{X}_{\text{vol}}$ are divergence-free vector fields on $\mathbb{R}^2$, the symplectic structure on the coadjoint orbit of ω at ω is given from Theorem 4.2 by

$$\Omega_\omega(L_{u_1}\omega, L_{u_2}\omega) = \int \omega(u_1, u_2) \, d\mu$$

$$= \sum_{i=1}^{N} \Gamma_i (dx \wedge dy)(u_1(x_i, y_i), u_2(x_i, y_i)).$$

This may be identified with the symplectic structure*

$$\Omega_{\Gamma_1, \ldots, \Gamma_N} = \sum \Gamma_i \, dx_i \wedge dy_i$$

on $\mathbb{R}^{2N}$ applied to the pairs of vectors $(u_1(x_i, y_i), u_2(x_i, y_i))$. In fact, if we consider $\mathbb{R}^{2N}$ with the symplectic structure above, then $\mathcal{D}_{\text{vol}}$ acts on it symplectically and its momentum map is precisely

$$V : (x_i, y_i) \mapsto \omega = \sum_{i=1}^{N} \Gamma_i \delta_{(x_i, y_i)} \, dx \wedge dy.$$

Notice that this is a diffeomorphism onto the coadjoint orbit.

Thus, *the variables (x_i, y_i) on $\mathbb{R}^{2N}$ are canonical (or Clebsch) variables for the motion of N vortices.* This Poisson map V clearly has only a trivial gauge group if the Γ_i's are distinct, otherwise the gauge group is a finite group of permutations.

Our construction is a special case of the example involving $\mathcal{M} = $ maps of P_{vol} to P_{sym} considered in section 5. Let P_{vol} be the set $\{1, \ldots, N\}$ with the ith point having mass Γ_i, and consider $P_{\text{sym}} = \mathbb{R}^2$, with $\omega_{\text{sym}} = dx \wedge dy$. Thus $\mathcal{M}$ is $\mathbb{R}^{2N}$ with symplectic structure $\Sigma \Gamma_i \, dx_i \wedge dy_i$. Now $\mathcal{D}_{\text{sym}}$ acts on $\mathcal{M}$ on the left with momentum map $\eta \mapsto \eta_* \omega_{\text{vol}}$. This is just the map V above.

We next consider the representation of the (kinetic energy) Hamiltonian for perfect fluids written in the above canonical variables. From $H(\omega) = \frac{1}{2} \int \langle \Delta^{-1} \omega, \omega \rangle \, d\mu$ (section 4) and the fact that $\Delta^{-1}(\delta_{(x_i, y_i)} \, dx \wedge dy) = -(1/2\pi) \log \|(x - x_i, y - y_i)\| \, dx \wedge dy$, the Green's function† for Δ, we get

$$H((x_1, y_1), \ldots, (x_N, y_N))$$

$$= -\sum_{i,j} \frac{1}{4\pi} \Gamma_i \Gamma_j \log \|(x_i - x_j, y_i - y_j)\|.$$

Of course the term with $i = j$ is infinite, corresponding to the infinite self-energy of a point vortex. If this term is removed we get the standard Hamiltonian picture of vortex motion (cf. Chorin and Marsden [18], p. 85). To summarize: *the usual Hamiltonian description of N vortices is just the restriction of the standard Euler equation Lie–Poisson Hamiltonian description to a particular coadjoint orbit, with the (infinite) self-energy terms ignored.* We did this in $\mathbb{R}^2$, but the description also works for bounded domains or curved surfaces (cf. Hally [24]). The necessary renormalization in the description of point vortices is related to difficulties in taking the limit $N \to \infty$ which are briefly discussed in the next section.

We conclude with some remarks on integrability and chaotic motions of vortices. For the Euler equations on $\mathbb{R}^2$, the Euclidean group E(2) acts on $\mathcal{D}_{\text{vol}}$ by composition on the left, commutes with right composition, and leaves the Hamiltonian invariant. The momentum map for the E(2) action

---

* This symplectic structure for point vortices is well known. Its derivation from the Poisson structure for smooth vorticities was also found by Morrison [47].

† Recall that Δ is the Laplace–DeRham operator; $\Delta(f \, dx \wedge dy) = (-\nabla^2 f) \, dx \wedge dy$.

is the total linear and angular momentum. The momentum map is right invariant, so it induces a momentum map on $\mathscr{X}_{\text{vol}}^*$; cf. Marsden and Weinstein [41], p. 127, Thm. 2. The momentum map on vortices is determined by the map of e(2) to $C^\infty(\mathscr{X}_{\text{vol}}^*)$ given by

$$\frac{\partial}{\partial x} \mapsto \int y\omega(x,y) \, dx \, dy = J_x,$$

$$\frac{\partial}{\partial y} \mapsto \int -x\omega(x,y) \, dx \, dy = J_y,$$

$$\frac{\partial}{\partial \theta} \mapsto \int -\tfrac{1}{2}(x^2+y^2)\omega(x,y) \, dx \, dy = J_\theta,$$

which satisfy the commutation relations

$$\{J_x, J_y\} = \Omega \stackrel{\text{def}}{=} \int \omega(x,y) \, dx \, dy,$$

$$\{J_x, J_\theta\} = J_y,$$

$$\{J_y, J_\theta\} = -J_x.$$

The vorticities which actually arise from momentum densities which vanish at infinity are those for which $\Omega = 0$; on this space the momentum map is Ad* equivariant.

On the full space of vorticities (i.e. all densities), Ω is a Casimir function, and the corresponding group which acts is the extension of E(2) (called the oscillator group), whose algebra is generated by $(J_x, J_y, J_\theta, \Omega)$ satisfying the commutation relations above. These quantities comprise an Ad*-equivariant momentum map for the oscillator group.

On point vortices,

$$\Omega = \sum_{i=1}^N \Gamma_i, \quad J_x = \sum_{i=1}^N \Gamma_i x_i, \quad J_y = -\sum_{i=1}^N \Gamma_i y_i,$$

and

$$J_\theta = -\tfrac{1}{2} \sum_{i=1}^N \Gamma_i (x_i^2 + y_i^2).$$

For $N = 3$ one can check that the motion is (completely) integrable in the sense that the (non-abeilan) reduced phase spaces are points. However one can also see that the dynamics of 3 point vortices is (completely) integrable by exhibiting 3 independent integrals in involution such as H, J, and $J_x^2 + J_y^2$. (This is a special case of the replacement of non-abelian by abelian complete integrability, as discussed by Mischenko and Fomenko [45].)

The general point of view presented here has the advantage that it extends to related, and perhaps more realistic, situations. For example, the motion of three vortices with cores (defined in section 8) is also completely integrable.

The motion of four vortices is generally believed to be chaotic. There are many papers, (see the references in Aref [2]) giving numerical evidence for this belief. Using a perturbation argument of Melnikov, Ziglin [62] outlines a proof of non-integrability. The proof involves the introduction of a fourth vortex ("the restricted four vortex problem") moving in the field of three vortices in stable triangular relative equilibrium (for the E(2) symmetry); cf. Palmore [51]. However, Ziglin's proof has some gaps involving exponentially small terms, as noted in Holmes and Marsden [28]. As shown in Synge [53], Aref and Pomphrey [3], and Aref [2], there is a different configuration of three vortices having a homoclinic orbit joining two saddle points in its reduced phase space. (The saddle points are configurations of three identical vortices on a line.) The methods of Holmes and Marsden [30] can now be applied—the result is almost surely that a nearby four vortex model (with Γ_4 small) will be chaotic in the sense of having Smale horseshoes in its dynamics. These facts suggest, but certainly do not prove, that two-dimensional Euler flow is not completely integrable.

8. Vortex cores

In this section we outline a modification of point vortices intended to model vortices with cores. This model has the advantage that it involves no renormalization, and its solutions converge as $N \to \infty$ to solutions of the Euler equations. Vortex cores have been successfully used in numerical computations by Chorin [14]; the analytical convergence proof is due to Beale and Majda [8] with important earlier work due to Hald. Our purpose is to indicate how this model fits into the Lie–Poisson picture and how, using it, one can gain geometric insight into "why" the scheme converges. We hope that such insight may inspire similar results in (geometrically related) plasma problems, which also have a Lie–Poisson description.

The motion of N vortex cores still forms a finite-dimensional Hamiltonian system. The idea is to cut off the logarithmic singularity in the Hamiltonian of the point vortex model. This cutoff is gradually removed as the number N of blobs gets large.

Let ψ be a function on $\mathbb{R}^2$ with integral 1 and let $\psi_\delta(z) = \delta^{-2}\psi(z/\delta)$, for $z \in \mathbb{R}^2$ and $\delta > 0$. Let

$$G_\delta(z) = \int G(z-z')\psi_\delta(z')\,\mathrm{d}z',$$

where $G(z) = (-1/2\pi)\log\|z\|$ is the Green's function for Δ. Consider the modified Lie–Poisson Hamiltonian system on $\mathscr{X}^*_{\mathrm{vol}}$ obtained by replacing Δ by Δ_δ, the operator whose kernel function is G_δ. Thus we consider

$H_\delta: \mathscr{X}^*_{\mathrm{vol}} \to \mathbb{R}$,

$$H_\delta(\omega) = \tfrac{1}{2}\int \langle \Delta_\delta^{-1}\omega, \omega \rangle \,\mathrm{d}\mu$$

(See equation ($H_{\mathrm{vorticity}}$) in section 4).

We consider the following three systems:

1) H_δ on the coadjoint orbit with N points: $\omega = \Sigma\, \Gamma_i \delta_{(x_i, y_i)}\, \mathrm{d}x \wedge \mathrm{d}y$;

2) H_δ on smooth vorticities; and

3) H on smooth vorticities.

Beale and Majda show that solutions of system 1, with initial conditions obtained by discretizing a smooth vorticity field, converge as $\delta \to 0$ and $N \to \infty$, with δ and N linked in a certain way, to the solutions of system 3 with the given initial condition. (It is a classical theorem of Wolibner [61] that there are smooth global solutions of the Euler equations in $\mathbb{R}^2$). This convergence includes particle paths as well as velocity fields. In other words, the convergence holds in material coordinates; i.e. on $T\mathcal{D}_{\mathrm{vol}}$ or $T^*\mathcal{D}_{\mathrm{vol}}$. How δ and N are linked depends on the norms used and on ψ. See Beale and Majda [8] for details.

We introduce the intermediate system 2 in order to make the following series of remarks:

a) With the cut off Hamiltonian H_δ, the vortex cores form a Hamiltonian system on $\mathbb{R}^{2n}$ and coincide with the cores defined by Beale and Majda.

b) For δ fixed, and N large (depending on δ) solutions of system 1 approximate solutions of system 2. This result is similar to that of Braun and Hepp [11]. (With a singular interaction it is easy to see directly that one cannot get convergence, by considering clouds of vortices passing through each other with close encounters).

c) For δ small, trajectories of H_δ with smooth initial conditions converge to those of H with the same conditions as $\delta \to 0$. In fact, it follows as in Ebin and Marsden [21] that H_δ and H generate smooth vector fields on $T^*\mathcal{D}_{\mathrm{vol}}$ (completed in suitable Sobolev topologies) that are C^1 close for δ small. Thus the convergence in this step merely results from elementary facts about trajectories of smooth vector fields on Banach manifolds.

d) Step c) does not involve N; δ and N are linked in step b).

The remarks above are not meant to replace the analytic estimates needed for the actual convergence proof. Rather, they are intended to show the overall structure of the method and to give a rather different argument for why it works.

9. Vortex patches

A vortex patch is a vorticity distribution in the plane which is the characteristic function of a region with smooth boundary (times $\mathrm{d}x \wedge \mathrm{d}y$ to regard it as a two form). Early work on stationary and steadily rotating patches was done by Kirchoff and Kelvin. We refer to Aref [2] and Burbea [13] for recent work and references. The Euclidean group and the breaking of its symmetry are an important part of this analysis. Numerical investigations of vortex patches have been made by Deem and Zabusky [20].

Here we are interested in the dynamics of vortex patches as a Hamiltonian system. We shall first show that the Lie–Poisson structure reduces to a symplectic structure used in the KdV equation.

The set of vortex patches supported on a set of fixed topological type and area forms a coadjoint orbit in $\mathscr{X}_{\text{vol}}^*$, since any two such patches are related by an area-preserving map. (This is proved as in Moser [50]; see Ebin and Marsden [21] p. 126.) For a vortex patch ω, let $\mathcal{O}_\omega$ be its coadjoint orbit. Let $v \in \mathscr{X}_{\text{vol}}$, with stream function ψ. Using our earlier notation,

$$v^\flat = \delta(\psi \, dx \wedge dy),$$

$$\psi \, dx \wedge dy = \Delta^{-1}\omega,$$

$$\omega = dv^\flat.$$

A tangent vector to $\mathcal{O}_\omega$ is represented by $L_v\omega$. Now we use theorem 4.2 to compute the symplectic structure:

Proposition 9.1. The symplectic structure Ω_ω on $T_\omega \mathcal{O}_\omega$, where ω is a vortex path associated with the set $M \subset \mathbb{R}^2$, is given by

$$\Omega_\omega(L_{v_1}\omega, L_{v_2}\omega) = \int_{\partial M} \psi_1 \, d\psi_2,$$

where ψ_1 and ψ_2 are the stream functions for v_1 and v_2, and ∂M is the boundary of M.

Proof. By theorem 4.2,

$$\Omega_\omega(L_{v_1}\omega, L_{v_2}\omega) = \int_M \omega(v_1, v_2) \, d\mu.$$

But on M, ω is the standard symplectic structure on the plane, so $\omega(v_1, v_2) = \{\psi_1, \psi_2\}$, the Poisson bracket.

Now $\{\psi_1, \psi_2\} \, d\mu = d(\psi_1 \, d\psi_2)$, a simple identity, so the result follows by Stokes' theorem. ∎

If M is diffeomorphic to a disc then ω is determined by the boundary loop ∂M. We observe that the symplectic structure in 9.1 is the same as that for the loop space in $\mathbb{R}^2$. (See Weinstein [56].)

In plasma physics, the analog of the vortex patch is called the water bag model†.

Let us briefly mention the evolution of the shape of the patch. Consider a patch near the unit disc whose shape is described in polar coordinates by

$$\tfrac{1}{2}(r^2 - 1) = \phi(\theta),$$

where $\int_0^{2\pi} \phi(\theta) \, d\theta = 0$ (so that the area is always π). In this representation the Poisson bracket corresponding to the symplectic structure in proposition 9.1 is

$$\{F, G\} = \int_0^{2\pi} \frac{\delta F}{\delta \phi} \frac{d}{d\theta} \frac{\delta G}{\delta \phi} \, d\theta.$$

Using this bracket and truncating the Hamiltonian $\tfrac{1}{2}\langle \Delta^{-1}\omega, \omega \rangle$ to third order in ϕ, one finds that the evolution equation for ϕ truncated at second order is

$$\phi_t = c_1 \phi_x + c_2 \mathscr{H} \phi + c_3 \phi \phi_x,$$

where c_1, c_2 and c_3 are constants and $\mathscr{H}$ is the Hilbert transform on the circle (convolution with $\tfrac{1}{2}\Sigma_{k \neq 0}(\operatorname{sgn} k) e^{ik\theta}$). It appears that the dispersion operator $\mathscr{H}$ is too weak to support travelling solitary waves without shocks; cf. the computations of Deem and Zabusky [20] and compare with the Benjamin–Ono equation, when the term $\mathscr{H}\phi$ is replaced by $\mathscr{H}\phi_{xx}$ (see Meiss and Pereira [43]).

We suspect that a Hamiltonian treatment of free boundary problems and surface waves for the

† For the Poisson–Vlasov equation, the analogue of 9.1 is

$$\Omega_f(\{\psi_1, f\}, \{\psi_2, f\}) = \int_{\partial M} \psi_2 (i_{X_{\psi_1}}\mu).$$

Euler equations themselves (see Miles [44]) will enable one to see how the Hamiltonian description of the KdV equation fits in with that for the Euler equations as described in this paper. What is missing is a suitable framework for taking limits of Hamiltonian systems. We hope that the ideas in Weinstein [59] will be of help in this direction.

10. Vortex filaments

The volution of a vortex filament in space, in the "self-induction approximation" is given by a motion at each point of the filament in the direction of the binormal, with velocity equal to the curvature (cf. Batchelor [7]). Hasimoto [25] showed that this motion is equivalent to the (completely integrable) nonlinear Schrödinger equation for the quantity

(curvature) $e^{i \int torsion}$,

which determines the filament up to a rigid motion. It is our purpose in this section to describe the Hamiltonian structure of the self-induction equation as *deduced* from that for the Euler equation; it remains to be seen how Hasimoto's transformation should be interpreted in our framework.

We consider our Lie–Poisson manifold $\mathscr{X}^*_{vol}$ on $\mathbb{R}^3$ and look at the set of vorticity distributions of the following form. Let C be a curve in $\mathbb{R}^3$ extending to infinity in both directions. (One can study closed loops in a similar way.) Let δ_C be the delta-function given by integration along C with respect to arc length. Let ω_C be the 2-form along C defined by $i_T dx \wedge dy \wedge dz$, where T is the unit tangent vector to C. Then if Γ is any constant, $\Gamma \omega_C \delta_C$ is the vorticity corresponding to C with strength Γ. (The constancy of Γ is equivalent to the vorticity being a *closed* generalized 2-form.)

The set of all such vorticities with a fixed Γ forms a coadjoint orbit $\mathcal{O}_\Gamma$ in $\mathscr{X}^*_{vol}$ and so it carries a symplectic structure Ω_Γ. The tangent space to $\mathcal{O}_\Gamma$ at a curve C consists of all vector fields normal to C, and for two such fields v and ω one finds that $\Omega_\Gamma(v, \omega) = \Gamma \int_C (T \times v) \cdot \omega \, ds$ using theorem 4.2. If we think of the space perpendicular to T as a copy of the complex numbers, then the operator $T \times$ corresponds to multiplication by $\sqrt{-1}$, and Ω_Γ is then equivalent to the symplectic structure relative to which the Schrödinger equation (linear or nonlinear) is Hamiltonian. (See Abraham and Marsden [1], p. 461.)

The Hamiltonian for the self-induction equations now turns out to be simply the arc-length functional on the curves C. Of course, this is only formal, since the curves C all have infinite arc length, and so one must renormalize the Hamiltonian somehow, such as by considering filaments asymptotic at infinity to a reference curve and then taking the difference of the two arc-length integrands. For closed loops the symplectic structure Ω_Γ is given by the same formula as above. We notice that in the coadjoint orbit corresponding to Γ, the loops C can have arbitrary lengths; i.e. *the stretching of vortex filaments is allowed*, although it does not occur for the arc length Hamiltonian flow.

Our approach and our discussion in section 8 make it clear how to introduce *vortex filaments with cores* (following ideas of Chorin [15], the cutoff is uniform along the filament and is not curvature dependent). Again one gets a Hamiltonian system, and the symplectic structure is still that of the Schrödinger equation. Vortex filaments (actually segments) with cores are useful in numerical work (see, for instance Chorin [15]) and there is a convergence theorem for them due to Beale Majda [8]. The convergence is valid for time periods for which smooth solutions for the Euler equations exist, just as in related algorithms for the three-dimensional Euler equations (cf. Ebin and Marsden [21] and Chorin et al. [17]). The geometry behind this convergence, which helps to explain why it works, is similar to that for point vortices with cores, discussed in section 8.

Acknowledgements

The work described here is an outgrowth of our work on plasmas, which was inspired by Phil

Morrison and Allan Kaufman. Conversations with Darryl Holm were important in our understanding of Clebsch variables. The hospitality of the Aspen Center for Physics made possible some useful discussions with Jim Meiss and Phil Morrison on vorticity equations. Finally, we thank Alex Chorin and Andy Majda for their helpful comments on vorticity algorithms.

References

[1] R. Abraham and J. Marsden, Foundations of Mechanics, second edition (Addison-Wesley, Reading, Mass. 1978).

[2] H. Aref, Integrable, Chaotic and Turbulent Vortex Motion in Two-Dimensional Flows. Ann. Rev. Fluid Mech. 15 (1983).

[3] H. Aref and N. Pomphrey, Integrable and chaotic motions of four vortices I. The case of identical vortices., Proc. Roy. Soc. Lond. A 380 (1972) 359–387.

[4] V. Arnold, Sur la géometrie differentielle des groupes de Lie de dimension infinie et ses applications a l'hydrodynamique des fluids parfaits, Ann. Inst. Fourier Grenoble 16 (1966) 319–361.

[5] V. Arnold, The Hamiltonian nature of the Euler equations in the dynamics of a rigid body and of an ideal fluid, Usp. Mat. Nauk. 24 (1969) 225–226.

[6] V. Arnold, Mathematical methods of classical mechanics, Graduate Texts in Math. No. 60 (Springer, New York, 1978).

[7] G.K. Batchelor, An introduction to fluid dynamics (Cambridge Univ. Press, London, 1970).

[8] J.T. Beale and A. Majda, Vortex methods II. Higher order accuracy in two and three dimensions, Math. Comp. (to appear).

[9] F. A. Berezin, Some remarks about the associated envelope of a Lie algebra, Funct. Anal. Appl. 1 (1967) 91–102.

[10] H. Born and L. Infeld, On the quantization of the new field theory, Proc. Roy. Soc. A. 150 (1935) 141.

[11] N. Braun and K. Hepp, The Vlasov dynamics and its fluctuations in the $1/N$ limit of interacting classical particles, Comm. Math. Phys. 56 (1977) 101–113.

[12] F.P. Bretherton, A note on Hamilton's principle for perfect fluids, J. Fluid Mech. 44 (1970) 19–31.

[13] J. Burbea, Motions of Vortex Patches, Lett. Math. Phys. 6 (1982) 1–16.

[14] A.J. Chorin, Numerical study of slightly viscous flow, Jour. Fluid Mech. 57 (1973) 785–796.

[15] A.J. Chorin, Vortex models and boundary layer instability, SIAM J. Sci. Stat. Comput. 1 (1980) 1–21.

[16] A.J. Chorin, The evolution of a turbulent vortex, Commun. Math. Phys. 83 (1982) 517–535.

[17] A. Chorin, T. Hughes, M. McCracken and J.E. Marsden, Product formulas and numerical algorithms. Comm. Pure and Appl. Math. 31 (1978) 205–256.

[18] A.J. Chorin and J.E. Marsden, A mathematical introduction to fluid mechanics (Springer Universitext, 1979).

[19] A. Clebsch, Über die Integration der hydrodynamischen Gleichungen J. reine angew. Math. 56 (1859) 1–10.

[20] G.S. Deem and N.J. Zabusky, Vortex waves: stationary 'V states', interactions, recurrence, and breaking. Phys. Rev. Lett. 40 (1978) 859–62.

[21] D. Ebin and J. Marsden, Groups of diffeomorphisms and the motion of an incompressible fluid. Ann Math. 92 (1970) 102–63.

[22] U. Fong and K.R. Meyer, Algebras of integrals, Rev. Colom. de Mathematicas IX (1975) 75–90.

[23] V. Guillemin and S. Sternberg, The moment map and collective motion, Ann. of Phys. 127 (1980) 220–253.

[24] D. Hally, Stability of streets of vortices on surfaces of revolution with a reflection symmetry, J. Math. Phys. 21 (1980) 211–217.

[25] R. Hasimoto, A soliton on a vortex filament, J. Fluid Mech. 51 (1972) 477–485.

[26] F.S. Henyey, Hamiltonian description of stratified fluid dynamics, preprint (1981).

[27] D. Holm and B. Kuperschmidt, Poisson brackets and Clebsch representations for magnetohydrodynamics, multifluid plasmas, and elasticity, preprint (1982).

[28] P.J. Holmes and J. Marsden, Horseshoes in perturbations of Hamiltonian systems with two degrees of freedom, Commun. Math. Phys. 82 (1982) 523–544.

[29] P.J. Holmes and J.E. Marsden, Melnikov's method and Arnold diffusion for perturbations of integrable Hamiltonian systems, J. Math. Phys. 23 (1982) 669–675.

[30] P.J. Holmes and J.E. Marsden, Horseshoes and Arnold diffusion for Hamiltonian systems on Lie groups, Ind. Univ. Math. J. (to appear).

[31] D. Kazhdan, B. Kostant, and S. Sternberg, Hamiltonian group actions and dynamical systems of Calogero type, Comm. Pure and Appl. Math. 31 (1970) 481–508.

[32] E.Z. Kuznetsov and A.V. Mikhailov, On the topological meaning of canonical Clebsch variables, Phys. Lett. 77A (1980) 37–38.

[33] E. Levich, The Hamiltonian formulation of the Euler equation and subsequent constraints on the properties of randomly stirred fluids, Phys. Lett. 86A (1981) 165–168.

[34] S. Lie, Theorie der Transformationsgruppen, second edition (B.G. Teubner, Leipzig, 1890). Reprinted by Chelsea, New York, (1970).

[35] G.M. Marle, Symplectic manifolds, dynamical groups and Hamiltonian mechanics, in Differential geometry and relativity. M. Cahen and M. Flato, eds. (D. Reidel, 1976).

[36] J. Marsden, Applications of global analysis in mathematical physics, (Publish or Perish, Berkeley, CA, 1974).

[37] J. Marsden, Well-posedness of the equations of a nonhomogeneous perfect fluid, Comm. P.D.E. 1 (1976) 215–230.

[38] J. Marsden, Lectures on geometric methods in mathematical physics, CBMS-NSF Regional Conference Series #37, SIAM (1981).

[39] J.E. Marsden, A group theoretic approach to the equations of plasma physics, Can. Math. Bull. 25 (1982) 129–142.

[40] J. Marsden, T. Ratiu and A. Weinstein, Semi-direct products and reduction in mechanics, Trans. Am. Math. Soc. (to appear).

[41] J. Marsden and A. Weinstein, Reduction of symplectic manifolds with symmetry, Rep. Math. Phys. 5 (1974) 121–130.

[42] J. Marsden and A. Weinstein, The Hamiltonian structure of the Maxwell–Vlasov equations, Physica 4D (1982) 394–406.

[43] J.D. Meiss and N. Pereira, Internal wave solitons, Phys. Fluids 21 (1978) 700–702.

[44] J.W. Miles, Hamiltonian formulations for surface waves, Applied Scientific Research 37 (1981) 103–110.

[45] A.S. Mishchenko and A.T. Fomenko, Generalized Liouville method of integration of Hamiltonian systems, Funct. Anal. Appl. 12 (1978) 113–121.

[46] H.K. Moffatt, Some developments in the theory of turbulence, J. Fluid Mech. 106 (1981) 27–47.

[47] P.J. Morrison, Hamiltonian field description of two-dimensional vortex fluids and guiding center plasmas, Princeton U. Plasma Physics Laboratory report PPPL-1783 (1981).

[48] P.J. Morrison, Poisson brackets for fluids and plasmas. Mathematical Methods in Hydrodynamics and Integrability in Dynamical Systems (La Jolla Institute, 1981), M. Tabor and Y.M. Treve, eds. AIP Conference Proceedings, 88 (1982) 13–46.

[49] P.M. Morrison and J.M. Greene, Noncanonical Hamiltonian density formulation of hydrodynamics and ideal magnetohydrodynamics, Phys. Rev. Letters. 45 (1980) 790–794.

[50] J. Moser, On the volume elements on a manifold, Trans Am. Math. Soc. 120 (1965) 286–294.

[51] J.I. Palmore, Relative equilibria of vortices in two dimensions, Proc. Natl. Acad. Sci. USA 79 (1982) 716–718.

[52] R.L. Seliger and G.B. Whitham, Variational principles in continuum mechanics, Proc. Roy. Soc. 305 (1968) 1–25.

[53] J.L. Synge, On the motion of three vortices, Can. J. Math. 1 (1949) 257–270.

[54] A.M. Vinogradov and B.A. Kuperschmidt, The structure of Hamiltonian mechanics, Russ. Math. Surveys 32 (1977) 177–243.

[55] A. Weinstein, Lectures on symplectic manifolds, CBMS Conf. Series No. 27, AMS (1977).

[56] A. Weinstein, Bifurcations and Hamilton's principle, Math. Zeit. 159 (1978) 235–248.

[57] A. Weinstein, Symplectic geometry. Bull. Am. Math. Soc. 5 (1981) 1–13.

[58] A. Weinstein, Gauge groups and Poisson brackets for interacting particles and fields. Mathematical Methods in Hydrodynamics and Integrability in Dynamical Systems (La Jolla Institute 1981), M. Tabor and Y.M. Treve, eds. AIP Conference Proceedings 88 (1982) 1–11.

[59] A. Weinstein, The local structure of Poisson manifolds, preprint (1982).

[60] A. Weinstein, Hamiltonian structure for drift waves and geostrophic flow preprint (1982).

[61] W. Wolibner, Un théorème sur l'existence du mouvement plan d'un fluide parfait homogène, incompressible, pendant un temps infiniment longue, Math. Zeit. 37 (1933) 698–726.

[62] S.L. Ziglin, Nonintegrability of a problem on motion of four point vortices, Soviet Math. Dikl. 21 (1980) 296–299.

[63] D.G. Ebin, Integrability of perfect fluid motion, Comm. Pure. Appl. Math. 36 (1983), 37–54.

[64] P.J. Olver, A nonlinear Hamiltonian structure for the Euler equations, J. Math, An. Appl. 89 (1982) 233–250.

BIFURCATION OF STATIONARY VORTEX CONFIGURATIONS

II. TOPOLOGY AND INTEGRABILITY

Julian I. PALMORE
University of Illinois, 1409 West Green Street, Urbana, Illinois 61801, USA

We study the dynamics of finitely many vortices in a circular disk and compare the integrability of this problem with that of Kirchhoff's problem of vortices in the plane. The effect of the topology of the phase space on the two Hamiltonian systems is compared. Our goal is to apply topological methods uniformly to investigate the flow of these dynamical systems.

1. Introduction

In a previous paper [1], we examined the dynamics problem of finitely many vortices in a circular disk and proved that *for any $n \geq 2$ and for any choice of positive circulations $\kappa_1, \ldots, \kappa_n$, for vortices in a disk, there are $n!/2$ families of stationary configurations in which bifurcations occur*. This result demonstrates the effect of the boundary via vortex images on the degeneracy of stationary configuration classes. There are two integrals in this problem: energy and angular momentum. These integrals are independent at almost all points of the configuration space for $n \geq 2$. The exceptional configurations where the independence of the integrals fails are precisely the stationary configurations.

For two vortices in a disk, for any choice of circulations such that $\kappa_1 \kappa_2 > 0$ or $\kappa_1 \kappa_2 < 0$, $\kappa_1 + \kappa_2 \neq 0$, and for each choice of angular momentum we show that there are either 1, 2 or 3 stationary configurations. For $\kappa_1 = -\kappa_2$ there are infinitely many stationary configurations for zero angular momentum. This particular family exists for the case where the 2 vortices are on the diagonal.

This result should be compared to that for Kirchhoff's problem for which one stationary configuration class exists for any choice of nonzero circulations such that $\kappa_1 + \kappa_2 \neq 0$.

By writing down the reduced phase space for two vortices in the disk, we show that each orbit is determined by a choice of energy and angular momentum; thus, this problem is integrable.

The topology of the reduced phase space is crucial to a calculation of the number of stationary configurations for a fixed choice of the circulations. The reduced phase space depends on the angular momentum so that these numbers may change without a bifurcation occurring within a family of stationary configurations. In the last section we compare some results for Kirchhoff's problem with those for vortices in a disk.

Bifurcation within a family of stationary configurations is a local phenomenon that must be taken into account in a global description of the flow. Integrability is a global phenomenon which depends on the existence of sufficiently many isolating integrals to partition the phase space. Superintegrable is the name given to the case where all orbits of the flow are periodic. An example of superintegrability in Kirchhoff's problem is the flow of two vortices having nonzero total circulation.

In contrast to the problem of two vortices in the disk, the three vortex problem for the disk has behavior analogous to aspects of the flow of four vortices of Kirchhoff's problem.

In [2] we observed that techniques which have been shown to be useful in studying celestial mechanics problems are of use in studying Kirchhoff's problem and the dynamics of vortices in a disk. In the last section we give examples of lower bounds on the numbers of stationary configuration classes for both Kirchhoff's problem and the vortices in a disk. These lower bounds arise also in the study of relative equilibria in celestial mechanics. Compare [3] and [4].

2. Topology of the reduced phase space

In this section we compare the structure of the reduced phase space of Kirchhoff's problem to the reduced phase space of the vortices in a disk. Let E^2 denote the Euclidean plane with inner product $\langle\,,\,\rangle$ and norm $\|\cdot\|$.

2.1. Kirchhoff's problem

Let $\kappa_1, \ldots, \kappa_n$ be nonzero circulations of n-vortices. Let $M \subset (E^2)^n$ be the linear subspace defined by $M = \{(x_1, \ldots, x_n) \in (E^2)^n \mid \sum \kappa_i x_i = 0\}$. Let $\Delta = \bigcup_{i<j} \Delta_{ij}$, $\Delta_{ij} = \{(x_1, \ldots, x_n) \in (E^2)^n \mid x_i = x_j\}$, denote the fat diagonal. This is the set of all configurations such that the positions of two or more vortices coincide.

The configuration space of Kirchhoff's problem is $M - \Delta$. The Hamiltonian function $H: M - \Delta \to \mathbb{R}$ is defined by

$$H((x_1, \ldots, x_n)) = -\sum_{i \neq j} \kappa_i \kappa_j \log\|x_i - x_j\|.$$

The domain $M - \Delta$ fixes the center of vorticity of all configurations at the origin. The angular momentum $I: M - \Delta \to \mathbb{R}$ is given by $I((x_1, \ldots, x_n)) = \frac{1}{2} \sum \kappa_i \|x_i\|^2$. The flow generated by H on $M - \Delta$ leaves the level set $I^{-1}(c)$ invariant. Thus, for each value c of the angular momentum, we may examine the flow on the level set $I^{-1}(c)$. A further reduction is made possible by the invariance of the Hamiltonian under rotations of the plane. By taking the quotient of $I^{-1}(c)$ by $SO(2)$ we obtain a manifold $I^{-1}(c)/SO(2)$ of dimension $2n - 4$. This quotient manifold is the reduced phase space of H at angular momentum value c.

For each choice of positive circulations $(\kappa_1, \ldots, \kappa_n) \in \mathbb{R}^n_+$, the reduced phase space is homeomorphic to $\mathbb{P}_{n-2}(\mathbb{C}) - \tilde{\Delta}_{n-2}$ where $\mathbb{P}_{n-2}(\mathbb{C})$ is complex projective space and $\tilde{\Delta}_{n-2}$ is the induced diagonal.

For each choice of circulations such that $(\kappa_1, \ldots, \kappa_k, -\kappa_{k+1}, \ldots, \kappa_n) \in \mathbb{R}^n_+$, $1 \leq k \leq n - 1$, a hyperbolic projective space results.

2.2. Vortices in a disk

Let $D^2 \subset E^2$ denote the open unit disk, $D^2 = \{x \in E^2 \mid \|x\| < 1\}$. The configuration space of the vortex problem in the disk is $(D^2)^n - \Delta$ where $\Delta = \bigcup_{i<j} \Delta_{ij}$, $\Delta_{ij} = \{(x_1, \ldots, x_n) \in (D^2)^n \mid x_i = x_j\}$, is the fat diagonal. The Hamiltonian for this problem is $H: (D^2)^n - \Delta \to \mathbb{R}$ given by

$$H((x_1, \ldots, x_n)) = -\sum_{i \neq j} \kappa_i \kappa_j \log\|x_i - x_j\|$$
$$+ \tfrac{1}{2} \sum_{1 \leq i,j \leq n} \kappa_i \kappa_j \log((1 - \|x_i\|^2)$$
$$\times (1 - \|x_j\|^2) + \|x_i - x_j\|^2).$$

The angular momentum $I: (D^2)^n \to \mathbb{R}$ defined by $I((x_1, \ldots, x_n)) = \frac{1}{2} \sum \kappa_i \|x_i\|^2$ is an integral of the flow.

The reduction in (2.1) applied to this configuration space yields the reduced phase space at angular momentum c as the quotient manifold $I^{-1}(c)/SO(2)$ of dimension $2n - 2$.

For each choice of positive circulations $(\kappa_1, \ldots, \kappa_n) \in \mathbb{R}^n_+$ and sufficiently small angular momentum $c(\kappa_1, \ldots, \kappa_n)$ the reduced phase space is homeomorphic to $\mathbb{P}_{n-1}(\mathbb{C}) - \tilde{\Delta}_{n-1}$ where $\mathbb{P}_{n-1}(\mathbb{C})$ is complex projective space of (complex) dimension $n - 1$ and $\tilde{\Delta}_{n-1}$ is the union of $n(n-1)/2$ codimension 1 complex projective subspaces. At large values of angular momentum, $I^{-1}(c)$ is no longer a sphere of dimension $2n - 1$.

2.3. Comparison of these two phase spaces

Consider Kirchhoff's problem where the configuration space is given as $(E^2)^n - \Delta$. The center of vorticity of each configuration is fixed in the plane. For each choice of positive circulations each reduced phase space has dimension $2n - 2$ and is homeomorphic to a complex projective space with diagonal removed. In the limit as angular momentum approaches 0, the reduced phase space for the vortices in a disk is homeomorphic to the reduced phase space for Kirchhoff's problem where the center of vorticity is *not* fixed at the origin.

3. Integrability of the flow

For one vortex in a circular disk, energy conservation fixes the distance of the vortex from the origin. Thus, all configurations are stationary. This situation is analogous to the behavior of two vortices in Kirchhoff's problem when the total circulation is nonzero. Both are examples of superintegrable systems.

For two vortices in a circular disk, there are two independent integrals of energy and angular momentum. Let h and c be regular values of the energy and momentum. The level sets $H^{-1}(h)$ and $I^{-1}(c)$ are 3-dimensional manifolds in $(D^2)^2$. The quotient spaces $H^{-1}(h)/SO(2)$ and $I^{-1}(c)/SO(2)$ are 2-dimensional manifolds which intersect either in tangent points or along 1-dimensional arcs. At the tangent points of the two level sets, the integrals of energy and angular momentum are dependent. *A stationary configuration is a configuration where the energy and angular momentum integrals are dependent.* The orbits other than stationary configurations are given by the 1-dimensional arcs. The system is integrable in that the relative orbits are determined by the values (h, c) of the integrals.

For a system of three vortices in Kirchhoff's problem the center of vorticity is fixed at the origin. The configurations of constant angular momentum c define a 3-dimensional manifold $I^{-1}(c)$ which can be reduced by taking the quotient by the SO(2) action. The reduced phase space $I^{-1}(c)/SO(2)$ is a 2-dimensional manifold as is $H^{-1}(h)/SO(2)$. The intersection of these two surfaces defines orbits of the flow. Thus, values (h, c) determine orbits as in the case of two vortices in a disk. Kirchhoff's problem of three vortices is integrable.

For three vortices in a disk the reduced phase space is homeomorphic to $\mathbb{P}_2(\mathbb{C}) - \Delta$ where the induced diagonal Δ is the union of three copies of $\mathbb{P}_1(\mathbb{C}) \approx S^2$. The Betti numbers of this reduced space are $\beta_0 = 1$, $\beta_1 = 2$ and $\beta_i = 0$ for $i \geqslant 2$. In the family of equilateral stationary configurations there is a value of angular momentum for which the member is degenerate. The family changes from minima to saddles of index 2 at this degeneracy. Therefore, in order to maintain the local topology of a minimum other critical points must appear. These other critical points will be minima of H and saddles having index 1.

4. Stationary configurations of two vortices

The stationary configurations are in a 1–1 correspondence with the zeros of a polynomial in two variables of degree 7. For any choice of nonzero circulations, this polynomial is defined as the numerator of the rational function $D\tilde{H}(x)(v) = 0$ where $x = ((\alpha, 0), (\beta, 0))$ and $v = ((1, 0), (1, 0))$. Here $\tilde{H}$ is the restriction of H to a reduced phase space. As a function of α, β, κ_1 and κ_2 this polynomial is given by

$$p(\alpha, \beta) = (\kappa_1 \alpha(1 - \alpha^2) + \kappa_2 \beta(1 - \beta^2))$$
$$\times (1 - \alpha^2)(1 - \beta^2) + \alpha\beta(\alpha - \beta)(1 - \alpha\beta)$$
$$\times (\kappa_1(1 - \beta^2) - \kappa_2(1 - \alpha^2)).$$

The results below on the local and global properties of the stationary configurations of two vortices in a disk are derived from an analysis of this polynomial. As a polynomial in β with coefficients in α, $p_\alpha(\beta)$ has degree 5.

The problem of finding stationary configurations reduces to that of finding the zeros of a polynomial. The degree of the polynomial bounds the number of stationary configurations.

Theorem 1. (Local behavior of stationary configurations.)

The curves defined by $p(\alpha, \beta) = 0$ in the neighborhoods of (a) $(0, 0)$, (b) $(-1, 0)$, (c) $(0, 1)$, (d) $(-1, 1)$, (e) $(1, 1)$ and (f) $(1, 0)$ are equivalent to
(a) $\kappa_1 \alpha + \kappa_2 \beta = 0$ (Kirchhoff's problem),
(b) $\beta = 4(\alpha + 1)^2$,
(c) $\alpha = 4(\beta - 1)^2$,
(d) $\kappa_2(\alpha + 1) + \kappa_1(\beta - 1) = 0$,
(e) $s = (\beta - 1)/(\alpha - 1)$ satisfies $\kappa_1 s^3 + 3\kappa_2 s^2 + 3\kappa_1 s + \kappa_2 = 0$ where if $-\kappa_1/\kappa_2 > 1$, then $s < 1$,
(f) $\beta = -4(\alpha - 1)^2$ (no solution for $\beta > 0$).

Proof. By rewriting

$$p(\alpha, \beta) = (\kappa_1 \alpha(1 - \alpha^2) + \kappa_2 \beta(1 - \beta^2))$$
$$\times (1 - \alpha^2)(1 - \beta^2) + \alpha\beta(\kappa_1(1 - \beta^2)$$
$$- \kappa_2(1 - \alpha^2))(\alpha - \beta)(1 - \alpha\beta)$$

as a polynomial in $(\alpha - \alpha_0, \beta - \beta_0)$ at each of the six points and evaluating the polynomial locally, we obtain the above results. The second part of (e) is found by writing $-\kappa_1/\kappa_2 = (1/s)(3s^2 + 1)/(s^2 + 3)$ and showing by differentiation that this function is monotone decreasing with s,

$$\frac{\mathrm{d}}{\mathrm{d}s}(-\kappa_1/\kappa_2)(s) = -s^{-2}(3s^2 + 1)/(s^2 + 3) < 0$$

for all s.

Theorem 2. (Global behavior of sotationary configurations.)

(a) For any choice of nonzero circulations κ_1, κ_2 the points $(-1, 0)$ and $(0, 1)$ are connected by a component of $p(\alpha, \beta) = 0$.
(b) For any choice of circulations $\kappa_1, \kappa_2 > 0$ the points $(-1, 1)$ and $(0, 0)$ are connected by a component of $p(\alpha, \beta) = 0$.
(c) For any choice of circulations $\kappa_1 < 0 < \kappa_2$ such that $\kappa_1 + \kappa_2 \neq 0$, the points $(1, 1)$ and $(0, 0)$ are connected by a component of $p(\alpha, \beta) = 0$.
(d) In case $\kappa_1 + \kappa_2 = 0$, the component of $p(\alpha, \beta)$ that connects $(0, 0)$ and $(1, 1)$ is the diagonal $\{(\alpha, \alpha)\}$.

(e) For any choice of circulations κ_1, κ_2 such that $\kappa_1 \neq -\kappa_2$ the configuration $(\alpha, -\alpha)$ where $\alpha^2 = \sqrt{5} - 2$ is stationary.

Corollary. For any choice of nonzero circulations κ_1 and κ_2, there are either 1, 2 or 3 stationary configurations for each choice of angular momentum.

For any choice of positive circulations the transition from 1 to 3 stationary configurations occurs at a degenerate stationary configuration.

Figs. 1 and 2 illustrate the stationary configurations schematically for positive circulations and for one positive and one negative circulation.

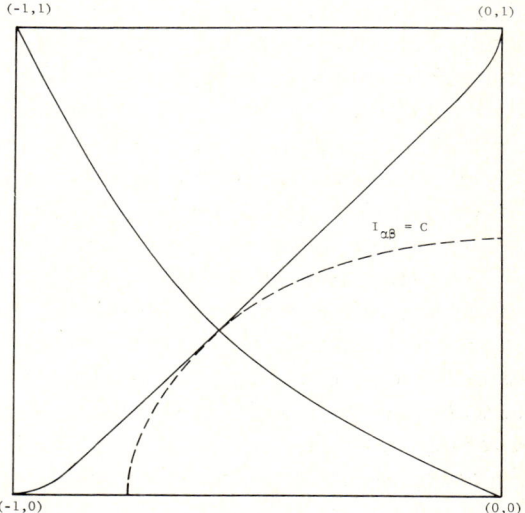

Fig. 1. Location of stationary configurations on the (α, β) plane for $0 < \kappa_1 < \kappa_2$.

5. Kirchhoff's problem and vortices in a disk

The bifurcation of stationary configurations occurs in Kirchoff's problem only for $n \geq 4$. An interesting example of bifurcation occurs for $n = 4$ with four vortices of equal circulation. Three of the

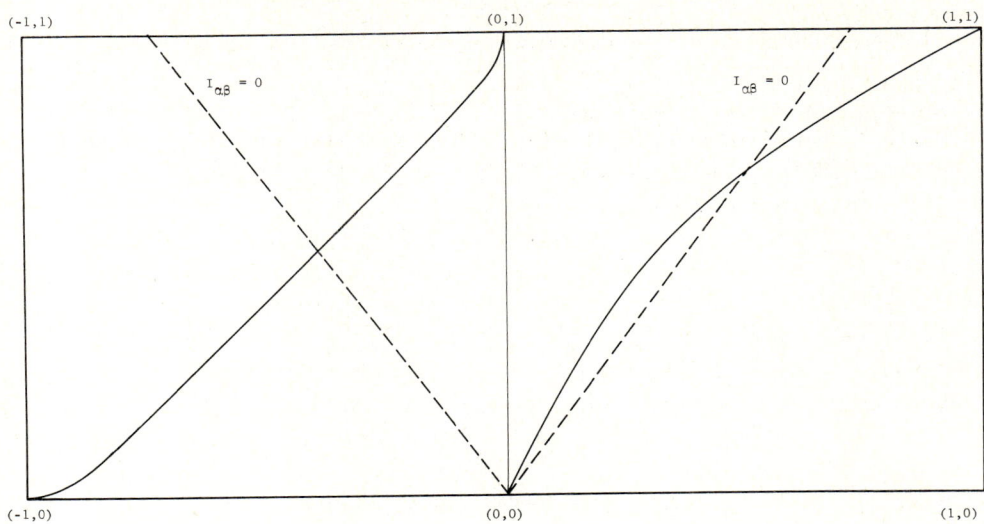

Fig. 2. Location of stationary configurations on the (α, β) plane for $\kappa_1 < 0 < \kappa_2$.

vortices are placed at the vertices of an equilateral triangle with the fourth vortex located at the center of the equilateral configuration. The rank of this configuration as a class of stationary configurations is equal to 2. The configuration has the maximum degeneracy possible for $n = 4$ vortices. This configuration represents 8 classes of stationary configurations. The collinear configurations represent 12 classes. The configurations of four vortices of equal strength located at the corners of a square represent 6 classes. We conclude that there are exactly 26 classes of stationary configurations of four vortices with equal circulation.

An example of lower bounds on the numbers of stationary configurations for Kirchhoff's problem and for vortices in a disk for arbitrary *positive* circulations for which the Hamiltonian is a non-degenerate function is given in table 1. Here μ_0 denotes a lower bound on the number of minima and μ_i, $i > 1$ denotes a lower bound on the number of saddles of index equal to i; χ_n is the Euler characteristic of the reduced phase space.

Acknowledgement

This research was supported in part by a grant from the National Science Foundation MCS81-21106.

TABLE I
Lower bounds on stationary configurations of vortices

n	μ_0	μ_1	μ_2	μ_3	μ_4	μ_5	χ_n
2	1						1
3	2	3					-1
4	6	16	12				2
5	24	90	120	60			-6
6	120	576	1080	960	360		24
7	720	4200	10080	12600	8400	2520	-120

References

[1] J.I. Palmore, Bifurcation of stationary vortex configurations, in Nonlinear problems: Present and Future, A. Bishop, D. Campbell and B. Nicolaenko, eds. (North-Holland, Amsterdam, 1982) pp. 127–131.

[2] J.I. Palmore, Relative equilibria of vortices in two dimensions, Proc. Nat. Acad. Sci. USA 79 (1982) 716–718.

[3] J.I. Palmore, Measure of degenerate relative equilibria, I, Ann. of Math. 104 (1976) 421–429.

[4] J.I. Palmore, Classifying relative equilibria, II, Bull. Amer. Math. Soc. 81 (1975) 489–491.

NONCANONICAL HAMILTONIAN FORMULATION OF IDEAL MAGNETOHYDRODYNAMICS

Darryl D. HOLM and Boris A. KUPERSHMIDT

Center for Nonlinear Studies, Los Alamos National Laboratory, Los Alamos, New Mexico 87545, USA

A noncanonical Poisson structure for ideal magnetohydrodynamics is presented and identified with a differential Lie algebra.

1. Introduction

Nonlinear hydrodynamics is still an active field of research in mathematical physics even though it has 250 years of history since the Bernoulli brothers and Euler. The most active areas in nonlinear hydrodynamics today are dynamical systems and Hamiltonian structures. In dynamical systems, much of the activity centers upon chaotic behavior of truncated modal expansions in convective flow. Such chaos is the main subject of this conference. In Hamiltonian structure studies, the main activity is to identify order – in the form of Lie algebras and related objects which underly the Lie–Poisson brackets responsible for fluid flows.

Noncanonical Poisson brackets for fluid systems first appear in Gardner's 1971 paper [1] on the Korteweg–de Vries equation. Since then, these noncanonical structures have proliferated; now they are known for a great many fluid dynamical theories.

In this paper we present a Poisson structure for ideal magnetohydrodynamics (MHD) in which the physical variables appear explicitly. Thus, the Galilean symmetries of MHD, for example, are realized as canonical transformations whose generators are physical quantities. In addition, the physical variables appear *linearly* in the Poisson bracket; which means that this bracket is intimately connected with a certain differential Lie algebra [see the next paper for details].

The Lie algebra given here and all of the other Lie algebras connected with Poisson brackets for hydrodynamically-related systems turn out to be semi-direct products of varying complexity (see, e.g. [2, 3]). From the calculational point of view, the presence of semi-direct products is an "experimental" observation, which is open to mathematical interpretations (see, e.g. [4]). The brackets themselves, though, can often be obtained by intuitive, physical reasoning [5] or even, sometimes, by trial and error [6].

In the ideal MHD model, electrically neutral plasma convects like an adiabatic fluid which carries an embedded magnetic field. The MHD fluid has mass density, ρ, and specific entropy, s. It moves through Euclidean space $\mathbb{R}^n$ with positions x_i and velocities v_j and carries an embedded magnetic field $B_{ij}(x, t)$. The magnetic field components B_{ij} are skew-symmetric and are derived from a vector potential, A_j, according to $B_{ij} = A_{i,j} - A_{j,i}$ with subscript notation also for partial derivatives. During convection, induced electrical currents flow: $J_i = -B_{ij,j}$ according to Ampere's Law. These induced currents oppose any change of magnetic flux through each co-moving surface. The resultant magnetic stresses alter the convective motion of the plasma by opposing bending of magnetic field lines.

In terms of momentum density $M_j = \rho v_j$, the MHD equations are:

$$\partial M_i/\partial t = -[M_i M_j/\rho + \delta_{ij}(p - 1/4\, \mathrm{Tr}\, B^2) - B_{ik} B_{kj}]_{,j}, \tag{1}$$

$$\partial \rho / \partial t = - M_{j,j}, \quad (2)$$

$$\partial s / \partial t = - \rho^{-1} M_j s_{,j}, \quad (3)$$

$$\partial B_{ij}/\partial t = -(B_{ik} M_k/\rho)_{,j} + (B_{jk} M_k/\rho)_{,i}. \quad (4)$$

Throughout, we sum on repeated indices. Eq. (1) is the hydrodynamic motion equation expressed in conservative form as the divergence of the stress tensor for MHD. In the stress tensor the fluid pressure p is determined as a function of ρ and s from a prescribed relation for the specific internal energy, $U(\rho, s)$, combined with the first law of thermodynamics,

$$dU = U_\rho \, d\rho + U_s \, ds = \rho^{-2} p \, d\rho + T \, ds, \quad (5)$$

where T is temperature. Eq. (2) expresses local mass conservation. Eq. (3) expresses the adiabatic flow condition, that no heat is exchanged between fluid elements as they convect. Eq. (4) is the Maxwell induction equation with electric field eliminated by Ohm's law, which is written for perfect electrical conductivity as $E_i = B_{ij} M_j / \rho$.

2. Poisson bracket relations

The Hamiltonian for the MHD system (1)–(4) is

$$H = \int d^n x [M^2/(2\rho) + \rho U(s, \rho) - 1/4 \, \text{Tr} \, B^2], \quad (6)$$

where $\text{Tr} \, B^2 = B_{ij} B_{ji}$. The Hamiltonian functional H is the sum of the kinetic energy, thermal energy, and magnetic energy of the fluid.

We introduce the following Poisson bracket defined over functionals $F[\rho, \sigma, M_i, A_j]$, where the variables $\sigma = \rho s$ and A_j are entropy density and magnetic vector potential, respectively,

$$\{F, G\} = \int d^n x \left[\rho \left(\frac{\delta F}{\delta M_j} \partial_j \frac{\delta G}{\delta \rho} - \frac{\delta G}{\delta M_j} \partial_j \frac{\delta F}{\delta \rho} \right) \right.$$

$$+ \sigma \left(\frac{\delta F}{\delta M_j} \partial_j \frac{\delta G}{\delta \sigma} - \frac{\delta G}{\delta M_j} \partial_j \frac{\delta F}{\delta \sigma} \right)$$

$$+ M_k \left(\frac{\delta F}{\delta M_j} \partial_j \frac{\delta G}{\delta M_k} - \frac{\delta G}{\delta M_j} \partial_j \frac{\delta F}{\delta M_k} \right)$$

$$- \left(\frac{\delta F}{\delta M_i} \frac{\delta G}{\delta A_k} - \frac{\delta G}{\delta M_i} \frac{\delta F}{\delta A_k} \right)(A_{k,i} - A_{i,k})$$

$$\left. + A_k \left(\frac{\delta F}{\delta M_j} \partial_i \frac{\delta G}{\delta A_i} - \frac{\delta G}{\delta M_k} \partial_i \frac{\delta F}{\delta A_i} \right) \right]. \quad (7)$$

The MHD eqs. (1)–(4) are then equivalent to the following bracket relations:

$$\dot{F} = \{H, F\}, \quad F \in \{\rho, \sigma, M_i, A_j\} \quad (8)$$

for Hamiltonian H given by eq. (6). The proof of (8) follows by comparison of the righthand side of eqs. (1)–(4) with terms in the identity below,

$$\{H, F\} = - \int d^n x \left\{ \frac{\delta F}{\delta \rho} (\partial_j M_j) + \frac{\delta F}{\delta \sigma} \partial_j (\sigma M_j / \rho) \right.$$

$$+ \frac{\delta F}{\delta M_k} \partial_j [M_j M_k / \rho$$

$$+ \delta_{jk}(p - 1/4 \, \text{Tr} \, B^2) - B_{ji} B_{ik}]$$

$$\left. + \frac{\delta F}{\delta A_k} [B_{ki} M_i / \rho + \partial_k (M_i A_i / \rho)] \right\}. \quad (9)$$

Thus the Hamiltonian H in (6) generates time evolution for MHD as a canonical transformation.

It is possible to transfer bracket (7) from **A**-space to **B**-space (see formula (55) of [7]). The resulting cumbersome bracket simplifies greatly for $n \leq 3$, when **B** can be treated not as a 2-form but as a vector (for $n = 3$) or scalar (for $n = 2$). This bracket in **B** space, for $n = 3$, was first found by Greene and Morrison [6].

The entire set of Galilean symmetries of the MHD system (1)–(4) may be realized as a subgroup of the canonical transformations defined in terms of the Poisson bracket (7). The generators of the Galilean transformations are expressible as physical quantities,

$$H = \int d^n x [M^2/(2\rho) + \rho U(s, \rho) - 1/4 \, \text{Tr} \, B^2],$$

$$P_i = \int d^n x M_i,$$

$$L_{ij} = \int d^n x (x_i M_j - x_j M_i), \tag{10}$$

$$G_i = -tP_i.$$

The functionals H, P_i, and L_{ij}, are respectively the total energy, kinetic momentum, and kinetic angular momentum of the MHD fluid.

As a result of the Poisson bracket (7) the generators (10) form a realization of the Lie Algebra of the Galilean group, viz.

$$\{H, G_i\} = P_i,$$
$$\{P_k, L_{ij}\} = P_j \delta_{ik} - P_i \delta_{jk},$$
$$\{G_k, L_{ij}\} = G_j \delta_{ik} - G_i \delta_{jk}, \tag{11}$$
$$\{L_{ij}, L_{kl}\} = \delta_{jk} L_{il} - \delta_{jl} L_{ik} + \delta_{il} L_{jk} - \delta_{ik} L_{jl},$$
$$0 = \langle H, P_k \rangle = \{H, L_{ij}\} = \{P_i, P_k\}$$
$$= \{P_i, G_k\} = \{G_i, G_k\}.$$

In addition, the following quantities Poisson-commute with all functionals defined over $\{\rho, \sigma, M_i, A_j\}$. That is, with

$$\mathcal{M} = d^n x \rho(x), \quad S = d^n x \sigma(x), \tag{12}$$

one finds that $\{\mathcal{M}, F\} = 0 = \{S, F\}$ for arbitrary $F[\rho, \sigma, M_i, A_j]$. So the total mass and entropy of the fluid each generate the identity transformation.

We have already seen in (9) that the total energy H generates time translations. We notice also that the kinetic momentum and angular momentum of the fluid generate spatial translations and rotations, respectively. Those results follow from the identities

$$\{P_k, F\} = -\int d^n x \left[\frac{\delta F}{\delta \rho} \rho_{,k} + \frac{\delta F}{\delta \sigma} \sigma_{,k} \right.$$
$$\left. + \frac{\delta F}{\delta M_l} M_{l,k} + \frac{\delta F}{\delta A_l} A_{l,k} \right], \tag{13}$$

$$\{L_{ij}, F\} = -\int d^n x (x_i \delta_{jk} - x_j \delta_{ik})$$
$$\times \left[\frac{\delta F}{\delta \rho} \rho_{,k} + \frac{\delta F}{\delta \sigma} \sigma_{,k} \right.$$
$$\left. + \frac{\delta F}{\delta M_l} M_{l,k} + \frac{\delta F}{\delta A_l} A_{l,k} \right]. \tag{14}$$

Note that for MHD the magnetic field plays no role in the canonical momentum and angular momentum. Finally we mention that the functional $G_i = -tP_i$ is the generator of Galilean boosts in the ith direction, cf. eq. (13). Thus all of the transformations in terms of the Galilean group are realized as canonical transformations in terms of physical variables with the Poisson bracket (7).

3. Lie-algebraic interpretation of the Poisson structure

We comment briefly on the mathematical origin of formula (7) for the Poisson bracket.

Recall that if $\mathfrak{G}$ is a finite-dimensional Lie algebra, with a basis $\{e_1, \ldots, e_n\}$ and structure constants $c_{ij}^k : e_i \Delta e_j = c_{ij}^k e_k$, then (smooth) functions on the dual space $\mathfrak{G}^*$ form a Lie algebra with the Poisson bracket

$$\{f, g\} = \sum_{ijk} \frac{\partial g}{\partial u_i} c_{ij}^k u_k \frac{\partial f}{\partial u_j}, \tag{15}$$

where u_i's are coordinates on $\mathfrak{G}^*$ in the basis dual to $\{e_i\}$. A formula analogous to (15) exists when one has a "differential" Lie algebra, that is, when c_{ij}^k are linear differential operators (examples of this sort first appeared in [8, 9]; more details can be found in [10]).

Let N be a (smooth) manifold, $C^\infty(N)$ be a ring of smooth functions, $\mathcal{D}(N)$ be a Lie algebra of vector fields on N (i.e., derivations of $C^\infty(N)$), $\wedge^i(N)$ be a $C^\infty(N)$ module of differential i-forms on N.

Consider $\mathbb{R}^p$ with coordinates $y_1, \ldots, y_p$. Denote by $\mathcal{D}^p(N)$ a Lie subalgebra of $\mathcal{D}(N \times R^p)$ consisting of vector fields X of the form

$$\left\{X = X' + \sum_{s=1}^{P} f_s \partial/\partial y_s, \quad X' \in \mathscr{D}(N), f_s \in C^\infty(N)\right\}.$$

Finally, denote by $\wedge^{i,P}(N)$ i-forms on N lifted to $N \times \mathbb{R}^P$ by pullback of the projection $N \times \mathbb{R}^P \to N$.

It is easy to see that $\mathfrak{G} = \mathscr{D}^P(N) \ominus \wedge^{i,P}(N)$ (direct sum of $C^\infty(N)$-modules) is a *Lie algebra* with respect to multiplication Δ given by

$$(X; \omega) \Delta (Y; v) = ([X, Y]; X(v) - Y(\omega)), \tag{16}$$

where, e.g., $X(v)$ is the Lie derivative of the i-form, v, with respect to the vector field, X.

Formula (7) now can be gotten from (16) for our $\mathfrak{G}$ by direct computation, when one takes $p = 2$, $i = n - 1$, $N = R^n = \{(x_1, \ldots, x_n)\}$ and denotes coordinates on $\mathfrak{G}^*$ in the following manner: ρ and σ are dual to $\partial/\partial y_1$ and $\partial/\partial y_2$, M_i's are dual to $\partial/\partial x_i$'s, $i = 1, \ldots, n$ and A_i's are dual to $\partial/\partial x_i \lrcorner (dx_1 \wedge \cdots \wedge dx_n)$, $i = 1, \ldots, n$.

The resulting bracket (7) necessarily satisfies the Jacobi identity and the other defining properties of a Poisson bracket (linearity and antisymmetry) because (7) has been constructed from a (differential) Lie algebra.

Thus, MHD fits into an algebraic Hamiltonian setting, directly in terms of the physical variables.

Acknowledgments

It is a pleasure to thank John Greene, for discussions of the Morrison–Greene bracket before its publication and J. Marsden, for his stimulating comments about the mathematical significance of the Morrison–Greene bracket.

This work was performed at the Center for Nonlinear Studies of the Los Alamos National Laboratory, sponsored by the United States Department of Energy.

References

[1] C.S. Gardner, J. Math. Phys. 12 (1971) 1548.
[2] J. Gibbons, D.D. Holm and B. Kupershmidt, Phys. Lett. 90A (1982) 281.
[3] D.D. Holm and B. Kupershmidt, Poisson Structures of Superfluids, Phys. Lett. 91A (1982) 425.
[4] J.E. Marsden, T. Ratiu and A. Weinstein, Semidirect Products and Reduction in Mechanics (preprint).
[5] J.E. Dzyaloshinskii and G.E. Volovick, Ann. Phys. 125 (1980) 67.
[6] P.J. Morrison and J.M. Greene, Phys. Rev. Lett. 45 (1980) 790; errata, ibid. 48 (1982) 569.
[7] D.D. Holm and B. Kupershmidt, "Poisson Brackets and Clebsch Representations for Magneto-hydrodynamics, Multifluid Plasmas, and Elasticity", Physica 6D (1983) 347.
[8] D.R. Lebedev and Yu.I. Manin, "Hamiltonian Operator of Gel'fand-Dikii and Coadjoint Representation of Volterra Group", Funk. Anal. Appl. 13:4 (1979) 40–46.
[9] D.R. Lebedev, "Benney's Long Waves Equations: Hamiltonian Formalism", Lett. Math. Phys. 3 (1979) 481–488.
[10] B.A. Kupershmidt, "On Dual Spaces of Differential Lie Algebras", these proceedings.

ON DUAL SPACES OF DIFFERENTIAL LIE ALGEBRAS

B.A. KUPERSHMIDT[†]
Center for Nonlinear Studies, Los Alamos National Laboratory, Los Alamos, New Mexico 87545, USA

We present a mathematical scheme which serves as an infinite-dimensional generalization of Poisson structures on dual spaces of finite-dimensional Lie algebras, which are well known and widely used in classical mechanics. These structures have recently appeared in the theory of Lax equations, long waves in hydrodynamics, and various other physical models: compressible hydrodynamics, magnetohydrodynamics, multifluid plasmas, elasticity, superfluid ^{4}He and ^{3}He–A, Ginzburg–Landau theory of superconductors, and classical chromohydrodynamics (the generalization of plasma physics to Yang–Mills interactions).

Introduction

The physical context of this paper and its relation to the rest of the proceedings has already been described in the previous paper. Therefore, here we will be quite succinct mathematically and aim for a greater degree of completeness than would be possible otherwise in a short proceedings article. Thus, we beg indulgence of the reader for the concentration of results and notation presented below. In fact, only the main theorems are given; details of their proofs may be found in [9].

1. Let A be a commutative algebra over a field $\bar{k}$ of characteristic zero, and let $\partial_l : A \to A$, $l = 1, \ldots, m$, be m mutually commuting derivations of A over $\bar{k}$.

Suppose $L = A^N$, $N \leq \infty$, is a differential algebra over $\bar{k}$. That means that if $\bar{X}_i = (0, \ldots, X_i, \ldots, 0)$ (nonzero element on ith-place), then a multiplication $*$ in L is given by

$$(\bar{X}_i * \bar{Y}_j)_k = \sum_{\sigma, \nu \in \mathbb{Z}_+^m} c_{ij,\sigma,\nu}^k \partial^\sigma X_i \cdot \partial^\nu Y_j, \quad c_{ij,\sigma,\nu}^k \in A, \quad (1.1)$$

where

$\mathbb{Z}_+ = \mathbb{N} \cup \{0\}$, $\partial^\sigma := \partial_1^{\sigma_1} \cdots \partial_m^{\sigma_m}$,

[†] Permanent address: Dept. of Mathematics, Univ. of Michigan Ann Arbor, Mich. 48109, USA.

for

$\sigma = (\sigma_1, \ldots, \sigma_m) \in \mathbb{Z}_+^m$.

If $\mathfrak{G}$ is a finite-dimensional Lie algebra over $\bar{k}$, then to any $h \in S(\mathfrak{G})$ (understood as a function on $\mathfrak{G}^*$) one assigns a derivation X_h of the ring $S(\mathfrak{G})$, and this map $\bar{\Gamma} : S(\mathfrak{G}) \to \mathrm{Der}(S(\mathfrak{G}))$ is *Hamiltonian*, meaning (see [1], ch. IV)

$$X_{\{h,f\}} = [X_h, X_f], \quad \forall h, f \in S(\mathfrak{G}), \quad (1.2)$$

where the *Poisson bracket* is defined by

$$\{h, f\} := X_h(f). \quad (1.3)$$

Here I describe what is an analog of this well-known map $\bar{\Gamma}$ in the differential case.

2. First we need an analog of "functions on L^*". Let $u_1, \ldots, u_N$ be differentially independent variables. Consider a differential ring $A_u := A[u_i^{(\sigma_i)}]$ with derivations ∂_l extended on A_u to act as $\partial_l(u_i^{(\sigma)}) = u_i^{(\sigma + 1_l)}$. If $X \in L$, $X = (X_1, \ldots)$, we denote

$$\langle u, X \rangle := \sum u_i X_i. \quad (2.1)$$

Denote $\mathrm{Im}\,\mathscr{D} = \Sigma_l \,\mathrm{im}\,\partial_l$ in A_u; we write $a \equiv b$ if $(a - b) \in \mathrm{Im}\,\mathscr{D}$.

Definition 2.1. A derivation $V: A_u \to A_u$ is called evolutionary if $[V, \partial_l] = 0$, $l = 1, \ldots, m$.

Note that $V(\text{Im } \mathcal{D}) \in \text{Im } \mathcal{D}$ and V is defined by its action on u_i's only: $V(u_i^{(\sigma)}) = \partial^\sigma V(u_i)$.

Define the matrix $B \in A_u[[\partial_1, \ldots, \partial_m]]$ by

$$\langle u, X * Y \rangle \equiv X^t B Y, \quad \forall X, Y \in L, \tag{2.2}$$

where "t" means "transpose" and elements of L are understood as vectors. From (2.2), (1.1) we get

$$B_{ij} = \sum_{k, \sigma, \nu} (-1)^{|\sigma|} \partial^\sigma \cdot u_k c_{ij, \sigma, \nu}^k \cdot \partial^\nu, \tag{2.3}$$

where $|\sigma| = \sigma_1 + \cdots + \sigma_m$.

Finally, we need "functional derivatives". For $P \in A_u$, the vector $\delta P / \delta \bar{u}$ is defined by $(\delta P / \delta \bar{u})_i = \delta P / \delta u_i$, where

$$\frac{\delta P}{\delta u_i} = \sum_\sigma (-\partial)^\sigma \left(\frac{\partial P}{\partial u_i^{(\sigma)}} \right). \tag{2.4}$$

We set $\delta P / \delta \bar{u}^t = (\delta P / \delta \bar{u})^t$.

Proposition 2.2. The matrix B is skew-adjoint iff multiplication in L is skew-commutative.

3. Thereafter we assume B (and L) to be skew. The matrix B generates a map $\Gamma = B\delta$ from A_u into $D^{\text{ev}}(A_u)$ (= evolution derivations of A_u) by

$$X_H(u_i) = \sum_j B_{ij} \left(\frac{\delta H}{\delta u_j} \right), \quad \forall H \in A_u. \tag{3.1}$$

We will write $X_H = B \delta H / \delta \bar{u}$.

Lemma 3.1. For any $P, Q \in A_u$,

$$\frac{\delta}{\delta \bar{u}} \left(\frac{\delta P}{\delta \bar{u}^t} B \frac{\delta Q}{\delta \bar{u}} \right) = D \left(\frac{\delta P}{\delta \bar{u}} \right) B \frac{\delta Q}{\delta \bar{u}}$$

$$- D \left(\frac{\delta Q}{\delta \bar{u}} \right) B \frac{\delta P}{\delta \bar{u}} + \frac{\delta P}{\delta \bar{u}} * \frac{\delta Q}{\delta \bar{u}},$$

where the Frechet derivative $D(\bar{R})$, $\forall \bar{R} \in (A_u)^N$, is a matrix (differential operator) defined by

$$D(\bar{R}) V \bar{u} = V(\bar{R}), \quad \forall V \in D^{\text{ev}}(A_u), \tag{3.2}$$

see e.g., [2], ch. I.

The proof is a straightforward analog of the computations in the lemma 7.14, ch. I of [2]

Remark 3.2. One can make A_u into an algebra using the Poisson bracket: $\{P, Q\} = X_P(Q)$. It is easy to see that, $\forall P, Q, R \in A_u$, $\{P, \{Q, R\}\} + $ c.p. $\equiv 0$ iff Γ is Hamiltonian, that is, $X_{\{P, Q\}} = [X_P, X_Q]$, $\forall P, Q \in A_u$.

Now I can formulate the main result.

Theorem 3.3. Γ is Hamiltonian iff L is a Lie algebra.

The proof follows from lemma 3.1.

Remark 3.4. In the case L is an ordinary Lie algebra, $B_{ij} = \Sigma_k c_{ij}^k u_k$ and the equations of trajectories of the field X_H, $H \in S(L) \in \text{``}C^\infty(L^*)\text{''}$, are familiar,

$$\dot{u}_i = \sum u_k c_{ij}^k \frac{\partial H}{\partial u_j}. \tag{3.3}$$

Thus X_H is tangent to the orbits of the coadjoint representation of a "group G" (whose Lie algebra is L). In the general differential situation, the notion of "orbits" is meaningless. That is why one needs to work algebraically. (The alternative would be either tight restrictions of the "compactness" type or pyramids of functional analysis.)

As a corollary, from the theorem 3.3 one gets

Proposition 3.5. The map $\Theta: L \to D^{\text{ev}}(A_u)$, $L \ni Y \mapsto X_{-\langle u, Y \rangle}$, is a Lie algebra homomorphism (assuming L is a Lie algebra).

4. Denote by $L_{(s)} = \{X \in L \mid X_i = 0, i \leq s\}$. Consider the matrix B^s defined by

$$B_{ij}^s = \sum_{k > s; \sigma, \nu} (-1)^{|\sigma|} \partial^\sigma \cdot u_k c_{ij, \sigma, \nu}^k \cdot \partial^\nu, \quad i, j > s. \tag{4.1}$$

Proposition 4.1. The matrix B^s defines a Hamiltonian structure (i.e., if H, F depend upon u_i with $i > s$ then the same is true for $\{H, F\}$) iff $L_{(s)}$ is a Lie subalgebra in L.

5. Suppose $L_1 = A^M$, $M \leq \infty$, is another algebra over $\bar{k}$, and let $\phi : L \to L_1$ is a map given by a linear differential operator. Denote by $A_v = A[v_j^{(v)}]$, $j \leq M$, the differential algebra which plays for L_1 the same role as A_u plays for L. We can define the map $\phi^* : A_u \to A_v$ by

$$\begin{cases} 1) \ \phi^* \partial_l = \partial_l \phi^*, \quad l \leq m; \\ 2) \ \langle \phi^* u, X \rangle \equiv \langle v, \phi(X) \rangle, \quad \forall X \in L. \end{cases} \quad (5.1)$$

Suppose L and L_1 are Lie algebras, thus corresponding matrices B_u and B_v define Hamiltonian structures. We say that ϕ^* is *canonical* if $\Gamma_v(\phi^* F)$ is ϕ^*-compatible with $\Gamma_u(F)$, $\forall F \in A_u$ (see motivations in [1], ch. IV).

Proposition 5.1. ϕ^* is canonical iff ϕ is a Lie algebra homomorphism.

6. Examples

6.1. Let $A[\xi] = \{\sum_{|\sigma| \leq M'} a_\sigma \xi^\sigma | a_\sigma \in A, \ \xi^\sigma := \xi_1^{\sigma_1} \ldots \xi_m^{\sigma_m}\}$ be a ring of differential operators and $A[\xi, \xi^{-1}]$ the corresponding ring of pseudodifferential operators. For the Lie algebra generated by elements of $A[\xi, \xi^{-1}]$ of negative order, the corresponding matrix B (for $m = 1$) was introduced by Manin ([2], ch. I) as a formal limit of Gel'fand–Dikii operators. The Lie-algebraic interpretation of that matrix (for $m = 1$) was given in [3].

6.2. Consider $C^\infty(\mathbb{R}^{2m}) = C^\infty(x, \xi)$ and its localization around subspace $\xi_1 = \cdots = \xi_m = 0$. The usual Poisson bracket on $\mathbb{R}^{2m} = T^*(\mathbb{R}^m)$ leads to the following algebraic set up: $L = \{\sum_{\mu \in \mathbb{Z}_+^m} X_\mu \xi^\mu | X_\mu \in A\} \leftrightarrow \{(X_\mu) | X_\mu \in A\}$.

The bracket on L is given by

$$(X * Y)_\mu = \sum_{\sigma + \tau = \mu + 1_l} [\sigma_l X_\sigma \partial_l(Y_\tau) - \tau_l Y_\tau \partial_l(X_\sigma)], \quad (6.1)$$

thus the corresponding Hamiltonian matrix B is

$$B_{\sigma,\tau} = \sum_{l=1}^m [\sigma_l u_{\sigma+\tau-1_l} \partial_l + \partial_l \tau_l u_{\sigma+\tau-1_l}]. \quad (6.2)$$

This matrix appears in m-D hydrodynamics. For $m = 1$ it was introduced in [4]; its Lie-algebraic interpretation is given in [5] (for $m = 1$).

6.3. In $\mathbb{R}^{r+1} = \{(x, \xi_1, \ldots, \xi_r)$ consider the Lie algebra of functions on $\mathbb{R}^{r+1}$, with the bracket

$$[f, g] = \sum_{\gamma=1}^r \left(d_\gamma \frac{\partial}{\partial \xi_\gamma} f \cdot \partial g - \partial f \cdot d_\gamma \frac{\partial}{\partial \xi_\gamma} g \right),$$

$$\partial = \partial_1 = \partial/\partial x. \quad (6.3)$$

Localization to the line $\xi = 0$ leads to the algebra $L = \{\sum_{\psi \in \mathbb{Z}_+^r} X_\psi \xi^\psi | X_\psi \in A, m = 1, \partial_1 = \partial/\partial x\}$, with Lie bracket

$$(X * Y)_\psi = \sum_{a+b=\psi+1_\gamma} (a_\gamma d_\gamma X_a \partial Y_b$$

$$- b_\gamma d_\gamma Y_a \partial X_b), \quad a, b \in \mathbb{Z}_+^r. \quad (6.4)$$

The corresponding matrix B is given by

$$B_{\phi\psi} = \sum_{\gamma=1}^r (\phi_\gamma d_\gamma u_{\phi+\psi-1_\gamma} \partial + \partial \psi_\gamma d_\gamma u_{\phi+\psi-1_\gamma}), \quad (6.5)$$

which is the multicomponent (for $r > 1$) generalization of the $m = 1$-case (6.2).

6.4. In $\mathbb{R}^{n+p}$ with coordinates $(y_1, \ldots, y_p; x_1, \ldots, x_n)$ consider Lie algebra $\bar{L}$ consisting of those vector fields $X \in \mathcal{D}(\mathbb{R}^{n+p})$ which commute with $\partial/\partial y_q$, $q = 1, \ldots, p$. The algebraic version of $\bar{L}$ would be the Lie algebra $\mathcal{D}(A) \bigoplus_{q=1}^p A \, \partial/\partial y_q$ acting on $A[y_1, \ldots, y_p]$ as derivations. The corresponding matrix B how yields Poisson bracket

$$\{F, G\} = \sum_{q=1}^p v_q \sum_{l=1}^n \left(\frac{\delta F}{\delta u_l} \partial_l \frac{\delta G}{\delta v_q} - \frac{\delta G}{\delta u_l} \partial_l \frac{\delta F}{\delta v_q} \right)$$

$$+ \sum_{l=1}^n u_l \sum_{i=1}^n \left(\frac{\delta F}{\delta u_i} \partial_i \frac{\delta G}{\delta u_l} - \frac{\delta G}{\delta u_i} \partial_i \frac{\delta F}{\delta u_l} \right), \quad (6.6)$$

which appeared in [6] (for $n = 3$, $p = 2$) in the context of hydrodynamics of compressible fluid.

6.5. On $C^\infty(\mathbb{R}^2)$ consider Lie algebra structure $[f, g] = (\alpha\xi\partial_\xi + \beta)f \cdot \partial g - (\alpha\xi\partial_\xi + \beta)g \cdot \partial f$, $\partial = \partial/\partial x$; $\alpha, \beta \in \mathbb{R}$. Localization on the line $\xi = 0$ leads to the following algebraic set up: $L = A^\infty = \{(X_i)\} \leftrightarrow \sum X_i \xi^i$, $m = 1$, with multiplication

$$(X * Y)_k = d_k \sum_{n+m=k} (X_n \partial Y_m - Y_m \partial X_n),$$

$$\partial = \partial/\partial x = \partial_1, \tag{6.7}$$

where $d_k = \alpha k + \beta$, and $\alpha, \beta \in \bar{k}$ are arbitrary. The corresponding matrix B [7]

$$B_{ij} = d_i u_{i+j} \partial + \partial d_j u_{i+j} \tag{6.8}$$

is the main source of integrable systems in two-space dimensions. Classification table of such systems (see, e.g. [7]) depends smoothly upon $(\alpha : \beta) \in \bar{k}P^1$. Despite this, one can prove the following

Theorem 6.7. The Lie algebra L with multiplication (5.12) admits a nonzero adinvariant (modulo Im ∂), bilinear (differential) form (not necessarily symmetric) iff $3\beta/\alpha \in \mathbb{Z}_-$:

$$(X, Y) = \sum_{i+j+3\beta\alpha^{-1}=0} X_i Y_j. \tag{6.9}$$

Remark. The existence of Hamiltonian structures on dual spaces of Lie algebras is not the privilege of the differential case only. Analogous results can be found: a) when one substitutes commuting automorphisms $\Delta_1, \ldots, \Delta_m: A \to A$ instead of derivations, as in [8]; b) and moreover, when both derivations and automorphisms are present. The corresponding matrices B behave naturally with respect to "continuum limits" $\Delta_l \to \exp(\lambda \partial_l)$ [8]. Details will be given elsewhere [9].

References

[1] B.A. Kupershmidt, "Geometry of Jet Bundles and the Structure of Lagrangian and Hamiltonian Formalisms", in Lect. Notes Math. no. 775 (Springer, Berlin, 1980) pp. 162–218.
[2] Yu. I. Manin, "Algebraic Aspects of Nonlinear Differential Equations", J. Sov. Math. 11 (1979) 1–22.
[3] D.R. Lebedev and Yu. I. Manin, "Hamiltonian Operator of Gel'fand-Dikii and Coadjoint Representation of Volterra Group", Funk. Anal. Appl. 13:4 (1979) 40–46.
[4] B.A. Kupershmidt and Yu. I. Manin, "Equations of Long Waves with a Free Surface II. Hamiltonian Structure and Higher Equations", Funk. Anal. Appl. 12:1 (1978) 25–32.
[5] D.R. Lebedev, "Benney's Long Waves Equations: Hamiltonian Formalism", Lett. Math. Phys. 3 (1979) 481–488.
[6] P.J. Morrison and J.M. Greene, "Nancanonical Hamiltonian Density Formulation of Hydrodynamics and Ideal Magnetohydrodynamics", Phys. Rev. Lett. 45:10 (1980) 790–794.
[7] B.A. Kupershmidt, "Deformations of Integrable Systems", Proc. Roy. Irish Acad.
[8] B.A. Kupershmidt, "On Algebraic Models of Dynamical Systems", Lett. Math. Phys. 6 (1982) 85–89.
[9] B.A. Kupershmidt, "Discrete Lax Equations and Differential-Difference Calculus", E.N.S. Lecture Notes.

CHAPTER 7

QUANTUM CHAOS

STOCHASTIC BEHAVIOR IN QUANTUM SCATTERING

Martin C. GUTZWILLER
IBM T.J. Watson Research Center, Yorktown Heights, N.Y. 10598, USA

A 2-dimensional smooth orientable, but not compact space of constant negative curvature with the topology of a torus is investigated. It contains an open end, i.e. an exceptional point at infinite distance, through which a particle or a wave can enter or leave, as in the exponential horn of certain antennas or loud-speakers. In the Poincaré model of hyperbolic geometry, the solutions of Schrödinger's equation for the reflection of a particle which enters through the horn are easily constructed. The scattering phase shift as a function of the momentum is essentially given by the phase angle of Riemann's zeta function on the imaginary axis, at a distance of $\frac{1}{2}$ from the famous critical line. This phase shift shows all the features of chaos, namely the ability to mimick any given smooth function, and great difficulty in its effective numerical computation. A plot shows the close connection with the zeros of Riemann's zeta function for low values of the momentum (quantum regime) which gets lost only at exceedingly large momenta (classical regime?) Some generalizations of this approach to chaos are mentioned.

1. Introduction

Classical mechanical systems manifest a wide variety of stochastic behavior where one would have believed that everything can be predicted in the long run rather simply from the initial conditions and the knowledge of a few periods. This phenomenon will cease to be the privileged domain of the specialists who were heard this week. Most physicists, and even their students, will become aware of classical chaos a few years from now, because the relevant concepts are largely agreed upon.

Quantum mechanics is another story. A number of trial investigations have been carried out, and a few have been reported in the last days. The general concepts seem still to be missing, however. It is safe to say that they will be more subtle because quantum mechanics stays away from singular mathematics and tries to smoothen any discontinuities. The price is much greater complexity, a Hilbert space where a Euclidean space of finite dimension used to do the job.

The first question to address is then: How does stochastic behavior show itself in a simple quantum-mechanical system? What are the "fractals" of q.m.? Only the study of different examples will tell. Considering the difficulty of the mathematics involved, general propositions are most likely to be either wrong or empty of content. On the other hand, it is quite legitimate to choose an individual example so as to insure its stochastic character on the basis of its classical limit. Such a case will be discussed in this contribution. Indeed, the goedesics in a space of constant negative curvature will guarantee that the corresponding waves can be expected to be equally stochastic if that is possible.

The simplest stochastic systems have either one degree of freedom and an external time-dependent perturbation or two degrees of freedom with a time-independent Hamiltonian. The latter seems more intriguing and unexpected. Closed systems of this kind were investigated by several authors, but I am not aware of any work on an open system. The difference is quite obvious in quantum mechanics: closed systems have a discrete spectrum, while open systems have a continuous one. In the first case, all the stationary states are bound, i.e. localized, while in the second case, a wave comes in from infinity and is scattered back out again. This occurs for all energies above a certain threshold, and the quantity of interest is the phase shift between incoming and outgoing wave as a function of momentum (or energy) and direction (if there is a choice).

The scatterer in the present case is most easily described as a box of the type which is dear to the heart of physicists, topologically a 2-torus, i.e. the surface of a doughnut or bagel. Such a box is ordinarily a closed system, but it will be provided with an entry-and-exit hatch through which a particle can get in or out. The width of this escape is negligible, however, compared to the size of the box, and a particle can get out only along one degree of freedom. A classical particle will in general not be able to gain that freedom, because the relevant initial conditions, although dense, form only a set of measure zero. Even a particle which comes in through the hatch will, generally, not get out again. Quantum mechanics is quite different in this respect: a particle which has gotten into the box from the outside will find its way out again. But there are also stationary states which will confine the particle to the inside of the box in perpetuity.

The box is put together from a piece of Lobachevski space, i.e. a Riemannian space of constant negative curvature, where certain boundaries are glued together (identified) so as to obtain a very smooth structure. The construction is quite elementary, and will be explained shortly so as to convey the whole picture. A special model for Lobachevski space is used where the geometry turns out to be a simple variant of Euclidean geometry. This is Poincaré's model which is based on the upper half of the complex plane and which allows simple sketches to be drawn on a piece of paper. Unfortunately, the whole box including its leak cannot be viewed as embedded in three-dimensional Euclidean space according to a theorem of Hilbert [1].

The incoming and outgoing waves are very simple functions of the coordinates in this model, and their phase can be calculated directly by taking advantage of the symmetries, exactly as in the traditional closed flat boxes. The phase shift is given essentially by the imaginary part of the logarithm (i.e. what is usually called the phase angle) for Riemann's zeta function. This neat and unexpectedly simple result could have been stated as a theorem without proof, or even as a postulate to be realized by some unknown physical system, but it is very pleasing to know its origin to be rather direct, and on a geometrical–physical foundation.

The stochastic elements in this system are, therefore, all hidden in this famous transcendental function, with its connection to prime numbers. Two of its features will be discussed in some detail, its ability to mimic other functions, and the difficulty of its computation. These two traits will be called its "chaos" and its "complexity." There are indications that Riemann's zeta function and its several cousins are present in many other scattering problems, whose classical analogs show stochastic behavior.

2. The Poincaré model

The space of constant negative curvature is described by the upper half H of the plane of complex numbers, $z = x + iy$ with $y > 0$. The distance ds between neighboring points z and $z + dz$ in H is given by the Riemannian metric

$$ds^2 = \frac{dx^2 + dy^2}{y^2}. \qquad (1)$$

The Gaussian curvature κ can be calculated from (1), and turns out to be -1. The radius of curvature has been used effectively to define a unit of length to which all other quantities will be referred [2].

The space H with the metric (1) is homogeneous; all points and all directions at a given point are equivalent. This equivalence is demonstrated explicitly by a transformation which carries $z = x + iy$ into $\zeta = \xi + i\eta$ through

$$\zeta = \frac{az + b}{cz + d}, \qquad (2)$$

with real numbers a, b, c, d which are normalized

by

$$ad - bc = 1. \tag{3}$$

These formulas yield immediately

$$d\zeta = \frac{dz}{|cz+d|^2}, \quad \eta = \frac{y}{|cz+d|^2}, \tag{4}$$

so that $(d\xi^2 + d\eta^2)/\eta^2 = (dx^2 + dy^2)/y^2$.

The composition of two transformations (2) follows the rules of matrix multiplication for the 2 by 2 matrices

$$\begin{pmatrix} a & b \\ c & d \end{pmatrix}. \tag{5}$$

The set of all these matrices forms a group G which is also called PSL(2, R), the Projective (since only the ratio occurs in ζ) Special (since its determinant = 1) Linear group of 2 by 2 matrices with Real elements.

The geodesics, i.e. the equivalent of straight lines, in this model are the Euclidean circles in H whose centers lie on the real axis. When two geodesics intersect at some point in H, the angle of intersection is the same as the Euclidean angle between the Euclidean circles. The apparent shapes of any object are therefore reminiscent of the real shapes. The distortion is no more than an optical illusion.

3. Plane waves

Consider an arbitrary trajectory $x(t), y(t)$ in H where t is the time. A particle with mass m on this trajectory has the classical kinetic energy

$$T_c = \frac{m}{2}\left(\frac{ds}{dt}\right)^2 = \frac{m}{2y^2}\left[\left(\frac{dx}{dt}\right)^2 + \left(\frac{dy}{dt}\right)^2\right]. \tag{6}$$

The corresponding expression in quantum mechanics for the wave function $\psi(x^1, x^2, t)$ in a Riemannian space of metric $ds^2 = g_{kl} dx^k dx^l$ is given by the integral

$$T_q = \frac{\hbar^2}{2m} \iint \sqrt{g}\, dx^1\, dx^2 \left[\sum_{kl} g^{kl} \frac{\partial \psi^*}{\partial x^k} \frac{\partial \psi}{\partial x^l} + \frac{\kappa}{4} \psi^* \psi \right], \tag{7}$$

to be integrated over all space. The determinant of the metric tensor g_{kl} has been called g, and its inverse is g^{kl} as usual. The Gaussian curvature κ (depending in general on x^1 and x^2) occurs in (7) with a factor $\frac{1}{4}$, in order to achieve a reasonable interpretation of quantum mechanics. $\hbar$ is Planck's quantum divided by 2π.

If a particle experiences no external forces, the kinetic energy determines its motion through the variational principle of Lagrange. The classical particle is then found to move along a goedesic with constant speed $ds/dt = v$, and has the energy $E = mv^2/2$. Schroedinger's equation follows from (7) in the form

$$i\hbar \frac{\partial \psi}{\partial t} = -\frac{\hbar^2}{2m} \frac{1}{\sqrt{g}} \frac{\partial}{\partial x^k}\left(\sqrt{g} g^{kl} \frac{\partial \psi}{\partial x^l}\right) + \frac{\hbar^2 \kappa}{8m} \psi. \tag{8}$$

In the space of constant negative curvature with the metric (1)

$$i\hbar \frac{\partial \psi}{\partial t} = -\frac{\hbar^2}{2m}(\Delta + 1/4)\psi,$$

with

$$\Delta = y^2\left(\frac{\partial^2}{\partial x^2} + \frac{\partial^2}{\partial y^2}\right). \tag{9}$$

in terms of the Laplacian Δ. In particular, a "stationary solution," i.e. a wave function of the form $\psi = \varphi(x, y) \exp(-i\omega t)$, satisfies

$$(\Delta + \tfrac{1}{4} + \lambda)\varphi = 0, \quad \text{with } \lambda = \frac{2mE}{\hbar^2}, \quad E = \hbar\omega. \tag{10}$$

A plane wave has its sources infinitely far away. At any one point, it is characterized by its wave vector and its direction. If one such wave is known, the others can be obtained from a coordinate transformation such as (2) in our case. Therefore, we start with the simplest case where ϕ depends only on y; the wave propagates in the vertical direction. One finds that

$$\varphi = y^{1/2+iw}, \quad \text{with } \lambda = w^2, \, E = \frac{\hbar^2 w^2}{2m}, \quad (11)$$

where w can be any real number.

The wave fronts are the lines $y = $ constant, and the distance s perpendicular to them is given by

$$s = \int_1^y \frac{dy}{y} = \log y. \quad (12)$$

Therefore, one can write the wave function as

$$\psi = \sqrt{y} \exp(iws - i\omega t). \quad (13)$$

The factor $y^{1/2}$ comes from maintaining the total flux in a vertical strip of constant width, such as $0 < x < 1$. The corresponding distance in the metric (1) is given by $\int_0^1 dx/y = 1/y$ which means that an effective narrowing as y increases has to be compensated by the factor $\sqrt{y}$ in (13).

Any other plane wave can be transformed into an expression like (11) by the transformation (2). Thus, one finds from (4) that

$$\phi = \eta^{1/2+iw} = \frac{y^{1/2+iw}}{|cz+d|^{1+2iw}}, \quad (14)$$

where c and d are real, not both zero because of (3). The wave fronts are obtained by assigning to the quantity $y/|cz+d|^2$ some constant value like $\exp(s)$ in analogy with (12). The corresponding points lie on a Euclidean circle which touches the real axis in $x_0 = -d/c$ and has the radius $\tfrac{1}{2}c^2 \exp(s)$.

An outgoing plane wave can be said to "start" at $y = 1$, and move up into H with increasing y, its wave fronts always parallel to the real axis. All other plane waves start at some circle which

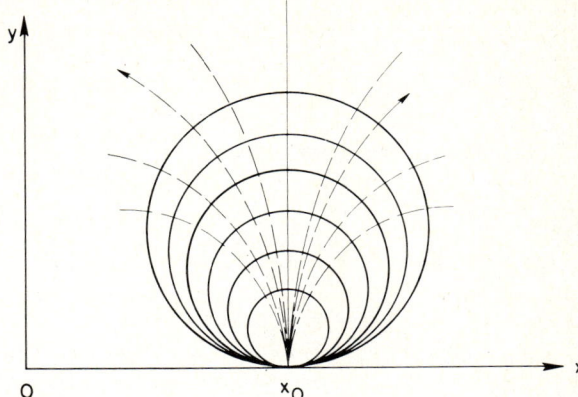

Fig. 1. Lines of equal phase (horocycles) for a "plane wave" originating in x_0, on the boundary of the space of negative curvature; particle trajectories (dashed) coming from the same point x_0.

touches the real axis at some arbitrary point x_0, of radius $\tfrac{1}{2}c^2$ for $s = 0$. Then, they contract by shrinking their radius while maintaining their point of contact x_0. The wave normals are Euclidean circles which hit the real axis in x_0, at a right angle. They are the geodesics which originate in x_0. The wave fronts are also called horocycles in the geometry of negative curvature. The situation is shown in fig. 1.

4. The leaky box

A "square" D is now cut out of the space of constant negative curvature. Its boundaries are geodesics, "straight" lines just as one would do in Euclidean space when defining a torus in a space of zero curvature. The boundaries are given in the Poincaré model H as follows:

(i) $\quad x = -1, \, 0 < y < \infty$;

(ii) $\quad (x - \tfrac{1}{2})^2 + y^2 = \tfrac{1}{4}, \, 0 < x < 1$;

(iii) $\quad x = 1, \, 0 < y < \infty$; $\quad\quad\quad\quad (15)$

(iv) $\quad (x + \tfrac{1}{2})^2 + y^2 = \tfrac{1}{4}, \, -1 < x < 0$;

fig. 2 shows the situation.

The torus results by identifying the points on opposite boundaries. This identification is made

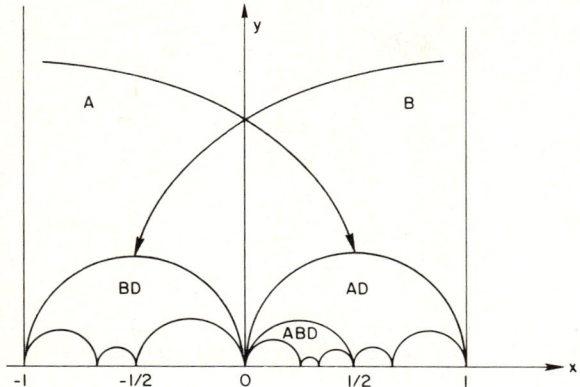

Fig. 2. Square-like domain D with the transformations A and B which identify opposite boundaries of D; copies of D by the transformations A, B, and AB.

with the help of a transformation (2) which sends opposite boundaries into each other. Thus, one uses the matrices

$$A = \begin{pmatrix} 1 & 1 \\ 1 & 2 \end{pmatrix}, \quad B = \begin{pmatrix} 1 & -1 \\ -1 & 2 \end{pmatrix}, \quad (16)$$

and the corresponding transformations (2) which are designated by the same letter. Boundary (i) gets transformed into (ii) by A, and (iii) is transformed into (iv) by B, as in fig. 2.

A geodesic (or other trajectory) may start inside D and hit one of the boundaries. Instead of continuing outside D, it is brought back into D by one of the transformations A, B, A^{-1}, or B^{-1}. Similarly, periodic boundary conditions are applied to the wave functions ψ or φ.

Another view of this situation is often equally useful, however. The geodesic or trajectory is allowed to continue outside of D; but as soon as it traverses one of the boundaries of D, it enters another copy of D. E.g. upon going beyond (ii), the geodesic enters the copy AD which results from D by transforming it with A. The corners -1, 0, 1 and ∞ are transformed into 0, $\frac{1}{2}$, $\frac{2}{3}$, and 1. The new boundaries are the Euclidean circles with centers on the real axis which connect the 3 pairs $(0, \frac{1}{2})$, $(\frac{1}{2}, \frac{2}{3})$, and $(\frac{2}{3}, 1)$ as sketched in fig. 2.

This process of generating new copies of D can be continued indefinitely. Any path on the torus is associated with a well-defined sequence of operations which is represented by a word in the four letters A, B, A^{-1} and B^{-1}, provided the path does not venture into one of the corners. That does not happen as long as the length of the path is kept bounded; the metric (1) shows any path to be infinitely long when it goes all the way to the real axis or to the point $y \to \infty$. This last feature allows us to say that our torus has a leak, a point which is infinitely far away. The width of this leak becomes the smaller, the further one advances into one of the corners, again because of the metric (1). When a particle moves with a finite energy, it will take an infinitely long time to move off into one of the corners. The plane wave (11) does the same thing for a wave function. Fig. 3 gives a schematic idea of such a surface.

The leaky torus is topologically different from the ordinary torus which physicists like to use in Euclidean space. The four corners of the domain D become one point as usual, but this point is now

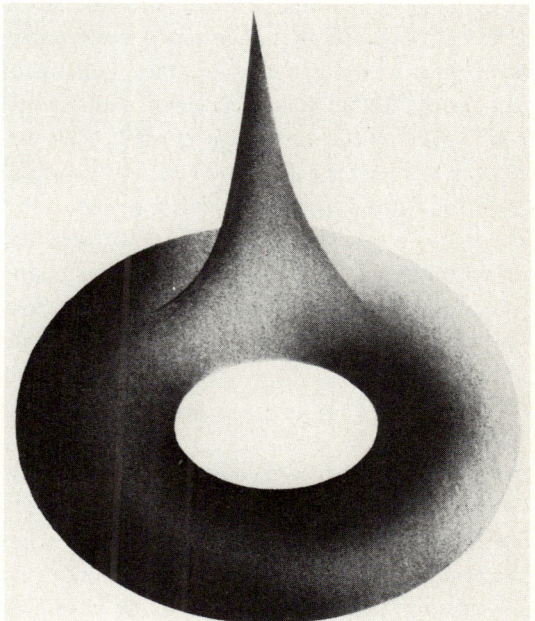

Fig. 3. Sketch of the leaky box to show its topology, embedded in Euclidean 3-dimensional space.

infinitely removed. A path which goes around this exceptional point cannot be contracted to zero, because that would require moving it over an infinite distance. Such a path is associated with the word $B^{-1}A^{-1}BA$, or some cyclic permutation, or inverse of this word. A straightforward calculation shows the matrix for this word to be

$$B^{-1}A^{-1}BA = K = \begin{pmatrix} -1 & -6 \\ 0 & -1 \end{pmatrix}, \tag{17}$$

which gives the transformation $\zeta = z + 6$, a translation parallel to the real axis by 6 units.

Every word in the four letters A, B, A^{-1}, and B^{-1} generates a new transformation as long as one eliminates the sequences AA^{-1}, BB^{-1}, $A^{-1}A$, or $B^{-1}B$. All these transformations together form a group Γ, the free group generated by A, B, A^{-1}, and B^{-1}. The copies of D under all these transformations completely fill H, covering each point exactly once (with proper allowance for the boundaries) [3].

The shift K can be visualized by redesigning D. The square D in fig. 2 is cut up into 12 congruent pieces which are then reassembled with the help of Γ. The new domain D' is contained between two vertical lines which are 6 units apart, and whose points are identified in the obvious manner. The "lower" part of the boundary consists of 6 arcs which are parts of Euclidean circles with radius 1 around the integer points on the real axis.

These 6 arcs are identified in pairs with the help of the matrices A and B, plus the matrix $C = (3, -1|1, 0)$. These 3 matrices satisfy the symmetric relations

$$ABC = I, \quad CBA = K. \tag{18}$$

The boundary conditions among the 6 arcs are given by the relations $A(-2 + e^{i\chi}) = 1 - e^{-i\chi}$, $B(2 + e^{i\chi}) = -1 - e^{i\chi}$, and $C(e^{i\chi}) = 3 - e^{-i\chi}$ where $\pi/3 < \chi < 2\pi/3$. If one cuts out the domain D', and glues together the two vertical boundaries, the lower boundary becomes essentially a hexagon where opposite sides are identified by the oper-

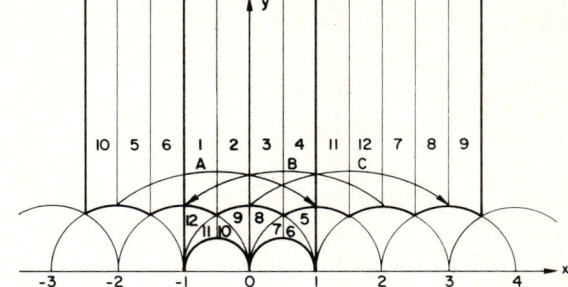

Fig. 4. Two equivalent versions. D and D', of the leaky box embedded in the upper half plane with the Poincaré metric (1).

ations A, B, and C. The torus is obtained in this last operation, while the escape hatch consists of the long vertical tube. It is easy to check that this construction leaves no singular points, except ∞.

The present domain should be distinguished from the better-known modular domain which is exactly 1/6 of D', such as the two partial domains with numbers 2 and 3 in fig. 4. If one glues together the outer boundaries of 2 and 3, and identifies symmetrically the two halves of the circular arc with radius 1 around 0, one finds a surface of genus 0, i.e. like a sphere, again with an excape to ∞. This surface has two singular points, $z = i$ and $z = e^{i\pi/3}$. It seems more satisfactory, however, to discuss a surface of constant negative curvature without any exceptional points besides ∞.

5. The incoming and the outgoing waves

An incoming wave is assumed to swoop down from $y = \infty$ in the form (13) without the time dependence

$$\varphi_{\text{inc}} = \alpha^{iw} y^{1/2 - iw}, \tag{19}$$

where $w > 0$ and $\alpha > 0$. The factor α^{iw} defines the place for zero phase to be $y = \alpha$; this is where the wave function is monitored on its way out after being scattered inside the box.

The incoming wave is transformed by the action of the group Γ as indicated in (14). These reflected waves are necessary in order to satisfy the boundary conditions on the 6 circular arcs of fig. 4. Thus one finds

$$\varphi = \alpha^{iw}\left[y^{1/2-iw} + \sum_{\gamma \in \Gamma}' \frac{y^{1/2-iw}}{|cz+d|^{1-2iw}}\right], \quad (20)$$

where the prime signals two exceptions. First, the identity of Γ does not appear because it is already contained in the incoming wave. Second, when γ is followed by the translation K, or some multiple of K, the expression (14) remains the same. Therefore, one element in Γ is necessary for the summation (19) from each left coset with respect to the translations in Γ.

When the matrix $(a, b|c, d)$ is multiplied by K on the right one finds $(a, b + 6a|c, d + 6c)$. The condition (3) implies that c and d are relatively prime, or, to use the nomenclature from number theory, their greatest common divisor $(c, d) = 1$. The question now arises: Which pairs of integers c and d are obtained as γ runs through the right cosets of Γ with respect to K? The answer comes from further study of Γ and is quite simple: every pair $(c, d) = 1$ occurs exactly once.

The geometry of the domain D (or D') has now been turned into an expression of pure arithmetic.

$$\varphi = \alpha^{iw}\left\{y^{1/2-iw} + \sum_{(c,d)=1}' \frac{y^{1/2-iw}}{|cz+d|^{1-2iw}}\right\}, \quad (21)$$

where the prime excludes the identity of Γ, i.e. $c = 0$, $d = 1$. One can think of the second term in (21) as a double summation; the outer sum goes over all integers $c \geq 1$, and the inner sum over all integers d from $-\infty$ to $+\infty$ which are prime to c. The convergence of (21) can be insured, if necessary, by giving w a positive imaginary part, and then considering the analytic continuation back to the real w-axis.

The full wave function is now known, and it is only necessary to extract the outgoing wave from the second term in (20) or (21). There are bound states on the leaky torus, i.e. eigenstates of the Laplacian whose amplitude decays exponentially as y increases toward ∞. Such states have a dependence on x, in contrast to (13) or (19), corresponding to an angular momentum around the nect of the leak in fig. 3. They are eliminated by integrating (21) over the width of the domain D', say from $x = -\frac{5}{2}$ to $x = \frac{7}{2}$. The first term in (21) simply gets multiplied with 6.

The second term is simplified by taking advantage of the right translations in Γ. They transform d into $d + 6c$, $d + 12c$ ..., $d - 6c$, $d - 12c$, ... Thus, they are properly accounted for if the integral in the second term is extended over x from $-\infty$ to $+\infty$, and the inner summation is restricted to $1 \leq d < 6c$. That leaves us with 6 equivalent contributions for any pair of c and d with $(c, d) = 1$ and $1 \leq d < c$, namely d, $d + c$, $d + 2c$, $d + 3c$, $d + 4c$, and $d + 5c$. With these preliminaries out of the way, the calculation of the phase shift can now be carried out.

The integral

$$\int_{-\infty}^{\infty} dx \frac{y^{1/2i-w}}{[(cx+d)^2 + c^2y^2]^{1/2-iw}}$$

$$= \frac{y^{1/2+iw}}{c^{1-2iw}} \cdot \frac{\Gamma(1/2)\Gamma(-iw)}{\Gamma(1/2-iw)} \quad (22)$$

results directly from Euler's B-function (cf. Whittaker and Watson, p. 254), and leads to

$$\int_0^6 dx\, \varphi = 6\alpha^{iw}\left\{y^{1/2-iw} + y^{1/2+iw}\frac{\Gamma(1/2)\Gamma(-iw)}{\Gamma(1/2-iw)}\right.$$

$$\left. \times \sum_{c=1}^{\infty}\sum_{(c,d)=1} \frac{1}{c^{1-2iw}}\right\}. \quad (23)$$

The first term is still the incoming wave, while the second term is exactly the outgoing wave whose amplitude is the sum over the integers c and their relative primes with the condition $0 < d < c$ [4].

6. Some elementary number theory

The evaluation of the summation over d in (23) requires the knowledge of all the integers d which are relatively prime to a given integer $c > 0$. The number of such integers d in the interval $0 < d < c$ is given by Euler's totient function $\phi(c)$. The explicit value of $\phi(c)$ can be found in every textbook on elementary number theory [5]. If c is decomposed into prime factors,

$$c = \prod_{j=1}^{k} p_j^{v_j}, \qquad (24)$$

where $v_j \geqslant 1$ and p_j are distinct primes, then one has

$$\phi(c) = \sum_{(c,d)=1} 1 = \prod_{j=1}^{k} \left(1 - \frac{1}{p_j}\right) p_j^{v_j}, \qquad (25)$$

for $c \geqslant 2$, while for $\phi(1) = 1$.

The first term in the summation of (23) becomes simply 1. The other terms are now written in terms of the decomposition (24)

$$\sum_{(c,d)=1} \frac{1}{c^{1-2iw}} = \prod_{j=1}^{k} \left(1 - \frac{1}{p_j}\right)(p_j^{v_j})^{2iw}. \qquad (26)$$

At this point, we use the unique decomposition of the integers into primes, and replace the summation over c by an independent summation over all the exponents v_j in (26). With a little reordering of terms the sum over c becomes

$$\prod_p \left[1 + \left(1 - \frac{1}{p}\right) \sum_{v=1}^{\infty} (p^{2iw})^v\right] = \prod_p \frac{1 - p^{2iw-1}}{1 - p^{2iw}}. \qquad (27)$$

The last expression can be written in terms of

$$\zeta(s) \equiv \sum_{n=1}^{\infty} \frac{1}{n^s} = \prod_p \left(1 - \frac{1}{p^s}\right)^{-1}. \qquad (28)$$

The first equality defines the Riemann zeta function, and the second equality gives Euler's expression for this function as an infinite product over all prime numbers. The definition is valid at first only in the half plane $\sigma > 1$, where $s = \sigma + it$, but it can be continued analytically without great trouble to the left (cf. the section on the complexity of the zeta function). The final expression for the reflected wave can be written compactly in terms of the combination

$$Z(s) = \Gamma\left(\frac{s}{2}\right) \pi^{-s/2} \zeta(s), \qquad (29)$$

which satisfies the crucial functional equation of Riemann

$$Z(1-s) = Z(s). \qquad (30)$$

(cf. Whittaker and Watson, chap. XIII [6].)

With all these technical improvements, we can now write

$$\int_0^6 dx\, \varphi = 6y^{1/2} \left\{ \left(\frac{y}{\alpha}\right)^{-iw} + \alpha^{+2iw} \frac{Z(1+2iw)}{Z(1-2iw)} \right.$$

$$\left. \times \left(\frac{y}{\alpha}\right)^{+iw} \right\}. \qquad (31)$$

The monitoring of the wave at $y = \alpha$ has been incorporated in writing everything in terms of the ratio y/α.

7. The scattering phase shift

The answer to our original problem could hardly be expected to be given in a more elegant form. The function $Z(s)$ is real on the real axis, so that $Z(1 - 2iw)$ is complex conjugate to $Z(1 + 2iw)$, and the factor in front of $(y/\alpha)^{iw}$ can be written as $\exp(i\beta)$ with real β. The flux of the reflected wave is equal and of opposite sign to the incoming flux. A particle which gets into the box at any given velocity will come out again.

The situation is much more intricate in classical mechanics. A geodesic line is given in the Poincaré model by an Euclidean circle whose center lies on the real axis. Such a circle is defined by its two endpoints ξ and η, both real, with the motion going from ξ to η. If ξ is a rational number, the geodesic originates in the leak, i.e. comes in from infinity;

otherwise, it wanders around the box in the past, but does not come in through the leak. An identical rule holds for η and the fate of the geodesic line in the future. Scattering trajectories are, therefore, dense among all trajectories, but they form a set of measure 0. Even a particle which has come in from ∞, will most probably not leave the box again; and symmetrically, a particle inside the box will be able to leave it only with vanishing probability.

Complications of this kind do not arise in quantum mechanics. The Hilbert space of all wave functions decays into two independent subspaces corresponding to scattering and to bound states, with one special state in between, so to speak, namely $\varphi = $ const whose square can be integrated since the total area is finite, but whose amplitude does not decay exponentially. Very little is known about the bound state spectrum, nor is it clear to what extent the wave function (20) contains any bound states. Selberg's trace formula can be extended so as to establish a direct relation between the discrete spectrum and the periodic geodesics, but the mathematical expressions are not as elegant as in the case of compact surfaces of constant negative curvature. All information about bound states was seemingly annihilated when φ was integrated over x in (23). The phase can now be written in detail as

$$\beta = 2w \log \alpha + \frac{1}{i} \log \left[\frac{\Gamma(\frac{1}{2} + iw)}{\Gamma(\frac{1}{2} - iw)} \pi^{-2iw} \right]$$
$$+ \frac{1}{i} \log \frac{\zeta(1 + 2iw)}{\zeta(1 - 2iw)}. \tag{32}$$

The first term simply indicates the need for the wave to cover the distance $\log \alpha$ before it gets reflected. The second term can easily be evaluated with the help of Stirling's formula (Whittaker and Watson, p. 252), and yields

$$\frac{1}{i} \log \left[\frac{\Gamma(\frac{1}{2} + iw)}{\Gamma(\frac{1}{2} - iw)} \pi^{-2iw} \right] \simeq 2w \left(\log \frac{w}{\pi} - 1 \right)$$
$$+ \frac{1}{12w} + \frac{7}{1440w^3} + . \tag{33}$$

This asymptotic formula for large w gives essentially correct values for w as small as 1, and contributes a smooth, monotonic dependence of β on w. The only sign of stochastic behavior comes from the last term in (32) which will be discussed for the rest of this paper as if it were the only important contribution to β.

Riemann's zeta function is mostly known to mathematicians who are interested in number theory. It has not come up so far in any problem with a physical background to my knowledge, although Hilbert seems to have proposed the idea of finding an eigenvalue value problem whose spectrum contains the zeros of $\zeta(s)$. As general background a few remarks $\zeta(s)$ may be of some interest.

8. The Riemann zeta function

Euler's product formula (28) makes it almost obvious that this mysterious object contains valuable information about prime numbers. Indeed, Hadamard and de la Vallée Poussin were able to prove in 1896 on the basis of (28) that the number of primes smaller than x goes asymptotically as $\int_1^x dx / \log x$. Two years later, Hadamard gave the first discussion of geodesics on a surface of negative curvature as an ergodic system, but he did not hint at any connection with the zeta function [7].

The great challenge is to prove Riemann's conjecture: $Z(s)$ as defined in (29) has all its zeros on the critical line $\sigma = \frac{1}{2}$. The function $Z(s)$ is real for $\sigma = \frac{1}{2}$ as follows directly from (30), and goes back and forth between positive and negative values as t increases from 0 in $s = \frac{1}{2} + it$. On the basis of extensive calculations, it has been shown that all zeros of $Z(s)$ occur for $\sigma = \frac{1}{2}$ as long as $t < 32{,}585{,}736.4$ in $s = \sigma + it$, giving the lowest 75,000,000 zeros [8]. It came out in these computations, however, that some zeros were exceedingly close to each other, and it is known that a double zero with $\sigma = \frac{1}{2}$ would destroy Riemann's conjecture. No proof seems to be in sight; but if one should be found, the density of prime number could be estimated much more closely.

Riemann's zeta function has only one singularity, a simple pole at $s = 1$ with residue 1. The last term in (32) becomes, therefore, equal to $-\pi$ as w goes to 0. Among the important steps in the work of Hadamard and de la Vallée Poussin was to show that $\zeta(s) \neq 0$ for $\sigma = 1$ and $t \neq 0$. The argument of the logarithm in the last term of (32) is different from zero, and the phase shift is well defined for all values of w.

If Euler's product (28) is inserted without worrying about convergence, one can write immediately

$$\sum_p \frac{1}{i}\left[\log\left(1 - \frac{1}{p}e^{2iw\log p}\right) - \log\left(1 - \frac{1}{p}e^{-2iw\log p}\right)\right]. \quad (34)$$

Each term in this sum is a real periodic function of w with the frequency $2 \log p$. Since $p \geq 2$, one can even expand the two logarithms and find the Fourier series

$$-\sum_{k=1}^{\infty} \frac{2}{kp^k} \sin(2wk \log p). \quad (35)$$

Wintner has shown that these expansions are legitimate [9]. The log of the zeta function is almost periodic on the line $\sigma = 1$ as follows: the mean value

$$\lim_{t \to \infty} \frac{1}{T} \int_0^T dt \, e^{i\lambda t} \log \zeta(1 + it) \quad (36)$$

exists for all $\lambda > 0$, and vanishes unless $\lambda = k \log p$ for some integer $k \geq 1$ and prime p. In that case, the value of (36) is given by $1/kp^k$, just as in the heuristic formula (35).

Wintner's result is important in view of the discussion in the next section. The zeta function will be shown to be about as chaotic as one might wish any smooth function to be. On the other hand, only a discrete set of frequencies suffices to bring about such behavior. But these frequencies are linearly independent. No linear combination of logarithms of prime numbers with integer coefficients can ever vanish. Otherwise the representation (24) of an arbitrary integer as a product of powers of primes would no longer be unique. All such linear combinations actually occur as frequencies in the definition (28) of $\zeta(s)$ if one writes each term in the sum as $n^{-\sigma}\exp(-it \log n)$, for constant σ and varying t.

9. Chaos in the Riemann zeta function

Since $\zeta(s)$ has no singularities in any vertical strip above the real axis, the phase shift $\beta(w)$ is a smooth function of w for arbitrary large w. The absolute value of $\zeta(s)$ grows exceedingly slowly, essentially as $\log \log t$ [10]. Nothing dramatic can possibly happen, although $\beta(w)$ never settles down to any kind of simple behavior because it contains an infinity of linearly independent frequencies.

The question then arises: How unruly is $\zeta(s)$ in the critical strip between Re $s = 0$ and Re $s = 1$? The answer was given recently by S. M. Voronin, and then made explicit by Axel Reich [11] in the following

Theorem. Let D be a disc in the complex s-plane, i.e. $D = \{s \in C : |s - s_0| < r\}$ and let D be contained in the strip $\frac{1}{2} \leq \text{Re } s \leq 1$. Consider an arbitrary non-vanishing holomorphic function $f(s)$ in D, i.e. an analytic function without singularities in D, or again an arbitrary power series in $s - s_0$ which converges in D. Then the set of integers L which is defined by the condition

$$L = \{l : \sup_{s \in D} |\zeta(s + il\Delta) - f(s)| < \epsilon\}, \quad (37)$$

with an arbitrary real $\Delta \neq 0$ and $\epsilon > 0$, has a positive density. The last term means that the numbers of integers $l \in L$ whose absolute value is smaller than N forms a non-vanishing fraction of N, as N goes to infinity.

The proof of this theorem is very involved, but

it uses two facts about $\zeta(s)$ in the critical strip, the presence of infinitely many independent frequencies and the conditional convergence of the series (28). The first item has two consequences. A storehouse of independent functions is available which can be used to approximate any other function, and the phase relations between these functions can be tuned arbitrarily closely to any given set of conditions by pushing the imaginary part of the argument in (37) far enough, i.e. by increasing l. The conditional convergence allows the absolute values of the independent functions to assume any necessary magnitude. The precise estimates are complicated and require a good deal of functional analysis.

Riemann's conjecture is avoided in this theorem by restricting $f(s)$ to have no zeros in D. With that exception, Riemann's zeta function turns out to be "the universal function." It can mimick any other function within a prescribed accuracy, and it does so with surprising ease. It suffices to check a simple linear set of points, and the test of approximation will succeed in a non-vanishing fraction of all cases.

This property is reminiscent of certain transcendental numbers like π. Their decimal (or binary) expansion contains any given finite sequence of decimals (or binaries), infinitely often. Also, the decimal expansion passes every conceivable test of randomness. This last feature must also be present in the zeta function: The holomorphic function $f(s)$ can be defined by a power series in $s - s_0$ whose coefficients have been selected by some random process, and properly normalized so as to insure the convergence in D. With this concept of randomness in function space, the zeta function is able to come arbitrarily close to any point there, with a well-defined return time which follows from the density of the set L.

10. The complexity of Riemann's zeta function

The stochastic behavior of any mathematical object can be recognized in the difficulty of its computation. The phase shift β as a function of the velocity w requires the calculation of $\zeta(s)$ on the line $s = 1 + it$, at the border of the critical strip $0 \leqslant \operatorname{Re} s \leqslant 1$. How easy is it to do the computations there?

The definition (28) together with the Euler–Maclaurin formula (cf. Whittacker and Watson, p. 127) leads to the asymptotic formula for $\zeta(s)$ as

$$\sum_{n=1}^{N} \frac{1}{n^s} + \frac{1}{s-1} N^{1-s} - \frac{1}{2N^s}$$
$$- \sum_{m=1}^{M} \frac{(-1)^m B_m}{(2m)!} \frac{s(s+1)\ldots(s+2m-2)}{N^{s+2m-1}}, \quad (38)$$

which involves the Bernoulli numbers B_m. They are rational numbers with large numerators and denominators and satisfy the inequality

$$\frac{2(2m)!}{(2\pi)^{2m}} < B_m < \frac{2(2m)!}{(2\pi)^{2m}} \left(1 - \frac{1}{2^{2m-1}}\right)^{-1}. \quad (39)$$

The mth term in (38) can, therefore, be estimated as given by $(|s|/2\pi N)^{2m-1}/\pi N^\sigma$. With $s = \sigma + it$ and σ constant, the remainder in (38) is of the same size as the first omitted term, i.e., $\approx (t/2\pi N)^{2M+1}/\pi N^\sigma$.

For the calculation of the phase shift β, one sets $\sigma = 1$ and $t = 2w$, and gets a very convenient and informative criterion for the use of (38) to a certain precision when w is large. Obviously when $N \approx w$, each additional term in the sum over m yields an improvement of the accuracy by $1/\pi^2 \approx 1/10$, i.e. one more decimal. The Bernoulli numbers have to be stored in the program which is not too bad when M is kept small. But the summation over n becomes rapidly prohibitive as w increases, which is just the interesting limit of approach to classical mechanics.

Riemann already had devised a more efficient method of evaluating $\zeta(s)$ in the critical strip, although he never published his results. They were deciphered by Siegel in left-over notes [12]. The functional relation (30) is used to give the following formulas: again with $s = \sigma + it$, define $\tau = t/2\pi$ and $N = [\sqrt{\tau}]$ where the bracket [ ] indicates the largest integer which does not exceed $\sqrt{\tau}$. Then one has again for $\zeta(s)$ asymptotically

$$\sum_1^N \frac{1}{N^s} + \chi(s) \sum_1^N \frac{1}{n^{1-s}} + (-1)^{N-1} C(s) S_N, \qquad (40)$$

with $\chi(s) = (2\pi)^s \cdot \sin(\pi s/2) \cdot \Gamma(1-s)/2$, and $C(s) = (-2\pi i t)^{1/2(s-1)} \, e^{-i/2(t+\pi/4)} \Gamma(1-s)$. The whole difficulty is now in S_N which is a power series in the two variables $\tau^{-1/2}$ and $\xi = 2(\sqrt{\tau} - N - \tfrac{1}{2})$. Its coefficients result from various recursion formulas, and from generating high derivatives of the function $\phi(z) = \cos \pi(\tfrac{1}{2}z^2 + 3\pi/8)/\cos \pi z$.

The summations over n are now shortened to go only as far as $\sqrt{w/\pi}$ which is a tremendous improvement over (38). The calculation of $\chi(s)$ and $C(s)$ is easy, but the correction term S_N becomes a great nuisance. It is difficult to generate and almost equally bothersome to store from tables, and it converges poorly [13]. Indeed, some of the more elaborate computations of the zeros in $\zeta(s)$ had a lot of trouble to decide whether certain pairs were separate or coalescing, because the value of $\zeta(s)$ between them was below the accuracy of S_N. The error estimates for S_N when τ is large are still incomplete. Clearly, the Riemann zeta function is much more difficult to evaluate than any of the standard functions of a complex variable which one encounters in mathematical physics. Although the formula (40) requires a much shorter summation than (38), the limit of large w can be computed only by using an ever large number of terms.

11. Discussion

This last section is concerned with some general conclusions which might be drawn from the special example of particle scattering in the earlier sections. In particular, some speculations will be made as to where similar behavior of the phase shift may be encountered. First, a few numerical results will be discussed, because they convey a much better impression about the stochastic character than the

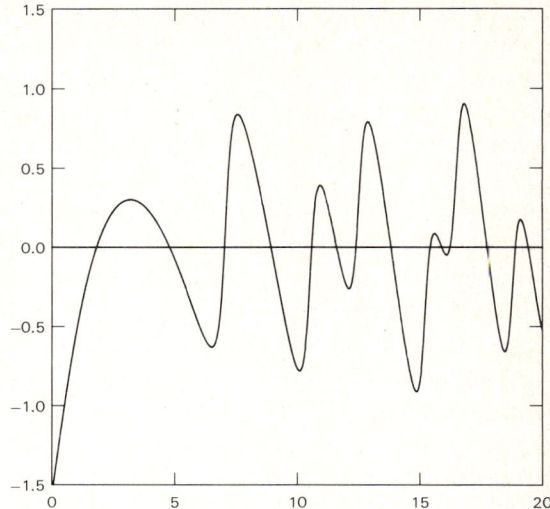

Fig. 5. Phase angle of $\zeta(1 + 2iw)$ as function of w in the range of $0 < w \leq 20$. The y-axis gives the angle in radians. The crossings of the w-axis with positive slope occur rather closely to $t/2$ where $\zeta(\tfrac{1}{2} + it) = 0$.

statements about chaos and complexity in the Riemann zeta function.

Figs. 5, 6 and 7 show the phase angle of $\zeta(1 + 2iw)$ as a function of w in 3 different ranges. In the lowest range, $0 < w \leq 20$, the various peaks and valleys can be understood in a rough manner. The steep rises with zero crossings at $w = 7.03$, 10.58, 12.40, 15.43, 16.22, 18.90, are directly related to the zeros of the zeta function on $s = \tfrac{1}{2} + it$ for $t = 14.13$, 21.02, 25.01, 30.42, 32.93, 37.59. Some simple periods are visible in fig. 5, although their assignment is not obvious. Their values correspond crudely to $\pi/\log p$ with $p = 2, 3$ as one would expect from (34).

The middle range, $100 < w \leq 120$, still shows a deceptively smooth function, although it contains some high frequencies. The detailed structure can still be related to the zeros of the zeta function, provided one considers not only the crossings with the w-axis but all the distinct points with a steep positive rise. The assignment of an exact location for the latter is somewhat arbitrary, but their numbers are correct. The tables of Hasselgrove show 23 zeros in this range, for an average distance

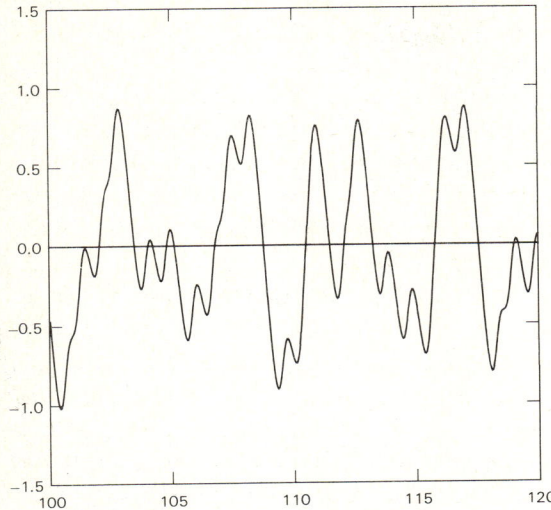
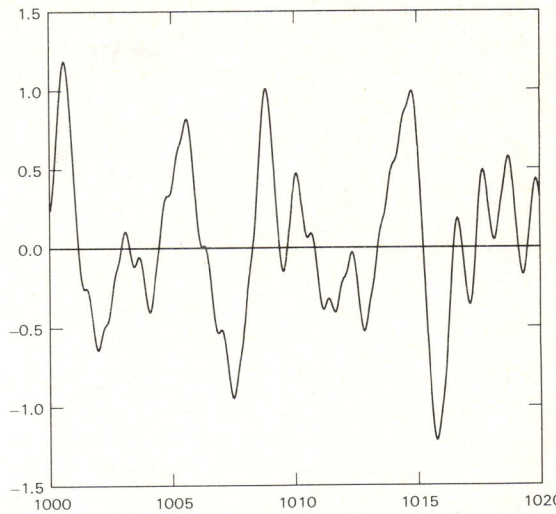

Fig. 6. Same as fig. 5, in the range $100 < w \leq 120$. The zeros of $\zeta(\tfrac{1}{2}+it)$ are now represented by the distinct places of positive slope, with only 2 missing out of 23.

Fig. 7. Same as fig. 5 and 6, in the range $1000 < w \leq 1020$. The correspondence with the zeros of $\zeta(\tfrac{1}{2}+it)$ is qualitative at best with $\tfrac{1}{3}$ of the places not recognizable any longer.

of 1.75. Since we are looking at $\zeta(s)$ at a distance of 0.5 from the critical line $\operatorname{Re} s = \tfrac{1}{2}$, it is not surprising to find a close correspondence.

Such an interpretation can be called quantum-mechanical, as opposed to an interpretation which is based on the original definition (28), or more particularly (34), and its offsprings (38) and (40). The integers represent the lengths of trajectories between entering and leaving the leaky box. Formula (38) requires N to go at least up to $t/2\pi$ for convergence, and leads to highest frequencies of $(\log N)/2\pi$. The mean distance between crests in the wavy structure is, therefore, found $2\pi/\log(t/2\pi)$ which is a well-known result for $\zeta(s)$. In terms of $t = 2w$, this yields $\pi/\log(w/\pi)$, whose value becomes ≈ 0.75 for $w = 120$, and accounts for fig. 6 in a general manner.

The use of (38) becomes prohibitive around $w = 500$, although (40) is not quite satisfactory as yet. The difference is easily noticed in the computed values, but falls below the resolution of the figures. The random quality of fig. 7 can hardly be denied. The tables indicate 37 zeros in the range $2000 < t < 2040$, but even an optimistic count of distinct places with positive slope in fig. 7 cannot go much beyond 25. Some smoothing is taking place.

The summation in (40) goes out to $N = [\sqrt{w/\pi}]$, and admits as shortest period $2\pi/\log(w/\pi)$. The remainder accounts for the finer structure and gives an idea of its complexity. The de Broglie wave length is $2\pi/w$, which is much smaller, so that fig. 7 ought really to be well into the quasi-classical regime. (The volume of the whole box is 2π.) And yet, with an average distance of 1.1 between zeros, the quantal structure should still be visible at our distance of $\tfrac{1}{2}$ from the critical line. As the density of zeros increases so slowly in the zeta function, one has to go to extremely large values of w before the quantal structure finally averages out. If one sets $2\pi/\log(t/2\pi) \approx \tfrac{1}{2}$, it follows that $t \approx 2\pi \exp(4\pi)$, and $w = t/2$ becomes 5000,000.

When classical mechanics is used to get an approximation for quantum mechanics, the trajectories have to be replaced by waves in the well-known manner: the Hamilton–Jacobi function $S(x,t)$ or $W(x,E)$ is used to get a wave function $\psi \sim \exp(iS(x,t))$ or $\varphi \sim \exp(iW(x,E))$. An amplitude factor is provided in order to maintain the density of flow. Such a wave which enters or exits through the leak in our box is represented in fig. 1 and is obviously represented by (14) where c and d are integers. Therefore, formula (21) can be looked upon as a classical approximation which happens to be exact. While Selberg's trace formula demonstrates the close connection between classical and quantal mechanics for bound states in a compact space of constant negative curvature, formula (20) does the same for scattering states.

The whole approach in this paper can be generalized rather trivially while staying among open surfaces of constant negative curvature and finite volume. The topology of fig. 3 can be preserved while changing the shape of the box. The construction of fig. 4 is modified by using a lower boundary which consists of 3 pairs of Euclidean circles with different radii, as long as the sum of the inner angles adds up correctly to 2π. It is not difficult to build models with more than 1 leak, so that the wave has a choice of where to escape. Finally, a box of higher genus can be chosen. In all these cases, the model has a finite set of parameters which can be varied continuously over certain ranges.

Does one always run into the Riemann zeta function for the scattering phase shift? The answer is clearly negative, unless the group of symmetries belongs to a special class called the arithmetic groups. So what remains of our analysis? Formula (20) remains valid, but not (21). One has now to deal with so-called Eisenstein series, and finds that many of the results can be generalized, but nothing as elegant as the phase shift in (31) is obtained.

By way of speculation, one can finally ask whether the Riemann zeta function shows up in other scattering problems. The answer is a resounding yes, although the connection is not directly in the phase shift. Consider the flat unit square with opposite sides identified and some kind of exponential horn attached in the middle through which a particle can enter or exit exactly as in the leaky box. The spherical waves from the center are now matched by as many copies in the whole square lattice, so that the boundary conditions on the unit square are satisfied. These waves are some kind of Bessel functions, of course, but their Mellin transforms with respect to the momentum are essentially sums over $(m^2 + n^2)^{-s}$ for m and n running through the integers. Expressions of this kind occur in the theory of algebraic numbers and lead eventually to modifications of Riemann's zeta function which are due to Dedekind and Hecke. Their behavior is just like $\zeta(s)$, including formulas like (28) and (30).

This connection with number theory is inherent in the work of Berry [14] on Sinai's billiard. Instead of a source or sink at the center of the square, an obstacle of circular symmetry is investigated and shown to give an ergodic system. The quantum-mechanical bound states are computed with the help of the KKR method which uses basically the Mellin transforms of the above-generalized zeta functions. As before, there are two cases to be considered. If the square is replaced by a parallelogram of different shape, there may not be a connection with number theory, and formula (28) does not arise. The functional equation (30) is still valid, however, and a Riemann–Siegel formula (40) can be derived to facilitate the computation. Again, the physicists have applied special cases of (40) when calculating certain sums in lattices, but they do not seem to have realized the relation to Riemann's zeta function and its chaotic properties.

Acknowledgements

It is a great pleasure to thank Professors Joan Birman, Harvey Cohn, Linda Keene, Burton Randol and Larry Schulman for a number of stimulating and instructive discussions.

References

[1] Hilbert, Transactions of the American Mathematical Society 2 (1901) 87–99.
[2] More details can be found in textbooks on complex variables and on differential geometry, but most of these presentations are highly selective and buried somewhere toward the end of the book. H.S.M. Coxeter, Introduction

to Geometry (Wiley, New York, 1961); W. Blaschke, Vorlesungen uber Differential Geometrie (Dover, New York, 1954); and its second edition by Blaschke and Reichardt (Springer, Berlin, 1960) give the essential ingredients.
[3] This group has been studied extensively by Harvey Cohn in connection with some problems in number theory, cf. Discontinuous Groups and Riemann Surfaces, L. Greenberg ed., Annals of Mathematics Studies No. 79 (Princeton Univ. Press, Princeton, 1974) p. 81–98.
[4] The calculation in this section is an obvious modification of a similar calculation which can be found in P.D. Lax and R.S. Phillips, Scattering Theory for Automorphic Functions (Princeton Univ. Press, Princeton, 1976) p. 170 ff.
[5] H. Rademacher, Lectures on Elementary Number Theory (Blaisdell, 1964) p. 18.
[6] E.T. Whittaker and G.N. Watson, A Course of Modern Analysis (Cambridge Univ. Press, Cambridge, 1946).
[7] J. Hadamard, "Sur le billiard non Euclidean," Soc. Sci. Bordeaux, Proc. Verbaux 1898, 147 (1898); J. Math Pure Appl. 4 (1898) 27.
[8] R.P. Brent, Mathematics of Computation 33 (1979) 1361–1372.
[9] A. Wintner, Duke Math J. 10 (1943) 429–440.
[10] E.C. Titchmarsh, The Theory of the Riemann Zeta-Function (Clarendon, Oxford, 1951).
[11] A. Reich, Arch. Math. 34 (1980) 440–451.
[12] H.M. Edwards, Riemann's Zeta Function (Academic Press, New York 1974) ch. 7.
[13] The coefficients for the computation of $\zeta(1 + it)$ are given in C.B. Hasselgrove, Tables of the Riemann Zeta Function (Cambridge Univ. Press, Cambridge, 1960).
[14] M.V. Berry, Annals of Physics 131 (1981) 163–216.

QUANTUM–CLASSICAL CORRESPONDENCE FOR THE FOURIER SPECTRUM OF A TRAJECTORY

Eric J. HELLER

Theoretical Division, T-12, Los Alamos National Laboratory, Los Alamos, NM 87545, USA

Using a displaced localized wavepacket (coherent state) as a quantum analog to a classical trajectory, we examine the Fourier spectrum of the expectation value of position X_t^Q, and compare it with the classical Fourier spectrum of position X_t. In both the quasiperiodic and chaotic regimes, a strong classical–quantum corrsespondence exists in the Fourier spectrum. However, the quantum spectrum has certain interesting features not present in the classical case.

In this paper, we shall address the question of classical–quantum correspondence of the Fourier spectrum of a classical variable. An example of such a variable from classical mechanics is

$$X_\omega = \frac{1}{\sqrt{2\pi}} \int_{-\infty}^{\infty} e^{i\omega t} X_t \, dt, \qquad (1)$$

where X_t is the x-coordinate as a function of time. In systems of interest, X is but one of several coordinates, and the dynamics may be integrable or chaotic. Such Fourier spectra are often used as qualitative indicators of chaos, because in the quasiperiodic region, only N fundamental frequencies $(\omega_{01}, \ldots, \omega_{0N})$ and their overtones and combinations may appear in X_ω:

$$X_\omega = \sum_m X_m \delta(\omega - \boldsymbol{m} \cdot \boldsymbol{\omega}_0), \qquad (2)$$

whereas in the chaotic domain, no such restrictions apply.

A quantum analog of eq. (1) is obtained by replacing X_t with X_t^Q, where $X_t^Q = \langle X \rangle_t = \langle \psi_t | X | \psi_t \rangle$. It is an interesting exercise to discover how the classical limit, eq. (2) is recovered from the quantum mechanics as $\hbar \to 0$, and how the classical chaos is reflected in the quantum spectra.

In order to correspond to the classical situation of setting all the initial conditions of the coordinates and momenta to some specified values, and "running a trajectory" to determine X_t, we take the initial wave-function ψ_0 to be as localized as possible in phase space. An ideal way to do this is with the coherent states (Gaussian wavepackets) $|g_0(X_0, P_0)\rangle$, which are minimum uncertainty states with

$$\langle g_0 | \boldsymbol{X} | g_0 \rangle = X_0,$$
$$\langle g_0 | \boldsymbol{P} | g_0 \rangle = P_0. \qquad (3)$$

Ehrenfest's theorem [1] tells us that X_t and X_t^Q will be nearly the same for some reasonable time, this time becoming longer as $\hbar \to 0$. This simple fact already implies a limited kind of quantum–classical spectral correspondence, because of a certain property of the Fourier convolution theorem which we might call "enveloping". Consider two trajectories, X_t and X_t^Q for which we may say

$$X_t \approx X_t^Q, \quad |t| \leq \tau. \qquad (4)$$

For $|t| > \tau$, suppose no claim is made as to the proximity of X_t and X_t^Q. Now consider

$$X_\omega = \frac{1}{\sqrt{2\pi}} \int_{-\infty}^{\infty} e^{i\omega t} X_t \, dt, \qquad (5)$$

with a similar equation for X_ω^Q, and

$$\bar{X}_\omega = \frac{1}{\sqrt{2\pi}} \int_{-\infty}^{\infty} e^{i\omega t} C_t X_t \, dt, \qquad (6)$$

with a similar equation for $\bar{X}_\omega^Q$, and where C_t has the property

$$\begin{aligned} C_t &\approx 1, \quad |t| < \tau, \\ C_t &\approx 0, \quad |t| > \tau. \end{aligned} \qquad (7)$$

With the conditions (4) and (7), we have immediately

$$\bar{X}_\omega \approx \bar{X}_\omega^Q. \qquad (8)$$

The Fourier convolution theorem dictates

$$\bar{X}_\omega = \frac{1}{\sqrt{2\pi}} \int d\omega' \, C_{\omega'} X_{\omega - \omega'}, \qquad (9)$$

with an analogous equation for $\bar{X}_\omega^Q$.

The Fourier transform of C_t, namely C_ω, has a spectral width of $\Delta\omega \approx 2\pi/\tau$. Eqs. (8) and (9) thus dictate that, whatever the differences in the detailed, high resolution X_ω and X_ω^Q, the smoothed (convoluted) versions $\bar{X}_\omega$ and $\bar{X}_\omega^Q$ have to agree with each other up to a resolution $\Delta\omega$, if X_t and X_t^Q agree up to a time τ, where $\Delta\omega\tau \approx 2\pi$.

In a loose sense, we may say that the spectrum X_ω must be consistent with its low resolution analog, $\bar{X}_\omega$, in which accumulations of intensity may be smoothed into bands at lower resolution.

Fig. 1 will help to illustrate the situation. The left sequence shows the quantum spectrum $\bar{X}_\omega^Q$ for τ increasing downward, the right sequence shows the classical spectrum, $\bar{X}_\omega$. For t longer than τ, X_t and X_t^Q start to deviate, and the quantum spectrum starts to show sub-bands not present in the classical spectrum. The classical spectrum simply narrows around the allowed Fourier components of the quasiperiodic motion.

So far, we have only required Fourier analysis and Ehrenfest's theorem, and a certain spectral

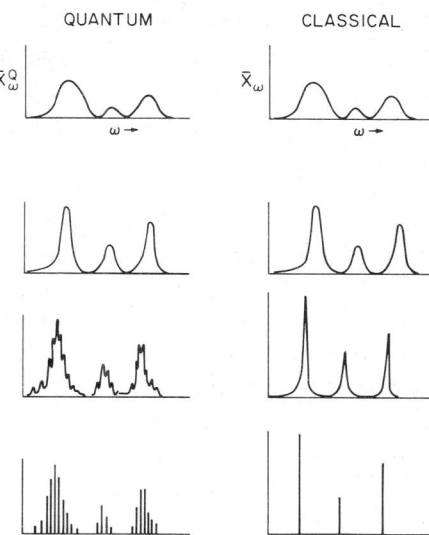

Fig. 1. In the left column we see the quantum spectrum and in the right, the classical, at increasing resolution as we move down the column. The classical and quantum spectra agree until the third case from the top. After that, the differences persist, showing the quantum spectrum to have band structure not present in the classical spectrum. This is the case in the classically quasiperiodic region.

similarity of $\bar{X}_\omega$ and $\bar{X}_\omega^Q$ has been noted. To understand the sub-structure present in the higher resolution quantum spectrum X_ω^Q, we must introduce a little quantum (or at least semiclassical) mechanics. Two features are present in each of the quantum clusters of lines: 1) The width of the cluster, 2) the number of peaks in each cluster.

Evidently, the width of the cluster, $\Delta\omega$, is directly related to the time, τ, after which the classical and quantum dynamics X_t and X_t^Q start to differ; i.e., $\Delta\omega \, \tau \approx 2\pi$. However, this observation is more tautology than anything else. What is needed is a basis for understanding or predicting the magnitude of τ or $\Delta\omega$.

Specializing to one degree of freedom we examine

$$X_t^Q = \langle g_t | X | g_t \rangle \qquad (10a)$$

$$= \langle g_0 | e^{iHt/\hbar} X e^{-iHt/\hbar} | g_0 \rangle \qquad (10b)$$

$$= \sum_{nn'} F_n^* X_{nn'} f_{n'} e^{-i(E_{n'} - E_n)t/\hbar}, \qquad (10c)$$

where $\langle g_0 | n \rangle = f_n^*$, etc., and the $|n\rangle$ are eigenstates. Also,

$$X_\omega^Q = \sum_{nn'} f_n^* X_{nn'} f_{n'} \delta(\omega - \Delta_{nn'}), \tag{11}$$

where $\Delta_{nn'} \equiv (E_{n'} - E_n)/\hbar$. The factors contributing to the overall appearance of X_ω^Q are the quantities f_n, E_n, and $X_{nn'}$. We examine each of these in turn.

In one degree of freedom for a confining (bound) potential, the main qualitative feature of the f_n's is a smoothly peaked pattern of intensity. The f_n's for a harmonic oscillator potential follow a Poisson distribution, tending to Gaussian as the displacement of $|g_0\rangle$ gets greater [2]. The overall energy spread is

$$\Delta E \approx \left(\frac{dV}{dx}\right)_{x=x_0} \sqrt{\frac{\hbar}{m\omega_0}}, \tag{12}$$

where the Gaussian g_0 is of the form

$$g_0(x) = \left(\frac{m\omega_0}{\pi\hbar}\right)^{1/2} \exp\left[-\frac{m\omega_0}{2\hbar}(x-x_0)^2\right]$$

and $V(x)$ is the potential energy. The distribution of f_n's always becomes Gaussian as $\hbar \to 0$, or as the displacements get large.

Now consider the energies E_n of the states. Let us take an energy near the center of the f_n distributions, call this E_p. The action of this state is

$$j_p = (p + \tfrac{1}{2})\hbar. \tag{13}$$

The hamiltonian H is expandable in terms of the action as

$$\begin{aligned}H(j) &= H(j_0) + \partial H/\partial j|_{j=j_0}(j-j_0) \\ &\quad + \tfrac{1}{2}\partial^2 H/\partial j^2|_{j=j_0}(j-j_0)^2 + \cdots \\ &= E(j_0) + v(j_0)(j-j_0)\end{aligned} \tag{14a}$$

$$+ \tfrac{1}{2}\left.\frac{\partial v}{\partial j}\right|_{j=j_0}(j-j_0)^2 + \cdots, \tag{14b}$$

where v is the classical frequency of motion. Applying these relations to the present situation we have, semiclassically,

$$E_n = E_p + \hbar\omega_p(n-p) + \frac{\hbar}{2}\omega_p'(n-p)^2 + \cdots, \tag{15}$$

where

$$H(j_p) \equiv E_p, \quad \omega(j_p) \equiv \omega_p, \quad \hbar\left.\frac{\partial\omega}{\partial j}\right|_{j=j_p} \equiv \omega_p'.$$

Note that because of the factor $\hbar$ in the definition of ω_p', $\omega_p' \to 0$, as $\hbar \to 0$, since $\omega(j)$ is a purely classical quantity. Thus eq. (15) is a statement of the Bohr Correspondence Principle, namely,

$$E_{n+1} - E_n \approx \hbar\omega_{\bar{n}}, \tag{16}$$

where $\omega_{\bar{n}}$ is the classical frequency of motion at the average value of the action

$$\bar{j} = \tfrac{1}{2}[(n+1+\tfrac{1}{2}) + (n+\tfrac{1}{2})]\hbar = (n+1)\hbar. \tag{17}$$

The last quantity we need to study is $X_{nn'}$. Standard semiclassical tricks (see the appendix) give

$$X_{nn'} \approx \int_0^{2\pi/\bar{\omega}} e^{i\Delta_{nn'}t} X_t \, dt, \tag{18}$$

where $\bar{\omega}$ is the frequency at the average action $\bar{j}$ and X_t is the classical trajectory. As $|n-n'|$ gets large, $X_{nn'}$ is expected to get small, since the Fourier frequency, $\Delta_{nn'} = (E_n - E_{n'})/\hbar \approx \bar{\omega}(n-n')$, gets large.

The spectral features corresponding to $n' - n = m$ are distinct for each m. Let us consider $m = 1$ first (the "fundamental"). For $n' = n + 1$ ($n' = n - 1$ is the negative frequency range), we have

$$X_\omega^Q = \sum_n f_n^* X_{nn+1} f_{n+1} \delta(\omega - \omega_p - \omega_p'(n-p+\tfrac{1}{2})). \tag{19}$$

For small $\hbar$, $f_n^* f_{n+1} \approx |f_n|^2$ since the f_n's vary smoothly, and from eq. (18) we have

$$X_{nn+1} \approx X_1, \tag{20}$$

where X_1 is the classical fundamental intensity. These observations yield, for this fundamental,

$$X_\omega^Q \approx X_1 \sum_n |f_n|^2 \delta(\omega - \omega_p - \omega_p'(n - p + \tfrac{1}{2})). \tag{21}$$

Already some interesting facts are emerging. First, since $\Sigma_n |f_n|^2 = 1$, the integral *intensity* of the quantum fundamental band equals the classical intensity of the fundamental. Second we can calculate the *width* of the quantum fundamental band of lines as follows: from eqs. (12) and (14) we have

$$\Delta E = \left(\frac{dV}{dx}\right)_{x=x_0} \sqrt{\frac{\hbar}{m\omega_0}} \tag{22}$$

$$= v \Delta j = \hbar \omega_p (n^+ - n^-)$$

$$\equiv \hbar \omega_p \Delta n,$$

where n^+ and n^- are the values of the quantum number n at the full width half maximum of the $|f_n|^2$ distribution. Eq. (22) gives

$$\Delta n = \left(\frac{dV}{dx}\right)_{x=x_0} \sqrt{\frac{\hbar}{m\omega_0}} \cdot \frac{1}{\hbar \omega_p}. \tag{23}$$

The frequency spread of the band is

$$\Delta \omega = h \left(\frac{d\omega}{dj}\right)_{j=j_p} \Delta n. \tag{24}$$

This gives, for $\Delta \omega$,

$$\Delta \omega = \sqrt{\frac{\hbar}{m\omega_0}} \left(\frac{d\omega}{dj}\right)_{j=j_p} \left(\frac{dV}{dx}\right)_{x=x_0} v_p^{-1}.$$

The time at which X_t and X_t^Q start to differ, by the arguments given above, is

$$\tau = 2\pi/\Delta\omega.$$

We see that

$$\tau \propto v_p, \left(\frac{dV}{dx}\right)_{x=x_0}^{-1}, \left(\frac{d\omega}{dj}\right)_{j=j_p}^{-1}, \omega_0^{1/2}, \hbar^{-1/2}.$$

That is, the correspondence between the classical and the quantal X_t and X_t^Q lasts longer if 1) the classical frequencies are higher; 2) the potential slope near the wavepacket is smaller; 3) the rate of change of classical frequency with action is smaller; 4) a larger ω_0 is used in the wavepacket; and 5) $\hbar \to 0$. All these trends will hold in the $\hbar \to 0$ limit. Now, from eq. (23) we see that $\Delta n \propto \hbar^{-1/2}$, i.e., the quantum number spread of the wave-packet is *increasing* as $\hbar \to 0$, and $\Delta \omega \propto \hbar^{1/2}$, i.e., the frequency spread of the fundamental quantum band of lines is *decreasing* as $\hbar^{1/2}$ as $\hbar \to 0$.

From eq. (21), we see that

$$X_\omega^Q \xrightarrow{\hbar \to 0} X_\omega, \tag{25}$$

since the intensity, the frequency, and the width of the overtone band all agree with the classical result as $\hbar \to 0$.

The analysis for an overtone band, $m = 2, 3, \ldots$ is much the same as for the fundamental band just presented.

The extension of the present result to two and more degrees of freedom is quite easy in the quasiperiodic domain of the classical mechanics. The results are the same; around each sharp classical fundamental, overtone, or combination line a quantum *cluster* of lines exist. The number of lines in the cluster increases as $\hbar^{-1/2}$ as $\hbar \to 0$, but the width of these cluster decreases as $\hbar^{1/2}$.

For an anharmonic n-dimensional system, a particular classical trajectory of a given energy E and action variables $j_1, j_2, \ldots, j_N$ has a given discrete set of fundamental frequencies $\omega_1, \ldots, \omega_N$. These frequencies, and their overtone and combinations, appear in X_ω. However, at the same energy, another set of actions $j_{1'}, j_{2'}, \ldots, j_{N'}$ would give in general a different set of classical frequencies, $\omega_{1'}, \ldots, \omega_{N'}$. In a certain sense, there are "missing" frequencies in a Fourier spectrum of any given quasiperiodic classical trajectory because the N actions $j_1, \ldots, j_N$, are conserved in the dynamics, preventing other frequencies from making an appearance. As $\hbar \to 0$ these statements apply to the quantum clusters of lines: The clusters bunch around the discrete allowed classical frequencies for the given actions $j_1, \ldots, j_N$.

In the classically chaotic regions the Fourier spectrum X_ω fills in, perhaps completely. No finite set of fundamentals can conspire to yield all the complexity in the spectrum, and indeed no set of actions is constant during the dynamics of a single trajectory. In a sense, the trajectory is able to sample all the frequencies now, since the actions are not conserved, thus the filling in of the Fourier spectrum X_ω. We now conclude with a few arguments to show that X_ω^Q has qualitatively the same behavior, namely many more "lines" in the spectrum, and approaching the classical continuum spectrum as $\hbar \to 0$.

A wavepacket $|g\rangle$ placed in a chaotic region of classical phase space still has a smooth energy envelope, but now far more "lines" appear under the envelope [3], because the chaotic wavefunctions of appropriate energy *all* sample the vicinity of the wavepacket. Thus, many more f_n's are nonzero in the chaotic region, compared to a similar wavepacket in a quasiperiodic domain, for the same number of degrees of freedom [3]. Furthermore, since the eigenstates $|n\rangle$ in the chaotic region are globally distributed, the matrix element $X_{nn'}$ will follow a random-like distribution as a function of n and n'; and the energies $E_{n'}$ likewise will not be derivable from any regular spacing formula such as eq. (15). Such a lack of systematic behavior in each of the quantities f_n, $X_{nn'}$ and $\Delta_{nn'}$ in eq. (11) implies no separation into fundamental, overtones, etc. is possible. Indeed, for fixed $\hbar$, eq. (11) evidently gives a very wild and random filling in of what were the "empty" regions between the clusters. This is because there are many, many possible $\Delta_{nn'}$ values, and nothing to distinguish them from each other because f_n, X_{nn}, etc. refuse to follow a "pattern". The only systematic aspect of the spectrum X^Q is that it does cut off at large frequencies, since the f_n's have to fall off at large n, corresponding to the finite energy width of the wave packet $|g\rangle$.

We thus have the qualitative trends:

–*Quasiperiodic regime*: Many "missing" frequencies.

–*Chaotic regime*: Filling in of the spectrum, no systematically missing lines.

These statements are true for both X_ω and X_ω^Q. They are in exact accord with the spectral features of molecular electronic transition discussed in earlier work [3]. Indeed, a quantiative criterion for the extent of quantum chaos was constructed from the observation that missing lines correspond to quasiperiodic motion [3].

In this paper, we have examined the correspondence between X_ω and X_ω^Q, using a displaced, initially localized a wavepacket to determine X_t^Q as an expectation value. We have found certain similarities and differences between X_ω and X_ω^Q. Earlier, Marcus and co-workers [4] had examined the Fourier spectrum of a single (but specially selected) classical trajectory and associated it directly with one and two quantum transitions in a vibrating molecule. This idea has its roots in the earliest days of quantum mechanics, when it was noticed that classical frequencies correspond to quantum energy level spacings (Bohr Correspondence Rule) (see eq. (16)). Using this idea, one is able to get intensities and frequencies of transition between states of similar quantum numbers [4] (i.e. large quantum jumps cannot be gotten so well with classical mechanics, unless $\hbar \to 0$). In any case, the viewpoint of refs. 4 is complementary to our own, in that we have used wavepackets to create an X_t^Q, and in so doing we rely on Ehrenfests' theorem for an $\hbar \to 0$ correspondence. Marcus and co-workers [4] have on the other hand relied more heavily upon the Bohr correspondence rule (eq. (14)). Both approaches provide insight into the relationship between classical and quantum mechanics.

Appendix A

A semiclassical, WKB wavefunction in one dimension is of the form

$$\psi_n^{\text{WKB}}(x) \approx \frac{1}{\sqrt{p_n(x)}} \exp\left[\frac{i}{\hbar} \int^x p_n(x')\, dx'\right], \quad (A.1)$$

where $p_n(x)$ is the momentum at position x of a classical particle of energy E_n. The matrix element $X_{nn'}$ reads

$$X_{nn'} = \int \psi_n^*(x) x \psi_{n'}(x)\, dx \quad (A.2)$$

$$\approx \int dx \cdot x \cdot \exp\left[-i/\hbar \int^x p_n(x')\, dx'\right.$$
$$\left. + \frac{i}{\hbar} \int^x p_{n'}(x')\, dx'\right] (\sqrt{p_n(x) p_{n'}(x)})^{-1}.$$

We now make the approximation

$$p_n(x') - p_{n'}(x') \approx \left.\frac{\partial p}{\partial n}\right|_{\bar{j}} (n - n') = \left.\frac{\partial p}{\partial j}\right|_{\bar{j}} \hbar(n - n'), \quad (A.3)$$

where

$$\bar{j} = \left(\frac{n + n'}{2} + \frac{1}{2}\right)\hbar.$$

Next, we note that

$$\left.\frac{\partial p}{\partial j}\right|_{j=\bar{j}} = \frac{m}{\bar{p}} \left.\frac{\partial E}{\partial j}\right|_{j=\bar{j}} = \frac{m\bar{v}}{\bar{p}}. \quad (A.4)$$

The integral in the exponent of (A.2) now reads

$$-\frac{i}{\hbar}(n - n')\hbar \int^x \frac{m\bar{v}}{\bar{p}(x)}\, dx = -\frac{i}{\hbar}(n - n')\hbar\bar{v} \int^t dt'$$

$$= i\Delta_{nn'} t. \quad (A.5)$$

Then if we approximate $\sqrt{p_n(x) p_{n'}(x)}$ by $\bar{p}(x)$, we have

$$X_{nn'} \approx \int e^{i\Delta_{nn'} t'} X_{t'}\, dt', \quad (A.6)$$

the desired result.

References

[1] See, for example, K. Gottfried, Quantum Mechanics (Benjamin, New York, 1966).
[2] See, for example, W.H. Louisell, Quantum Statistical Properties of Radiation (Wiley, New York, 1973).
[3] E.J. Heller, J. Chem. Phys. 72 (1979) 1337; M.J. Davis, E.B. Stechel, and E.J. Heller, Chem. Phys. Lett. 76 (1980) 21; E.J. Heller and M.J. Davis, J. Phys. Chem. 86 (1982) 2118.
[4] D.W. Noid, M.L. Koszykowski and R.A. Marcus, J. Chem. Phys. 67 (1977) 404, M.L. Koszykowski, D.W. Noid and R.A. Marcus, J. Chem. Phys. 86 (1982) 2113; and references therein.

INDEX OF CONTRIBUTORS

AUBRY, S.   240

BISHOP, A.R.   259

CRUTCHFIELD, J.P.   201

EPSTEIN, I.R.   47

FARMER, J.D.   153
FAUVE, S.   73
FESSER, K.   259
FEIGENBAUM, M.J.   16
FOWLER, A.C.   126

GIBBON, J.D.   126
GLASS, L.   89
GREBOGI, C.   181
GUCKENHEIMER, J.   105
GUEVARA, M.C.   341

HAUCKE, H.   69

HELLER, E.J.   356
HOLM, D.D.   330
HOLMES, P.   111
HUERRE, P.   135

KUPERSHMIDT, B.A.   330, 334

LANFORD, O., III   124
LAROCHE, C.   73
LIBCHABER, A.   73
LOMDAHL, P.   259

MacKAY, R.S.   283
MAENO, Y.   69
MANDELBROT, B.B.   224
MARSDEN, J.   305
McGUINESS, M.J.   126
MOON, H.T.   135

OTT, E.   153, 181

PACKARD, N.H.   201
PALMORE, J.I.   324

REDEKOPP, L.G.   135
ROUX, J.-C.   57
RUELLE, D.   40

SHENKER, S.J.   301
SHRIER, A.   89
SIGGIA, E.D.   302
SMITH, C.W.   85
SWINNEY, H.L.   3

TEJWANI, M.J.   85
TRULLINGER, S.E.   259

WEINSTEIN, A.   305
WHITLEY, D.   111

YORKE, J.A.   153, 181